# ATLAS OF THE LAND AND FRESHWATER MOLLUSCS OF BRITAIN AND IRELAND

# ATLAS OF THE LAND AND FRESHWATER MOLLUSCS OF BRITAIN AND IRELAND

MICHAEL KERNEY

1999

Published by Harley Books
(B. H. & A. Harley Ltd)
Martins, Great Horkesley,
Colchester, Essex CO6 4AH, England

in association with

The Conchological Society of Great Britain and Ireland

Published April 1999
Text © Michael Kerney
Maps © The Conchological Society of Great Britain and Ireland
and the Biological Records Centre, Monks Wood
Cover photo of *Perforatella subrufescens* (Brown snail) © D. G. Rands

All rights reserved. No part of this publication may be
reproduced, stored in a retrieval system, or transmitted, in any form
or by any means, electronic, mechanical, photocopying, recording or
otherwise, without prior permission of the publisher.

Text set in Ehrhardt by Saxon Graphics Ltd, Derby
Printed and bound at the University Press, Cambridge

British Library Cataloguing-in-Publication Data.
A catalogue record for this book is available from the British Library

ISBN 0 946589 48 8

# CONTENTS

|  | Page |
|---|---|
| PREFACE | 7 |
| EARLY HISTORY OF RECORDING | 8 |
|   The foundations | 8 |
|   Conchological Society Census | 9 |
| THE MAPPING SCHEME | 11 |
|   Origin | 11 |
|   Field procedure | 11 |
|   Completeness of survey | 11 |
|   Older records | 12 |
|   Fossil records | 13 |
|   Tetrad surveys | 13 |
| FACTORS INFLUENCING DISTRIBUTIONS | 14 |
| HISTORY OF THE BRITISH FAUNA | 16 |
|   Origins | 16 |
|   The Lateglacial period | 16 |
|   The Early Postglacial period | 16 |
|   The problem of Ireland | 17 |
|   The Late Postglacial period: the role of man | 17 |
| THE FUTURE | 19 |
| DISTRIBUTION MAPS AND SPECIES ACCOUNTS | 21 |
|   Explanation of maps and accompanying notes | 21 |
|   Totals recorded in 10km squares | 23 |
|   Gastropoda: Prosobranchia | 24 |
|     Neritidae | 24 |
|     Viviparidae | 25 |
|     Valvatidae | 27 |
|     Pomatiasidae | 30 |
|     Hydrobiidae | 31 |
|     Truncatellidae | 38 |
|     Bithyniidae | 39 |
|     Assimineidae | 41 |
|     Aciculidae | 43 |
|   Gastropoda: Pulmonata | 44 |
|     Ellobiidae | 44 |

|   |   |
|---|---|
| Physidae | 48 |
| Lymnaeidae | 51 |
| Planorbidae | 58 |
| Ancylidae | 72 |
| Acroloxidae | 74 |
| Succineidae | 75 |
| Cochlicopidae | 80 |
| Pyramidulidae | 84 |
| Vertiginidae | 85 |
| Chondrinidae | 102 |
| Pupillidae | 103 |
| Valloniidae | 107 |
| Enidae | 112 |
| Punctidae | 114 |
| Discidae | 116 |
| Arionidae | 119 |
| Vitrinidae | 133 |
| Zonitidae | 136 |
| Milacidae | 149 |
| Boettgerillidae | 153 |
| Limacidae | 154 |
| Agriolimacidae | 161 |
| Euconulidae | 165 |
| Ferussaciidae | 168 |
| Clausiliidae | 169 |
| Testacellidae | 175 |
| Bradybaenidae | 178 |
| Helicidae | 179 |
| Bivalvia | 207 |
| Margaritiferidae | 207 |
| Unionidae | 209 |
| Sphaeriidae | 214 |
| Dreissenidae | 236 |

MAPS OF ENVIRONMENTAL FACTORS 237
   January mean temperature 238
   July mean temperature 239
   Annual rainfall 240
   Calcareous rocks 241
   Maximum altitude 242
   Sulphur dioxide 243

LIST OF LOCALITIES CITED IN THE TEXT 244
   Maps of national grids and of vice-counties 247

LIST OF RECORDERS 250

BIBLIOGRAPHY 251

INDEX TO SPECIES 258

# PREFACE

In 1961 the Conchological Society launched its scheme to map the land and freshwater Mollusca of the British Isles, using a grid of 10km squares. The botanists were the pioneers in this. But the history of organized recording of molluscs goes back much further, to 1876, when a Leeds solicitor, William Denison Roebuck, began systematically to file under their localities and botanical vice-counties records securely authenticated by referees – an idea derived from the successful operation of the Botanical Exchange Club and inspired by the work of the pioneer biogeographer Hewett Cottrell Watson. Roebuck's 'Census' must be regarded as the real foundation of the present work.

By its very nature such an enterprise can never be complete. Nevertheless for no other invertebrate group does so clear a picture emerge. Against these maps – a record of a small but unusually representative section of our fauna as it exists during the late 20th century – future change can be measured, and, it may be, in some cases controlled.

An immense debt, unquantifiable in economic terms, is owed to the many people who collected information over the past thirty-seven years; their names are given at the end of this book. A few whose records were submitted indirectly are unknown to me, and for their omission I apologise. For specialist advice I am grateful to Miss Stella Davies and to Dr L. Lloyd-Evans for help with slugs and with Succineidae respectively, to Dr R. C. Preece, who checked and amplified information on geological history, and to Professor J. G. Evans for numerous unpublished fossil records from archaeological sites. The illustrations have been culled from a variety of sources, and here I am indebted in particular to Dr A. Wiktor and Dr Beata Pokryszko (Wrocław), Dr J. L. van Goethem (Brussels), Professor A. Piechocki (Łódź), Professor E. Gittenberger (Leiden) and Mr G. Riley for kindly allowing me to reproduce drawings which originally appeared elsewhere. A number of fine drawings were also made specially for this book by Mr D. W. Guntrip.

The computer work was done at the Biological Records Centre at Monks Wood, Huntingdon, by Mr H. Arnold, Dr M. Telfer, Mrs Wendy Forrest and Mrs Val Burton, under the direction of Mr P. T. Harding; Dr Telfer was responsible for producing the maps and for efficiently resolving a number of technical problems. I am indebted also to Mr Harding for making available the additional maps showing environmental and other factors. The primary mapping data are at present stored in the Department of Zoology, The Natural History Museum, London, and I am grateful to the Museum authorities, and in particular Dr P. B. Mordan, for providing me with congenial accommodation and research facilities during the preparation of this book.

The Conchological Society acknowledges financial help from the Royal Society towards the cost of publication.

London,  MICHAEL KERNEY
February 1999

# EARLY HISTORY OF RECORDING

## THE FOUNDATIONS

Land and freshwater molluscs have attracted scientific attention in Britain since the 17th century. Christopher Merrett in his *Pinax rerum naturalium Britannicarum* of 1666 – the first attempt at a complete catalogue of the plants, animals and minerals of the British Isles – lists six kinds: four terrestrial and two aquatic. Establishing their identity is not easy, but the Roman snail (*Helix pomatia*) is among them, as are probably species of *Cepaea* and *Anodonta*. Twelve years later Martin Lister published a fundamental account of the group in his *Historiae Animalium Angliae* which includes descriptions as well as rough though recognizable engravings of some thirty-four species. It is evident that Lister was aware that certain snails had limited ranges in the British Isles; for example, he knew that *Pomatias elegans* was a southern species, common in France and in Kent, and drew attention to the isolated colonies he had discovered in the north of England, giving the exact locations, at Burwell woods, Lincolnshire, and at Oglethorpe, near Tadcaster, Yorkshire.

Over the next few decades several naturalists availed themselves of Lister's book to compile local catalogues. Outstanding among these is the Revd John Morton's remarkable account in his *Natural History of Northampton-shire* (1712) in which he lists all the species he had encountered over the preceding twenty or so years, carefully noting their locations, habitats and relative frequencies. A few, like *Acanthinula aculeata* and *Vallonia pulchella*, he recognized as 'wholly new and undescribed' and furnished precise descriptions (Kerney, 1987).

Nothing comparable was achieved for almost a century: most 18th-century works on British Mollusca are derivative in character and show little interest in the land and freshwater species. An exception is a curious account of 1784 of minute shells from Faversham in Kent discovered by a local bookseller, George Walker. Walker was the first to detect several uncommon British species, notably *Acicula fusca* and *Segmentina nitida* (Walker, 1784; Kerney, 1976b).

Revival began with the great work of George Montagu, *Testacea Britannica* (1803, 1808), full of original descriptions and observations and undertaken, it appears, entirely independently of contemporary pioneer work in France. Its size and cost, and somewhat awkward layout, must however have limited its usefulness. The first manual devoted exclusively to the land and freshwater species was that by William Turton (1831), an idiosyncratic though serviceable little book which undoubtedly gave a stimulus to a popular study of the group. About this time began again the flow of local lists, of varying merit but all helping to provide the necessary groundwork on which country-wide distribution studies might be based. An annotated checklist of species recognized as British, remarkably modern in its approach, was published by Joshua Alder in 1837; J. E. Gray's excellent revision of Turton's book three years later benefited greatly from this. In the present context Gray's *Turton* is notable also as including what may be regarded as the first attempt at a national distribution register, showing in parallel columns those species, 128 in number, reliably known from different parts of the British Isles. A blank column is left for the reader's own observations (Gray, 1840, pp. 28–32).

Thereafter the literature grew rapidly. A milestone of the mid-Victorian years was the publication in 1862 of the first volume of John Gwyn Jeffreys' *British Conchology*. Attractively written yet authoritative, it helped to codify taxonomy in a way strongly felt up to the first World War and even beyond. J. W. Taylor's sumptuous *Monograph of the land & freshwater Mollusca of the British Isles* began to appear in parts in 1894. Its aim was to synthesize all available information, including the geographical distribution of each species. Though expensive, the names of over 200 subscribers appear in the first volume. Over-ambitious in its conception, the work was never to be completed.

Apart from these two major publications, a feature of the Victorian and Edwardian years was the appearance of a considerable number of popular identification guides, like those of Dixon & Watson (1858), Reeve (1863), Tate (1866), Harting (1875), Rimmer (1880), Adams (1884), Williams (1888), Step (1901), Swanton (1906), Stubbs (1907) and Horsley (1915). That they satisfied a real demand among naturalists is shown by the fact that in several cases reprints or new editions were called for. Indeed, innumerable local lists and species notes published during this period bear witness to a widespread interest in land and freshwater shells. To a large extent this reflects the aesthetic attraction the group held for collectors, especially of the colourful and highly variable Helicidae with their endless named varieties; nevertheless a good deal of permanent scientific value was also achieved.

## Conchological Society Census

The Conchological Society of Great Britain & Ireland was founded in Leeds in 1876. The study of geographical distribution was an interest from the start. A 'Census' scheme was organized by the Leeds solicitor William Denison Roebuck (1851–1919), based closely on the system employed by the Botanical Exchange Club (Roebuck, 1881; Kerney, 1967). The geographical units of recording were the well-known vice-counties proposed by H. C. Watson in 1859 for Great Britain (Dandy, 1969) and extended to Ireland by C. C. Babington and by R. L. Praeger (1896). Under this system records were entered into MS ledgers and the increments published annually in the *Recorder's Report* in the *Journal of Conchology*. The first entry is dated 12 October 1876 (*Unio tumidus* collected by Miss F. M. Hele from the River Frome, Stapleton, Gloucestershire, v.c. 34). No records were accepted, even of the commonest and most distinctive species, unless accompanied by voucher specimens submitted to the Society's referees (in the early years Roebuck himself or J. W. Taylor). Cumulative, vice-comital Censuses were published in tabular form, the first within a decade (Taylor & Roebuck, 1885). Roebuck (1921) and Ellis (1951) included in their editions complete sets of small maps which give a useful rough idea of the range of each species within the British Isles. Some vice-comital maps had been used earlier in Taylor's *Monograph*, the first (for *Testacella haliotidea*) in 1902. Books on two individual groups have also included sets of vice-comital distribution maps: for the genus *Pisidium* (Woodward, 1913) and for the slugs (Quick, 1960).

In Ireland recording took a somewhat independent course. An excellent beginning was made by William Thompson (1840) who tabulated the species known to him from different areas of the country. He was aided by a small band of now mostly forgotten local naturalists like, for example, the Revd B. J. Clarke of Portarlington, Co. Leix, whose observations on slugs were much ahead of their time. Yet over the next half century interest within Ireland faded away almost completely. Roebuck at first experienced much difficulty in obtaining Irish information for his Census. Then, in 1892, R. F. Scharff of the National Museum, Dublin, published a summary of existing knowlege, using Babington's twelve botanical 'provinces' to indicate the range of each species, albeit very roughly. A period of intense activity followed, lasting until 1914, which supplied material for a flow of articles to the *Irish Naturalist*. The fieldwork was done mainly by four people: R. J. Welch (working mainly in Ulster), R. A. Phillips (working in the southern half of Ireland; v.cs H1– H21), P. H. Grierson and A. W. Stelfox. By 1911 Stelfox was able to publish a comprehensive account of the Irish fauna, accompanied by a set of vice-comital maps based on the 'typomap' system.

After 1918 there is no doubt that interest in recording faded. Partly this was because the object for which the Census had been established had largely been achieved, but it is also a reflection of changes in society as a whole and of a new generation turning away from the interests of its Victorian predecessors – a phenomenon analysed by David Allen in his *The Naturalist in Britain* (1976). Membership of the Conchological Society fell from 350 in 1913 to just over 200 in the late thirties (a low point of 150 was touched in 1948). Only the enthusiasm of Professor A. E. Boycott, Recorder from 1919 to his death in 1938, maintained the Census scheme in operation. Boycott, a biologist of eminence, demonstrated the value of methodical recording by the use to which he put Census data in his classic papers on the ecology of non-marine Mollusca (Boycott, 1934, 1936a). The publication of A. E. Ellis's *British Snails* in 1926 also helped to keep alive interest in the group during the lean inter-war years.

The vice-comital system continues to the present day. It remains useful in providing some control over large volumes of data submitted by field workers of widely differing abilities. In the words of Boycott (*in* Roebuck, 1921), 'the uniformity of truth – or of error as the case may be – is thus considerable'. The most recent edition of the Census (in tabular form only) appeared in 1982 (Kerney, 1982a).

Nevertheless it has long been apparent that the

system has two grave drawbacks. It is cartographically crude, and secondly it makes no allowance for changes in distribution with time, bringing together records made over a period of well over a century. A remarkably early attempt at greater precision was a survey of *Vertigo alpestris* and *V. pusilla* in the southern Lake District, using a dot map to show precise 'find-spots' (Dean & Kendall, 1909). Locality maps, though superficially attractive, nevertheless have limitations of their own and are retrograde in the sense that they abandon the concept of the standardized recording unit, and hence of comparability of survey, both temporal and spatial. It was not until the 1950s that the vital principle of the equal-area grid as a basis for biological mapping was to be accepted (Clapham *in* Lousley, 1951; Walters, 1954).

# THE MAPPING SCHEME

## Origin

In 1961 the Conchological Society launched a scheme to map the non-marine Mollusca of the British Isles on a grid-square basis, following the lead given by the Botanical Society. A provisional atlas appeared fifteen years later (Kerney, 1976a).

A number of favourable factors assured the success of the new scheme. The existing vice-comital Census provided a secure framework, backed by a set of published maps and a valuable checklist of species recognized in Britain (Ellis, 1951). Several illustrated guides were in print from which most species could be identified from field characteristics alone, without dissection. Map-making facilities were made available to the Society, initially through the good offices of Dr F. H. Perring at the BSBI maps office in Cambridge, and later at the newly founded Biological Records Centre at Monks Wood. Lastly a small but enthusiastic body of field workers could be appealed to through a national organization with a now rapidly rising membership.

## Field procedure

Some idea of the procedure by which the information was brought together may be given here. As with other schemes, the basic mapping unit adopted was the 10km square of the British or Irish national grids, the smallest practical unit at this level. For the bulk of the work a specially printed 'field' or 'general' card was used, listing the fauna in abbreviated form (Fig. 1). The initial aim was to record at least 60 per cent of the estimated number of species living in each square. For large areas where there were no resident conchologists, especially in Ireland and Scotland, recording expeditions were organized. A small, experienced group with a vehicle and armed with litter-sieves and other equipment could with practice adequately work from four to six 10km squares in a day. This demanded careful route-planning. An examination of the map usually suggested the most profitable stopping points, the aim being to cover a spectrum of habitats – woodland, rocks, marsh, open water, waste ground and so forth. River crossings were often excellent places in this respect.

Empty shells present a particular problem with molluscs. They are a useful source of information but unless fresh must be interpreted with caution, at least on calcareous soils where they may persist for many years after death or even be fossils washed out of geological deposits. There is a similar difficulty in assessing the age of bleached, shelly debris left on river banks after winter floods. As a precaution, records known to be based solely on such evidence have been assigned to the 'before 1965' category. Nevertheless, the maps for certain species, especially in limestone areas, undoubtedly show some distortion for this reason. Examples are the maps for *Pomatias elegans*, *Pupilla muscorum*, *Helicella itala* and *Helicigona lapicida* – all receding species whose current ranges appear slightly exaggerated.

## Completeness of survey

As the first map shows, a good coverage of the British Isles has been achieved. Thinly worked areas of course remain, especially in Scotland and Ireland, but nevertheless the differences in species numbers recorded do broadly reflect true variations in regional species diversity ascribable to climate, topography or geology. These variations are very wide: some of the poorest 10km squares (for example in the central Scottish highlands) probably contain fewer than ten species; the richest contain over a hundred. The distribution patterns may therefore be accepted as real, except in a few cases involving recent segregates, exceptional minuteness or elusiveness, or special difficulties of identification. Limitations of this kind are noted in the text where appropriate.

From the point of view of systematic mapping at whatever scale, a relative evenness of survey is more

ATLAS OF MOLLUSCS

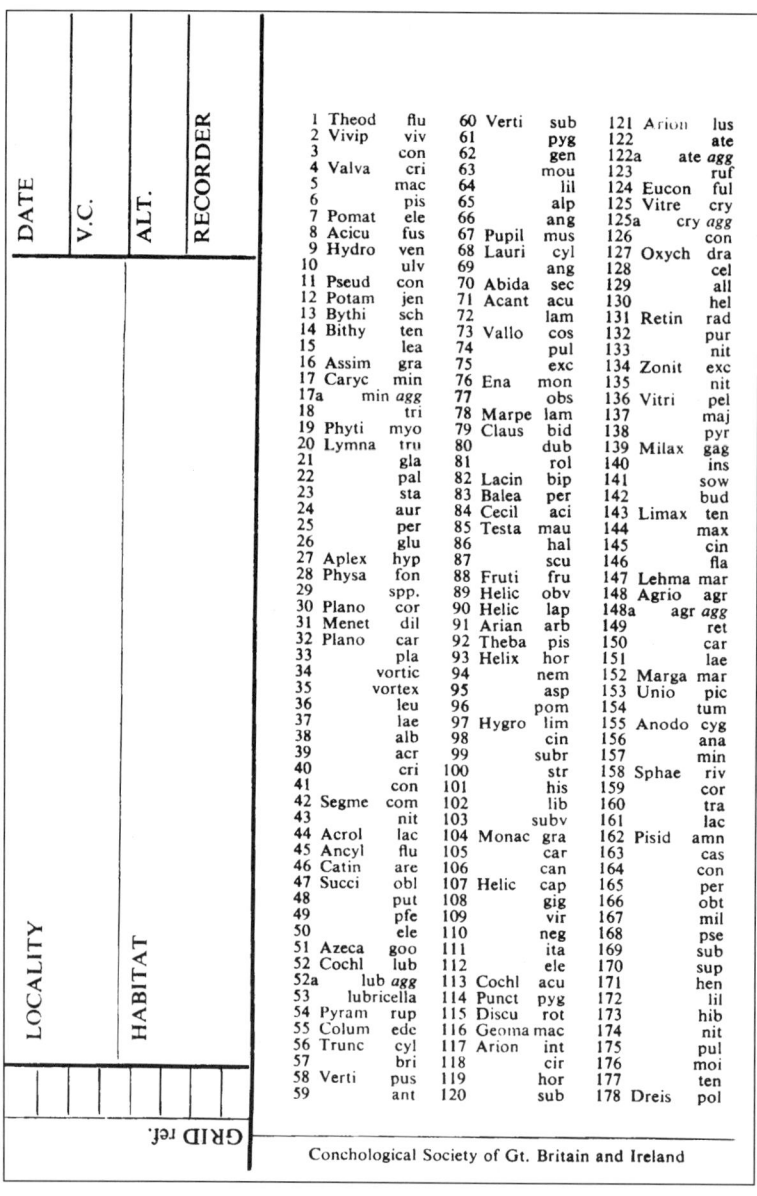

Figure 1   The original field card used to launch the Conchological Society's mapping scheme in 1961. The nomenclature follows Ellis (1951).

important than completeness (in any case an unachievable ideal). For this reason, the primary organized surveys of the 1960s and '70s were extensive rather than intensive in character. No attempt was generally made to work grid-squares exhaustively, or to locate all potential rarities; in practice these would turn up sporadically by the laws of chance, indicating their status clearly by a sparser patterning of dots on the maps.

A measure of the effectiveness of the mapping programme was the number of species new to the British fauna which came to light: *Vitrea subrimata* (1966), *Sphaerium solidum* (1973) and *Vertigo genesii* (1979) are notable examples. Taxonomic research was also stimulated, leading to important revisions in certain groups, for example in the genus *Arion*.

OLDER RECORDS

An important objective of mapping is to illumine change. Older literature, collections and MS sources have therefore been searched for all records not confirmed at 10km-square level since 1965. Published sources have included all papers in the *Journal of Conchology* and the majority of those in British and Irish local journals over the past century and a half.

An extremely valuable manuscript source, from which some thousands of records have been taken, is the series of 182 quarto species notebooks compiled by W. D. Roebuck and J. W. Taylor in the preparation of Taylor's unfinished *Monograph* (1894–1921), now preserved in the Department of Zoology, The Natural History Museum. The thirty or so MS ledgers of the Conchological Society's Census, the basis of the older system of vice-comital recording, have also been abstracted. Personal notebooks and collection catalogues include those of L. E. Adams (1854–1945), A. E. Boycott (1877–1938), J. E. Cooper (*c*.1870–1944), L. W. Grensted (1884–1964), D. K. Kevan (1895–1968), L. B. Langmead (?–1964), A. Loydell (?– 1910), C. Oldham (1868–1942), R. A. Phillips (1866–1945), A. W. Stelfox (1883–1972), R. J. Welch (1859–1936) and C. E. Wright (*c*. 1849–1926).

Nomenclatural and taxonomic changes have of course introduced difficulties in interpreting this mass of older work, especially where nominal species have later been split into two or more segregates, as in *Carychium*, *Columella*, *Vitrea*, *Euconulus* or *Arion*; many such records have unfortunately had to be discarded in the absence of voucher material. Virtually all records of *Pisidium* made prior to about 1910 are similarly unreliable.

There are also pitfalls in the gridding of old and often vague locality data, familiar to anyone who has attempted such work; for this and other reasons a small number of records shown in the 1976 *Atlas* have been deleted, usually without comment.

## Fossil records

Fossil records of Lateglacial and Postglacial age have been plotted on the maps where they fall outside modern ranges, i.e. where no current or historical record exists for a particular 10km square. They come from a variety of sources. The many pioneering papers of A. S. Kennard (1870–1948) have proved indispensable; the Kennard collection is now housed in The Natural History Museum.

In more recent years a considerable literature on the subject has developed, for example dealing with stratified shells from archaeological sites, and this too has been abstracted. Much use has also been made of unpublished information, and here I am especially indebted to Dr Richard Preece and Dr David Holyoak for allowing me access to the comprehensive site-lists contained in their forthcoming book on the history of the British and Irish mollusc fauna.

Unlike recent mapping data, fossil records are very unevenly distributed over the British Isles, with a heavy bias towards southern and eastern England and to areas of calcareous rock where suitable geological deposits are most common. Few sites in Scotland or Ireland have yet been properly investigated.

## Tetrad surveys

A tetrad ($2 \times 2$km square) map showing the distribution of *Abida secale* in Gloucestershire was published nearly thirty years ago (Long, 1970). The first tetrad survey recording all species present within a given area was that for a single 10km square around Penistone in South Yorkshire (Lloyd-Evans, 1975). More recently several county tetrad surveys have been undertaken in England and Wales, notably in Bedfordshire, Berkshire, Cardigan, Devon, the Isle of Wight, Kent, Leicestershire, Northamptonshire, Somerset, Suffolk and Surrey; atlases for the Isle of Wight and Suffolk have been published (Preece, 1980; Killeen, 1992). Unsuspected local patterns and relationships obscured by mapping at a coarser level inevitably emerge, especially for so-called common species. A virtue of such schemes is that a good coverage can be achieved within a relatively short period of time (say 5–10 years) giving a snapshot view, as it were, against which future changes can be measured (Kerney, 1982b). Such schemes have therefore an important part to play in conservation strategies.

# FACTORS INFLUENCING DISTRIBUTIONS

Within the limitations of the cartographic system employed, knowledge of distributions is now very complete. What do the patterns signify? The maps of environmental factors at the end of this book will suggest some answers. A few, very general comments are made below; others are contained in the notes accompanying the distribution maps.

Some species are nearly ubiquitous but the majority are restricted in various ways. As might be expected, ecologically catholic species like *Lymnaea peregra* or *Deroceras reticulatum* usually have the widest distributions, whereas demanding species tend to be localized.

Lime is an important need for most molluscs. The richest faunas are therefore found on calcareous soils or in areas of hard water. Calcium carbonate compensates to a high degree for climatic disadvantage, so that isolated patches of limestone in the north or at high elevations may have surprisingly rich faunas.

The clearest example of a calcicole snail is *Pomatias elegans*, which requires not less than five per cent free calcium carbonate in the soil. Though limited northwards by climatic and other factors, its distribution in southern Britain exactly follows the chalk and limestone belts, as well as some areas of coastal shell-sand in the extreme west. Other evident calciphiles are *Pyramidula rupestris*, *Pupilla muscorum*, *Cochlodina laminata*, *Cernuella virgata*, *Helicella itala* and *Helicigona lapicida*. Conversely, the calcifuge snail *Zonitoides excavatus* is restricted to areas of neutral and acid soils. Among freshwater molluscs, a continuum is apparent on the maps from evidently hardwater species (*Bithynia tentaculata*, *Lymnaea stagnalis*, *Planorbis carinatus*, *Anisus vortex*), through those which are moderately calciphile (*Valvata cristata*, *Hippeutis complanatus*), to species which are indifferent (*Lymnaea peregra*, *Ancylus fluviatilis*, *Pisidium casertanum*); one species, *Lymnaea glabra*, suggests an aversion to carbonates. Exact relationships with water hardness should however not be looked for; soft waters are frequently enriched in species by the compensatory effect of other factors, especially high pH values.

Some geologically-based relationships are mysterious. The common and ecologically undemanding 'weed' species, *Trichia striolata*, avoids the Triassic clays and sandstones of the English midlands, for reasons which remain unclear.

Relationships may be evident with climate. This is perhaps surprising, especially within so small an area as the British Isles, since to invertebrates it is the nature of the immediate habitat and its microclimate which is generally of overriding importance. In practice, nevertheless, equal-area mapping allows certain broad climatic effects to reveal themselves.

Prevailing humidities influence the distribution of several snails, including *Leiostyla anglica*, *Arion subfuscus*, *Spermodea lamellata*, *Zonitoides excavatus*, *Limax cineroniger*, *Lehmannia marginata* and *Perforatella subrufescens*. It is not uncommon furthermore for species of open habitat in the west to become increasingly restricted to humid, sheltered places in the climatically drier south-east: *Ashfordia granulata* is an excellent example. High humidities may account for the scarcity of a number of xerophiles of open calcareous ground in otherwise suitable limestone habitats of the oceanic west, for example *Pupilla muscorum* and *Vallonia costata*.

Most British molluscs will withstand prolonged cold in hibernation. A few however are frost sensitive. *Cochlicella acuta* is perhaps the clearest example: it is abundant chiefly along western coasts but occurs also inland in southern Ireland and – very occasionally – in England as transient populations vulnerable to winter frosts. The Lusitanian snail *Ponentina subvirescens* and the Mediterranean *Theba pisana* are likely also to be frost tender. Cold-intolerant species to a lesser degree are *Pomatias elegans*, *Acicula fusca*, *Lauria cylindracea* and *Helix aspersa*: their European distributions are markedly Atlantic/oceanic, not transgressing to the north-east beyond the mean

January isotherm for about +1°C. An analogy can be drawn with such plants as common ivy (*Hedera helix*) and holly (*Ilex aequifolium*).

Low summer temperatures are probably important in limiting northward and westward penetration by a number of central European snails, such as *Ena montana*, *Monacha cartusiana*, *Helicodonta obvoluta* and *Helix pomatia*, which just reach Britain at the edges of their ranges. This group of species is absent or poorly represented in Ireland.

In general, there are many species of broadly southern European distribution whose British and Irish limits are likely to include some element of thermal control. The failure of *Cepaea nemoralis* in northern Scotland is a good example. On the other hand, a few species appear to require coolness for their survival, notably the northern and western glacial relicts *Vertigo lilljeborgi*, *V. modesta*, *V. genesii*, *Deroceras agreste*, *Pisidium conventus* and *P. lilljeborgii*. Their continental distributions are primarily montane and Scandinavian.

Nevertheless, it is evident that distributions are rarely controlled by single factors, climatic or otherwise, but result from several factors acting together. *Pomatias elegans*, for example, requires limestone soils, friable in texture, in relatively warm situations, whereas *Zonitoides excavatus* is restricted to acid woodland primarily in the oceanic west. Such various requirements are not always easily separable. By a geological accident, to complicate matters, most of our limestones are in the climatically favoured south and east, whereas the harsher highland zone is largely non-calcareous. Human pressures are also generally at their most intense in the driest and warmest parts of the British Isles, favouring xerophiles and 'weed' species at the expense of species of marsh and woodland.

There is a further difficulty. Existing factors alone, singly or in combination, cannot explain distributions fully; many are the products also of complex past events, both natural and human. The history of the fauna must therefore be considered.

# HISTORY OF THE BRITISH FAUNA

## Origins

Apart from slugs, molluscs leave good fossils and consequently the geological history of the group in Britain is excellently documented.

There is some doubt as to whether any molluscs survived in the British area from the last interglacial (Ipswichian) period through the intense cold of the last glacial (Devensian) period, which began about 70,000 years ago. A few hardy species may have done so in favoured refugia in the unglaciated zone in the south but most of our fauna (like our flora) results from re-immigration from the continental mainland as deglaciation proceeded.

## The Lateglacial period

By the end of the somewhat milder Lateglacial period (c. 15,000–10,000 years ago) at least 29 freshwater molluscs and about 27 land molluscs had re-established themselves in the British area. The terrestrial species reflect the open nature of the landscape. Most of the evidence comes from the chalk and limestone tracts of southern England. Here the assemblages are typified by a mixture of hardy, ecologically catholic species (*Cochlicopa lubrica*, *Punctum pygmaeum*, *Vitrina pellucida*, *Nesovitrea hammonis*, *Euconulus fulvus*, *Arianta arbustorum*) and species of scree and grassland (*Vertigo pygmaea*, *Abida secale*, *Pupilla muscorum*, *Vallonia costata*, *Helicella itala*, *Trochoidea geyeri*). Distinctively woodland snails are absent. In marshes we find *Catinella arenaria*, *Cochlicopa nitens*, *Oxyloma pfeifferi*, *Columella columella*, *Vertigo antivertigo*, *V. genesii* and *V. geyeri*.

Aquatic molluscs are known from deposits of Lateglacial age in several parts of the British Isles. Most are familiar British species but, among the bivalves, two have Arctic or boreo-Alpine distributions (*Pisidium conventus*, *P. lilljeborgii*) and one (*P. vincentianum*) is extinct here. In small lake basins left by the ice *Gyraulus laevis* often abounds.

The high proportion of freshwater species, slightly outnumbering the terrestrial, may be explained by their great powers of passive dispersal, coupled with the rapid warming up of shallow water-bodies due to the high position of the sun in our latitudes, in contrast to polar regions today.

## The Early Postglacial period

With the opening of the Postglacial proper (Preboreal and Boreal periods), steeply rising temperatures brought about a rapid enrichment of the molluscan fauna. A few Lateglacial species became extinct (e.g. *Cochlicopa nitens*, *Columella columella*) or were reduced thereafter to the status of relicts (e.g. *Vertigo genesii*). As forests developed, species requiring shade and shelter spread. *Carychium tridentatum*, *Aegopinella nitidula* and *Clausilia bidentata* became common very early in the Postglacial. A Scandinavian forest-snail, *Discus ruderatus*, occurred widely in England between about 9,500 and about 8,000 BP and was then replaced by the more southern *Discus rotundatus*. The extinct *Nesovitrea petronella* occasionally occurred also.

By the deciduous forest optimum (the Atlantic period, c. 7,500–5,000 BP), woodland faunas showed their greatest diversity and were characterized by the presence throughout lowland Britain of species of now restricted (often westerly or oceanic) distribution. *Spermodea lamellata* and *Leiostyla anglica* are good examples. Conversely, terrestrial species of dry, open habitats, such as *Pupilla muscorum*, *Abida secale*, *Vallonia costata*, *V. excentrica* and *Helicella itala*, all common in the bare landscapes of the Lateglacial, were largely suppressed, surviving only on sea coasts and other refuges which escaped a dense tree-cover; two, *Trochoidea geyeri* and *Helicopsis striata*, failed to survive the period of forest dominance and became extinct in the British Isles.

The tally of 'native' species – i.e. those which reached the British area from the continental mainland by their own powers of passive dispersal –

was by now complete. There is fossil evidence of some 48 freshwater species and about 67 terrestrial species living at this period in southern Britain. All of these survive today. It is likely also that at least eighteen of our 29 species of slugs had reached us by this time, though this cannot be proved. A large element of chance was doubtless involved, since earlier interglacials of the Pleistocene show rather different associations of species living here.

## The problem of Ireland

Ireland presents a special problem. As with nearly all groups of animals and plants, the Irish molluscan fauna is impoverished compared with that of Britain. In general terms, this is usually explained by the fact that the land connection in the Irish Sea was submerged earlier than that in the much shallower English Channel as sea-levels rose at the end of the ice age. Less time was therefore available for immigration from refugia, which in any case were further away. Nevertheless the detailed make-up of the Irish fauna does not accord particularly well with our present knowledge of the actual sequence of recolonization. For example, snails like *Oxychilus cellarius* or *Acicula fusca*, now common in Ireland, appear on fossil evidence to have arrived quite late in the British Isles, seemingly after the refilling of the Irish Sea. Conversely, some early (Lateglacial) immigrants into England, like *Abida secale*, do not occur in Ireland at all.

The problem perhaps is not so much explaining the poverty of the Irish fauna as explaining its relative richness in the face of geological evidence. Equally difficult to account for is the presence of a few species which live in Ireland but not in Britain. The most famous of these is the Kerry slug, *Geomalacus maculosus*.

## The Late Postglacial period: the role of man

During the second half of the Postglacial period the fauna has been progressively affected by the activity of man. His role can hardly be exaggerated: with a few exceptions nearly everything we now see in the landscape of the British Isles is the result of human interference. The faunal changes brought about have been of two kinds: first, alterations in the distribution of native species as a result of habitat change; and secondly, the introduction from abroad, for the most part accidentally, of a considerable number of additional species, which now include some of our commonest snails. Evans (1972) gives a detailed account of these changes, based largely on evidence from archaeological excavations.

Forest clearance was begun in earnest by Neolithic settlers in the fourth millenium BC. At first this took place mainly on the lighter soils, especially on the chalk and limestone uplands and on the alluvial floodplains of some lowland rivers. With the introduction first of metal tools and later of effective wheeled ploughs, the heavier soils were progressively cleared, largely for arable farming. The process was well advanced by the beginning of the Christian era and was largely complete by the end of the Middle Ages.

The most obvious result of this process has been the restriction of the woodland element. Though few molluscs are entirely confined to woodland, many require humidity and shade: in consequence, their distributions have become discontinuous. Several have poor powers of passive dispersal, so that their presence may be evidence of a continuity of forest growth from early in the Postglacial. Useful indicators are *Spermodea lamellata* and *Phenacolimax major*, and *Malacolimax tenellus*, a slug which seems never to occur in plantations; it is a true relict of the forest optimum.

Clearance favoured the spread of species of grassland and open ground, allowing them to re-expand from their refugia in the wake of agriculture. Common Lateglacial snails, like *Pupilla muscorum*, *Vertigo pygmaea*, *Helicella itala* and the three species of *Vallonia*, became common once more, especially on base-rich soils. In general, xerophiles were favoured at the expense of hygrophiles, an effect accentuated in the climatically drier, eastern parts of the British Isles.

An important effect of agriculture was the drainage of wetlands. All marsh species have suffered in consequence, but more especially those with narrow hydrological requirements and an intolerance of disturbance. The clearest example is *Vertigo angustior*, a snail once abundant in lowland Britain but now an endangered rarity.

In spite of such losses, it is important to emphasize that the overall impact of man on the natural fauna has been to enrich and diversify it. Deciduous forest, deeply shaded and relatively uniform over wide areas, has given way to an intricate mosaic of arable land, grassland, scrub, stone walls, hedgerows, managed woodland and so forth. All these new habitats speedily acquired their own characteristic assemblages of molluscs by the natural processes of passive dispersal. In areas of calcareous bedrock, poor, deeply-leached soils were reactivated by ploughing, bringing fresh carbonate to the surface once more.

Stone walls provide important secondary refuges for a number of snails (*Pyramidula rupestris*, *Vertigo alpestris*, *Clausilia dubia*, *Balea perversa*, *Helicigona lapicida*). New aquatic habitats were created by the digging of ponds and marsh drains. The Norfolk Broads, which are flooded mediaeval peat-diggings, give refuge to some of our rarest freshwater species. Of particular importance was the creation of the canal system from about 1750 onwards – a man-made network which today is undoubtedly the richest repository of freshwater molluscs in Britain. The pattern of canals and canalized rivers is closely mirrored by the distribution maps of several species: *Viviparus viviparus*, *Bithynia leachii*, *Unio tumidus* and *Sphaerium rivicola* are the best examples.

With the possible exception of *Margaritifera auricularia*, no species appears to have been lost through human activity; on the other hand, no fewer than thirty species are certain or probable introductions. The process began in the Neolithic and has continued to to present day. Some remain restricted to gardens and other disturbed and artificial habitats but many have spread widely to become among our commonest snails. *Helix aspersa*, a Romano-British arrival from the Mediterranean region, is a striking example. Other such introductions include probably *Oxychilus draparnaudi*, the common *Tandonia* species, *Cecilioides acicula*, *Balea biplicata*, the *Testacella* species, *Candidula intersecta*, *C. gigaxii*, *Cernuella virgata*, *Cochlicella acuta*, *Monacha cartusiana*, *M. cantiana*, *Theba pisana* and *Helix pomatia*. The first notifications of some nineteenth- and twentieth-century introductions, now well established, are: *Trochoidea elegans* (1890), *Hygromia limbata* (1917), *Deroceras panormitanum* (1931), *Hygromia cinctella* (1950), *Boettgerilla pallens* (1972) and *Paralaoma caputspinulae* (1985). Among freshwater species are *Dreissena polymorpha* (1824), *Potamopyrgus antipodarum* (1852), *Musculium transversum* (1856), *Menetus dilatatus* (1869) and *Ferrissia wautieri* (1977).

# THE FUTURE

Distributions are not static. Long-term climatic change will continue to modify distribution patterns, especially for those species at the limits of their thermal tolerances. Some accidentally introduced species, like *Potamopyrgus antipodarum* or *Boettgerilla pallens*, appear to be spreading largely by their own powers of passive dispersal, unaided by man. Other species, like the arionid slugs, show cyclic population explosions or regressions the causes of which are obscure. But, overwhelmingly, it is human activity which is responsible for the changes we now see.

Armed with knowledge of the past history of the British mollusc fauna and the pressures now acting upon it, how should we proceed in the future? Our fauna cannot be compared in biogeographical interest with that, say, of a tropical rain forest or remote oceanic island. We have no endemic species. In a few cases, populations survive *in situ* in habitats first colonized at the end of the ice age or very early in the Postglacial. Some of these species, like *Catinella arenaria* or *Vertigo genesii*, are in danger of extinction and a need to protect their habitats is generally accepted, especially when a similar threat to survival is evident in other European countries. Secondary, man-made habitats can also be important refuges – sometimes the only refuges – for rare native species. But should we attempt to control the activities of man more generally in order to preserve distribution patterns which, for the most part, themselves result from several millenia of random human interference?

The broad answer must surely be that intervention is justified when overall diversity is diminishing, a point certainly now reached in many parts of the British Isles. The causes are well understood.

First, agriculture. In the past the effect of farming was, on the whole, to increase diversity. This has mostly ceased to be the case. The scale of modern farming – especially arable farming – tends instead to remove shelter and dry the landscape. Deciduous woods, hedgerows and managed wetlands have very largely lost their economic basis. Limestone grasslands, once maintained by sheep grazing, have disappeared from many lowland counties, causing a loss of such snails as *Pupilla muscorum* and *Helicella itala*. Farm ponds and ditches are becoming rare, as the maps for *Aplexa hypnorum* and *Anisus leucostoma* show. Equally serious is the excessive use of pesticides, and of nitrate and phosphate fertilizers, which leach out into watercourses, encouraging algal growths which remove light and oxygen. Silage slurry has a similar effect. Bottom sediments become foul and anaerobic, killing the bivalves. Several species, notably *Myxas glutinosa* and *Segmentina nitida*, are seriously threatened for reasons of this kind.

Second, urbanization and industrialization. The physical effects of the industrial revolution surround us. Rural landscapes have been invaded by houses, factories and roads. Many species have declined in consequence, even if urban areas can provide interesting secondary refuges, for example in gardens or in city-centre canals. More serious is the effect of pollutants. This has long been apparent where toxic wastes were discharged directly into rivers, killing or impoverishing the freshwater fauna. Less obvious but more insidious and widespread are the gaseous and other airborne emissions of industry and power generation from coal. Already by the 1870s the disappearance of helicids from begrimed areas of the Black Country and in south Lancashire was ascribed to this cause. The deleterious effect of $SO_2$ on lichen floras in England and Wales has been mapped and studied in detail (Hawksworth & Rose, 1970; Ferry, Baddeley & Hawksworth, 1973). Many land snails that graze on hard surfaces such as rock faces, stone walls and tree trunks have certainly been affected by atmospheric pollution, either directly, or indirectly through damage to lichens and other epiphytes on which they feed. The widespread extinction of *Balea perversa* in central and eastern England has convincingly been explained in this way (Holyoak, 1978a) and several other maps in this Atlas (*Pyramidula rupestris*, *Lehmannia marginata*, *Clausilia*

19

*bidentata, C. dubia, Helicigona lapicida*) reveal what is probably a comparable effect. Recent evidence also suggests that existing levels of airborne pollution are likely gradually to raise the acidity of soils in those areas poorly buffered by naturally occurring carbonate, most seriously in the highland zone. This may lead to a loss of terrestrial shelled species from old non-calcareous western woodlands, and be therefore especially damaging to such snails as *Spermodea lamellata*. A corresponding lowering of pH in fresh waters naturally poor in carbonates in these same areas would impoverish the aquatic fauna – a process amply demonstrated in Scandinavian lakes and capable of being monitored by the disappearance of molluscs with different pH thresholds (K. A. Økland, 1979; J. Økland, 1990).

Conservation is no longer therefore only a matter of protecting rarities in isolated refuges but raises much wider problems concerning our surroundings as a whole. At the most general level, climatic change caused by the production of greenhouse gases may have universal effects; that we should reduce our dependence on fossil fuels seems inescapable. A reversion to less intensive systems of farming seems also necessary if the intricacies of our landscapes are to survive. Historical awareness nevertheless brings its own problems. Is the fauna of a piece of limestone grassland, say, the result of forest clearance and long human usage, as 'important' as that of a surviving fragment of primary woodland? Should we allow particular habitats naturally to revert or evolve, or deliberately freeze them at some arbitrary stage, like ancient buildings? And what should our attitude be towards introductions? If we protect the rare *Monacha cartusiana*, a 'weed' of cultivation introduced by Neolithic farmers, should we not also protect the equally rare *Trochoidea elegans*, a Victorian equivalent?

Though no absolute answers can be given to such questions, two general principles may be suggested. First, the desirability of maintaining faunal variety, at every level from the national to the quite local. And secondly, a respect for habitats not merely as repositories for this or that species, but as embodying a record of long-continued historical change, both natural and man-made.

# DISTRIBUTION MAPS AND SPECIES ACCOUNTS

Explanation of maps and
accompanying notes

A map is included for all terrestrial and freshwater species, whether native or introduced, living in open habitats in the British Isles. A few semi-marine snails of saltmarsh and brackish water in the families Hydrobiidae, Assimineidae, Truncatellidae and Ellobiidae have additionally been mapped. Also included are maps for those few species which have become extinct in the British Isles since the last ice age. Greenhouse aliens are excluded, as are also in most cases greenhouse finds beyond the ranges of otherwise naturalized species. Chance 'casuals' and a few introduced species briefly established in the past but which now seem to be extinct (e.g. *Cernuella neglecta* at Luddesdown, Kent, *c*.1915–22) have similarly been disregarded.

The mapping unit is the 10km square of the British or Irish national grids (or, in the case of the Channel Islands, the international UTM grid). The symbols used are as follows:

- ●    Records made in or after 1965
- ○    Records made prior to 1965 only; the majority of these belong to the period 1880–1914
- +    Fossil occurrences (Lateglacial and Postglacial) lying outside modern ranges

An illustration accompanies each map. It is hoped that they may be found helpful. Nevertheless this book is not intended to serve as an identification guide and for some families (e.g. the arionid slugs) the drawings are manifestly inadequate for this purpose. An approximate measurement is given for the size of the shell at maturity – either height or breadth, whichever is the greater. In the case of slugs, the measurement is for the extended body-length when crawling. A reference is also given, usually to one of the following works, abbreviated as follows:

| | |
|---|---|
| Ellis | A. E. Ellis, *British Snails*. Oxford, 1926 (revised edition with unchanged pagination, 1969) |
| Ellis (Bivalves) | A. E. Ellis, *British freshwater bivalve Mollusca*. Linnean Society of London, 1978. |
| K. & C. | M. P. Kerney & R. A. D. Cameron, *A field guide to the land snails of Britain and north-west Europe*. Collins, 1979 (reprinted 1987, 1994). |
| Macan | T. T. Macan, *A key to the British fresh- and brackish-water gastropods*. Freshwater Biological Association, 1949 (several later reissues). |

Other references will be found in the bibliography.

*Nomenclature*

Molluscan taxonomy is far from settled. In the absence of a generally agreed international checklist, the familiar if somewhat conservative nomenclature of the 1976 *Atlas* has been retained with a few modest changes where favoured by current opinion. Taxonomic order has been left largely unaltered. Common synonyms are given, mainly names employed in 20th-century British literature. Also included in the synonymy are a few recently introduced names which may gain wider currency. For innumerable older synonyms, Kennard & Woodward (1926) should be consulted.

English names are mainly from Ellis (1926) or from Victorian popular manuals. Very few of these names are genuinely vernacular and most are evidently back-formations from the Latin names of the species (in many cases now changed). Though hardly used in serious scientific literature they are included here for their historical interest.

The explanatory text is normally divided into three paragraphs, dealing respectively with habitat; with history and status; and with range outside the British Isles. A fourth paragraph is occasionally appended dealing with problems of taxonomy or related matters.

*Habitat*

Unless specified to the contrary, this refers to the British Isles only and may sometimes be different elsewhere. Much of the information comes from the classic papers of Boycott (1934, 1936a), modified or amplified from more-recently published work and from personal experience. A few additional references are given to papers describing specific occurrences within the British Isles, usually of the rarer species.

*History and status*

Species are categorized as either *native* (those which recolonized the British area from refugia on the continental mainland following the retreat of the ice) or *introduced* (accidentally or – much more rarely – deliberately brought in by man from the Neolithic period onwards).

Fossil occurrences are denoted as follows:

Lateglacial (Lgl): fossils known from the closing stages of the ice age, prior to about 10,000 bp.

Early Postglacial (E.Pgl): earliest known certain fossils from the period subsequent to 10,000 BP but prior to the first widespread evidence of prehistoric (Neolithic) agriculture about 5,000 BP (roughly the Preboreal, Boreal and Atlantic periods, covering the forest optimum of the Postglacial).

Late Postglacial (L.Pgl): earliest known certain fossils from within the period of prehistoric and later agriculture, subsequent to about 5,000 BP (roughly the Sub-boreal and Subatlantic periods).

For a much fuller account of the geological history of each species, see Preece & Holyoak (in preparation).

Comment is made also on historical or recent changes in distribution. It is important to note that statements like 'no evidence of significant change' should be viewed in the context of a national survey; mapping on a finer scale (e.g. by tetrads) or over a briefer period may reveal evidence of flux or decline otherwise masked. For the thirty or so species believed to be nationally at risk and included in the British Red Data Book (RDB; Bratton, 1991), the degree of threat is indicated as follows:

Extinct: not located in the wild during the past 50 years.

Endangered: now extremely localized and extinction possible if causal factors continue. Species now occurring in only one 10km square are also assigned to this category.

Vulnerable: declining and likely soon to move into the Endangered category if causal factors continue.

Rare: populations small or localized but where the degree of threat is relatively low.

Insufficiently Known: suspected to belong to one of the above categories but precise information lacking.

Such ratings of course need to be periodically reviewed in the light of increasing knowledge or changing circumstances; a few published in 1991 are already open to question. No corresponding Red Data Book has yet been published for Ireland. Separate RDB ratings are however suggested in a few cases where the situation differs significantly from that in Great Britain.

*Range*

A brief diagnosis of total world range is given. Distribution maps for the terrestrial species in western and central Europe will be found in Kerney, Cameron & Jungbluth (1983).

# MAPS OF MOLLUSCAN DISTRIBUTION

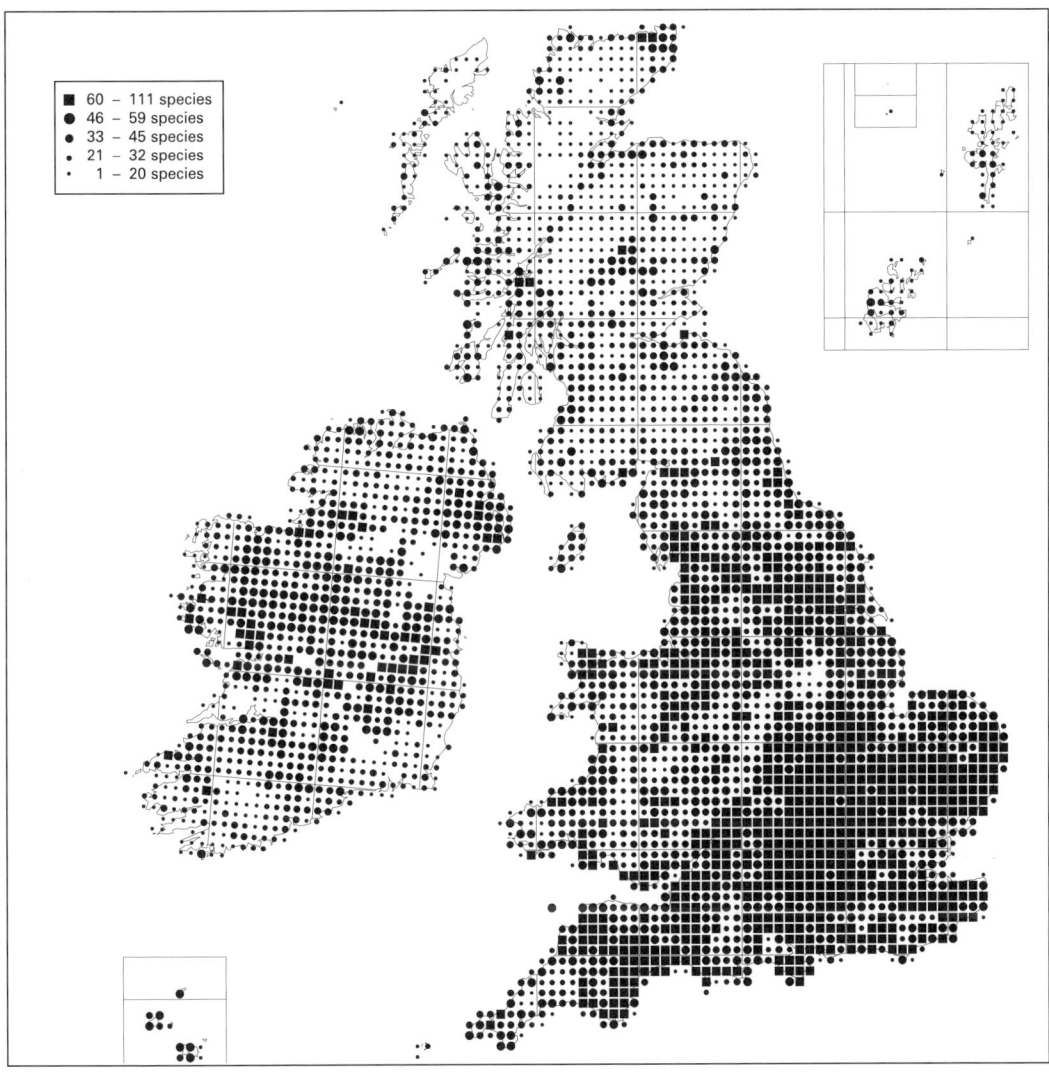

## TOTALS RECORDED IN 10 KM SQUARES

This map shows the number of species recorded since 1965 in each 10km square of the British and Irish national grids. Most parts of the British Isles have now been adequately sampled, and therefore in a rough way the map also reflects variations in the numbers of species living in different areas. The south-eastern lowlands are evidently richer in species than the cooler, non-calcareous highlands of the north and west. This difference tends however to be exaggerated by the relatively greater attention paid by recorders to southern Britain. Even within England and Wales there are considerable inequalities in the recording coverage: some counties (like Nottinghamshire and Staffordshire) remain poorly known, whereas others (like Devon, Cardigan, Bedfordshire and Suffolk) have been mapped in great detail. It is important to note also that clusters of old (pre-1965) records shown on certain maps may signify only a lack of recent fieldwork in those areas, and not the disappearance of the species in question. This is especially true in Scotland and Ireland. In Ireland, Co. Louth (v.c. H31) has hardly been visited by malacologists since it was intensively studied by P. H. Grierson in 1904–5. In south-east Scotland (roughly v.cs 77–87) most of the detailed freshwater recording was done by D. K. Kevan of Edinburgh in the years 1930–60; the apparent decline of many common freshwater species in this area (e.g. *Physa fontinalis*, *Lymnaea truncatula*, *Bathyomphalus contortus*, *Pisidium casertanum*, *P. milium*, *P. subtruncatum*) is therefore illusory. Biases like these have been taken into account in the interpretive notes which follow.

NERITIDAE: *THEODOXUS*

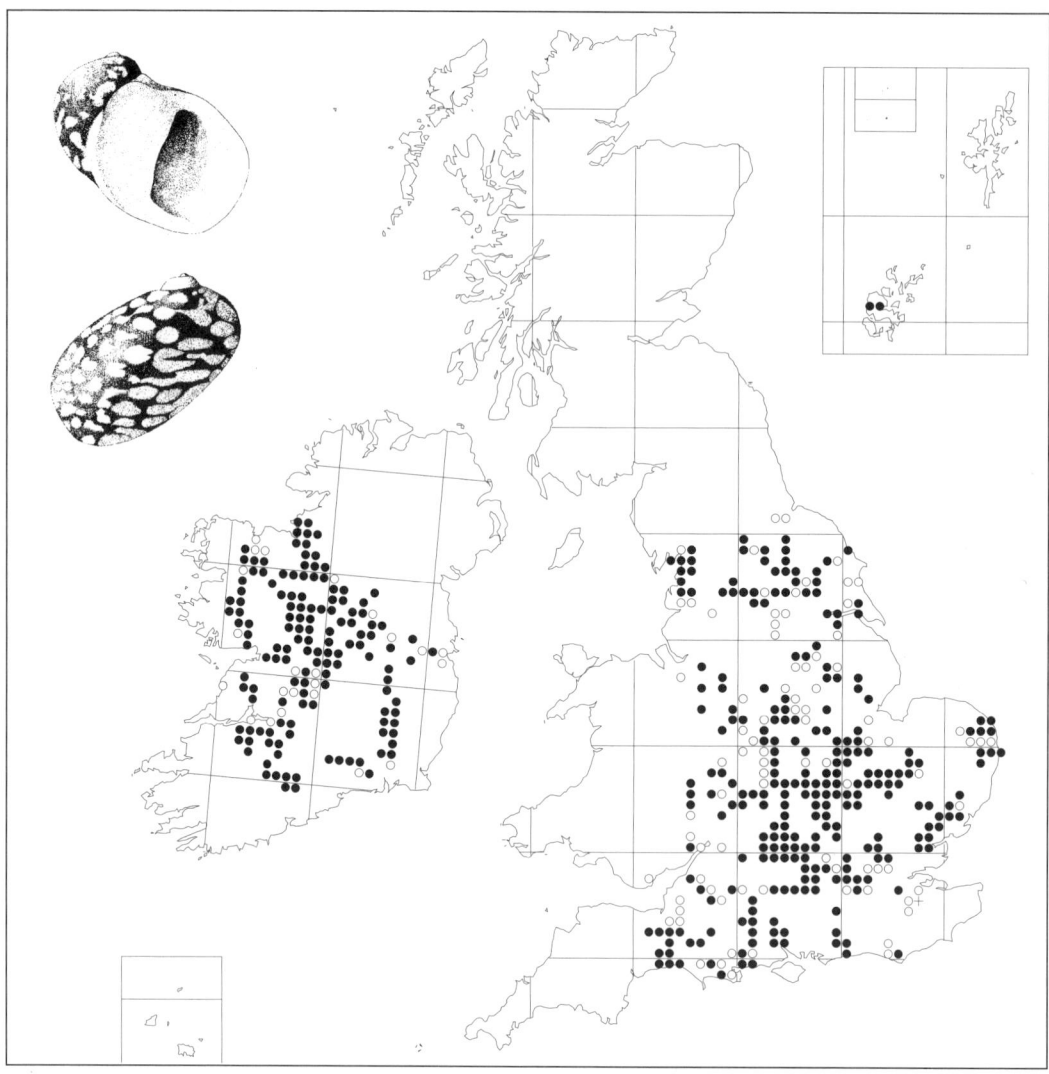

**Theodoxus fluviatilis** (L., 1758) (*Neritina fluviatilis* (L.))

River nerite   Size: 9–13mm

A species restricted to well-oxygenated waters rich in lime, living in both slow-flowing and fast-flowing rivers, in canals and in the wash-zone of large calcareous lakes (notably in Ireland, and in the very isolated populations in Loch of Harray and Loch of Stenness, Orkney; Boycott, 1936b). It likes hard surfaces for grazing, on pebbles, rocks or the masonry of bridges. It is tolerant of brackish water, commonly up to 2.5 parts per thousand NaCl but much higher salinities have occasionally been reported. The robust shells washed out to sea may sometimes be picked up on beaches.

Probably native (poorly dated or L.Pgl fossils only). *Theodoxus* has declined in some areas due to habitat loss, and in seriously polluted canals and rivers, but is not seriously threatened. It will tolerate mild organic pollution (Lucey *et al.*, 1992).

W. Palaearctic; in Europe to 64°N in Sweden.

(Ellis: 69; Macan: 13)

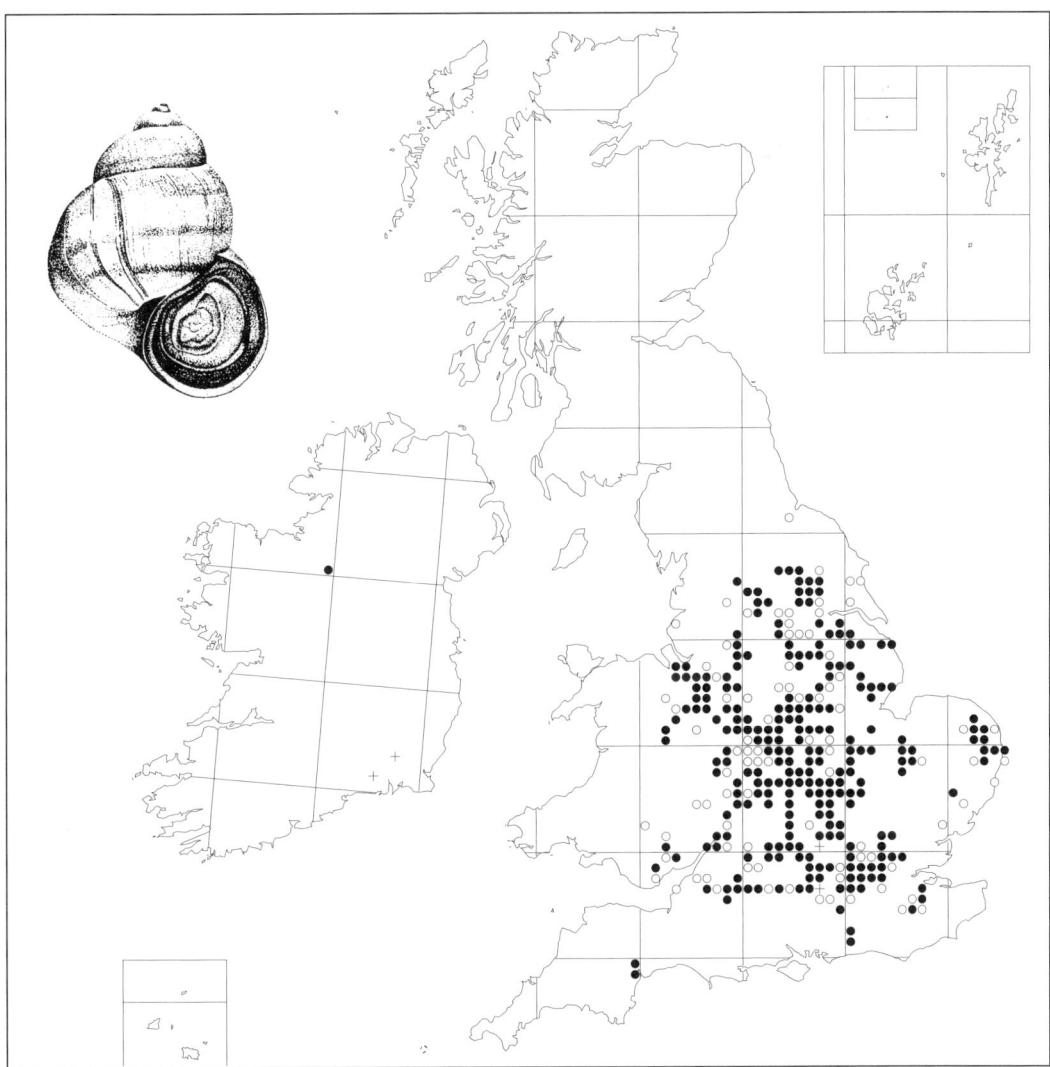

**Viviparus viviparus** (L., 1758) (*Paludina vivipara* (L.))

Common river snail    Size: 25–35mm

A calciphile, restricted to large, deep bodies of still or slowly moving, well-oxygenated water in major lowland rivers, canalized waterways and canals (it is a characteristic species of the English canal system). It is not normally found in small isolated ponds, unless introduced by man. *Viviparus* is a suspension feeder, favouring a muddy bottom in which it lies for long periods with the mouth uppermost.

Probably native (L.Pgl fossils only). In England the 'canal basin' distribution of this species owes much to man. Though not in serious decline, some losses may be ascribed to habitat change or pollution. In Ireland the status of *V. viviparus* is unclear. Shells of uncertain age have been dredged from the Rivers Suir and Barrow (McMillan & Stelfox, 1962). A population in a pond in Co. Leitrim may be a recent introduction (Cotton, 1996).

European (not in S. Europe): widespread in lowland areas to 60°N in Scandinavia.

(Ellis: 85; Macan: 13)

VIVIPARIDAE: *VIVIPARUS*

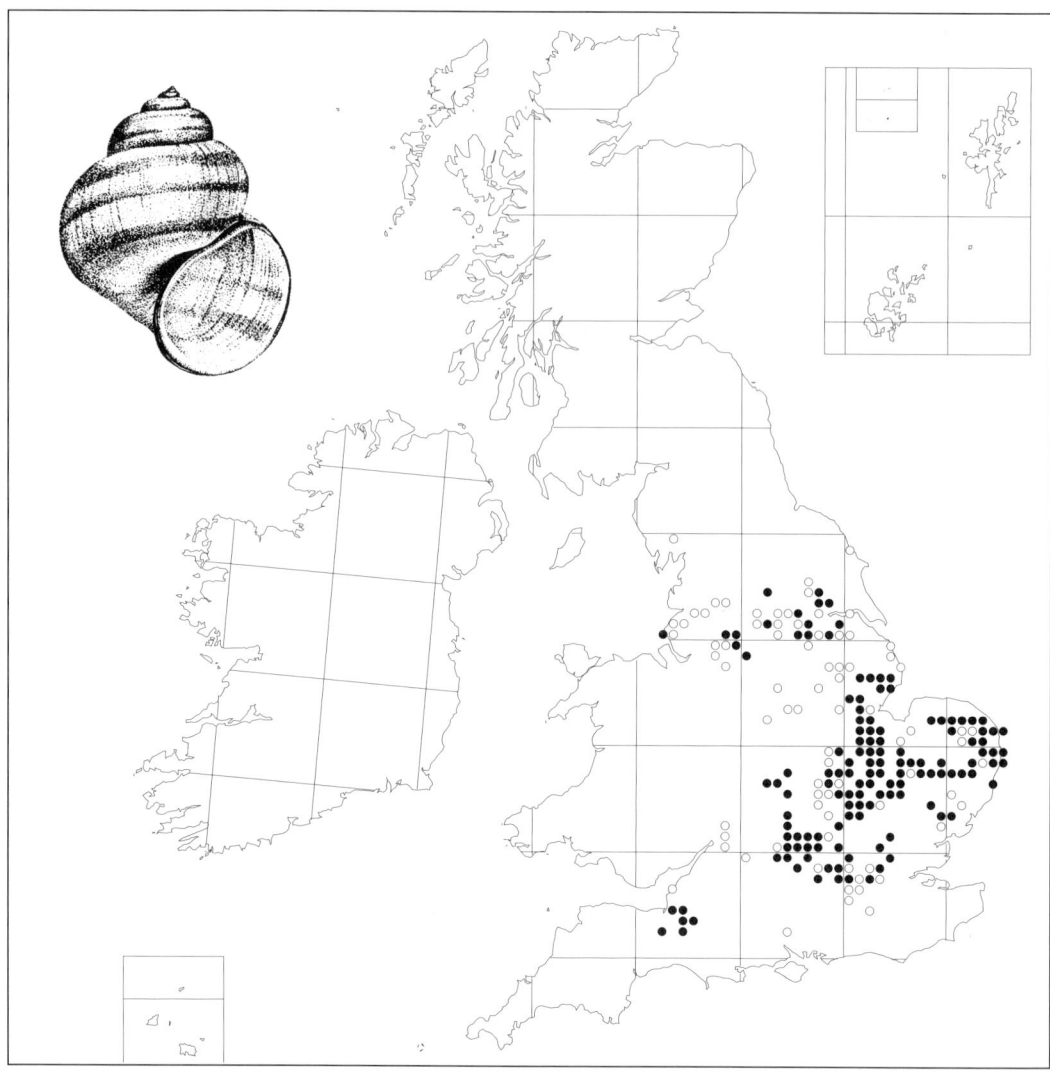

**Viviparus contectus** (Millet, 1813) (*Paludina contecta* Millet, *Viviparus fasciatus* of British authors, non Müller)

Lister's river snail    Size: 30–40mm

This species favours broadly similar habitats to *Viviparus viviparus*: large bodies of well-oxygenated, hard water over muddy substrates in major lowland rivers, canals and drainage ditches. It is more frequent in still water than *V. viviparus* and is less associated with canals, often occurring in lakes and fenland dykes. The two species may however be associated. Occasionally introduced successfully into garden ponds.

Probably native (poorly dated Pgl fossils only). This species shows a clearer decline than *V. viviparus*, especially in the north of England and around London. Nevertheless it is still common in its strongholds in the East Midlands and in East Anglia. The isolated occurrences in the Lake District (Wise Fen Tarn, Scales Tarn) are deliberate introductions of the 1940s; they have not been confirmed in recent years.

European: Sweden to 58°N, absent from southern Europe.

(Ellis: 85; Macan: 13)

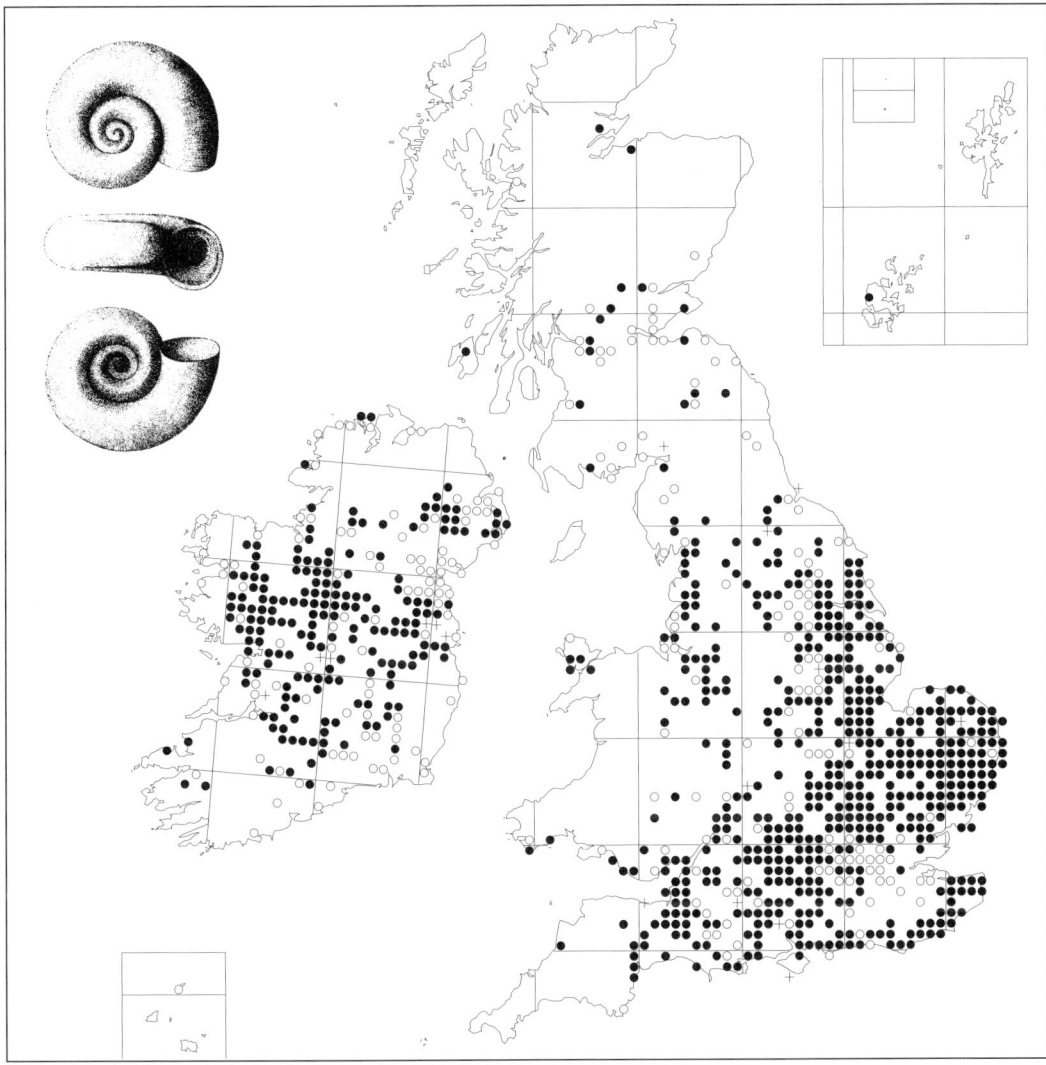

**Valvata cristata** Müller, 1774

Flat valve snail    Size: 3.0–3.5 (occ.–4.0)mm

A species restricted to well-oxygenated, slowly flowing or still water, with a strong preference for richly vegetated places on muddy substrates. It is often common in the quiet clean backwaters of lowland rivers, or among emergent vegetation at the shallow edges of ponds, lakes and drainage ditches. A diverse fauna is usually associated.

Native (Lgl). There is little evidence of significant recent national change. Nevertheless the habitat type is a vulnerable one and there have been local extinctions through pollution and the embanking of waterways. This is especially true around London.

Palaearctic; nearly throughout lowland Europe but absent from northern Scandinavia.

(Ellis: 89)

VALVATIDAE: *VALVATA*

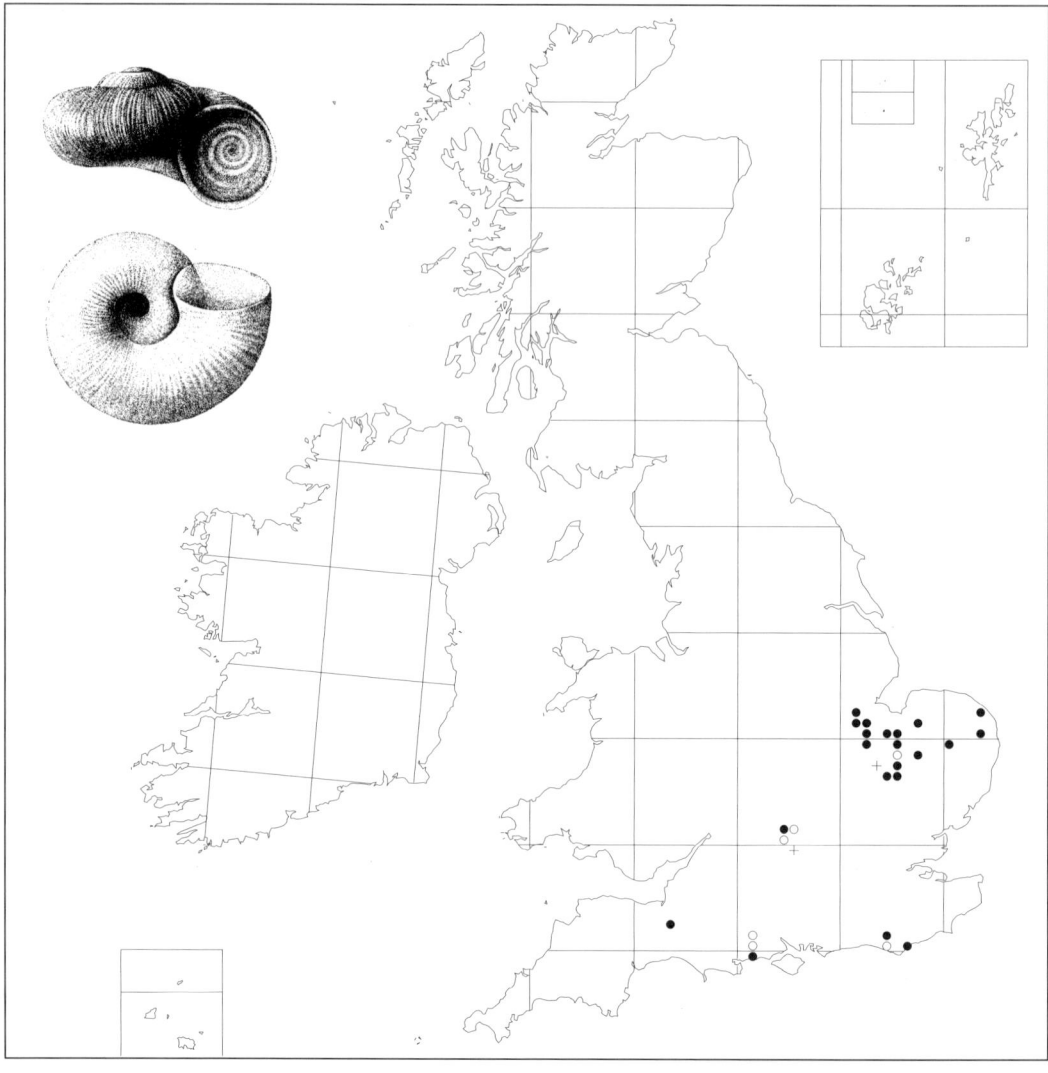

**Valvata macrostoma** Mörch, 1864 (*V. pulchella* of many Continental authors)

Size: 3.5–5.0mm

A rare calcicole restricted to drainage ditches in marshland levels and river floodplains. It lives in stagnant or slowly moving water in well-vegetated places with a good diversity of species. Other rare molluscs may be associated, notably *Anisus vorticulus*, *Segmentina nitida* and *Pisidium pseudosphaerium* (Ellis, 1931).

Probably native (Pgl fossils of uncertain date only). This very local mollusc is still common at a few sites – notably in the southern Fenland and the Pevensey Levels – but is declining. Most of the populations in the area north of Peterborough recorded in the 1960s and early '70s appear now to be extinct. Near Oxford it has not been seen since about 1975. Surviving populations are at risk through habitat loss by disturbance or eutrophication (Hingley, 1979). RDB category: Vulnerable.

C. and N. European; exact range somewhat uncertain owing to confusion with other species.

Care is needed in identifying *V. macrostoma*. Juveniles or depressed forms of *V. piscinalis* are commonly mistaken for it, and some literature records are consequently unreliable.

(Ellis: 88)

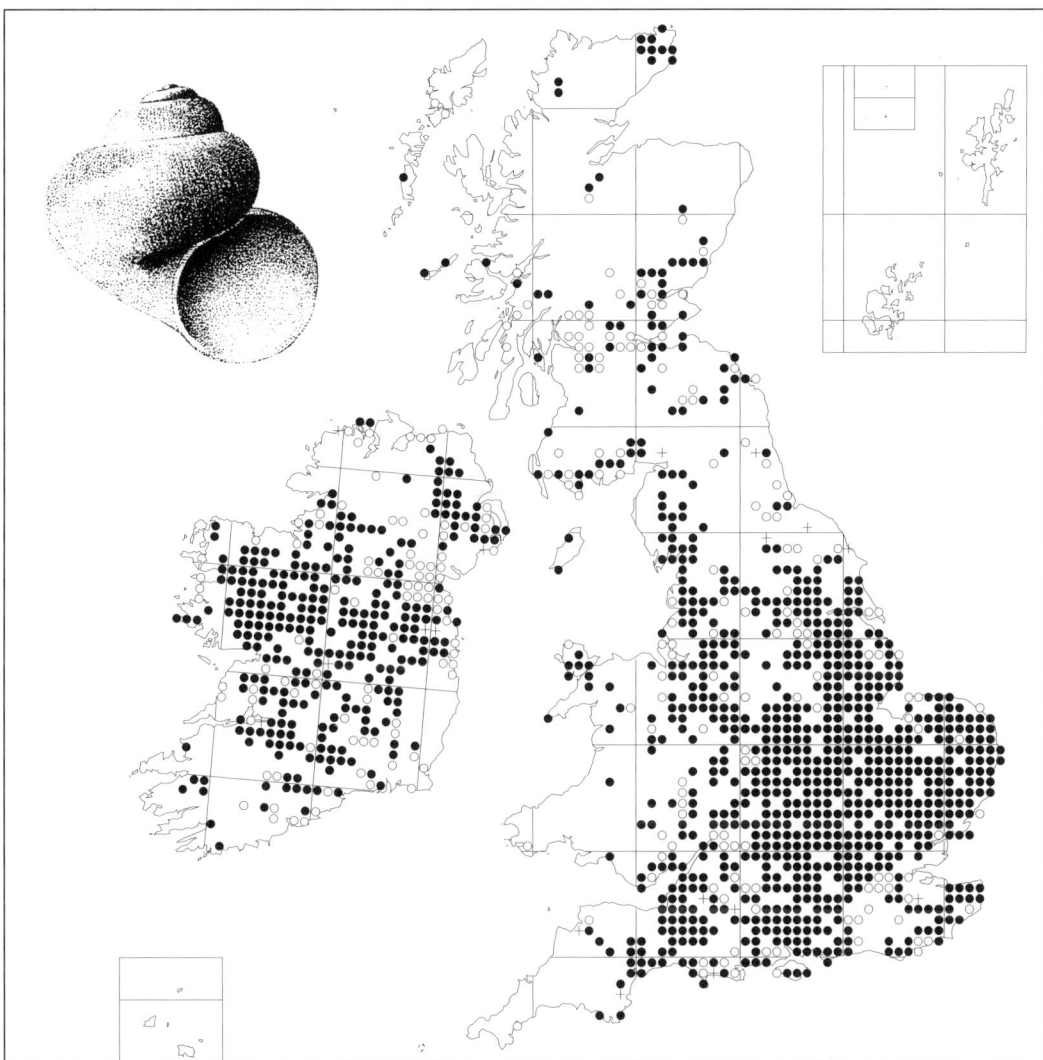

**Valvata piscinalis** (Müller, 1774)

Common valve snail   Size: 4.5–6.0mm

A species common in larger bodies of slowly flowing or still water, with a preference for muddy or silty substrates: rivers, canals, lakes, large permanent running ditches (but not usually in closed ponds). It is tolerant of soft water. In major lakes it extends in small numbers to considerable depths (commonly 3–10m, occasionally to 50m or more), well below the range of other freshwater gastropods. *V. piscinalis* can form an important part of the diet of fish.

Native (Lgl). No evidence of any regional change.

Palaearctic; common throughout Europe, becoming sporadic in mountain areas of Britain and Scandinavia.

(Ellis: 87; Macan: 15)

POMATIASIDAE: *POMATIAS*

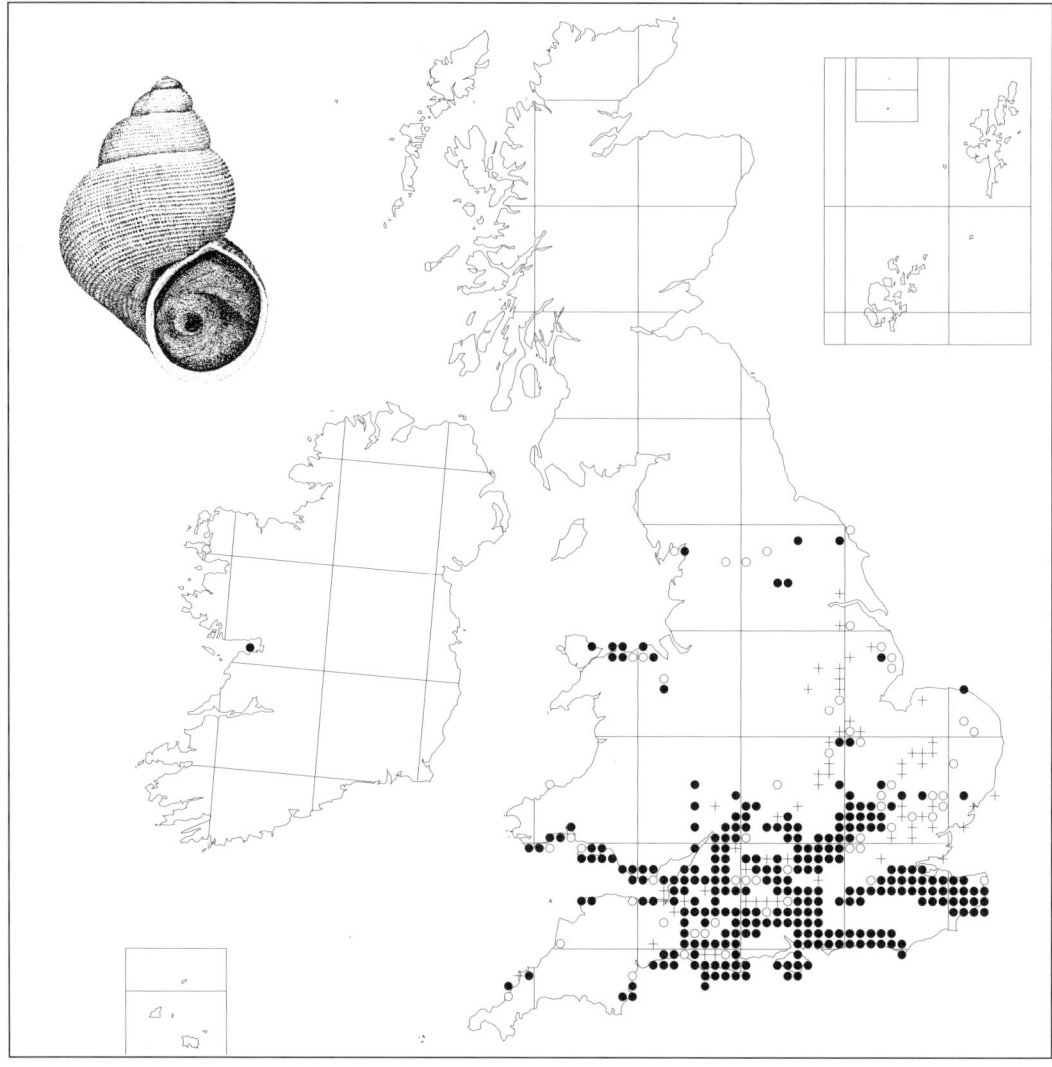

**Pomatias elegans** (Müller, 1774) (*Cyclostoma elegans* (Müller))

Round-mouthed snail    Size: 12–16mm

A snail locally common in hedge banks, open woods, old quarries, cliffs and maritime grassland, always on loose rubbly soils in which it can burrow and hibernate. It is a strict calcicole, found only on chalk, limestone or other highly calcareous rock, including glacial deposits in East Anglia and shell-sand in Cornwall (Boycott, 1921).

Native (E.Pgl). This species has declined strongly over much of central and eastern England, an area of relatively low winter temperatures where Postglacial fossils show it once to have been common (Kerney, 1972). Relict populations are here at risk from intensive farming and several have been destroyed in recent years, notably in East Anglia and in Lincolnshire, by the ploughing-out of hedge banks and similar uncultivated refuges. In Leicestershire it was last seen about 1850 and in Warwickshire in 1916. *P. elegans* is under less pressure further south and west where it is still abundant in appropriate habitats. The apparently unique Irish population in the Burren was discovered only in 1976 (Platts, 1977). RDB category (Ireland only): Endangered.

Mediterranean and W. European, north to the British Isles and east to C. Germany; in Scandinavia only in Denmark.

(Ellis: 73; K. & C.: 53)

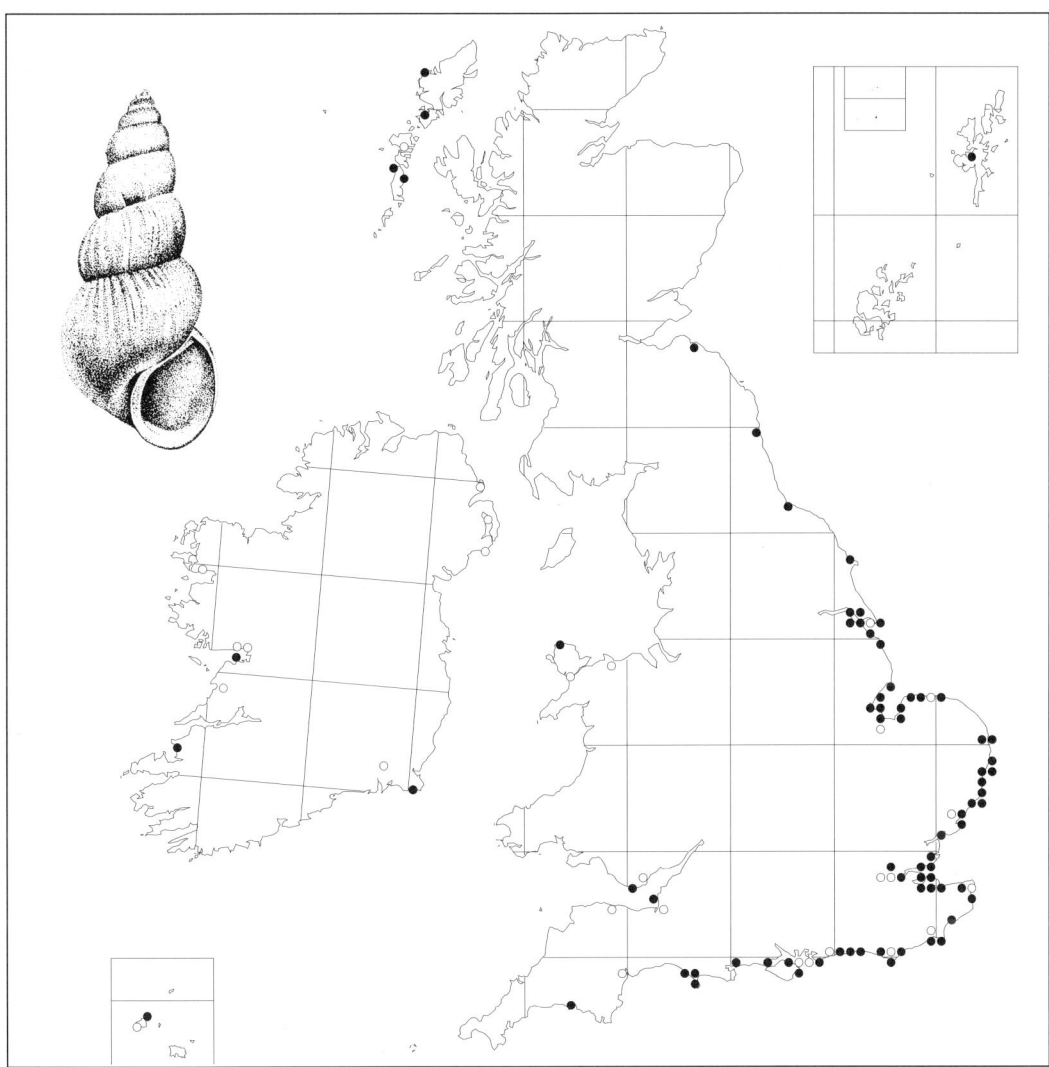

**Hydrobia ventrosa** (Montagu, 1803) (*Paludestrina ventrosa* (Montagu), *Ventrosia ventrosa* (Montagu))

Spire snail    Size: 4–6mm

Inhabits water of low to moderate salinities (most commonly 5–25 parts per thousand NaCl) in quiet estuaries, ponds behind shingle bars, and lagoons and drainage ditches in coastal marshes. Unlike *H. ulvae*, it lives more or less permanently submerged, on bottom mud or in aquatic vegetation (especially *Enteromorpha* and *Potamogeton*). *H. ventrosa* prefers more sheltered places than *H. ulvae*, usually non-tidal and lacking a direct connection with the open sea (Muus, 1963; Bishop, 1976; Cherrill & James, 1985).

Native (E.Pgl). *H. ventrosa* shows some local decline, its habitats being more vulnerable than those of *H. ulvae*, but is not seriously threatened.

W. European and Mediterranean coasts, north to Iceland and to 63° in the Baltic.

This species is not easily distinguishable on shell characters from *H. neglecta*, to which a few of the mapped occurrences may belong.

(Ellis: 77; Bishop, 1976)

HYDROBIIDAE: *HYDROBIA*

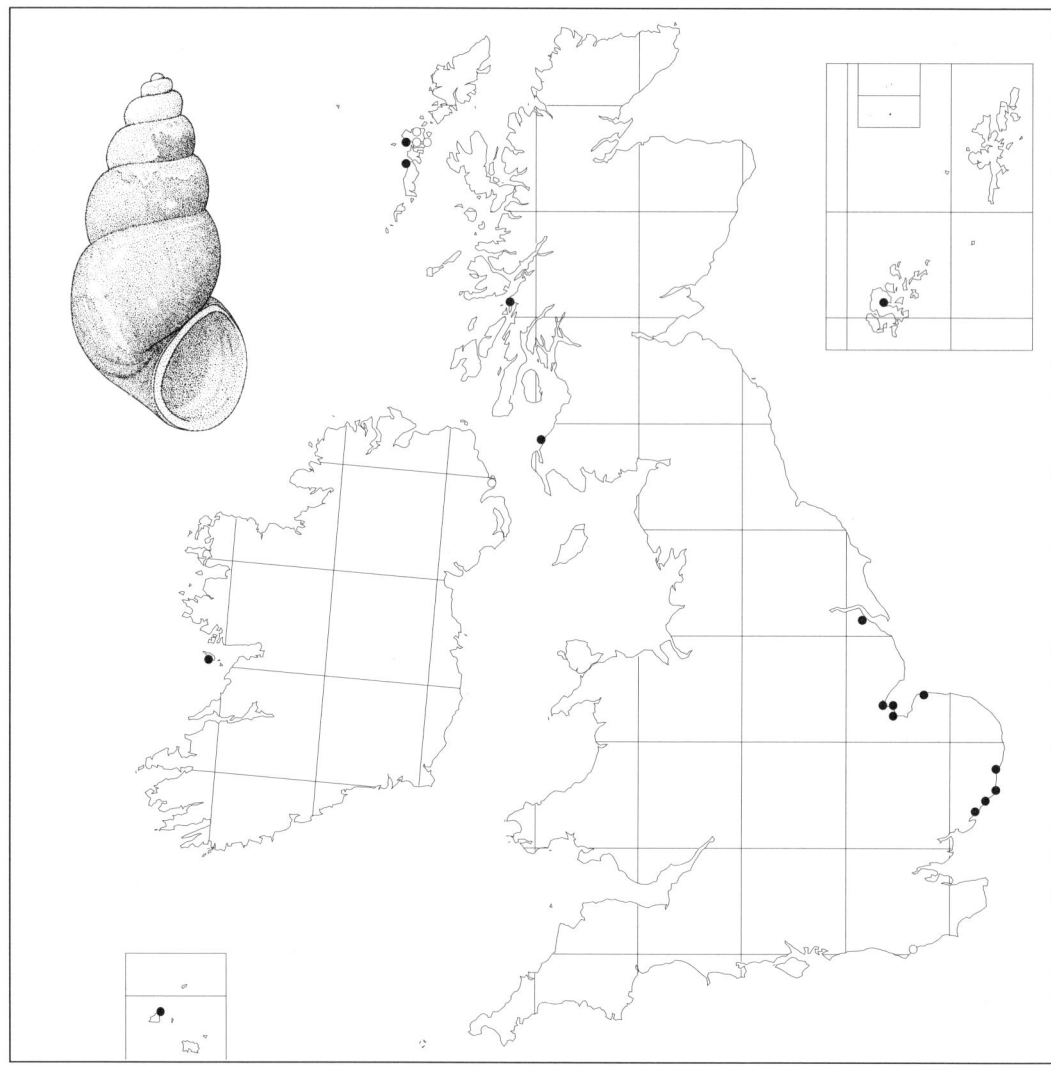

**Hydrobia neglecta** Muus, 1963

Size: 3–4mm

Inhabits rather strongly brackish water (commonly 10–33 parts per thousand NaCl) in sheltered lagoons and marsh drains. It tends to prefer salinities intermediate between those suitable for *H. ventrosa* and *H. ulvae*. It occurs in permanently submerged (non-tidal) situations with such plants as *Enteromorpha*, *Zostera* and *Potamogeton*, in this resembling *H. ventrosa* rather than *H. ulvae* (Muus, 1963; Bishop, 1976; Cherrill & James, 1985).

Probably native (no certain Pgl fossils). There is no clear evidence for decline. Though underrecorded, the species is undoubtedly much more local than *H. ventrosa* and to that extent vulnerable in the small number of known sites.

W. European coasts from France to Denmark.

This species was first described only in 1963. It is not easily distinguishable on shell characters from *H. ventrosa* with which it was previously confused, and care is needed in identification.

(Muus, 1963; Bishop, 1976)

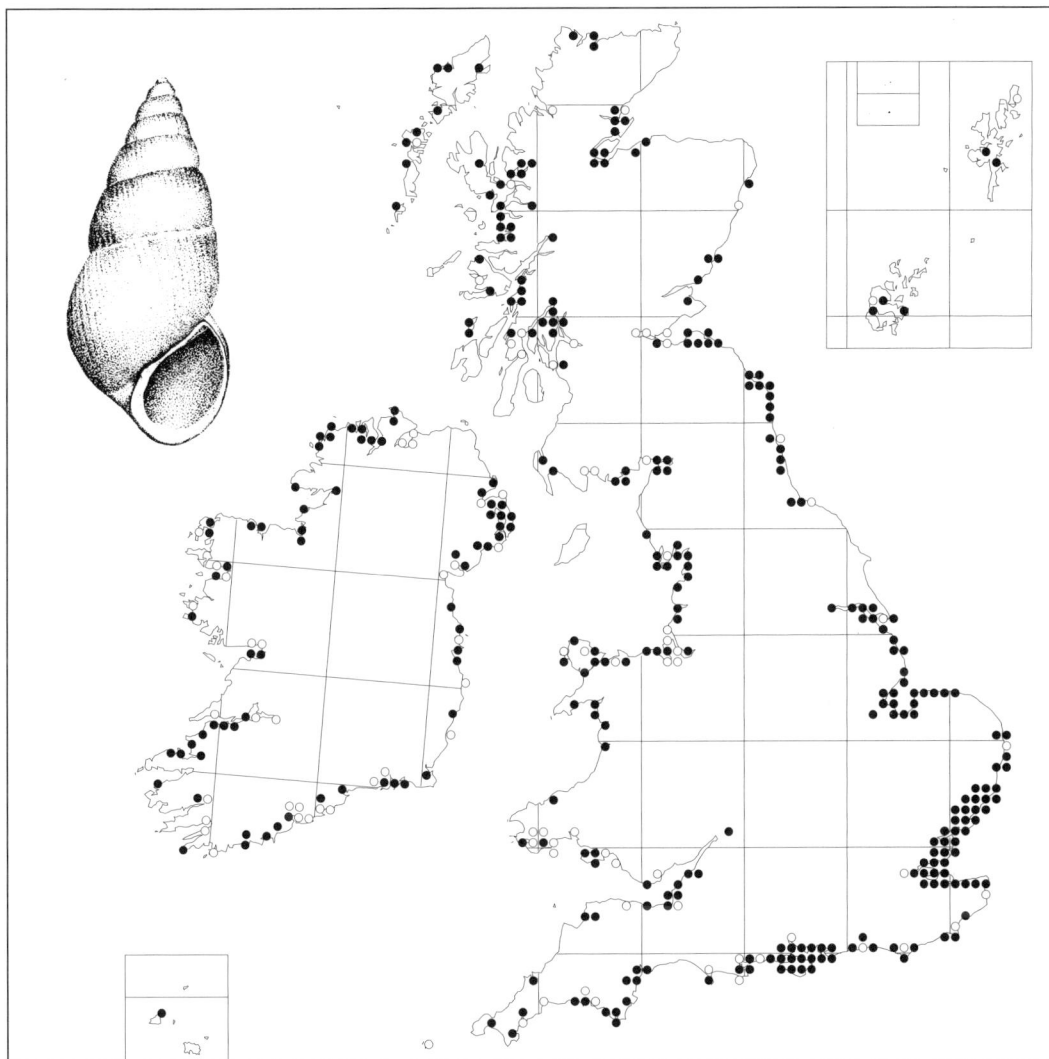

**Hydrobia ulvae** (Pennant, 1777) (*Sabanaea ulvae* (Pennant), *Peringia ulvae* (Pennant), *Hydrobia* (or *Paludestrina*) *stagnalis* of British authors)

Laver spire snail   Size: 4–6 (occ.–9)mm

A species restricted to brackish or salt water (commonly 10–33 parts per thousand NaCl; occasionally in hypersaline conditions) in estuaries, intertidal mudflats and saltmarshes. It is commonest within the upper half of the intertidal zone, where at low tide it may occur in countless millions on exposed muddy or silty surfaces, sometimes associated with marine molluscs like *Cerastoderma* and *Scrobicularia*. It is mainly a detritus feeder, but will also graze directly on seaweeds such as *Ulva* (laver) (Muus, 1963; Bishop, 1976; Cherrill & James, 1985).

Native (E.Pgl). No evidence of regional change.

W. African and European coasts, north to Arctic Russia.

(Ellis: 76; Macan: 17)

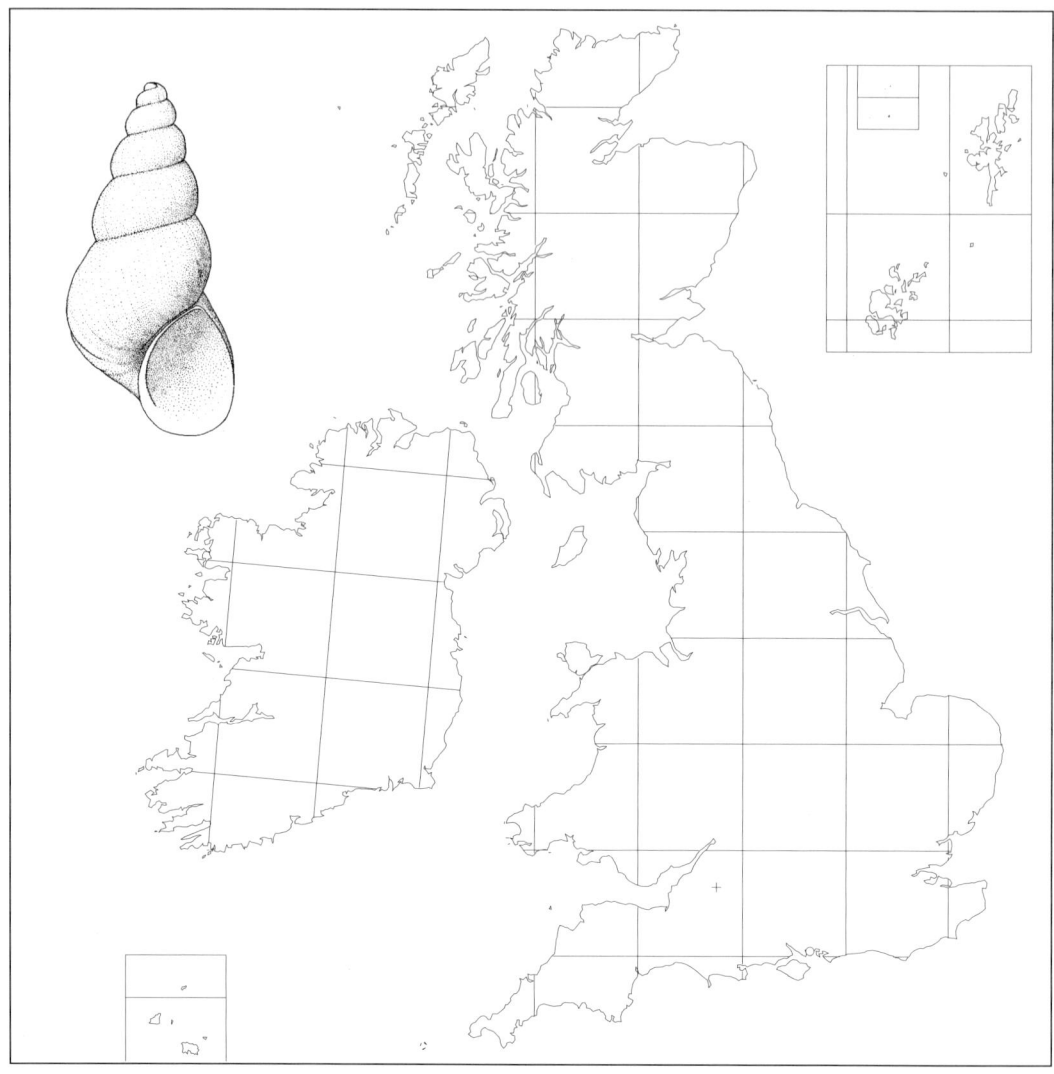

**Heleobia stagnorum** (Gmelin, 1791) (*Hydrobia stagnorum* (Gmelin), *Semisalsa stagnorum* (Gmelin))

Size: 3.5–5.0mm

In the Netherlands this species lives in a few non-tidal lagoons and drainage ditches near the sea, in brackish water of low salinity (2.5–8.0 parts per thousand NaCl). It is usually associated with *Hydrobia ventrosa*.

Native? *H. stagnorum* has been recognized at only two sites in Britain. At Farlington marshes, near Portsmouth, Hants, dead shells occur in the bottom sediment of a brackish lagoon, from which the species appears to have died out about 1960–70. Numerous shells originally identified as *Hydrobia ventrosa* are known also from deposits of presumed Roman age in the Great Bath at Bath, Som. In the Netherlands, *Heleobia stagnorum* is declining and the few remaining sites are threatened by urban pollution and by changes in salinity brought about by coastal control schemes.

Poorly known; the Netherlands, eastern Germany and probably S. France. Related species occur around the Mediterranean and in Asia Minor.

(Bank *et al.*, 1979; Bank & Butot, 1984)

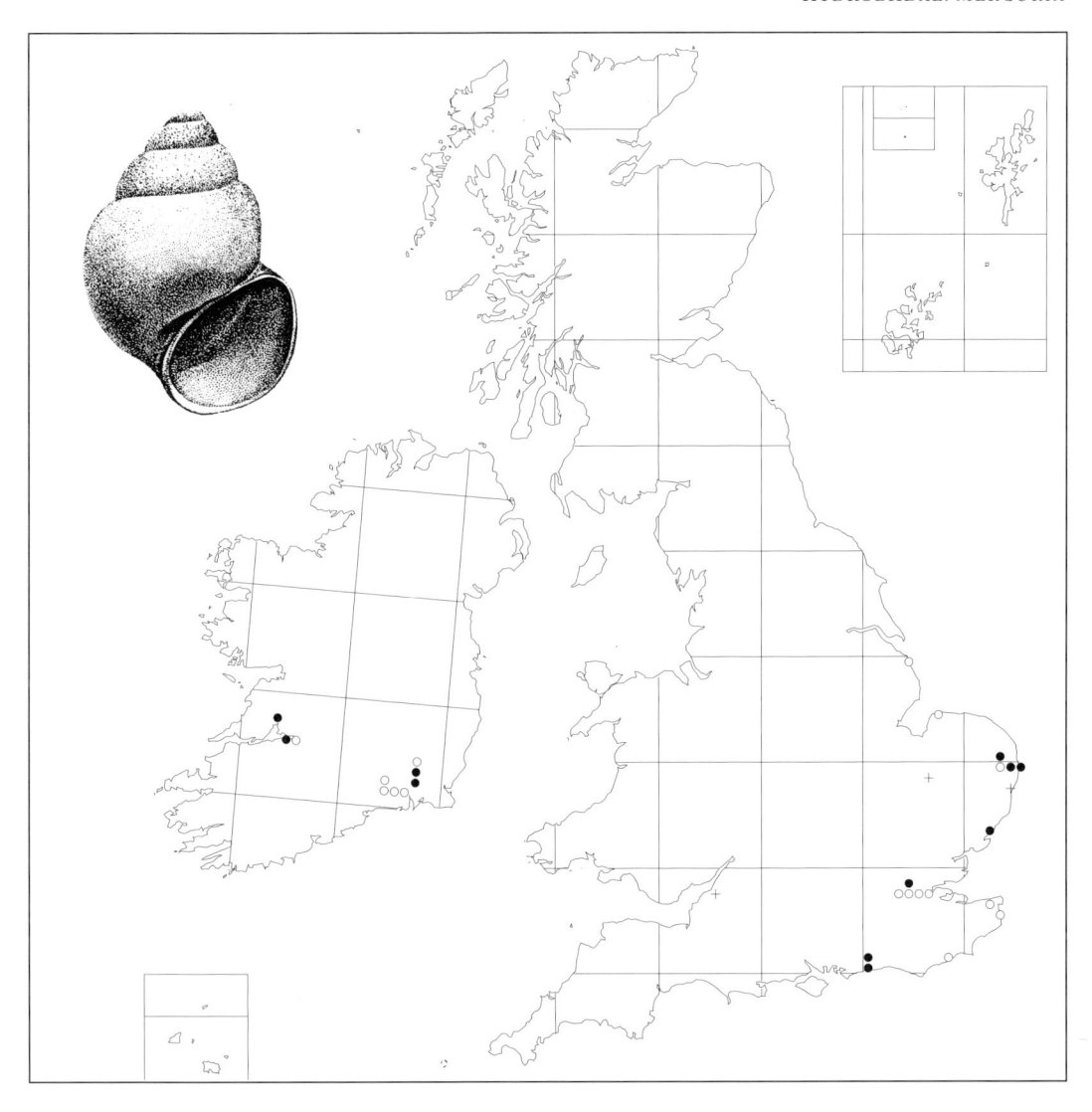

**Mercuria confusa** (Frauenfeld, 1863) (*Amnicola confusa* (Frauenfeld), *Pseudamnicola confusa* (Fr.), *Paludestrina confusa* (Fr.), *Hydrobia similis* and *H. anatina* of older British authors)

Swollen spire snail   Size: 3–4mm

Restricted to nearly fresh water (1–5 parts per thousand NaCl) in river estuaries and in tidal ditches and pools. It prefers soft muddy substrates exposed at low tide, in quiet sheltered situations at the base of emergent vegetation such as *Phragmites* or *Carex*. It is usually associated with freshwater molluscs only, not with other brackish-water species like *Hydrobia* or *Ovatella* (Oldham, 1922a; Holyoak, 1983). Occasionally it is found with *Assiminea grayana*.

Probably native (poorly dated or L.Pgl fossils only). This is a declining species, at risk by reason of its narrow ecological requirements and by being at the edge of its geographical range. The largest surviving British populations are probably those in Oulton Broad, Suffolk. Though formerly common along the lower Thames, it was considered extinct there by 1896 owing to industrial pollution (Castell, 1962). However, in 1984 a small population was found in Barking Creek, Essex (Harris, 1985). The northernmost records, at Thornham, Norfolk, and Saltfleet, Lincs, date from 1960 and 1913 respectively. All populations are at risk from pollution and river-management schemes affecting salinities. RDB category: Endangered.

W. Mediterranean and Atlantic coasts, north to the British Isles and the Netherlands.

(Ellis: 81; Macan: 17)

HYDROBIIDAE: *POTAMOPYRGUS*

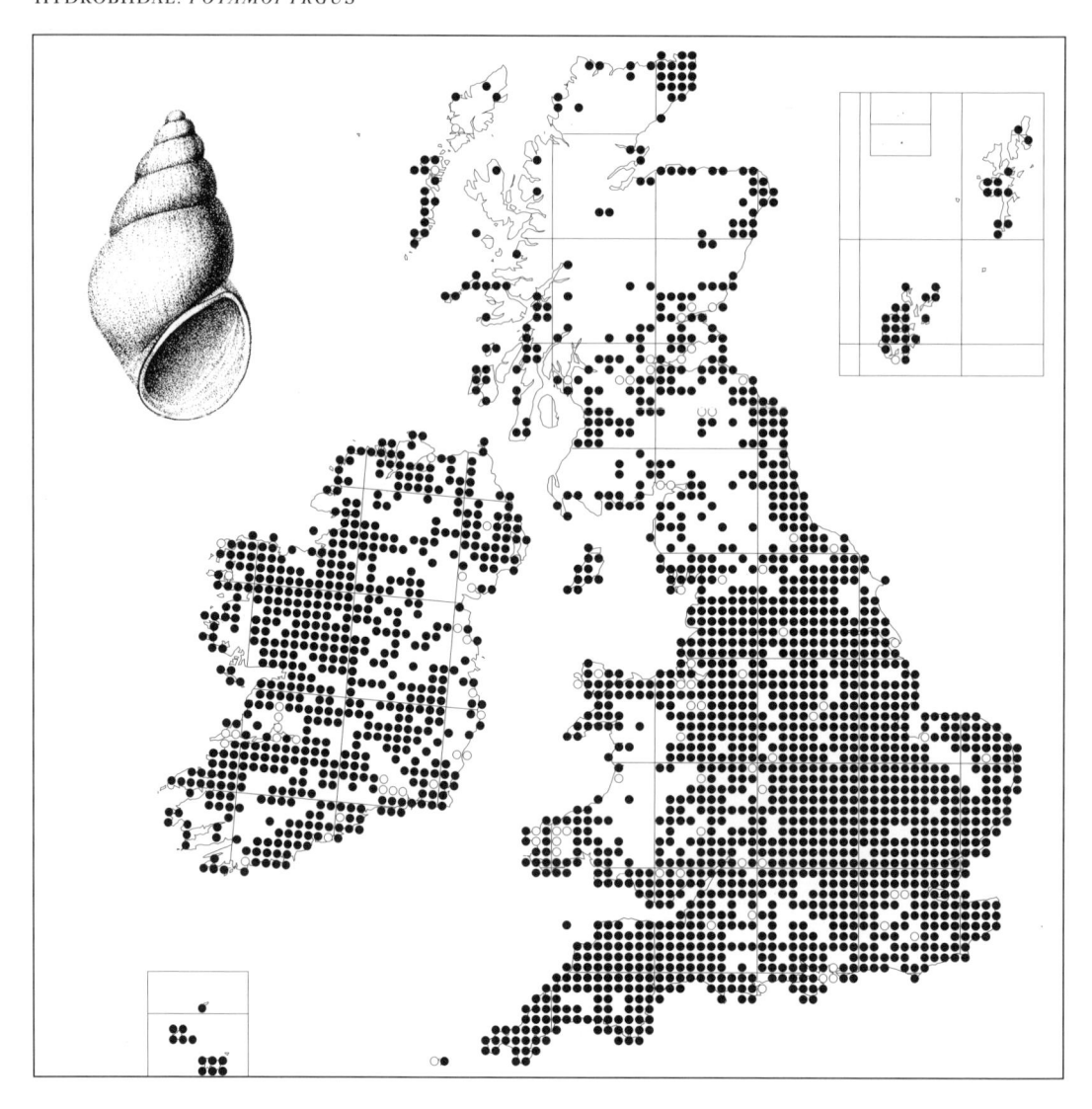

**Potamopyrgus antipodarum** (Gray, 1840)
(*Hydrobia jenkinsi* E. A. Smith, *Potamopyrgus jenkinsi* (E. A. Smith))

Jenkins's spire snail    Size: 4.0–5.5mm

Common in flowing water of all kinds, hard or soft, in rivers, canals and streams, or in brackish ditches and lagoons (up to 17 parts per thousand NaCl). It also lives in caves, springs and roadside trickles, associated with few or no other molluscs. It has been reported inside water mains. In still water it is rare. A detritus feeder, it is virtually indifferent to the nature of the substrate; it can abound in poorly vegetated places, on stones or bare mud. Newly established colonies often show enormous populations, generally declining sharply after a few seasons. The parthenogenic mode of reproduction favours dispersal. It is very hardy, and tolerant of mild pollution.

Introduced (supposed fossil records are erroneous). It was first noted in England at Grays, Essex, before 1852. By the 1880s it was well established in the Thames marshes below London, and was then discovered at Sandwich, Kent (1891), Exminster, Devon (1892), Dudley, Staffs (1893) and Lewes, Sussex (1894). In Ireland it was first found near Coleraine, Co. Londonderry, in 1893. By 1930 it was widespread in inland waters. It is now one of our commonest freshwater molluscs and still extending its range in parts of northern England, Wales and Scotland.

Europe north to the Shetlands, S. Scandinavia and the Baltic. Original home probably New Zealand (Winterbourn, 1972; Ponder, 1988).

(Ellis: 78; Macan: 17)

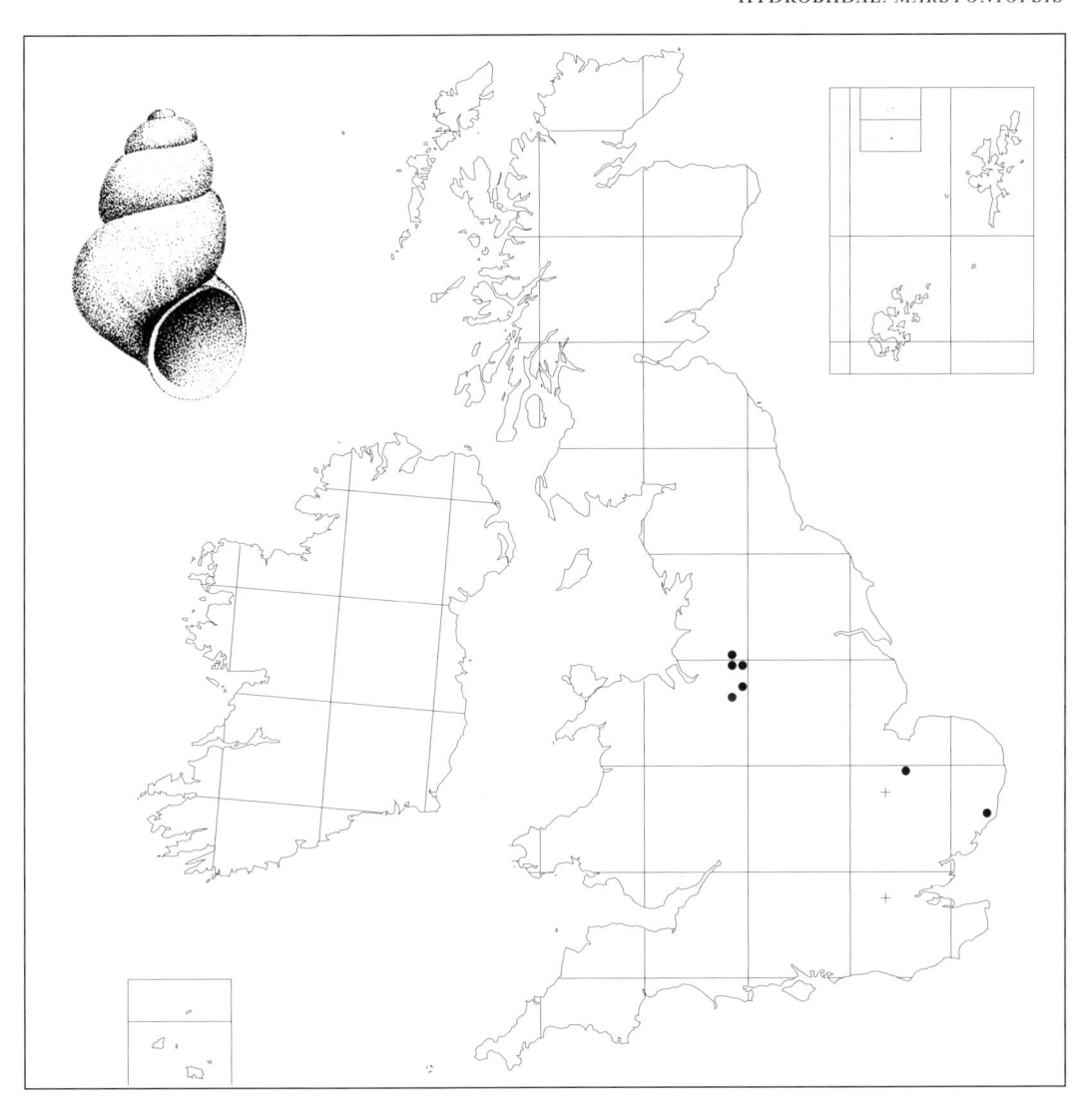

**Marstoniopsis scholtzi** (A. Schmidt, 1856)
(*Bythinella scholtzi* (Schmidt), *B. steinii* (Martens), *Paludestrina taylori* E. A. Smith, *Amnicola taylori* (E. A. Smith)).

Taylor's spire snail    Size: 2.5–3.0mm

Found in still or slow-moving water in canals (Manchester area) or in lowland rivers (Great Ouse, Norfolk; R. Deben, Suffolk). At the East Anglian sites only small numbers of fresh dead shells have been recovered, derived from unlocalized habitats. In canals it appears to prefer quiet, weedy conditions over a muddy bottom, often among *Glyceria maxima*.
  Native? (L.Pgl fossils only). *M. scholtzi* may be native in S.E. England where a fossil probably of Roman age is known from deposits of the R. Thames (Preece & Wilmot, 1979). In the north it is probably introduced. Near Manchester it was first noted in 1900 (canals at Droylsden and Dukinfield; Jackson & Taylor, 1904) and in Stirling in 1931 (timber-seasoning ponds by canal at Grangemouth; Waterston, 1934). Apart from the disappearance of the Grangemouth colony in the 1950s, there is no evidence that populations have declined significantly in recent years, and in the Manchester canals they are seemingly resistant to mild industrial pollution. The species must however be considered at some risk because of its rarity. RDB category: Rare.
  N. European: France and Britain eastwards to N. Germany, Poland and S. Scandinavia.

(Ellis: 81; Macan: 17)

TRUNCATELLIDAE: *TRUNCATELLA*

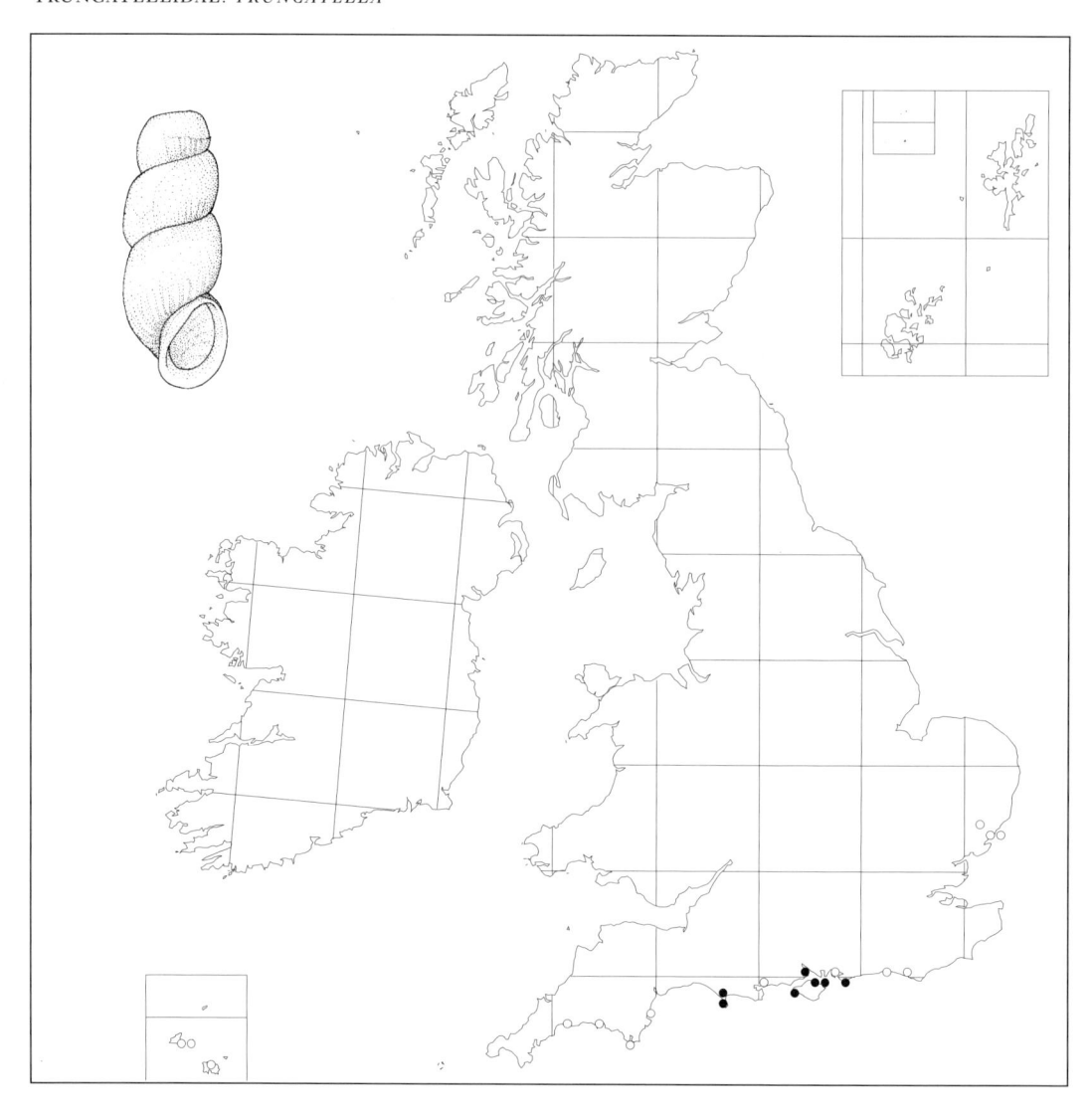

**Truncatella subcylindrica** (L., 1758) (*T. truncatula* (Draparnaud), *Acmea subcylindrica* (L.))

Looping snail    Size: 3.5–5.0mm

This is a semi-marine snail. Usually it lives out of water, under stones and driftwood in muddy places at high-tide level or in saltmarshes, among seablite (*Suaeda maritima*) and sea purslane (*Halimione portulacoides*). It is difficult to find alive and is more often recorded as bleached shells in tidal debris washed up in brackish lagoons and estuaries.

Probably native. Though doubtless underrecorded, *Truncatella* may be declining. It is likely to be vulnerable as a mollusc of specialized habitat at the edge of its geographical range. At its northern limit on the Suffolk coast it was last reported in 1893. RDB category: Rare.

Atlantic and Mediterranean coasts of Europe, eastwards to the Black Sea.

(Fretter & Graham, 1978: 137)

BITHYNIIDAE: *BITHYNIA*

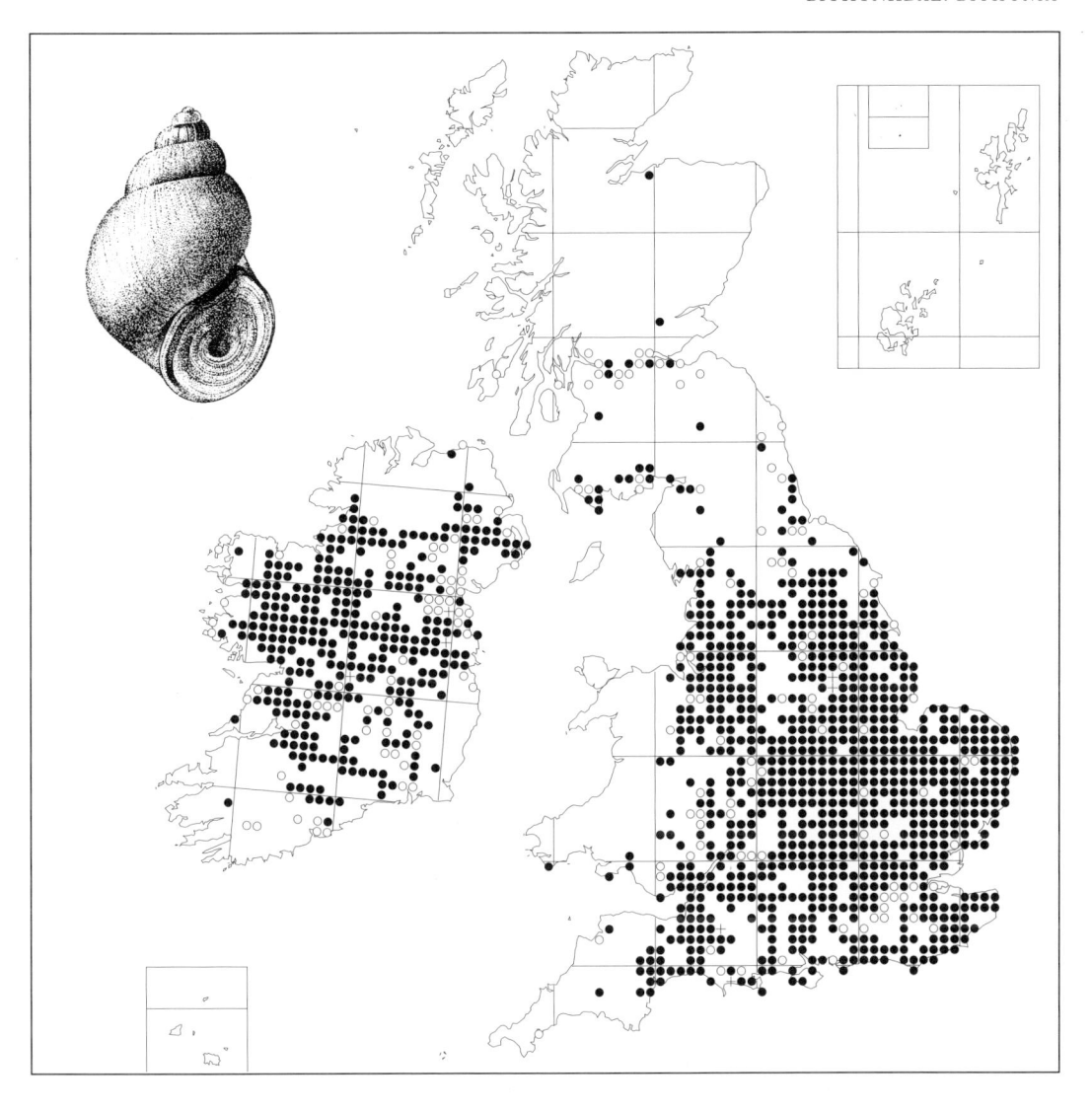

**Bithynia tentaculata** (L., 1758)

Common Bithynia    Size: 8–13 (occ.–16)mm

A common species in large bodies of slow-moving, well-oxygenated hard water in lowland rivers, canals, drainage dykes and lakes. It is rare in small closed ponds. It particularly favours muddy-bottomed situations where there are dense growths of aquatic plants (especially the moss *Fontinalis antipyretica*).

Native (Lgl). This species remains common in suitable habitats and gives no evidence of significant regional decline.

Palaearctic: lowland Europe, eastwards to W. Siberia and Kashmir. Introduced in N. America.

(Ellis: 82; Macan: 15)

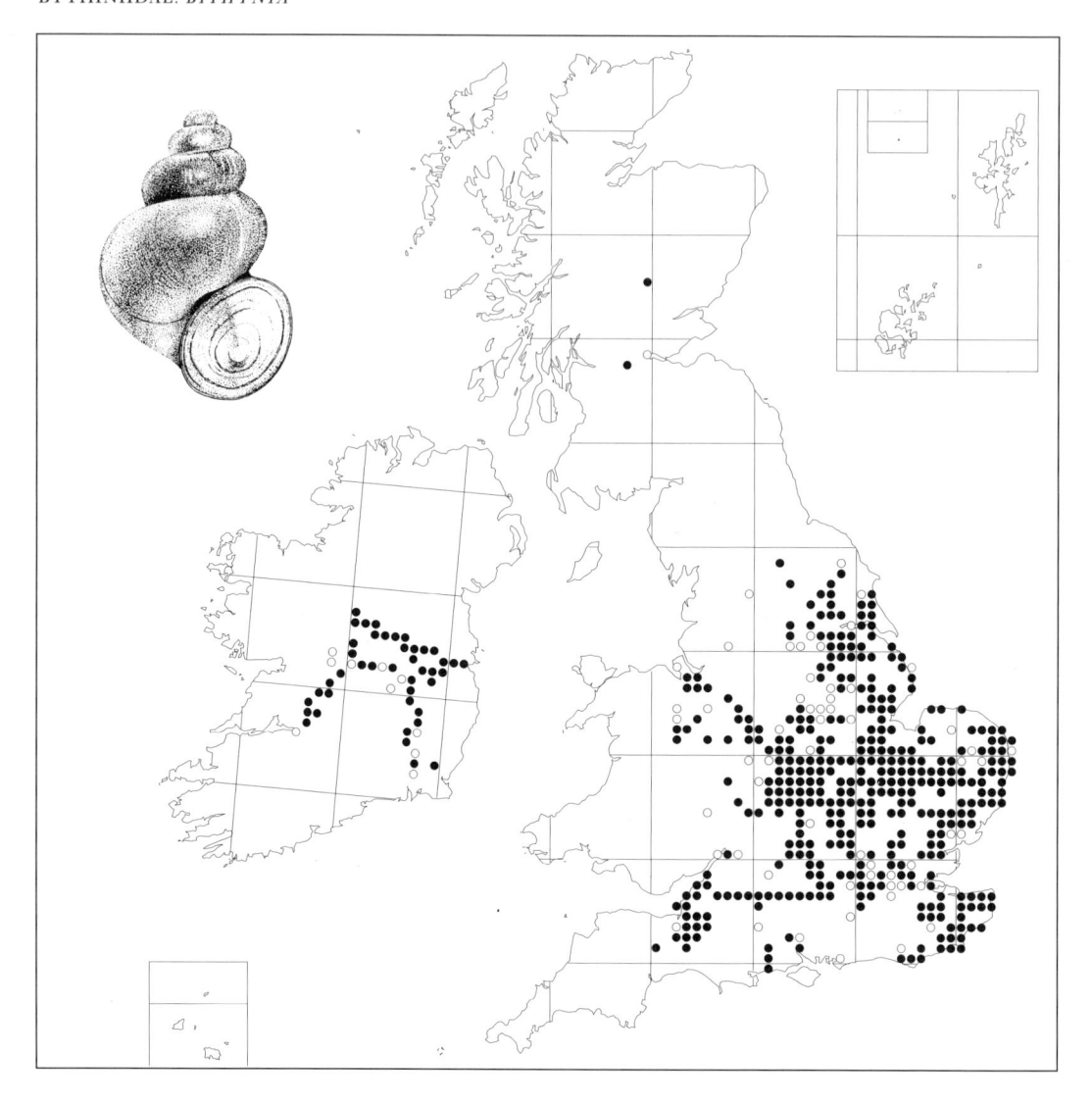

**Bithynia leachii** (Sheppard, 1823)

Leach's Bithynia   Size: 5–7mm

Generally found in similar places to *B. tentaculata* (which is nearly always associated with it) but more local and restricted to the best calcareous lowland habitats with a high diversity of species, especially in canals, canalized rivers and marshland drainage dykes. It is very rare in lakes and ponds.

Native (E.Pgl). This species is still frequent over most of its English range though it shows signs of decline locally. Protecting its canal habitats in Scotland (Forth & Clyde Canal) and in Ireland (Royal Canal, Grand Canal) may be important for its future in those countries. The northernmost Scottish station, a large artificial pond near Pitlochry, Perthshire, is anomalous and probably due to chance introduction.

Palaeartic; in Europe mainly confined to the N. European plain (to 61°N in Sweden).

(Ellis: 83; Macan: 15)

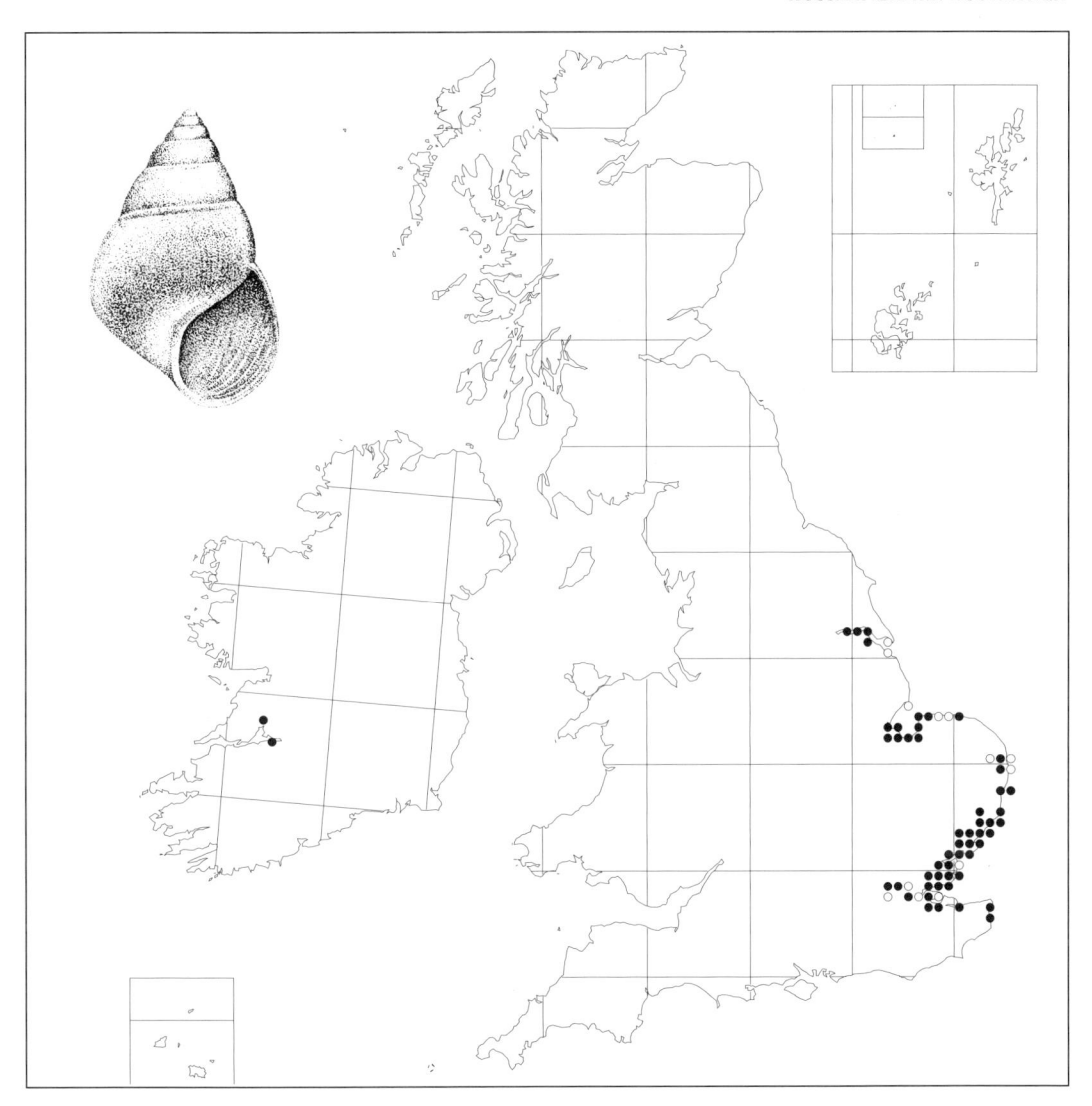

## Assiminea grayana Fleming, 1828

Dun sentinel   Size: 4–6mm

Locally abundant in saltmarshes and estuaries at or just above high-tide level, either in brackish pools or more frequently out of water, crawling on wet mud or on sedges (it is more a terrestrial than an aquatic mollusc). It can often be found under stones or pieces of driftwood lying in marshy ground.

Native? (L. Pgl fossils only). *A. grayana* is abundant where it occurs and gives no evidence of significant recent change. It was discovered in the Shannon estuary in Ireland only in 1991 and may perhaps be a recent immigrant in that country (Colville, 1992).

W. European coasts, mainly around the southern North Sea.

(Ellis: 72; Macan: 15)

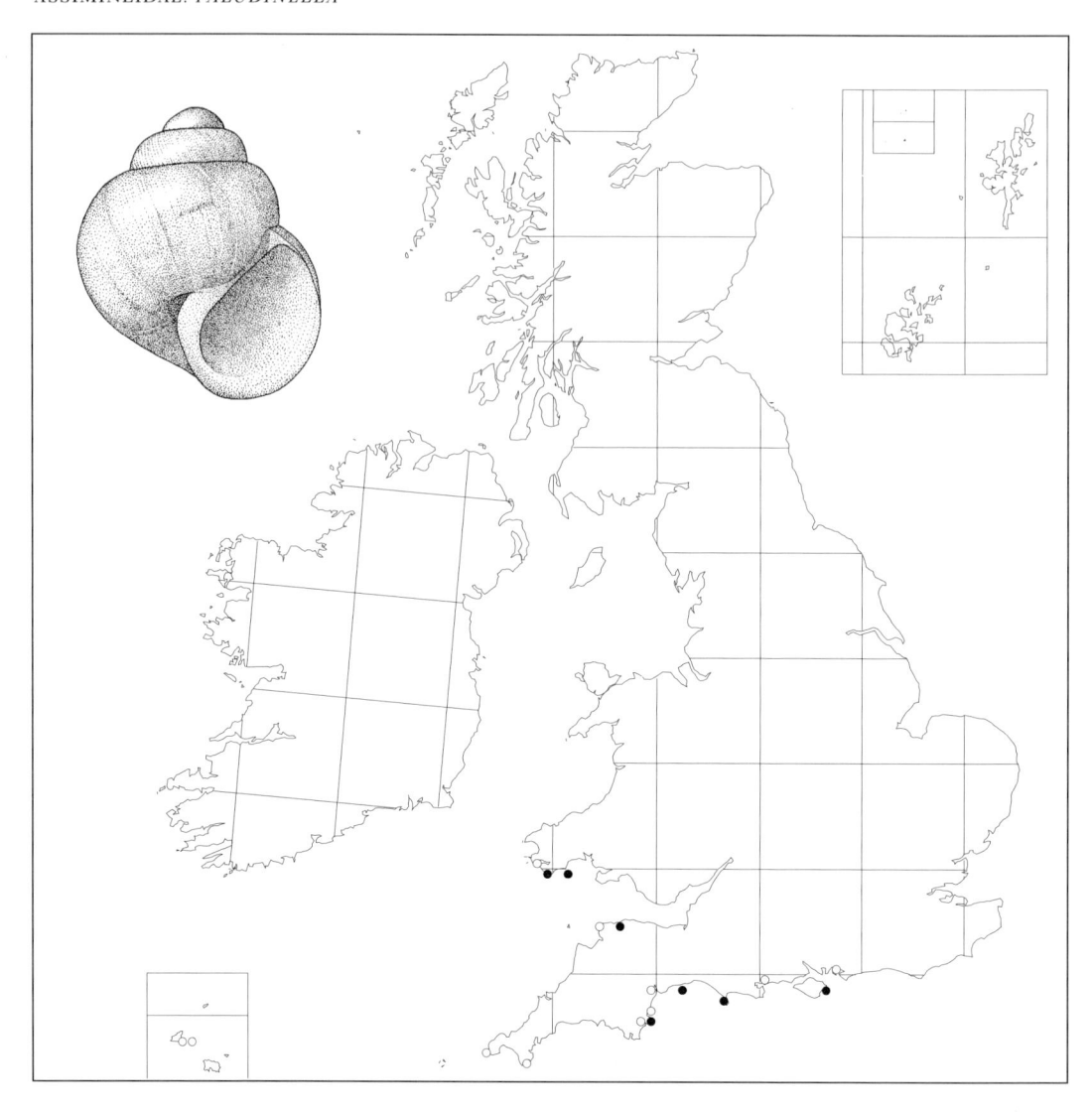

**Paludinella littorina** (delle Chiaje, 1828) (*P. littorea* (Forbes & Hanley), *Assiminea littorina* (delle Chiaje))

Size: 1.7–2.0mm

This little-known species inhabits sheltered places on coasts and in estuaries around high-tide mark, usually out of water. It lives under stones and in shingle, in rock crevices, sea caves and more rarely among maritime plants and debris from saltmarsh vegetation. It favours less muddy situations than the closely related *Assiminea grayana*. It is commonly associated with *Ovatella myosotis* or *Leucophytia bidentata* (Light, 1986; 1998).

Probably native. Though likely to be underrecorded, *P. littorina* may be declining. Many records date from the 19th century or are based on dead shells. Sites where it has recently been found alive are in the Fleet (Dorset), Whitecliff Bay (IoW), Woody Bay (N. Devon), and a few places in S. Devon and in Pembrokeshire. It is a species potentially vulnerable in Britain from being at the edge of its geographical range. RDB category: Rare.

W. Mediterranean and Atlantic coasts, northwards to the English Channel.

(Fretter & Graham, 1978: 148)

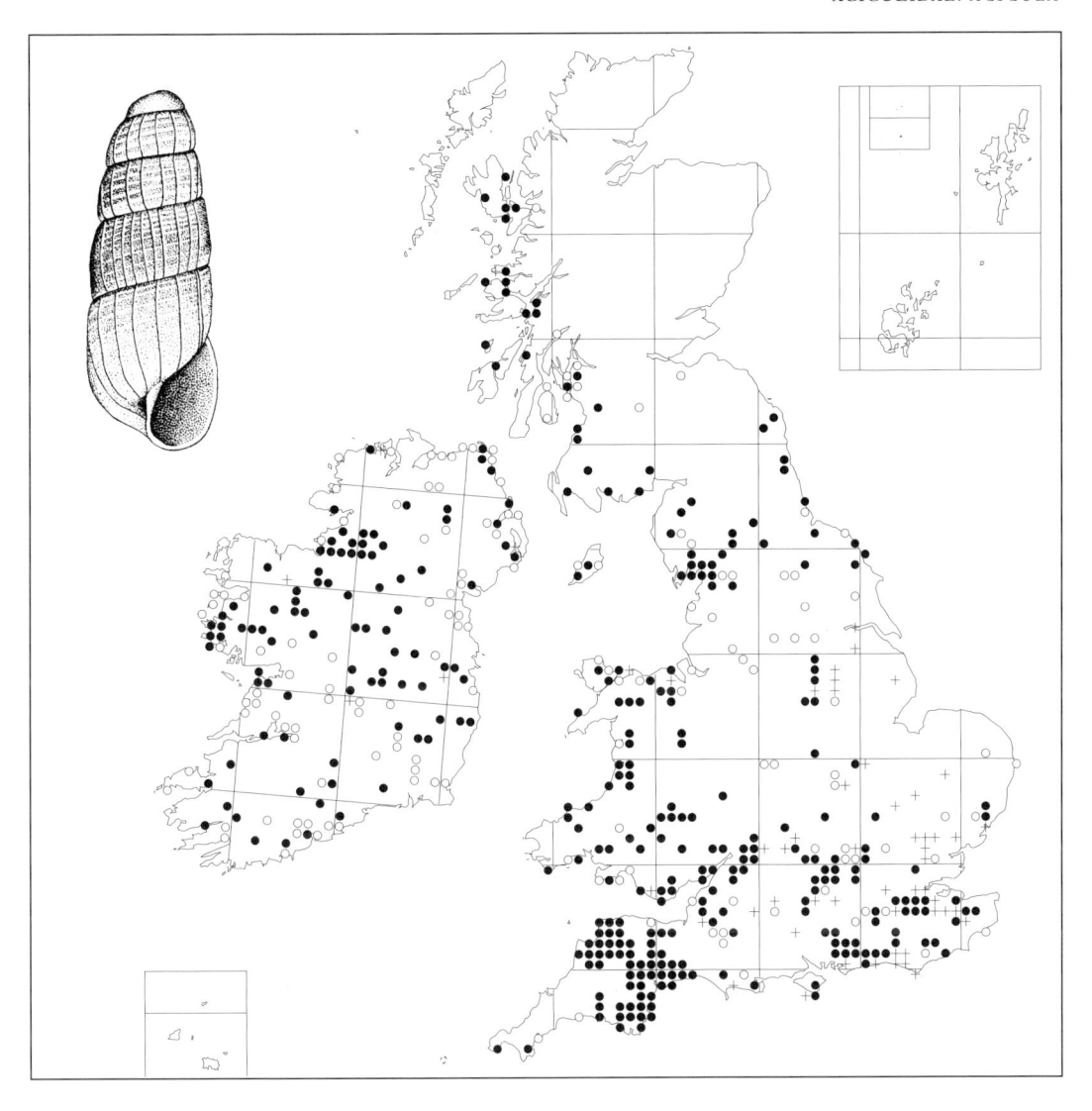

**Acicula fusca** (Montagu, 1803) (*Acme lineata* of British authors, *non* Draparnaud)

Point snail   Size: 2.2–2.5mm

A snail found in moss and leaf litter, mainly in old, undisturbed deciduous woodland. It is tolerant of non-calcareous soils. In well-drained chalk or limestone woods it usually occurs only in damp hollows or around flushes not affected by summer droughts. In Ireland and in western Britain it is sometimes found additionally in relatively exposed situations, such as sea cliffs and mossy roadside banks.

Native (E.Pgl). Owing to its minute size this species is somewhat underrecorded, especially in Ireland. But locally it is retreating as a result of intensive farming and the destruction or replanting of deciduous woodland. *Acicula* is intolerant of human disturbance. The decline is most evident in central England and East Anglia, perhaps reinforced here by the fall in winter temperatures since the thermal optimum of the Postglacial (cf. *Pomatias elegans*).

N.W. European: British Isles, France, Belgium, W. Germany.

(Ellis: 75; K. & C.: 54)

ELLOBIIDAE: *CARYCHIUM*

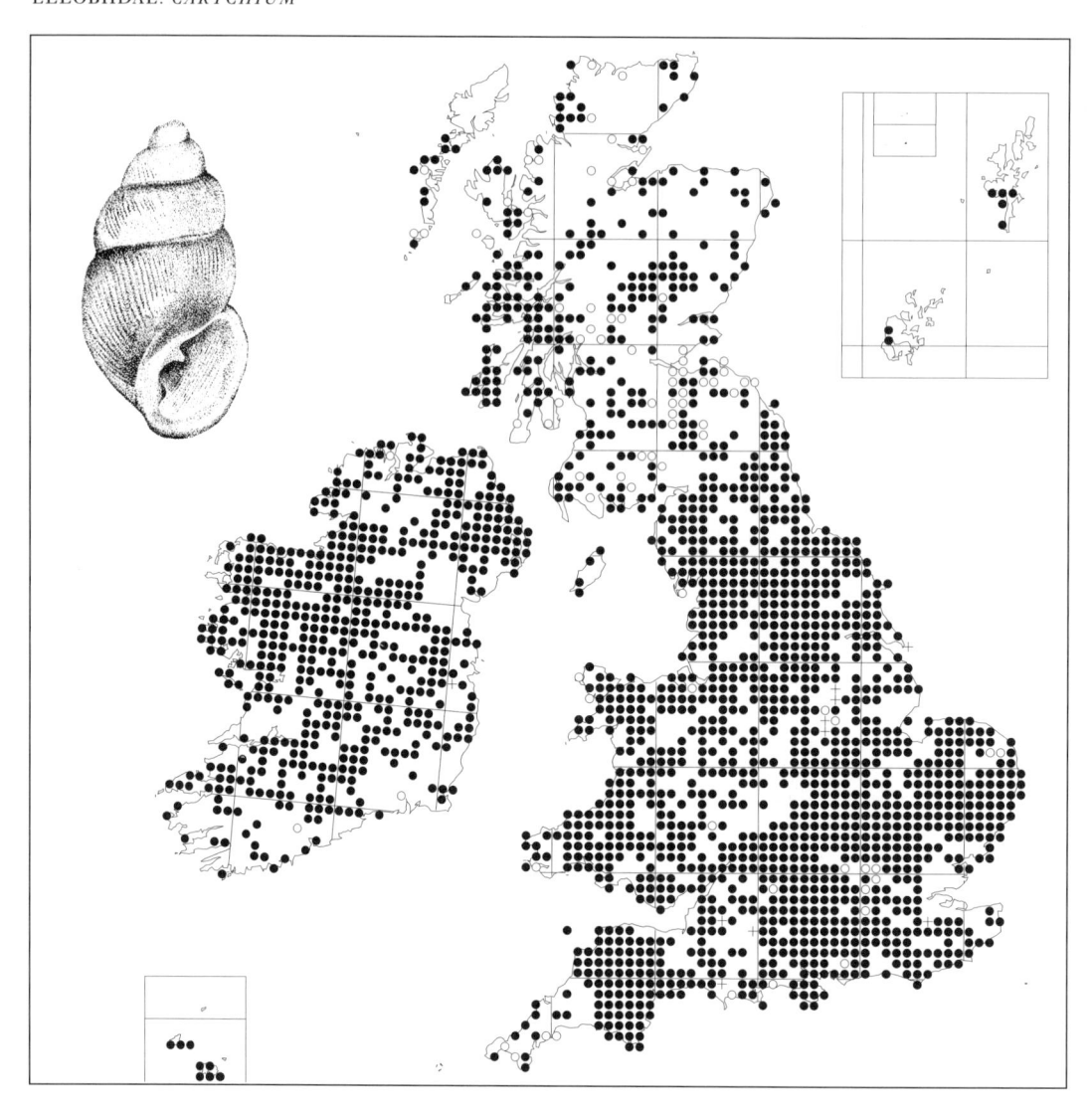

**Carychium minimum** Müller, 1774

Herald snail; sedge snail   Size: 1.6–1.9mm

Common in wet places generally: fens and marshes, water meadows, dune slacks and moist woods. It is virtually amphibious and can survive prolonged winter flooding. It is much commoner in woods in oceanic areas (especially in Ireland) than in eastern Britain, where it tends to be replaced by *C. tridentatum* (Watson & Verdcourt, 1953).

Native (Lgl). *C. minimum* remains common throughout and shows no evidence of regional change. It is relatively tolerant of human disturbance.

European and Siberian: uncommon and mainly coastal beyond 60°N.

(K. & C.: 57)

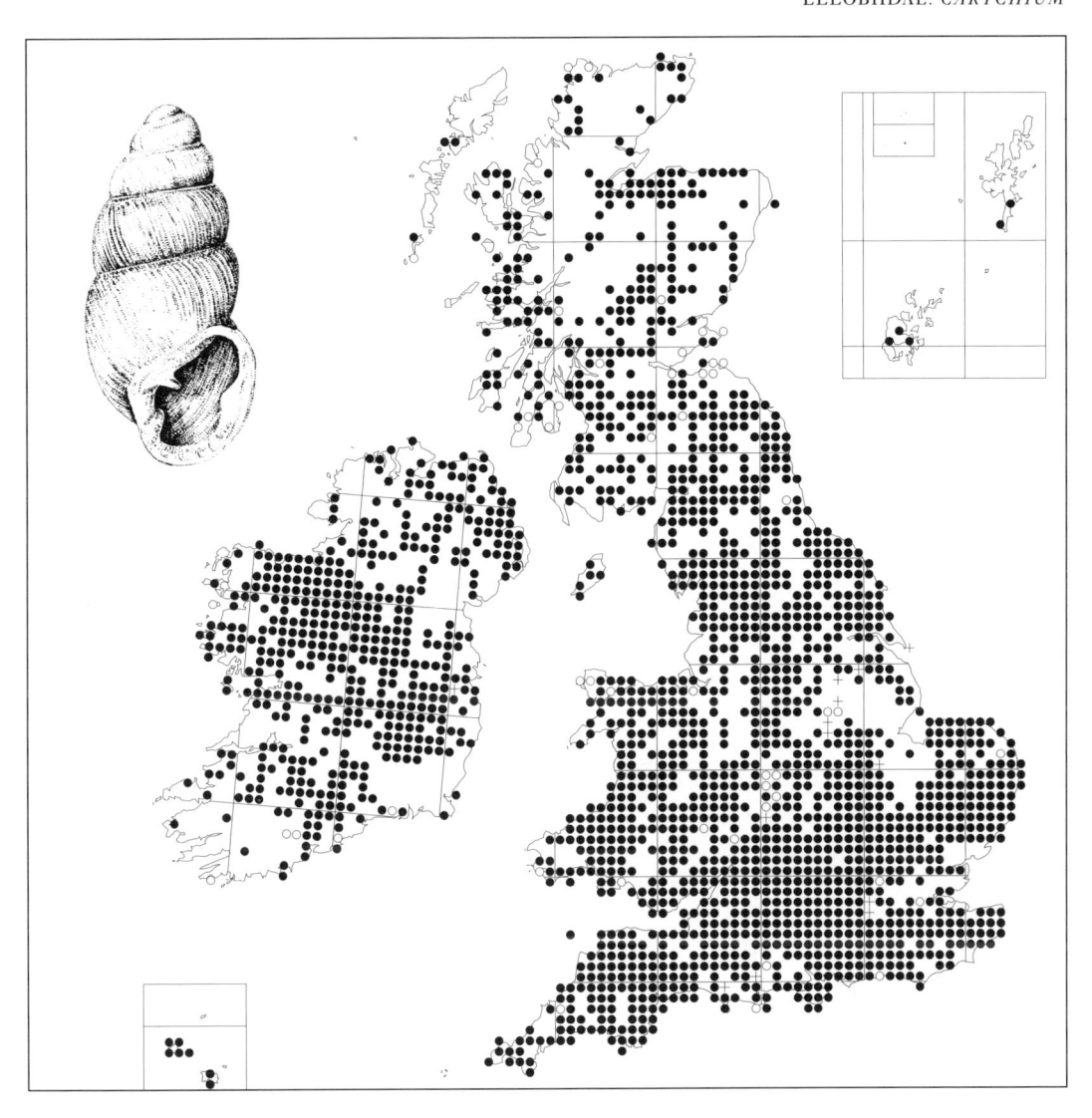

**Carychium tridentatum** (Risso, 1826) (*C. minimum* var. *tridentata* Risso)

Slender herald snail   Size: 1.8–2.0mm

A species characteristic of relatively moist, sheltered, well-vegetated places – woods, hedge banks or damp grassland – especially on base-rich soils. It sometimes occurs in marshes, but is less frequent than *C. minimum* in really wet places. It also extends into much drier habitats, such as leaf litter in well-drained chalk and limestone woods. The two species are however often found together (Watson & Verdcourt, 1953).

Native (E.Pgl). *C. tridentatum* remains common nearly everywhere and shows no evidence of regional change.

European: beyond 60°N uncommon and mainly coastal.

(K. & C.: 58)

ELLOBIIDAE: *OVATELLA*

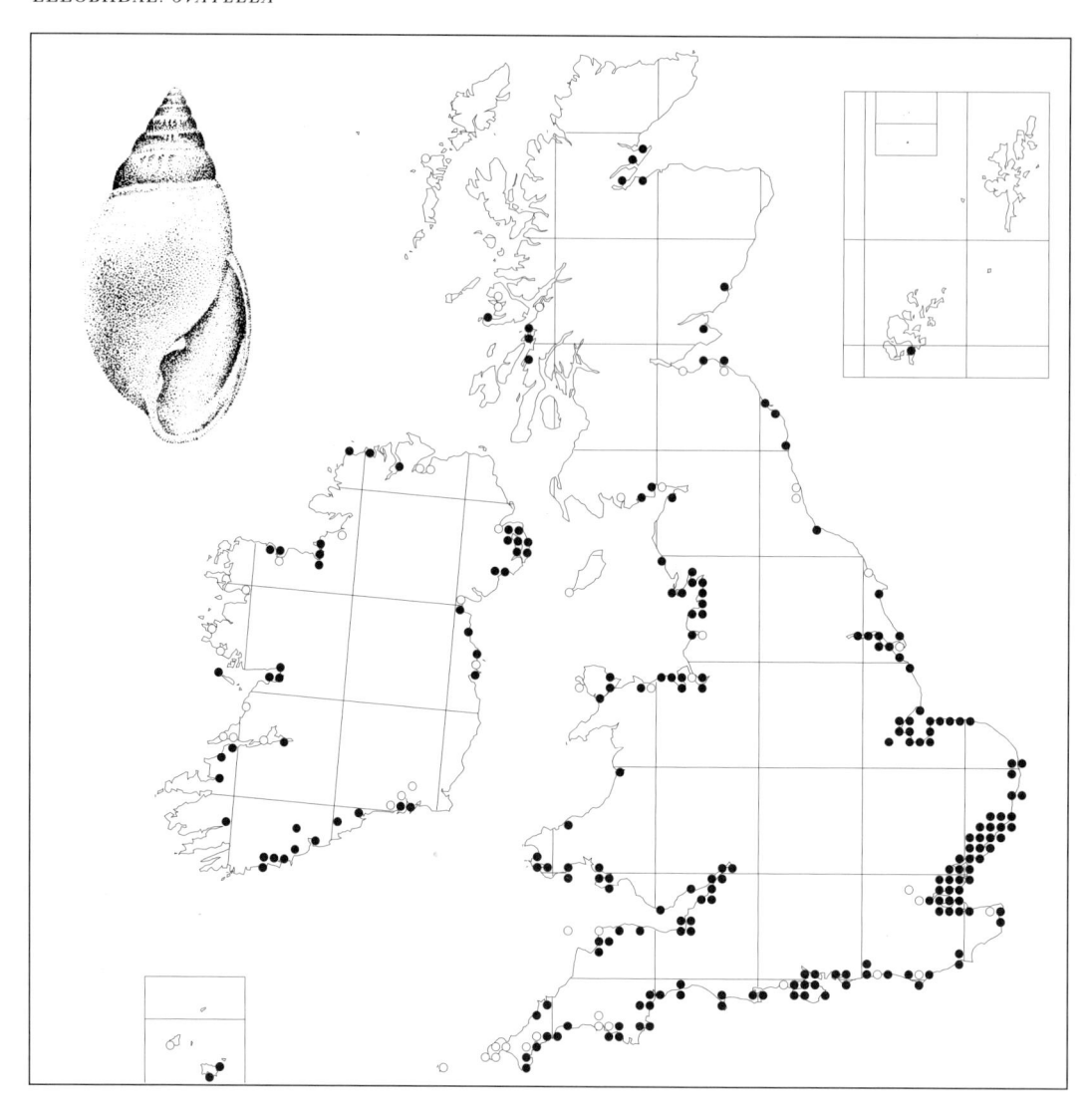

**Ovatella myosotis** (Draparnaud, 1801) (*Phytia myosotis* (Draparnaud), *Ovatella denticulata* (Montagu))

Mouse-eared snail   Size: 6–8mm

A maritime species found in muddy, sheltered places at high-tide level in brackish estuaries and saltmarshes, often under driftwood and other flotsam. It lives mostly out of water and is more a terrestrial than an aquatic snail. Less often it inhabits shingle or the crevices of rocks in exposed situations – a habitat more typical of *Leucophytia bidentata*. Bleached shells are sometimes common in tidal debris.

Probably native (L.Pgl fossils only). *Ovatella* remains common in suitable situations and gives no evidence of regional change.

Mediterranean and W. European coasts, north to the British Isles and Denmark.

Note that the open-coast form (var. *denticulata* Montagu) is sometimes considered to be a separate species.

(Ellis: 96; Macan: 37)

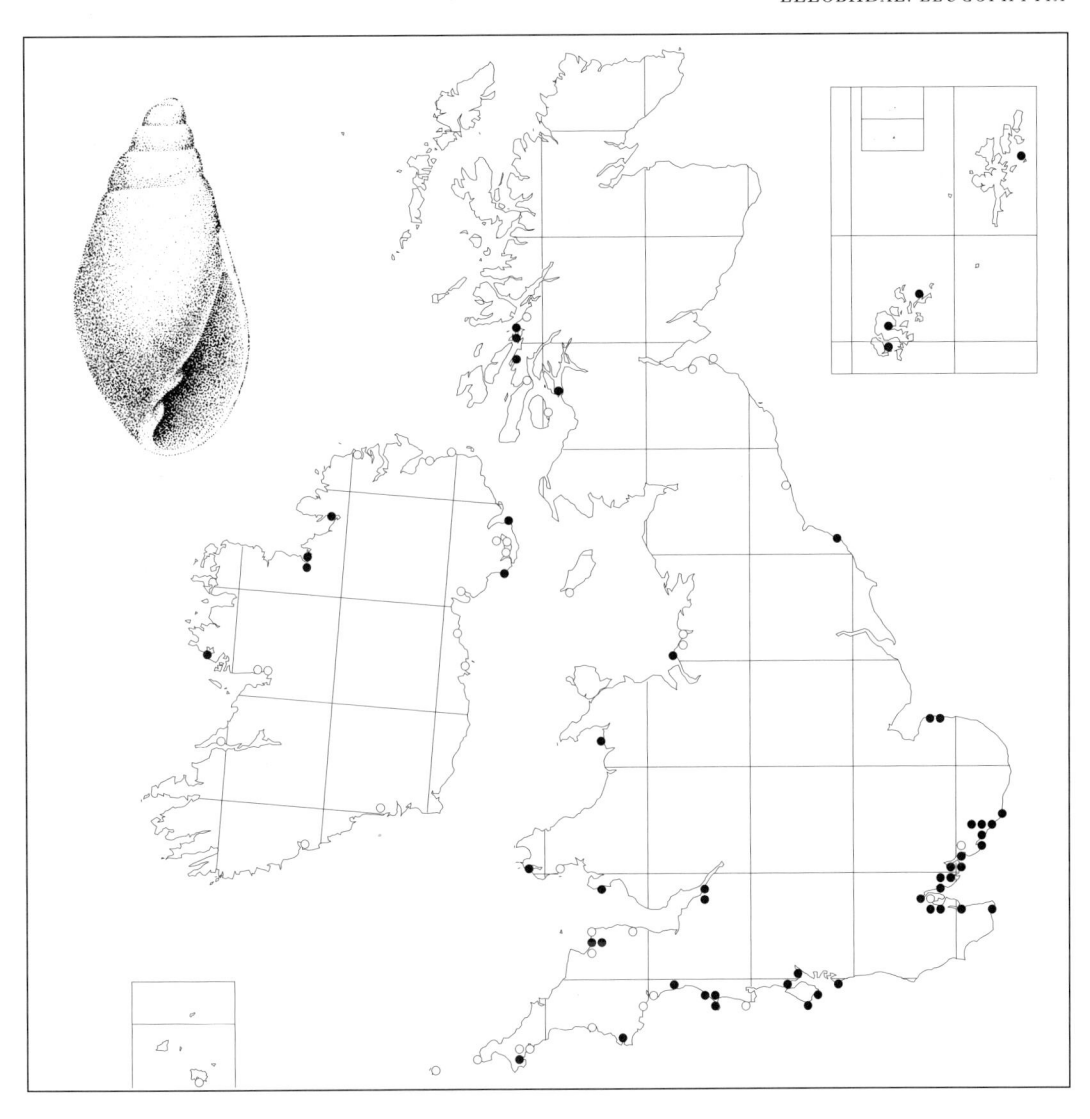

**Leucophytia bidentata** (Montagu, 1808) (*Leuconia bidentata* (Montagu), *Auriculinella bidentata* (Montagu))

Size: 5–6mm

A semi-marine species, living typically on open shores, around high-tide mark among shingle and the crevices of rocks. Less commonly it may be found in more sheltered situations in estuaries or in tidal lagoons. Like *Ovatella myosotis* it is a terrestrial rather than an aquatic mollusc.

Probably native. There is no evidence of significant recent change.

Mediterranean & W. European coasts, north to the British Isles.

(Ellis: 94; Macan: 37)

PHYSIDAE: *APLEXA*

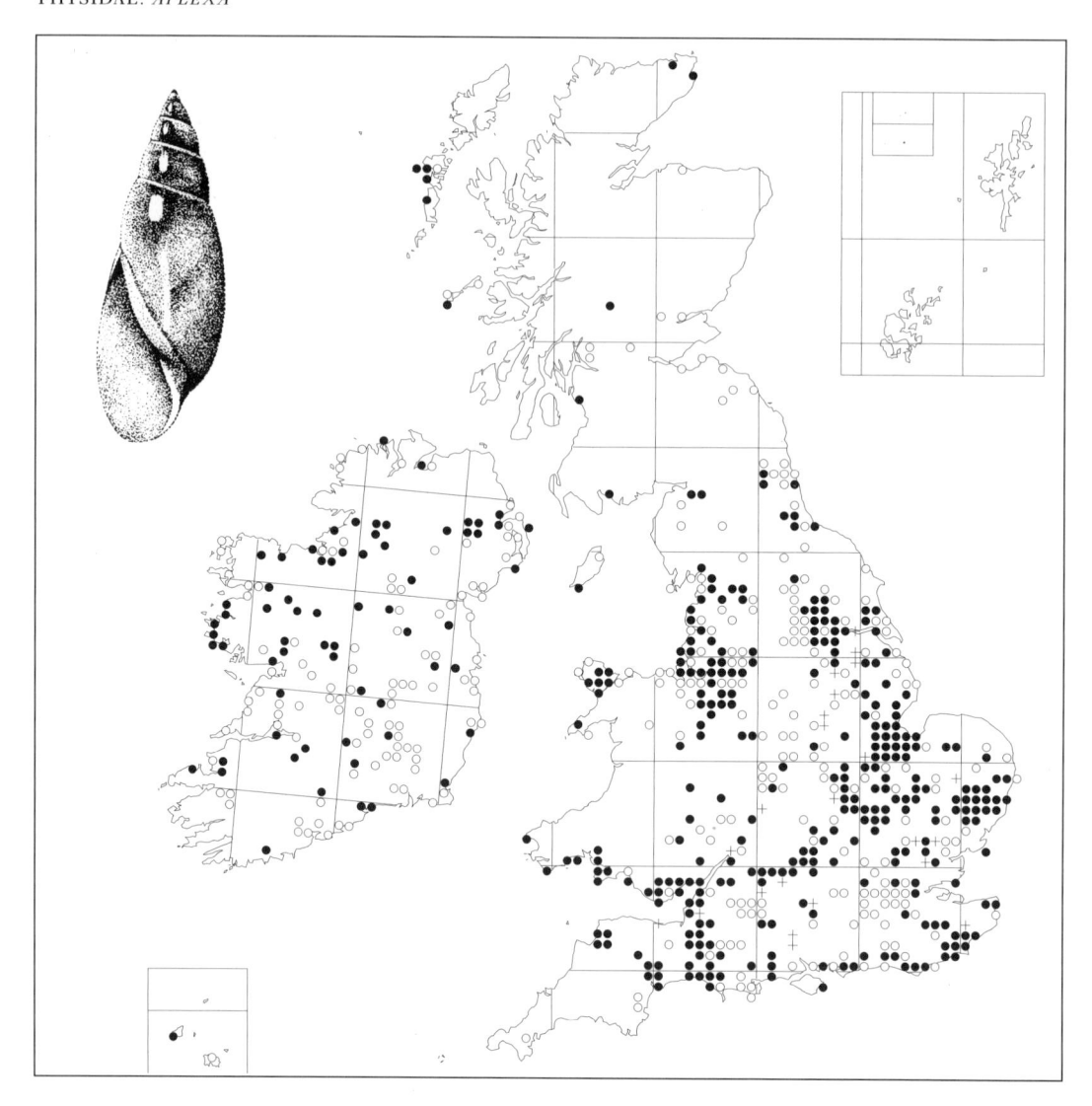

**Aplexa hypnorum** (L., 1758) (*Physa hypnorum* (L.))

Moss bladder snail    Size: 9–13mm

A species typical of swampy pools and ditches, usually associated with *Lymnaea truncatula*, *Anisus leucostoma* or *Pisidium personatum*; conversely it is rare in 'good' freshwater habitats with a high diversity of species. It can survive periodic desiccation. In lowland areas it is often found in weed-choked roadside ditches or in small grassy ponds, especially among floating sweet-grass (*Glyceria fluitans*).

Native (Lgl). This species has declined during the 20th century through the infilling of farm ponds and ditches; this is especially noticeable around London and in the Midlands where at 10km-square level there are certainly some extinctions. In Ireland it is underrecorded, though here also it is probably rarer than formerly.

European, to about 65°N in Scandinavia.

(Ellis: 115; Macan: 21)

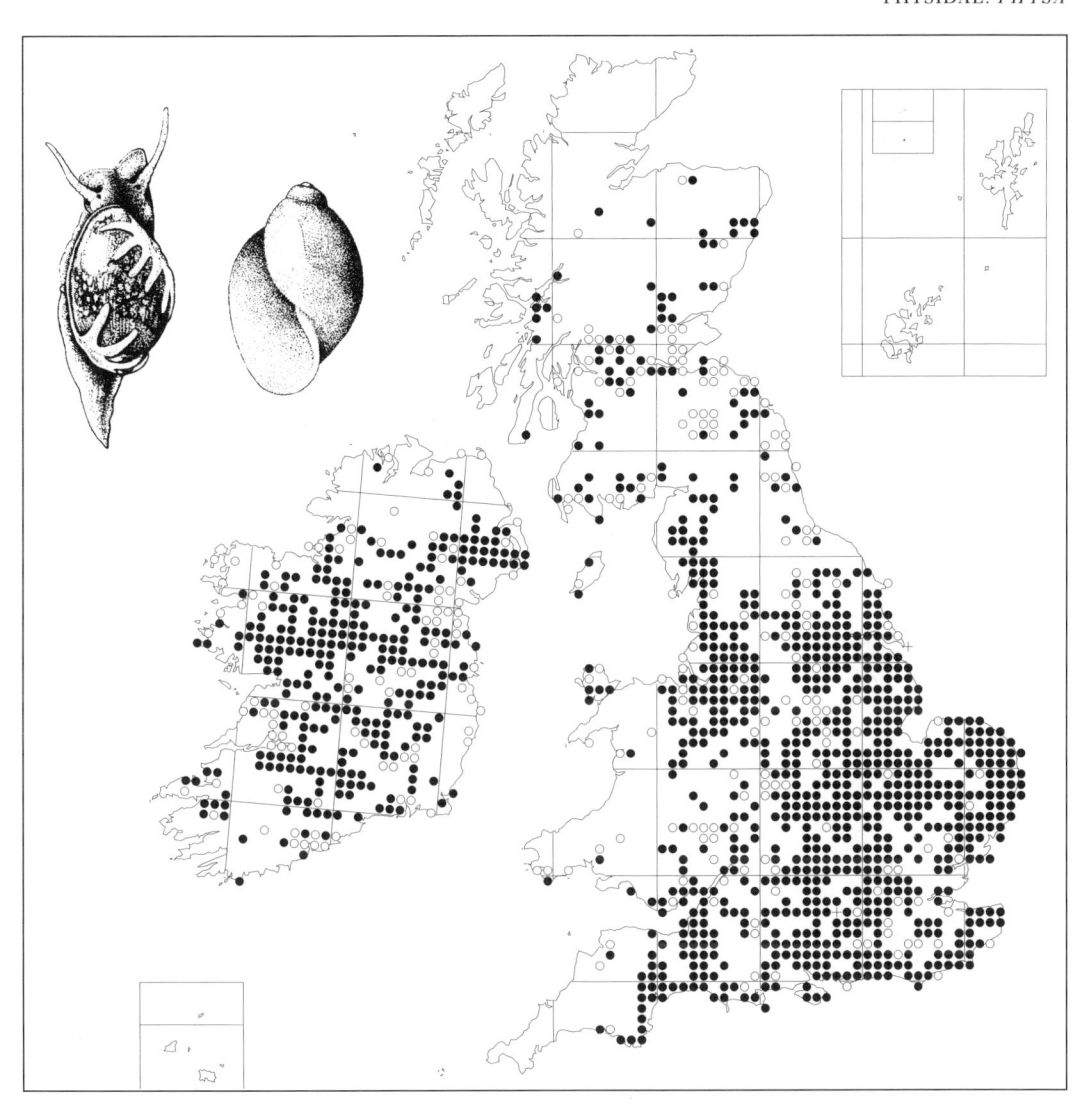

PHYSIDAE: *PHYSA*

**Physa fontinalis** (L., 1758)

Common bladder snail   Size: 8–12mm

Common in bright, clean, running water (hard or soft), in lowland rivers, streams, canals or drainage ditches. It is not infrequent in lakes but very rare in small closed ponds and absent from places subject to desiccation. It is a good indicator of unpolluted conditions. It favours dense growths of aquatic weed, often in quite strong currents.

Native (Lgl). There is no evidence of any significant regional change, though the species may show local decline because of water pollution; in dirty rivers and canals it is often replaced by introduced species of *Physa*.

European: widespread in lowland areas north to 63° in Scandinavia.

(Ellis: 115; Macan: 21)

PHYSIDAE: *PHYSA*

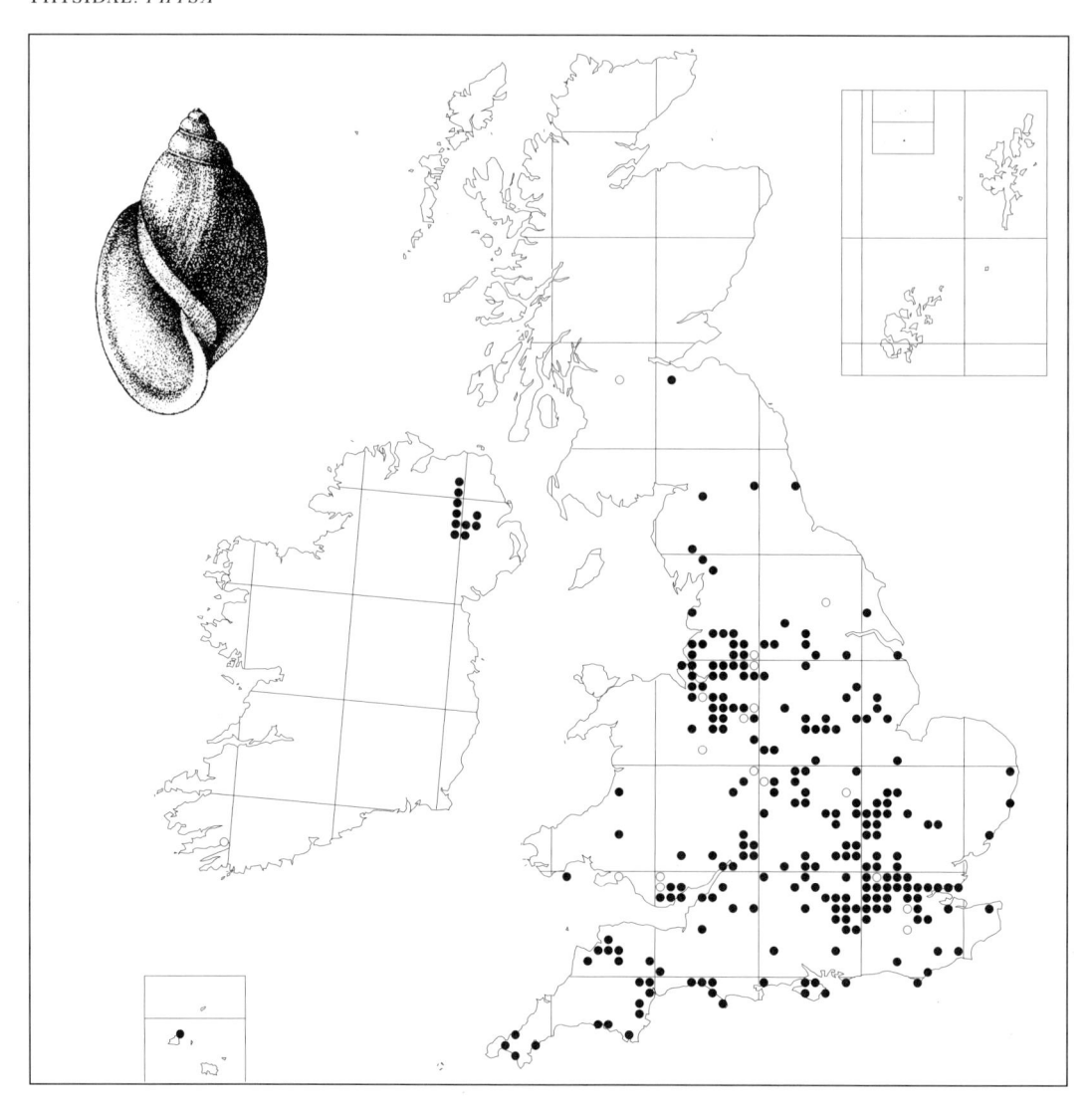

**Physa** spp. (introduced species)

Size: 10–18mm

Alien *Physa* are locally common in quiet or slowly moving water in lowland rivers, canals, lakes, ponds and reservoirs; they are frequent also in ornamental garden ponds and lily tanks. Some at least are tolerant of de-oxygenation and can be found in habitats affected by industrial effluents or damaged by eutrophication.

Introduced (no reliable fossil records). Alien *Physa* were first noted in England early in the 19th century but have spread greatly in recent years, no doubt through human activity.

Probably originally N. American, but spread by man in many parts of the world.

There is much confusion about the identity of these species, sometimes placed together in the genus *Physella*. Usually they are named on shell characters either as *Physa acuta* Draparnaud (described from southern Europe) or as *P. heterostropha* Say (described from N. America). Populations from S. Wales (Caerphilly and Whitchurch) and from N. Ireland (Lough Neagh) have been shown to be identical with the common N. American species *P. gyrina* Say on the basis of their genital anatomy (Dean, 1920; Anderson, 1996). It is likely that *P. gyrina* is widespread in the British Isles, although, like all *Physa*, the shell is very variable and is an unsafe guide to identification. Whether *P. acuta* or *P. heterostropha* also occur in the British Isles is still unproven. It should be noted that some authors believe that the European *P. acuta* may itself be an old (pre-1800) introduction from N. America.

(Ellis: 117; Macan: 21)

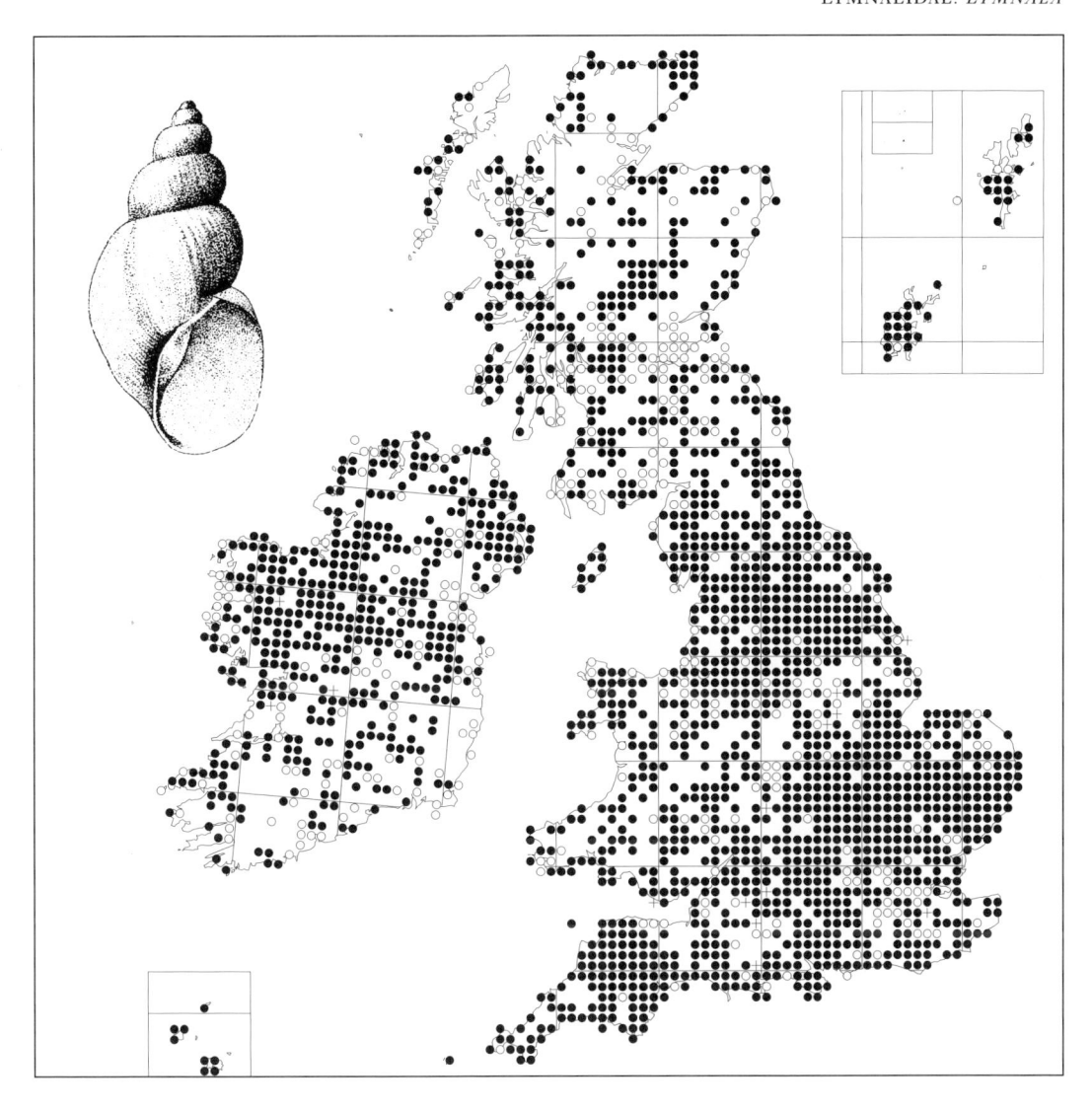

**Lymnaea truncatula** (Müller, 1774) (*Galba truncatula* (Müller))

Dwarf pond snail   Size: 7–12mm

This snail, notorious as the intermediate host of the sheep liver-fluke *Fasciola hepatica*, inhabits marshy grassland, shallow ephemeral ponds, roadside trickles, flushes and dune slacks, living mostly out of water. It is commonly associated with *Pisidium personatum* and with hygrophile land species. It tolerates disturbed or poorly vegetated places, such as bare mud in marshes or wet rocks. It occurs on most soils though avoiding highly acid ground.

Native (Lgl). *L. truncatula* remains common throughout the British Isles, in spite of many attempts at local control or eradication in areas of sheep farming. In parts of the highland zone it has invaded poor, acidic marginal land as a result of soil enrichment.

Holarctic: throughout Europe to beyond the Arctic Circle.

(Ellis: 100; Macan: 23)

LYMNAEIDAE: *LYMNAEA*

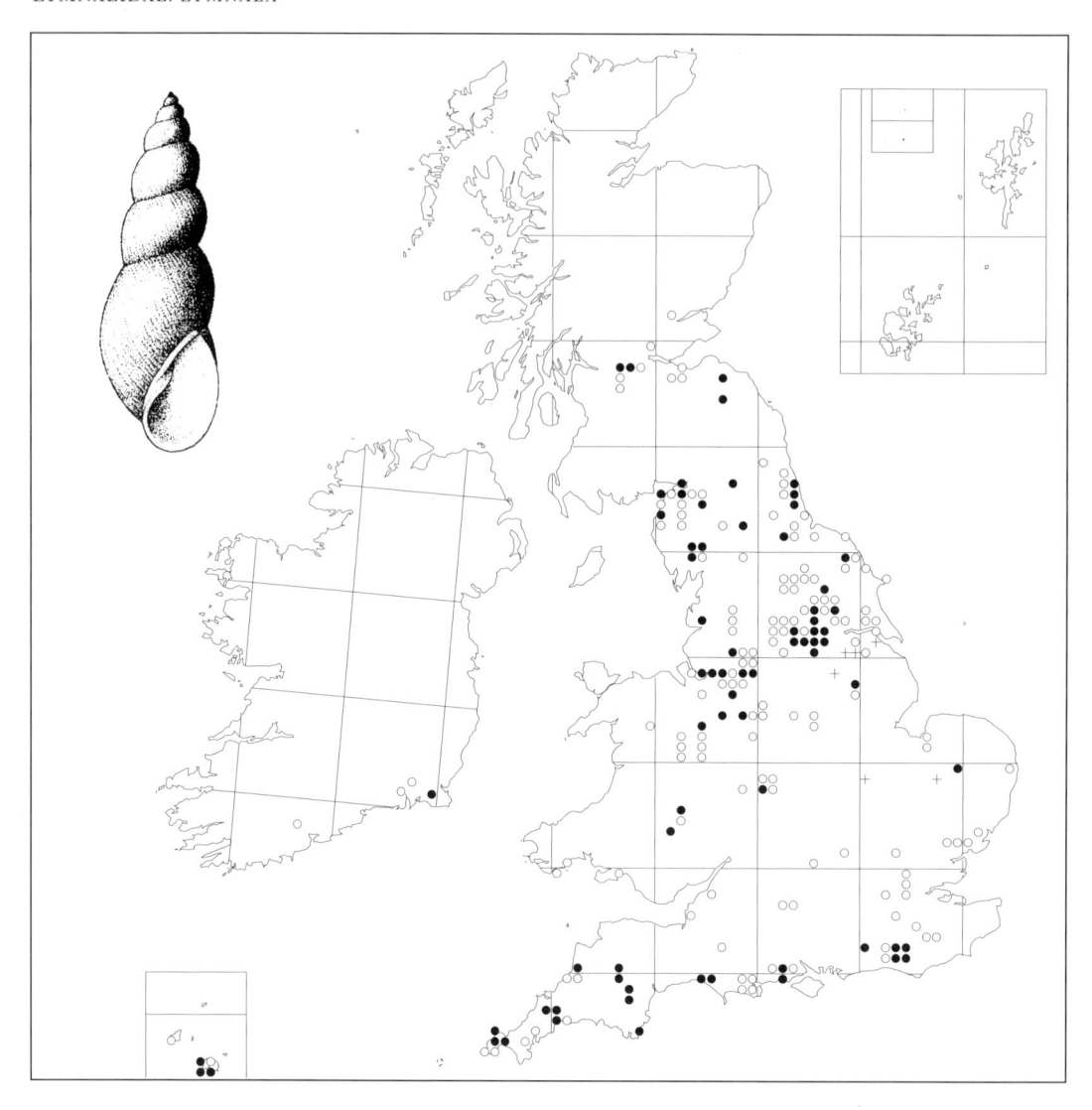

**Lymnaea glabra** (Müller, 1774) (*Stagnicola glabra* (Müller), *Omphiscola glabra* (Müller))

Mud pond snail   Size: 12–20mm

Lives in water low in nutrients in ponds and ditches or around seepages, especially in places which dry out periodically and where the aquatic flora is poor. Occasionally it is found in larger bodies of water such as swampy drainage dykes. Usually it lives either alone or with a few other species tolerant of desiccation (*L. truncatula*, *Anisus leucostoma*, *Pisidium casertanum*, *P. personatum*); it is not found in habitats with a high molluscan diversity. Most sites are on uncultivated land on acid, sandy or gravelly soils, such as heaths and commons.

Native (E. Pgl). This species is underrecorded but is certainly declining, its habitats eroded by drainage and agricultural encroachment. Eutrophication by fertilizers has undoubtedly destroyed some populations. In many parts of lowland England (e.g. around London) only 19th-century records exist, though it still remains fairly common in parts of S.W. England and south Yorkshire. In Ireland it has always been rare and the only population noted in the last fifty years (Shelmaliere Commons, Co. Wexford) was destroyed by farm drainage in 1980 (Hurley, 1981). RDB category: Vulnerable (in Ireland: Endangered).

W. European, to about 61°N in Scandinavia.

(Ellis: 111; Macan: 23)

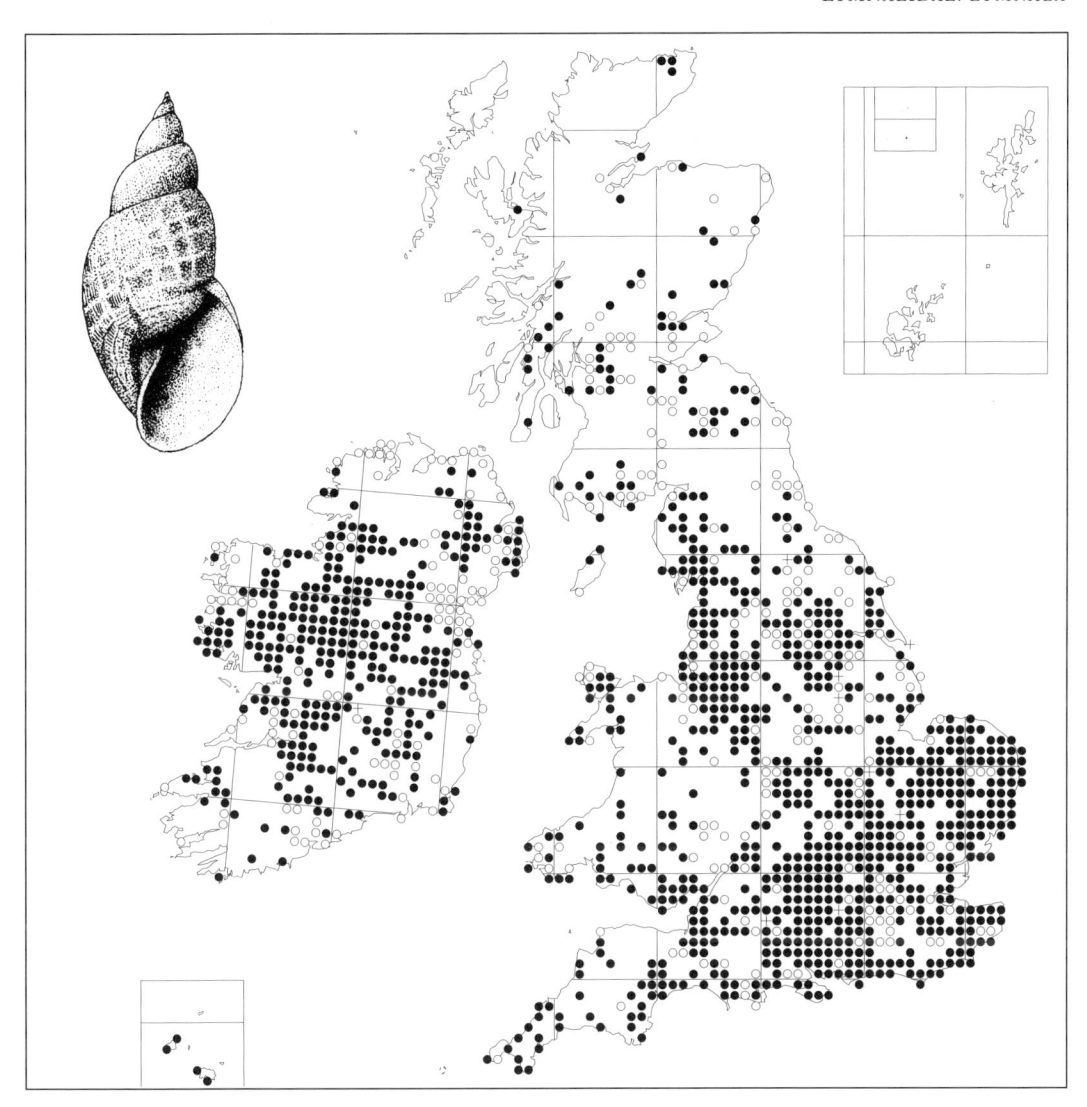

**Lymnaea palustris** (Müller, 1774) (*Stagnicola palustris* (Müller))

Marsh pond snail   Size: 15–24mm

This mainly lowland species lives in stagnant or slowly moving water, hard or soft. It is typical of swamps, shallow drains and ditches choked with aquatic or emergent vegetation, including places liable to dry up in the summer. More rarely it inhabits open water in ponds, lakes, rivers and canals.

Native (Lgl). *L. palustris* is still widespread, though it shows some local decline in parts of central and eastern England as a result of agricultural drainage.

Holarctic: in Europe to 70°N in Norway.

Further research is needed on the systematics of this taxon in the British Isles. Many Continental authorities now regard European '*palustris*' as a complex of two or more related species within the genus or subgenus *Stagnicola*.

(Ellis: 102; Macan: 23)

LYMNAEIDAE: *LYMNAEA*

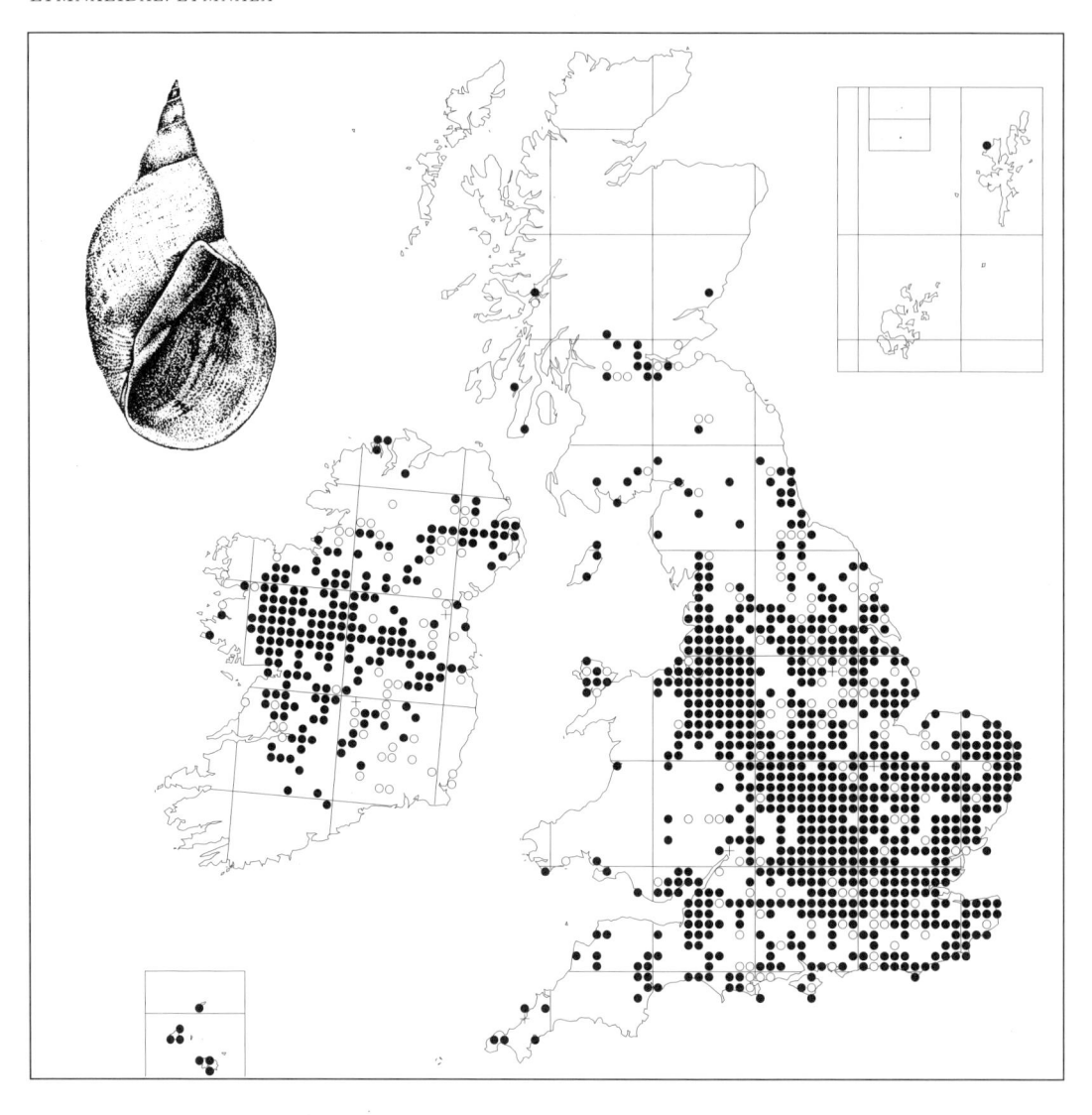

**Lymnaea stagnalis** (L., 1758)

Great pond snail   Size: 35–45 (occ.–50)mm

The great pond snail lives in large permanent bodies of clean, hard water. It is typical of slow-flowing lowland rivers, canals, drainage dykes and lakes, usually in places with a good aquatic flora and a diversity of molluscan species. It is uncommon in small closed ponds, though it is often successfully introduced into garden ponds and large aquaria. The juveniles frequently leave the water and may be found attached to emergent vegetation.

Native (E. Pgl). This species is widespread in suitable habitats and shows no sign of significant regional change. Many isolated colonies are known or suspected to be deliberate 20th-century introductions, for example in the Channel Islands, Cornwall, Cardigan, Cumberland, the Shetlands and the Isle of Man.

Holarctic; nearly throughout lowland Europe.

(Ellis: 103; Macan: 23)

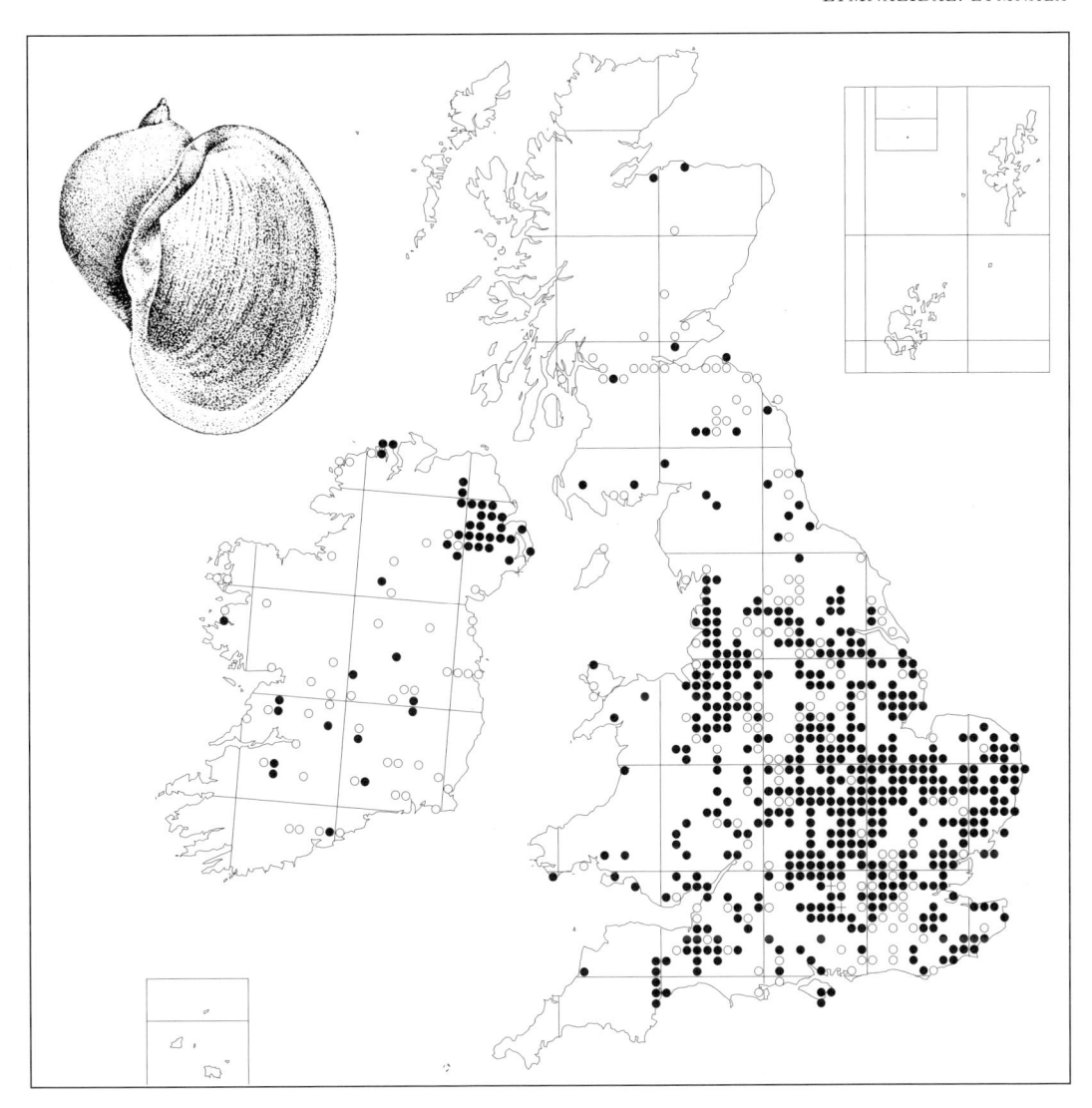

**Lymnaea auricularia** (L., 1758) (*Radix auricularia* (L.))

Ear pond snail    Size: 15–32mm

A species restricted to the best lowland aquatic habitats: sizable bodies of still or slowly moving, hard water in large rivers, canals, drainage dykes and lakes, usually in places with a good aquatic flora. It is rare in small closed ponds, though sometimes introduced successfully into ornamental ponds in gardens and parks.

Native (E.Pgl). There is no clear evidence of any significant regional decline affecting this species, though it may disappear locally through pollution and de-oxygenation. In Ireland, where there are few modern records outside the Lough Neagh basin, it is possibly underrecorded.

Palaearctic: nearly throughout Europe, but rare in the highland zone.

(Ellis: 110; Macan: 25)

LYMNAEIDAE: *LYMNAEA*

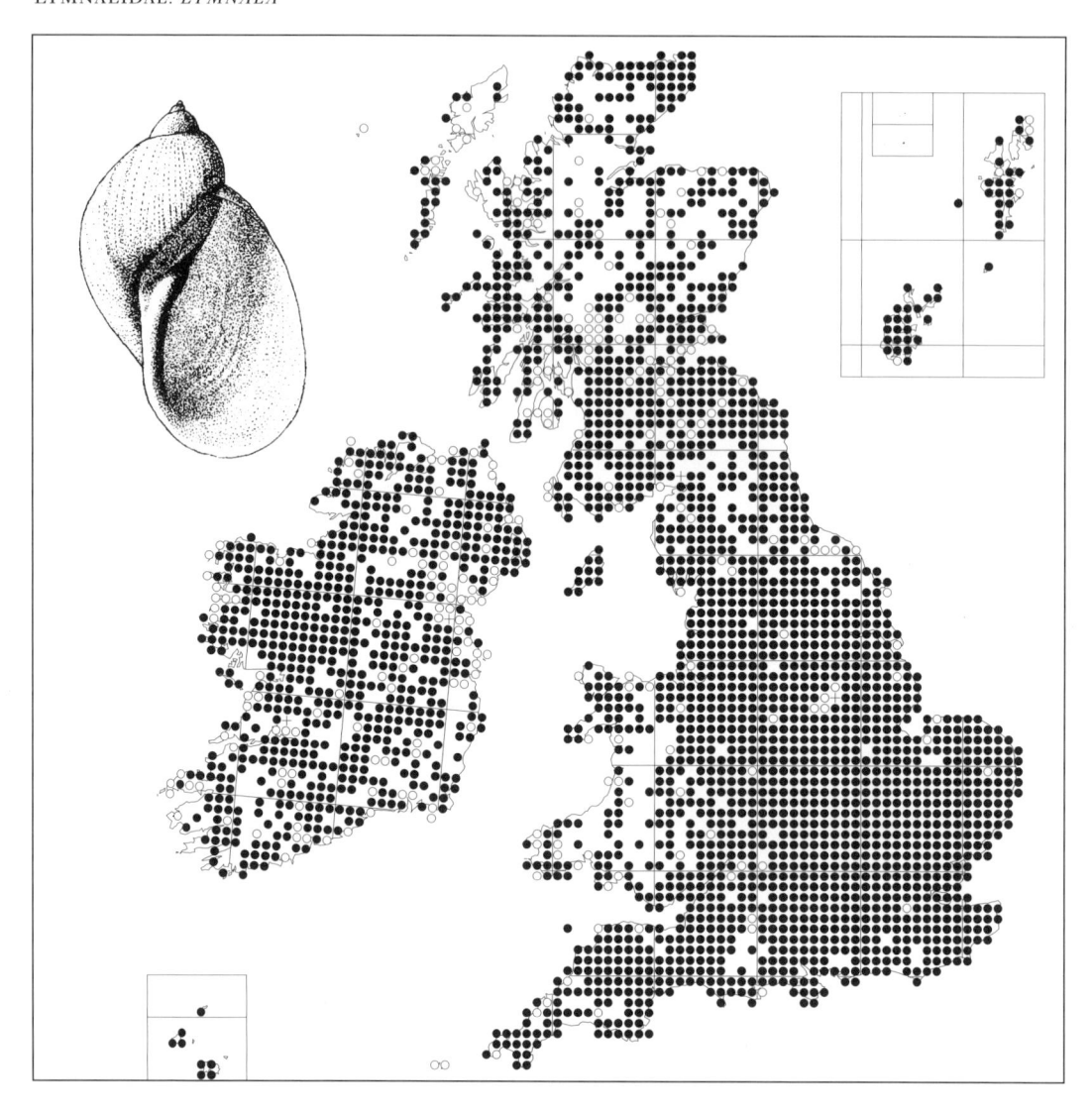

**Lymnaea peregra** (Müller, 1774) (*Radix peregra* (Müller), *Lymnaea ovata* (Draparnaud))

Common (or wandering) pond snail    Size: 12–28mm

This ubiquitous species occurs in aquatic habitats of all kinds, in hard or soft water alike, from the richest rivers and canals to the meanest ephemeral ponds and ditches where it may be the only gastropod present. It can live on almost bare, muddy or stony bottoms with little obvious plant-life and can withstand seasonal desiccation. It extends from sea-level (it is tolerant of brackish water) to over 700m in the Scottish and Welsh highlands.

Native (Lgl). No evidence of any local or regional change. It is a rapid colonist of new, man-made habitats and is tolerant of mild pollution.

Palaearctic: common throughout Europe.

*L. involuta* Thompson, from mountain lakes in Co. Cork and Co. Kerry, and *L. burnetti* Alder, from Loch Skene, Dumfries, are probably forms of *L. peregra*. Their peculiar characters appear to be due to a combination of harsh environment and genetic isolation (Boycott *et al.*, 1932; Boycott, 1938; Diver *et al.*, 1939; Ballantine & Bradley, 1963). The common form *L. ovata* (Draparnaud) has also been considered by some authorities to be a distinct species.

(Ellis: 104; Macan: 25)

LYMNAEIDAE: *MYXAS*

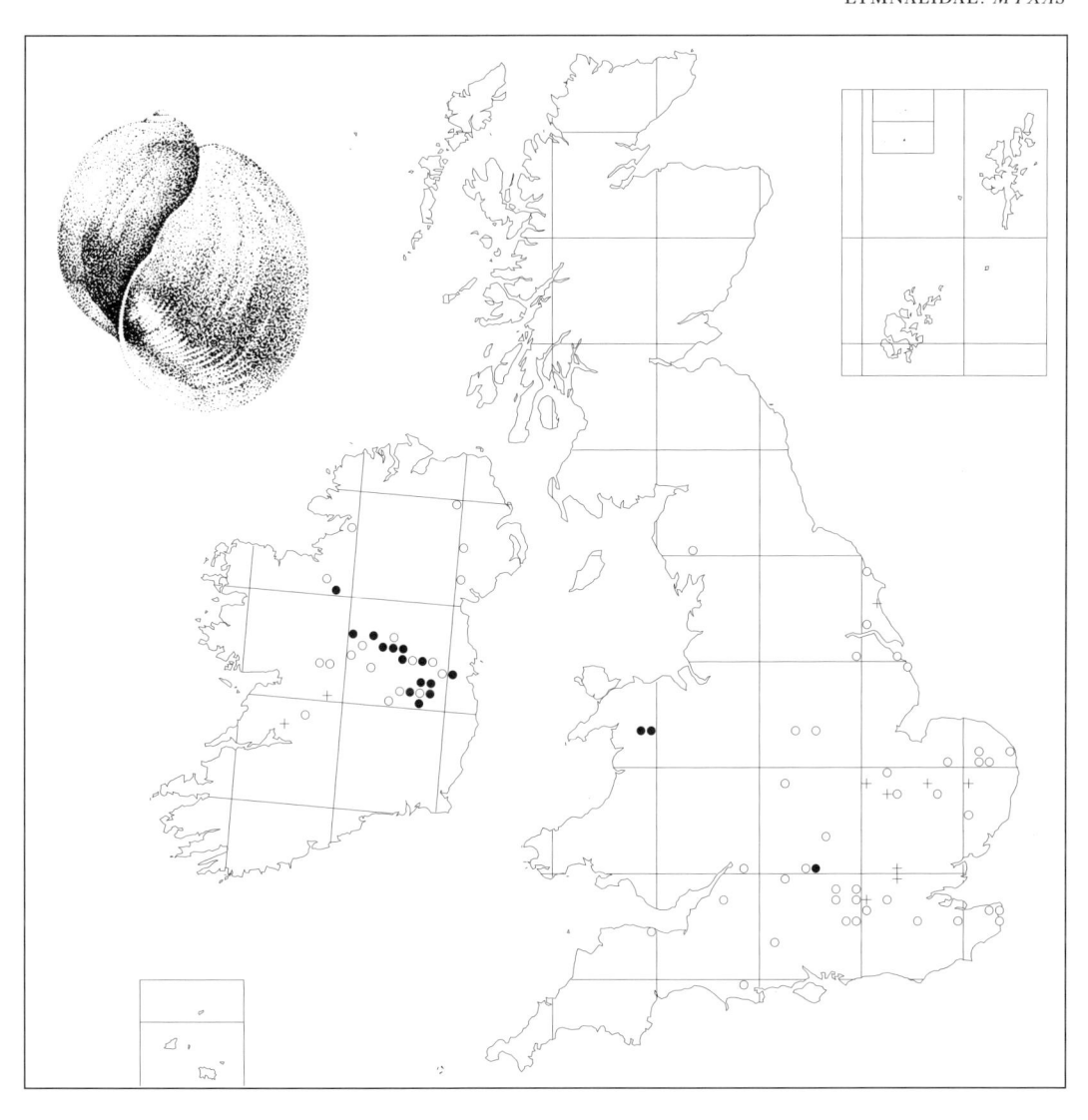

**Myxas glutinosa** (Müller, 1774) (*Lymnaea glutinosa* (Müller))

Glutinous snail   Size: 12–15mm

Restricted to spacious bodies of quiet, very clean water in slow rivers, canals, drainage ditches and lakes. Most occurrences are in hard water, notable exceptions being Windermere (calcium 6–8 ppm) and Bala Lake (calcium 2 ppm). It avoids turbid or weed-choked places. It likes firm substrates: in the Irish canals it is typically found on stones or on the masonry of locks and bridges rather than on the vegetation. In Windermere and in Bala Lake it is recorded living on stones lying in firm mud in about a metre of water.

Native (Lgl). Though reported over the past 150 years from about 35–40 localities in England and Wales, this is now a rare species. The most recent British records are: Deal (Kent) *c*. 1930; Chislet Marshes (Kent) 1930; R. Yare (Norfolk) 1940; Windermere (Westmorland) 1957; Fleet (Hants) (shell) 1969; Kennington (Oxon.) 1990 (Walter *et al.*, 1991); and Bala Lake (Llyn Tegid) (Merioneth) 1998. The reason for the decline of *Myxas* is unclear, but it may be unusually sensitive to disturbance and to pollution (including enrichment from arable land). In Ireland it was common in the 1970s in a few places in the Royal and Grand Canals; its survival in that country may depend on protecting these sites, especially in the now abandoned and degenerating Royal Canal. RDB category: Endangered (in Ireland: Vulnerable).

N. European: between the Alps and the Arctic Circle (Finland to 71°N), but everywhere very local.

(Ellis: 112; Macan: 25)

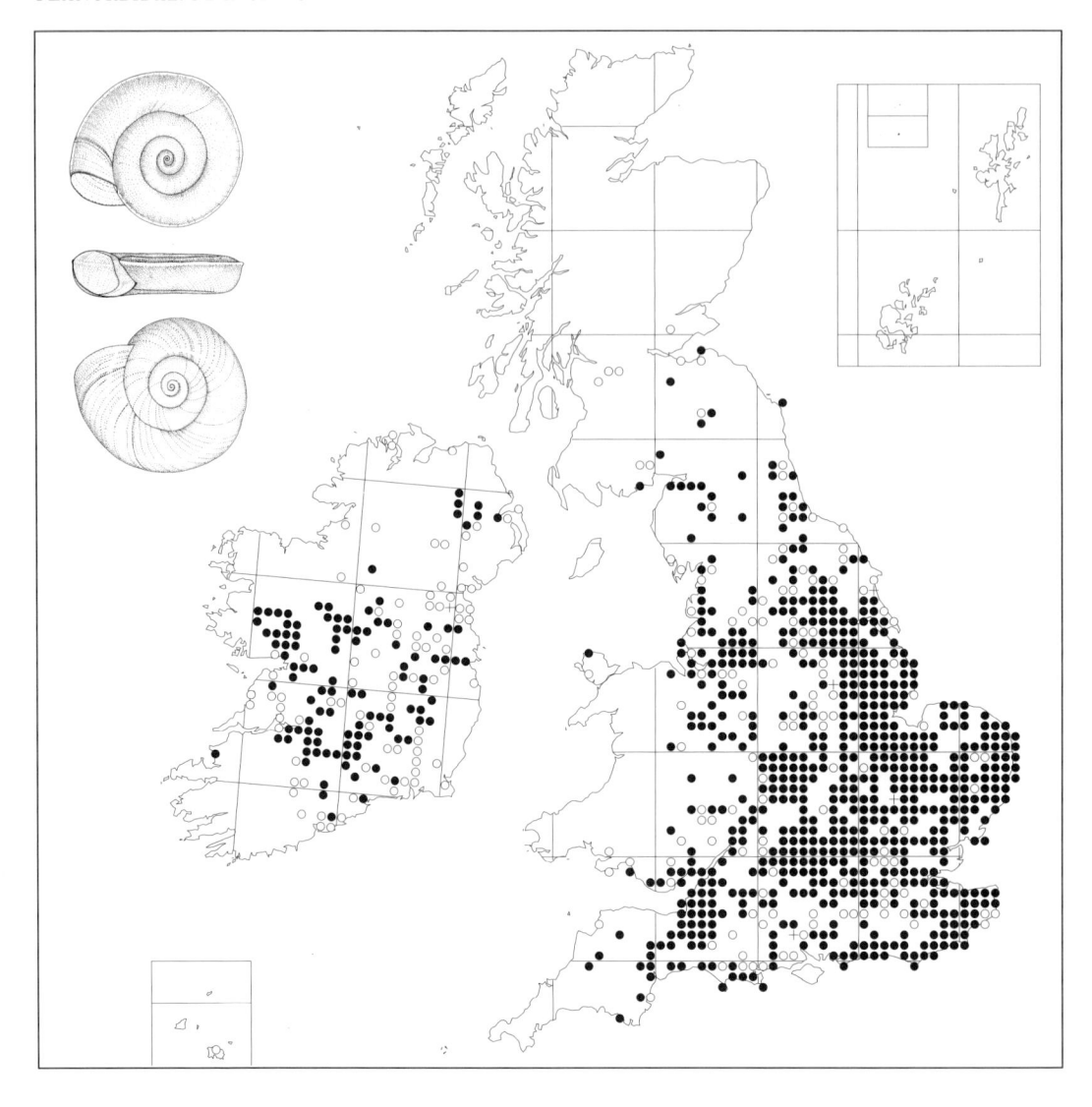

**Planorbis planorbis** (L., 1758) (*P. umbilicatus* Müller, *P. marginatus* Draparnaud, *P. complanatus* of some British authors, *non* Linnaeus)

Margined ram's-horn    Size: 12–17mm

Found in all kinds of well-vegetated aquatic habitats of lowland type – slow rivers, canals, lakes and closed ponds – but especially characteristic of shallow pools and swampy ditches liable to dry up in the summer and where it is often associated with *Lymnaea palustris* or *Anisus leucostoma*. It prefers hard water.

Native (E.Pgl). *P. planorbis* shows no evidence of any serious regional decline, but is becoming scarcer in some areas through habitat destruction.

European & W. Asiatic; in Scandinavia to 63°N.

The morphological distinction between *P. planorbis* and *P. carinatus* is not always clear-cut. Some forms of *P. carinatus* (especially when young) are often mistaken for the present species, which may therefore be somewhat overrecorded.

(Ellis: 119; Macan: 31)

PLANORBIDAE: *PLANORBIS*

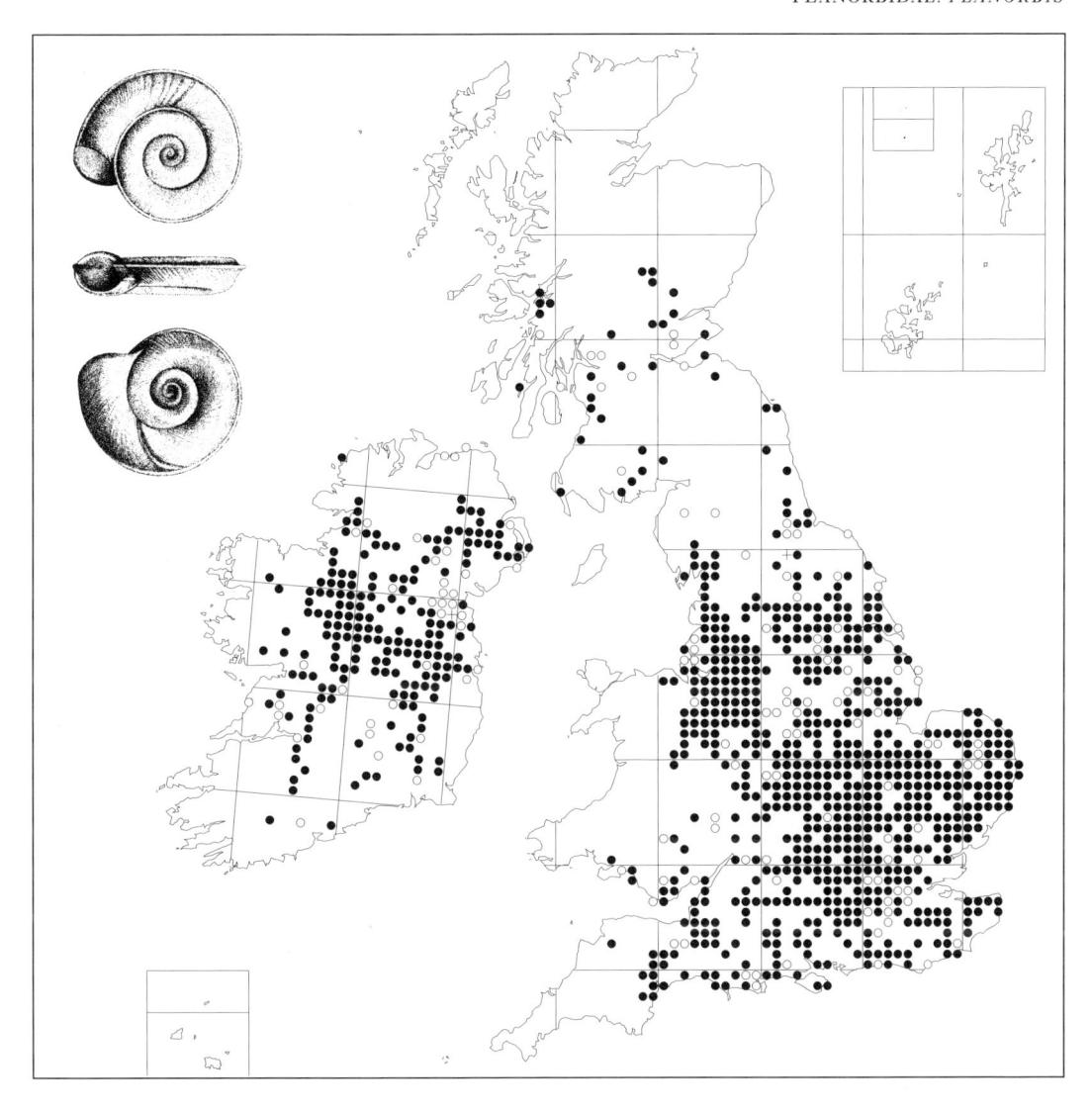

**Planorbis carinatus** Müller, 1774

Keeled ram's-horn    Size: 12–15mm

This common calciphile lives in a wide variety of permanent, well-vegetated aquatic habitats of lowland type such as slow rivers, canals, running ditches, lakes and ponds. A diverse fauna is usually associated. *P. carinatus* avoids marshy places subject to desiccation, where *P. planorbis* may replace it.

Native (E.Pgl). No evidence of any significant regional change.

European: in Scandinavia to 63°N.

(Ellis: 120; Macan: 31)

PLANORBIDAE: *ANISUS*

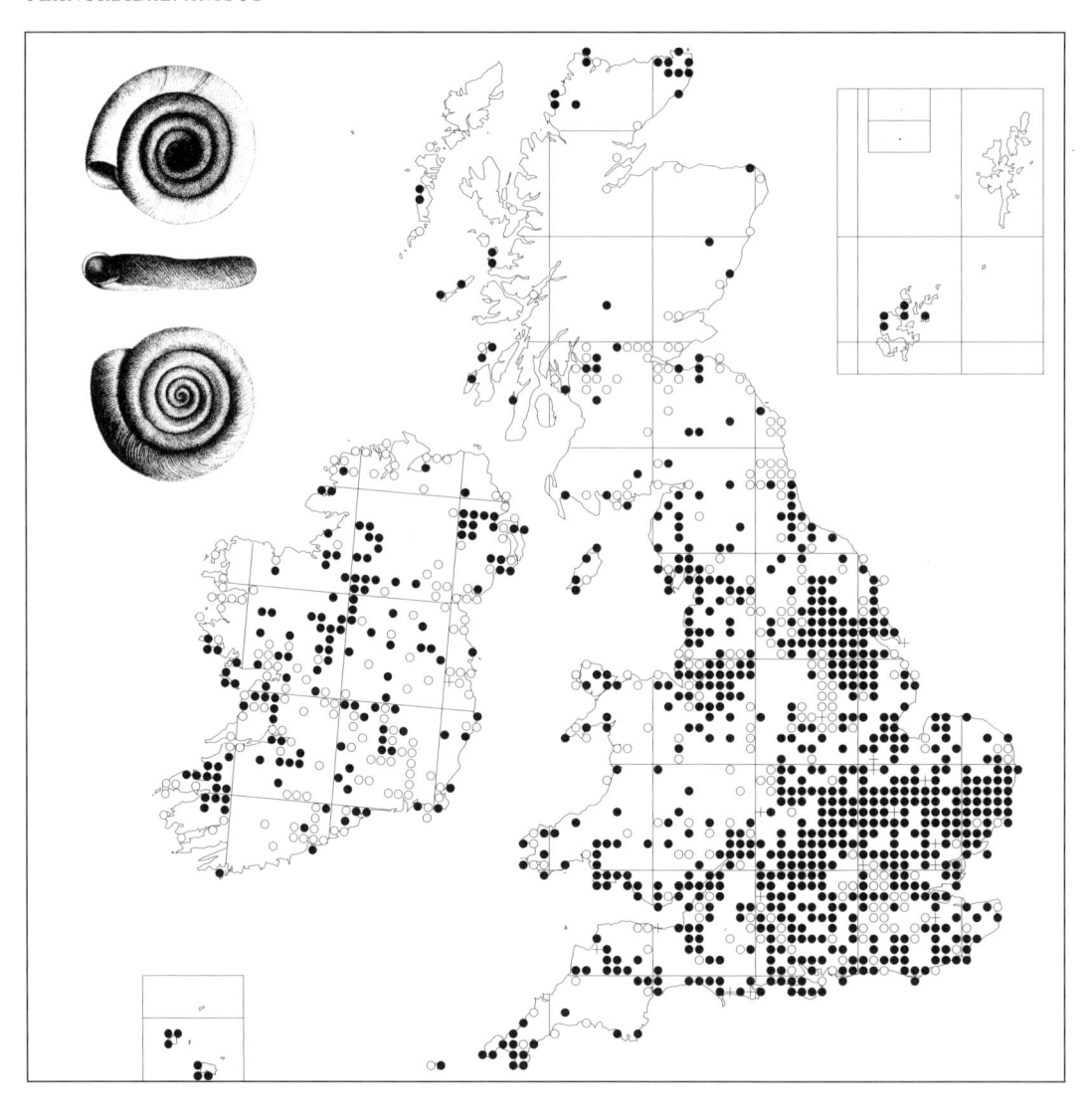

**Anisus leucostoma** (Millet, 1813) (*Planorbis leucostoma* Millet, *P. spirorbis* of British authors)

Button (or white-lipped) ram's-horn    Size: 5–8mm

Found in a variety of aquatic habitats – rivers, canals, lakes and ponds – but most typical of swampy pools and ditches, especially those drying up in the summer; common associates are *Aplexa hypnorum*, *Lymnaea truncatula*, *L. palustris* and *Pisidium personatum*. It is mainly a lowland species, with some preference for hard water.

Native (Lgl). *A. leucostoma* shows some decline in heavily farmed areas due to land drainage. In Ireland it is probably underrecorded.

European and W. Asiatic; in Sweden to 63°N.

The form *spirorbis* (in which the whorls are more broadly coiled and relatively fewer in number) is regarded by many Continental authorities as a separate species. Typical narrow-whorled *leucostoma* is the commoner form throughout the British Isles.

(Ellis: 123; Macan: 33)

PLANORBIDAE: *ANISUS*

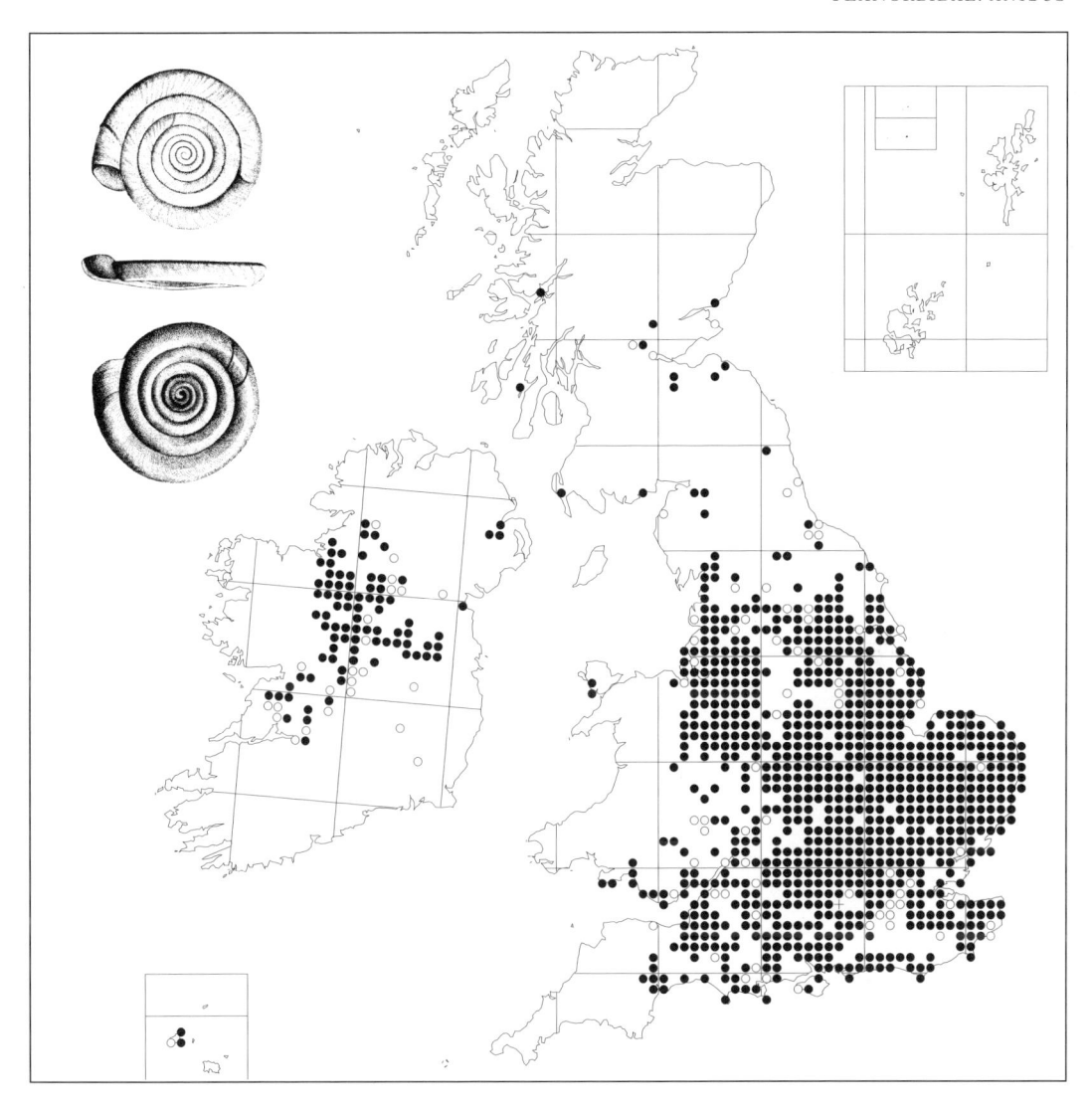

**Anisus vortex** (L., 1758) (*Planorbis vortex* (L.))

Whirlpool ram's-horn   Size: 7–10mm

A common species in the better kinds of lowland aquatic habitats: clean, well-oxygenated, hard water with plenty of weed in slow rivers, canals, lakes and drainage ditches. It is rare in small closed ponds and, unlike *A. leucostoma*, never occurs in places subject to desiccation.

Native (E.Pgl). *A. vortex* gives no evidence of significant regional decline.

European and W. Asiatic; mainly in lowland areas, reaching to the Arctic Circle in Finland.

Note that scattered old records exist for further north in eastern Scotland as far as Caithness (Loch Hempriggs, near Wick). These may be errors for *A. leucostoma* and have been omitted.

(Ellis: 121; Macan: 33)

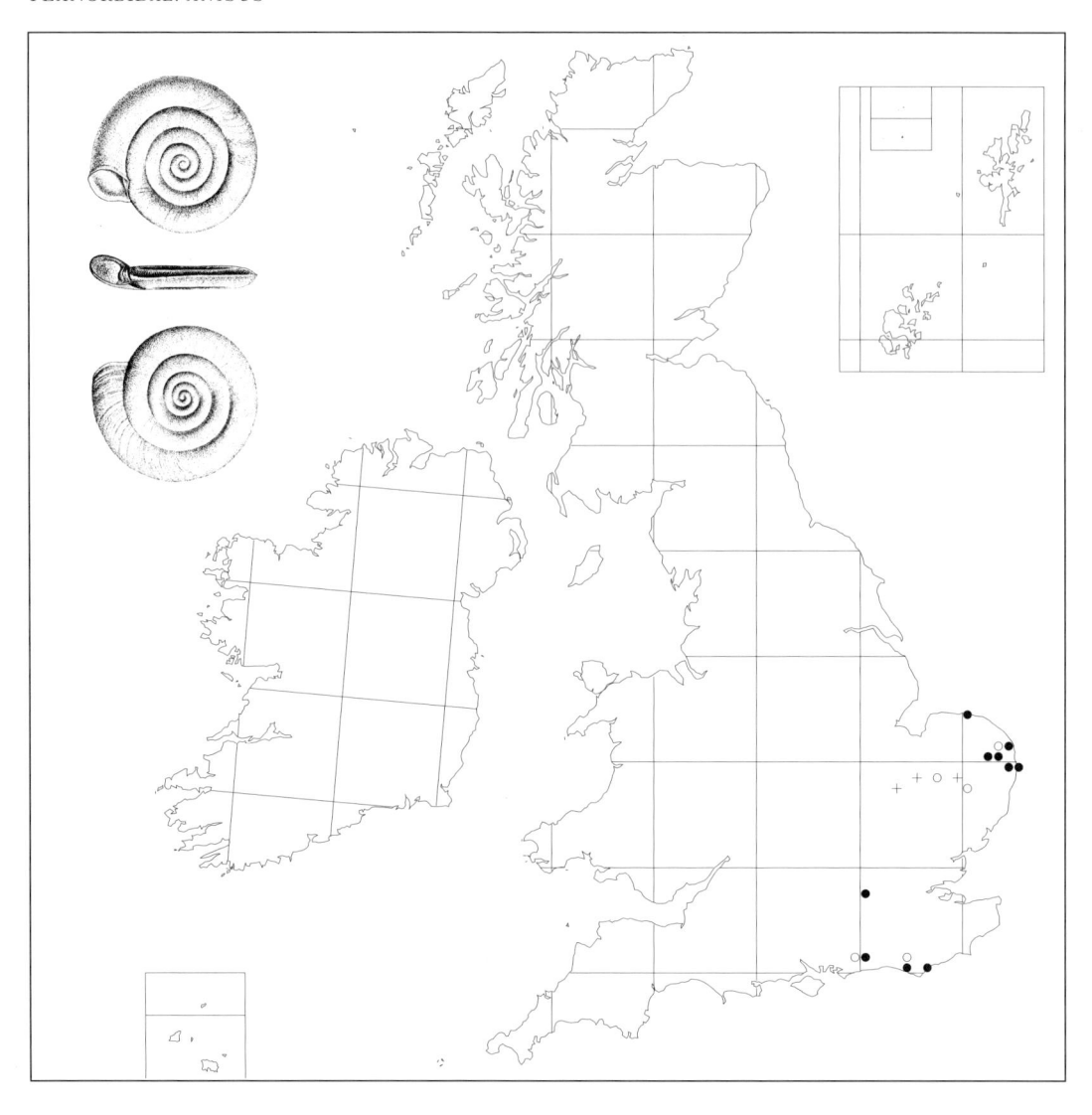

**Anisus vorticulus** (Troschel, 1834) (*Planorbis vorticulus* Troschel)

Size: 3.5–6.0mm

A rare species found chiefly in marsh drains in levels, in clean, still water with a dense aquatic flora, such as *Lemna trisulca*, *Hydrocharis* and *Ceratophyllum*. It favours ditches with little emergent vegetation and often floats on the surface among *Lemna*. A diverse mollusc fauna is often present and other rarities, most notably *Valvata macrostoma*, *Segmentina nitida* or *Pisidium pseudosphaerium*, may occur with it (Ellis, 1928, 1931, 1932; Long, 1968).

Probably native (poorly dated Pgl fossils only). This very local species has been seen in recent years at a few sites in Norfolk, in the lower Waveney valley (E. Suffolk), in the Pevensey and Lewes Levels and Amberley Wild Brooks (Sussex), and in a pond near Staines (Middx.) (Kerney & May, 1960) where it was last noted in 1976 and may now be extinct. Old records and Postglacial fossils show that it formerly lived also in W. Suffolk and Cambridgeshire. All surviving populations are potentially or actually threatened by drainage, overfrequent dredging and eutrophication (Hingley, 1979; Willing & Killeen, 1998). RDB category: Vulnerable.

Central and S. European; in Scandinavia known with certainty only from Denmark.

(Ellis: 122)

PLANORBIDAE: *BATHYOMPHALUS*

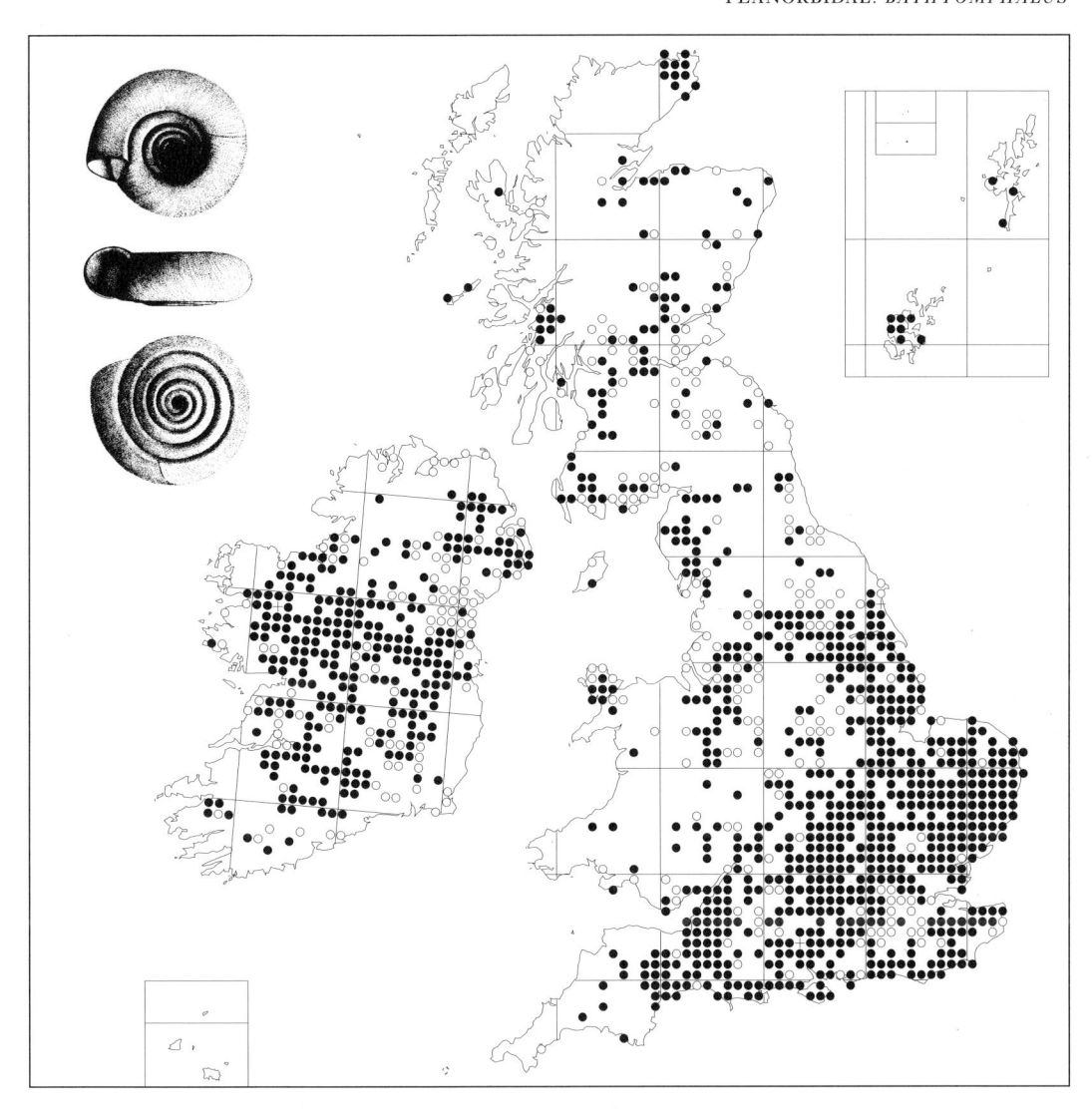

**Bathyomphalus contortus** (L., 1758) (*Planorbis contortus* (L.))

Twisted ram's-horn    Size: 4.5–6.0mm

Lives in hard or soft water in a variety of aquatic habitats: rivers, canals, lakes, small closed ponds and swampy ditches. It is often common in thick growths of weed in bright, running water, but equally it may be found in stagnant drains associated with few or no other gastropods. It avoids places subject to seasonal desiccation.

Native (Lgl). This species gives no evidence of regional change.

Palaearctic; throughout Europe to 69°N in Finland.

(Ellis: 129; Macan: 29)

PLANORBIDAE: *GYRAULUS*

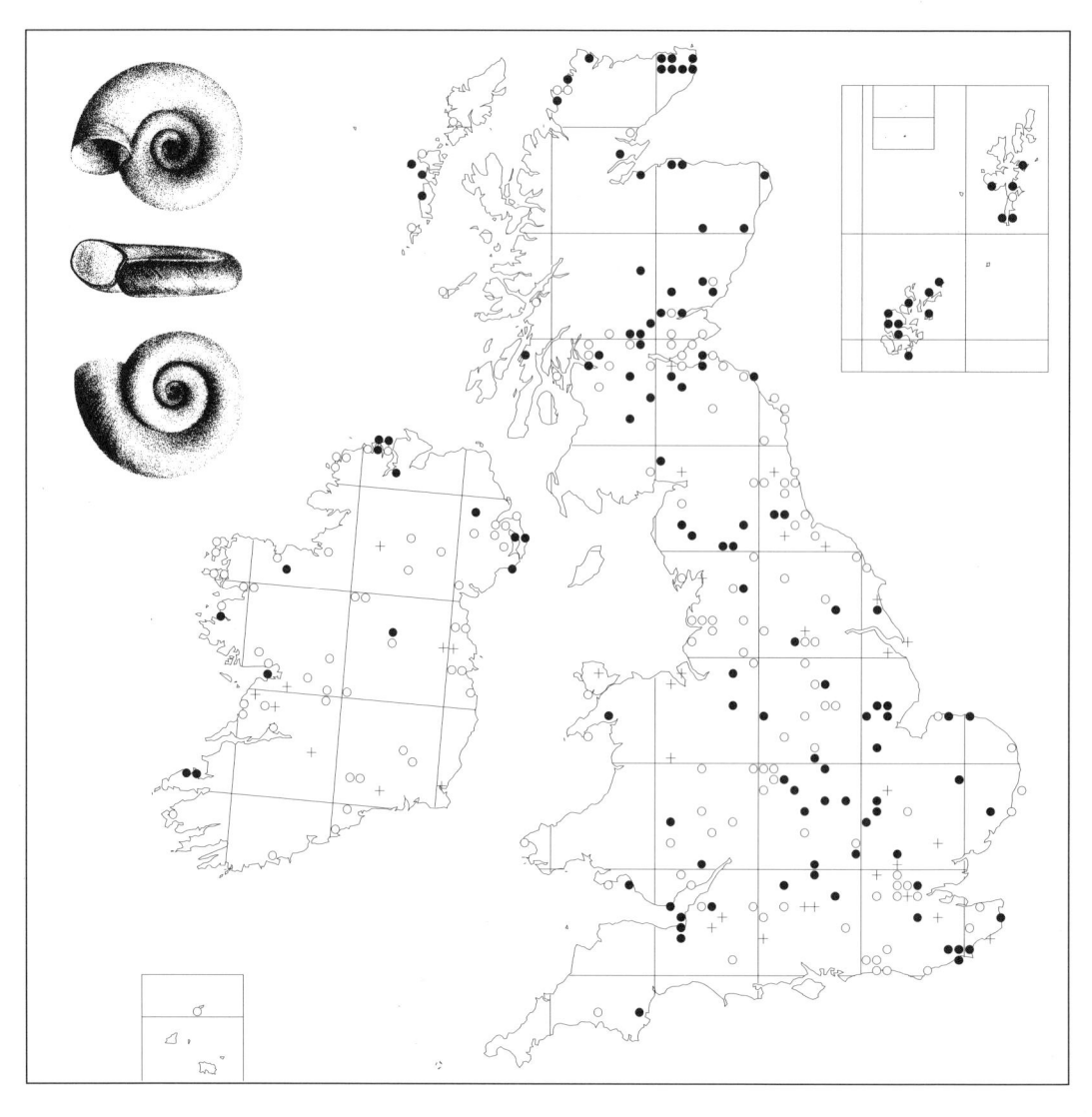

**Gyraulus laevis** (Alder, 1838) (*Planorbis laevis* Alder, *P. glaber* of older British authors)

Smooth ram's-horn    Size: 4.5–6.0mm

Very local in clean, quiet water, hard or soft, in lakes and ponds, usually among weeds but sometimes on bare, bottom mud or stones. In northern or western areas it typically occurs in shallow natural lakes close to the sea, often among sandhills. It will tolerate slightly brackish water. In lowland England it has frequently been noted in artificial habitats, such as flooded gravel pits and quarries, newly formed reservoirs and ornamental lakes in public parks. Exceptionally it is found in canalized rivers and drainage dykes.

Native (Lgl). *G. laevis* established itself in the British Isles immediately after the retreat of the ice and is a characteristic fossil in Lateglacial and early Postglacial lake marls. Its present distribution is partly relict, but results also from an ability to colonize 'raw', man-made habitats of an ephemeral type; probably it possesses good powers of passive dispersal. Though uncommon it shows no clear sign of overall decline.

Palaearctic (probably Holarctic); local throughout Europe, extending to beyond the Arctic Circle in Finland.

(Ellis: 126)

PLANORBIDAE: *GYRAULUS*

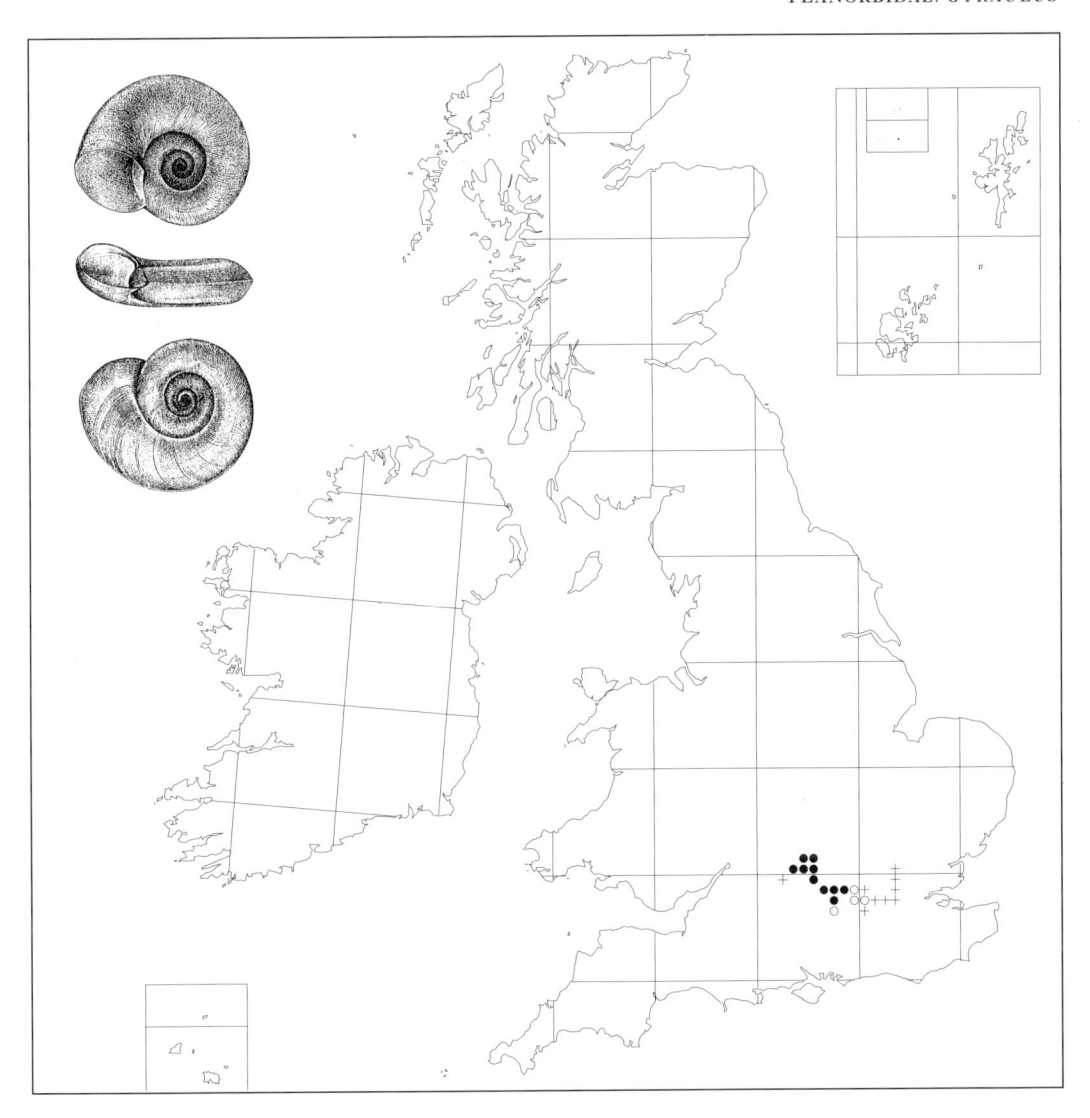

**Gyraulus acronicus** (Férussac, 1807) (*Planorbis acronicus* Férussac)

Thames ram's-horn   Size: 6–10mm

In the upper Thames and its tributaries this species lives in quiet stretches and backwaters, among weeds and stones (Cooper, 1924). It is found more often as dead shells in flood debris than living.

Native (E.Pgl). In the earlier part of the Postglacial period *G. acronicus* had a slightly wider range than today, living additionally in the lower Thames and in the River Lea. There is no clear evidence that the decline continues, but more needs to be known about this species, which is rarely collected alive. RDB category: Vulnerable.

C. and N. European; mainly Alpine and northern, to 71°N in Finland (but see note below).

The taxonomy of the genus *Gyraulus* is complex and the identity of this particular British form with the species known under the same name in mountain lakes in Scandinavia and the Alps is not firmly established.

(Ellis: 125)

65

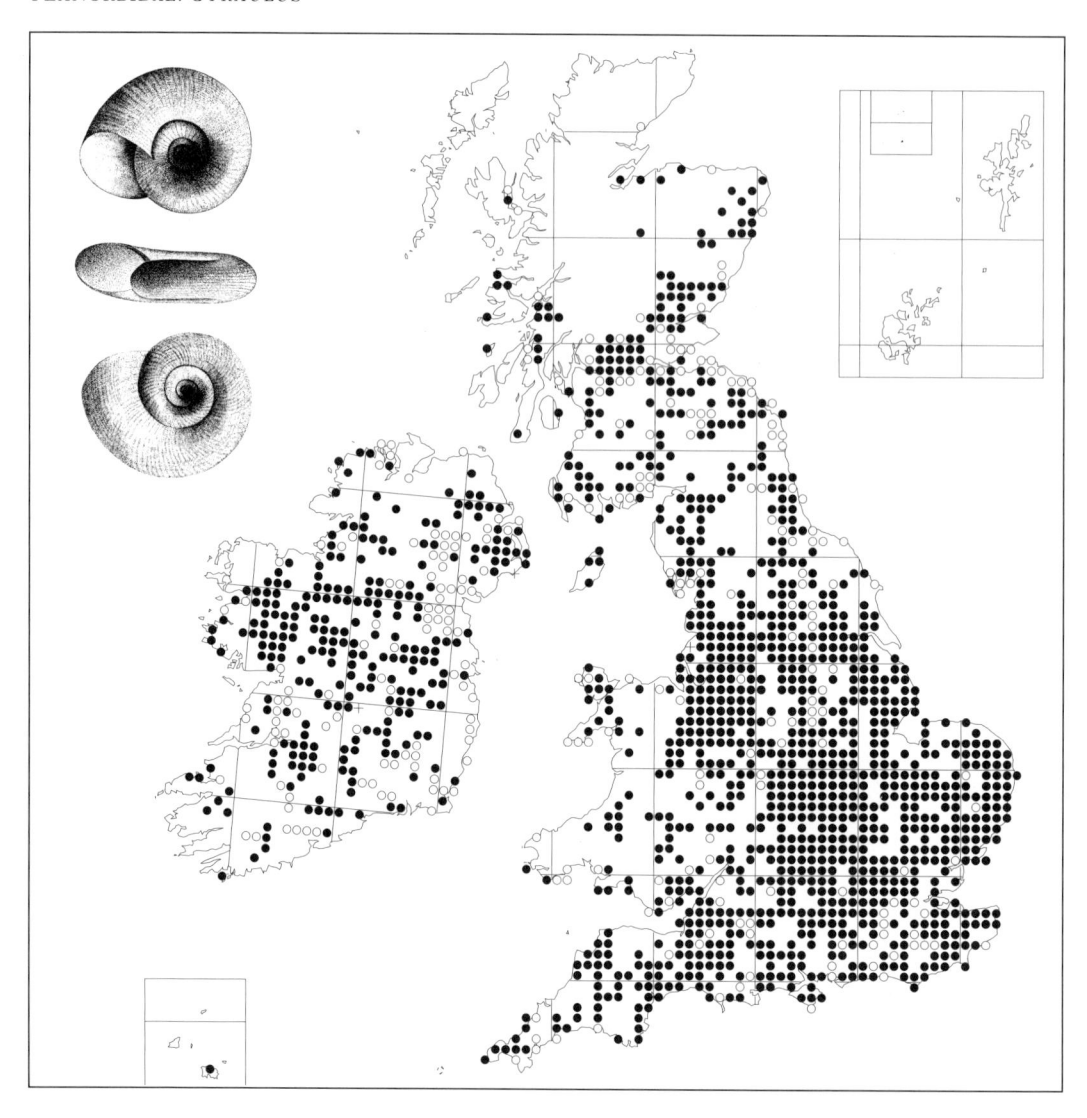

**Gyraulus albus** (Müller, 1774) (*Planorbis albus* Müller)

White ram's-horn    Size: 5–8mm

This is a successful and adaptable species found in most kinds of aquatic habitats apart from those subject to seasonal desiccation. It lives in stagnant or flowing water, hard or soft, in rivers, canals, lakes, ponds and ditches. It is the commonest planorbid in rivers and streams in upland areas.

Native (E.Pgl). *G. albus* shows no evidence of any regional change. It is tolerant of mild pollution.

Holarctic; in Europe to 65°N in Finland.

(Ellis: 124; Macan: 35)

PLANORBIDAE: *GYRAULUS*

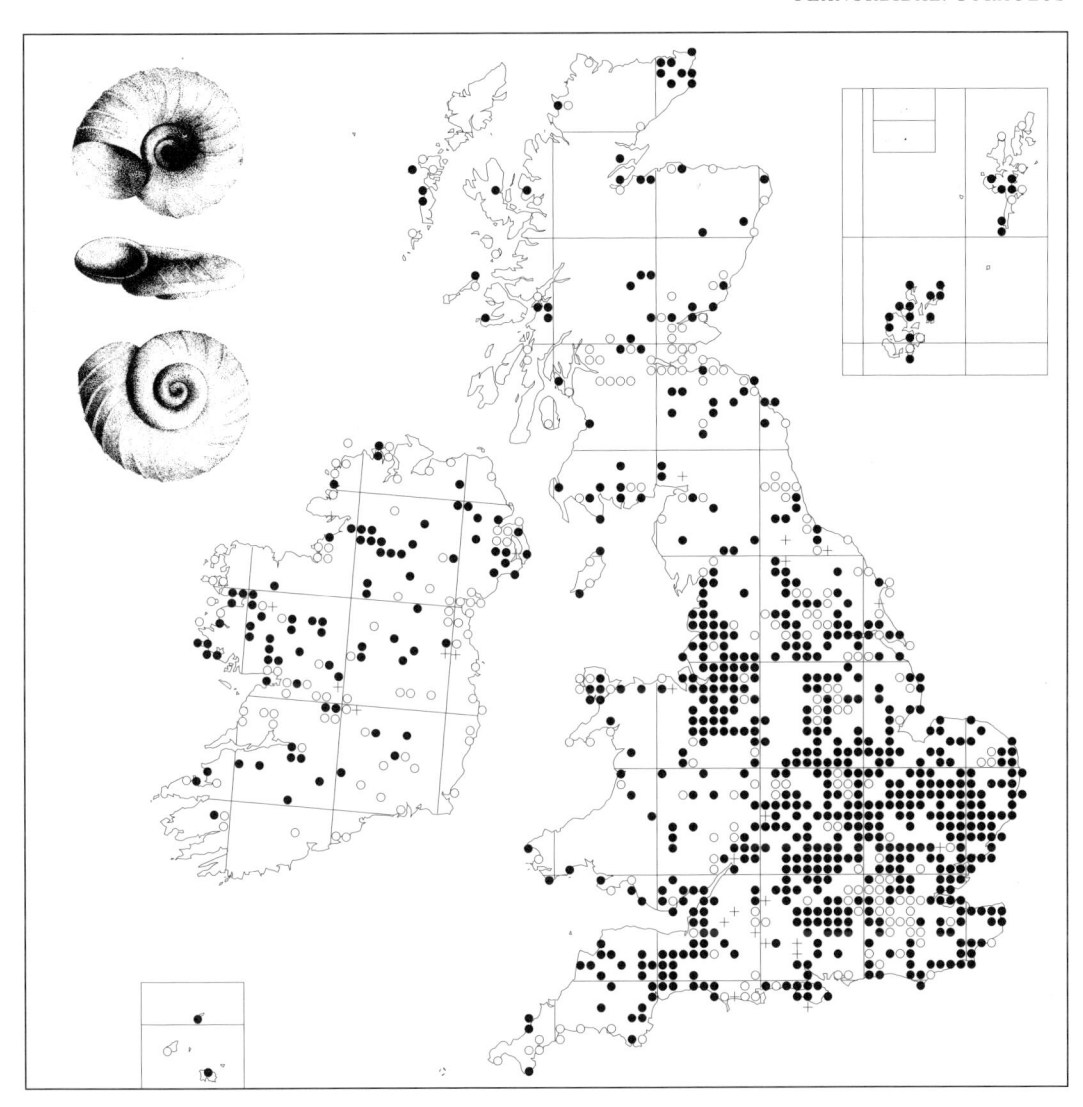

**Gyraulus crista** (L., 1758) (*Armiger crista* (L.), *Planorbis crista* (L.), *P. nautileus* (L.))

Nautilus ram's-horn   Size: 2–3mm

This minute species lives in most kinds of lowland aquatic habitats apart from those liable to dry up: quiet rivers, canals, lakes, ponds, weedy ditches. Though calciphile it is tolerant of soft water. It has good powers of passive dispersal and is frequent in tiny isolated ponds with few molluscan associates, but is found equally in the richest habitats.

Native (Lgl). *G. crista* gives no evidence of significant regional change. It is frequently overlooked owing to its minute size and is hence somewhat underrecorded, especially in the north and west.

European: nearly throughout Europe but sparse north of about 62°.

(Ellis: 127)

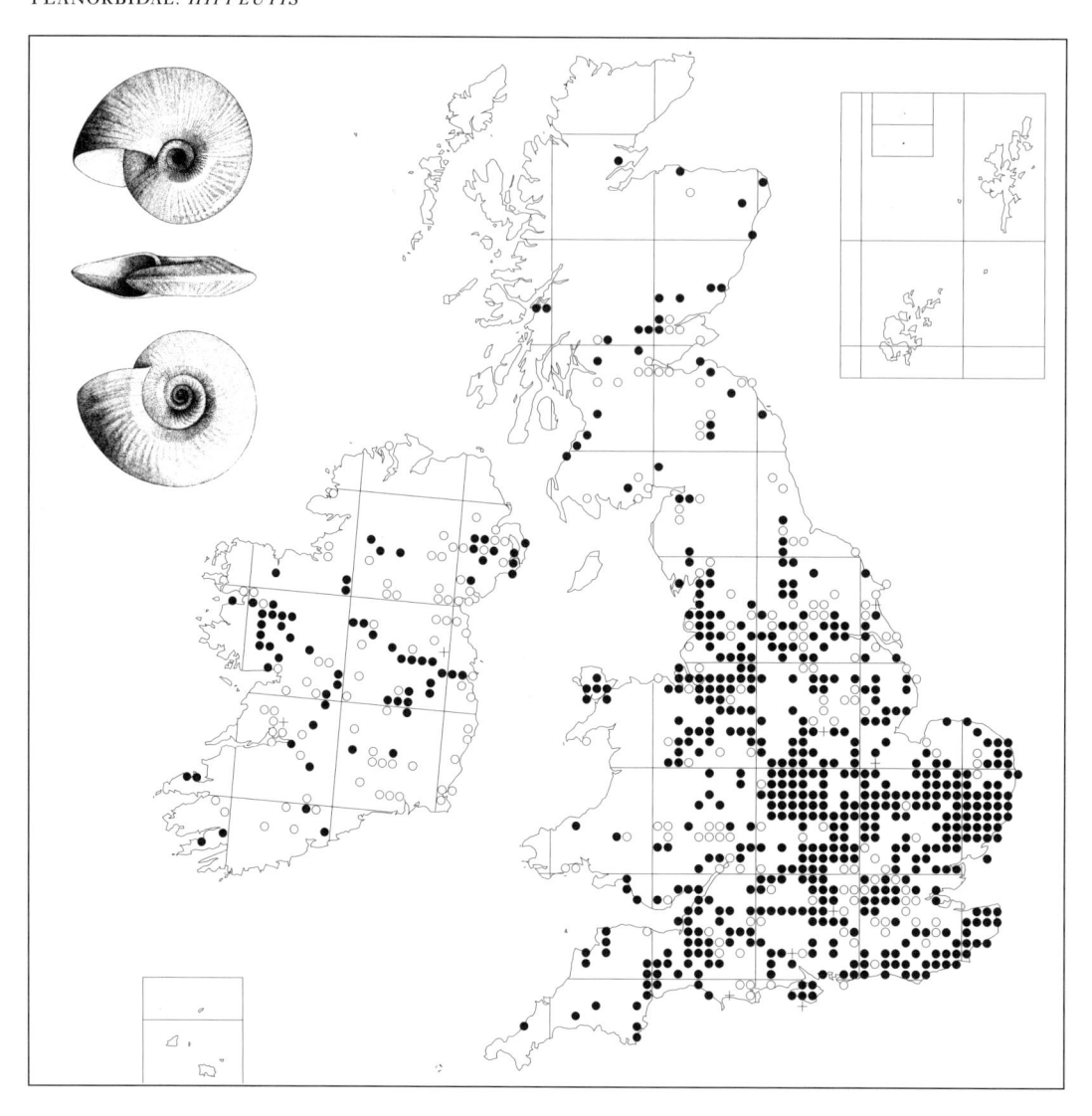

**Hippeutis complanatus** (L., 1758) (*Planorbis complanatus* (L.), *P. fontanus* (Lightfoot), *Segmentina complanata* (L.))

Flat ram's-horn   Size: 3–5mm

Inhabits well-vegetated places in quiet or slowly moving water in a variety of lowland habitats, such as canals, rivers, lakes and drainage ditches. It is however especially characteristic of small closed ponds, in which it is often associated with *Gyraulus crista* or *Acroloxus lacustris*. It is a calciphile, and intolerant of desiccation.

Native (Lgl). No evidence of significant regional change. In Ireland it is underrecorded.

European and W. Asiatic; widespread in lowland areas north to 63° in Finland.

(Ellis; 129; Macan: 31)

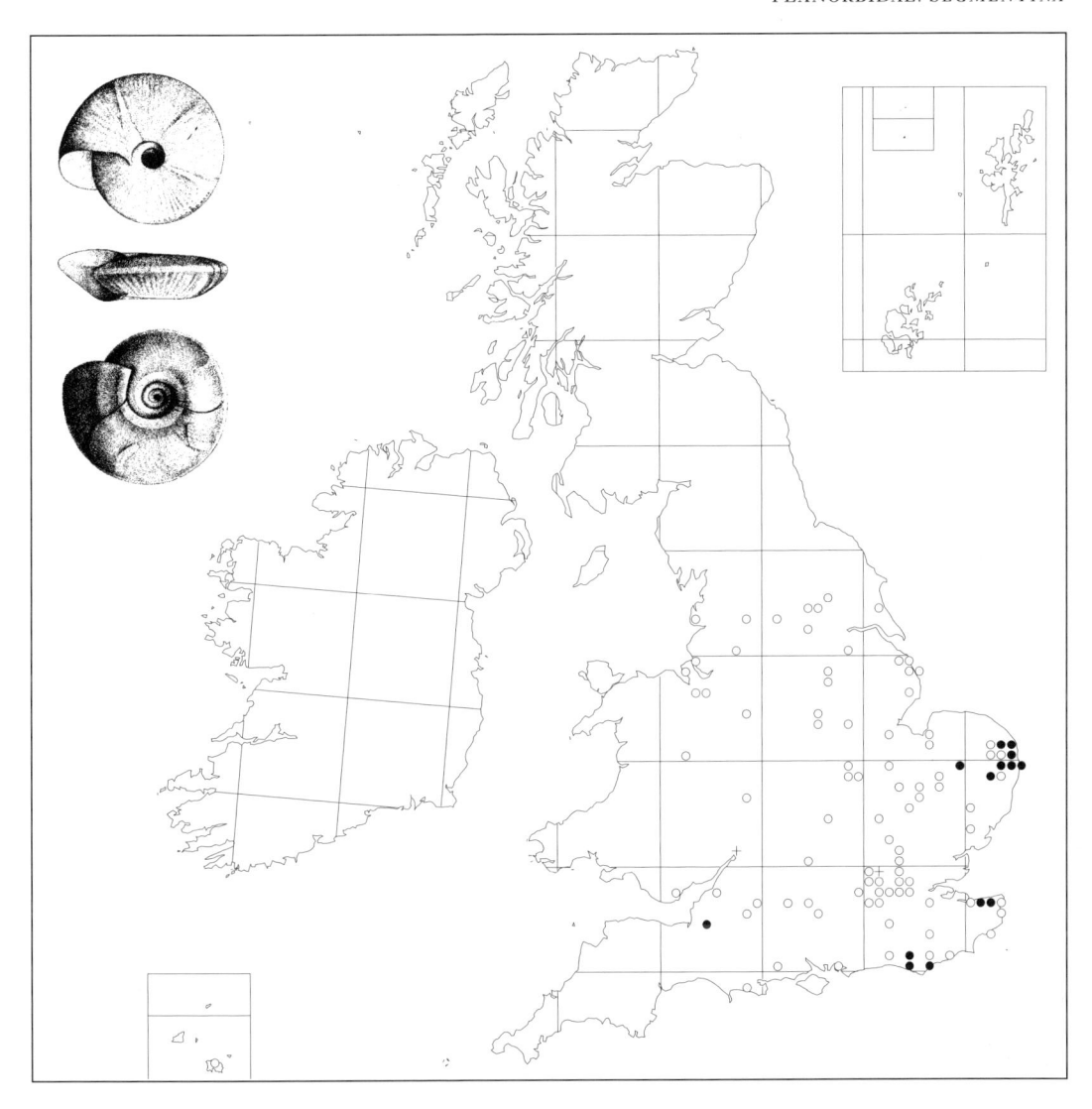

**Segmentina nitida** (Müller, 1774) (*S. lineata* Fleming)

Shiny ram's-horn    Size: 4.0–5.5mm

Occurs today mostly in drainage ditches in marsh levels, in clean, hard water in densely vegetated places among thick growths of, for instance, *Glyceria* and *Lemna trisulca*. The associated fauna is generally rich and may include other rarities such as *Valvata macrostoma*, *Anisus vorticulus* and *Pisidium pseudosphaerium*. Formerly it was known also from lakes and ponds, often situated on former floodplains or in areas of reclaimed marshland. At Thompson Common (Norfolk) it lives uniquely in water-filled glacial hollows.

  Native (E.Pgl). Very local, and declining. It is now extinct over most of England, surviving in a few places in Norfolk and Suffolk (notably the lower Waveney valley), in Kent (Monkton and Westbere Marshes), in Sussex (Pevensey and Lewes Levels) and at one site in the Somerset Levels. In the 19th century it was common around London (most recent record: Walton-on-Thames, 1952). Some last records elsewhere are: Staffs., 1844; Wilts., 1864; Montgomery, 1887; Lancs., 1895; Notts., 1910; Lincs., 1928; Cambs., 1931; Yorks., 1956; Hants, 1964. Reasons for this catastrophic decline are unknown, but over-frequent dredging, the lowering of water-levels and pollution have all played some part (Hingley, 1979; Killeen & Willing, 1997). It is perhaps significant that surviving populations are in areas of traditional grazing where phosphate and nitrate enrichment remains low.

  Palaearctic; local throughout lowland Europe, to southernmost Scandinavia (62°N in Finland).

(Ellis: 130; Macan: 31)

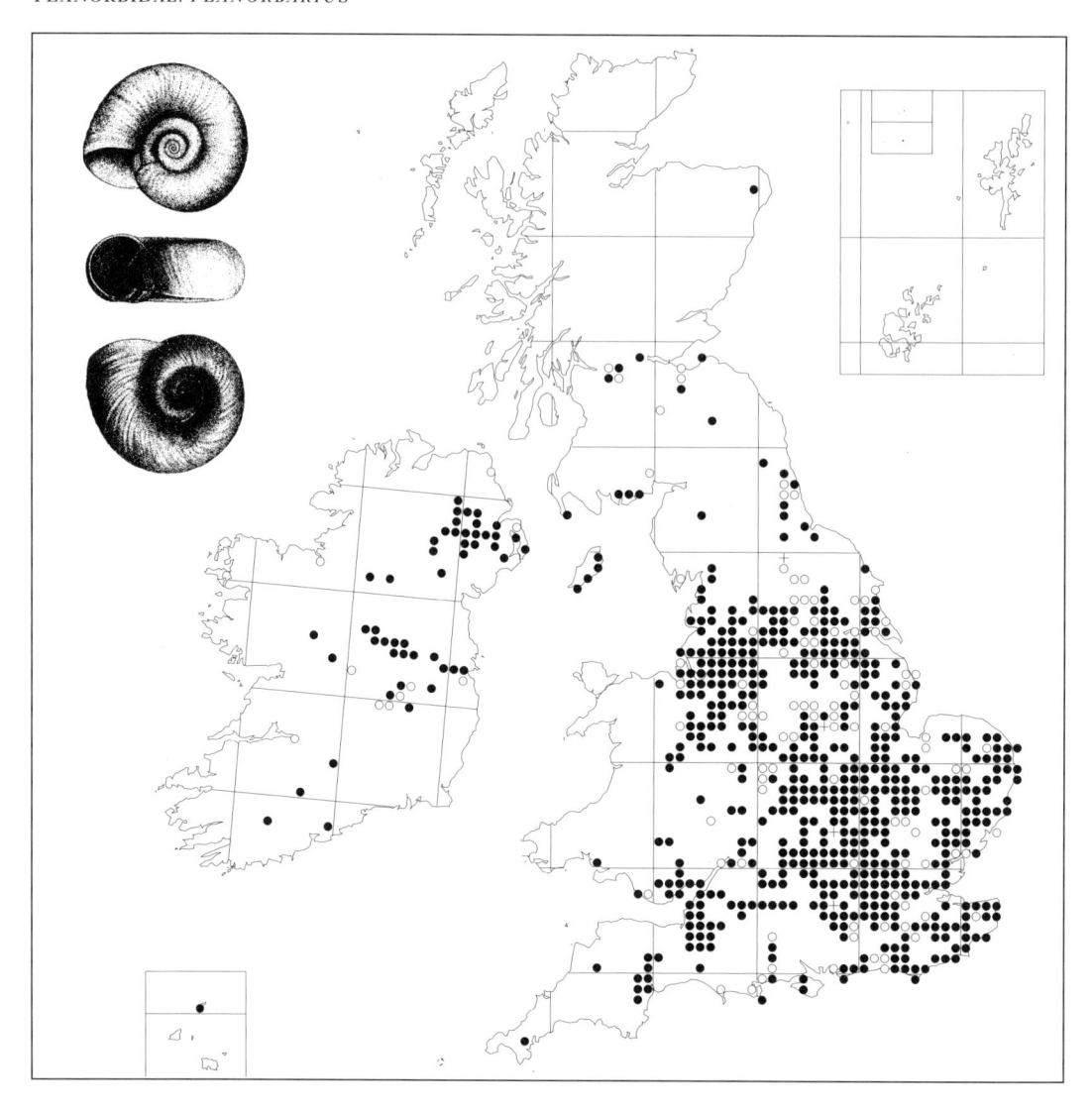

**Planorbarius corneus** (L., 1758) (*Planorbis corneus* (L.))

Great ram's-horn   Size: 25–35mm

This large species normally inhabits sizable bodies of quiet or slowly moving water, in canals, rivers, lakes and drainage ditches. It can tolerate places choked with rotting vegetation and where the bottom is anaerobic. It has poor powers of passive dispersal and is not normally found in small isolated water bodies, though often successfully introduced into such places, and may flourish in ornamental lakes, garden ponds and large aquaria.

Probably native (L.Pgl. or poorly dated fossils only). This is a species which is extending its range, both naturally along waterways and by deliberate introduction; several of the more outlying occurrences are certain or probable introductions (McMillan, 1955). In Ireland its spread has taken place within the present century. In Ulster, where it was first noted only in 1930, it is now a common species in the Lough Neagh basin.

European and W. Asiatic; in Europe extending to 64°N in Finland. The species has been widely dispersed by man.

(Ellis: 118; Macan: 29)

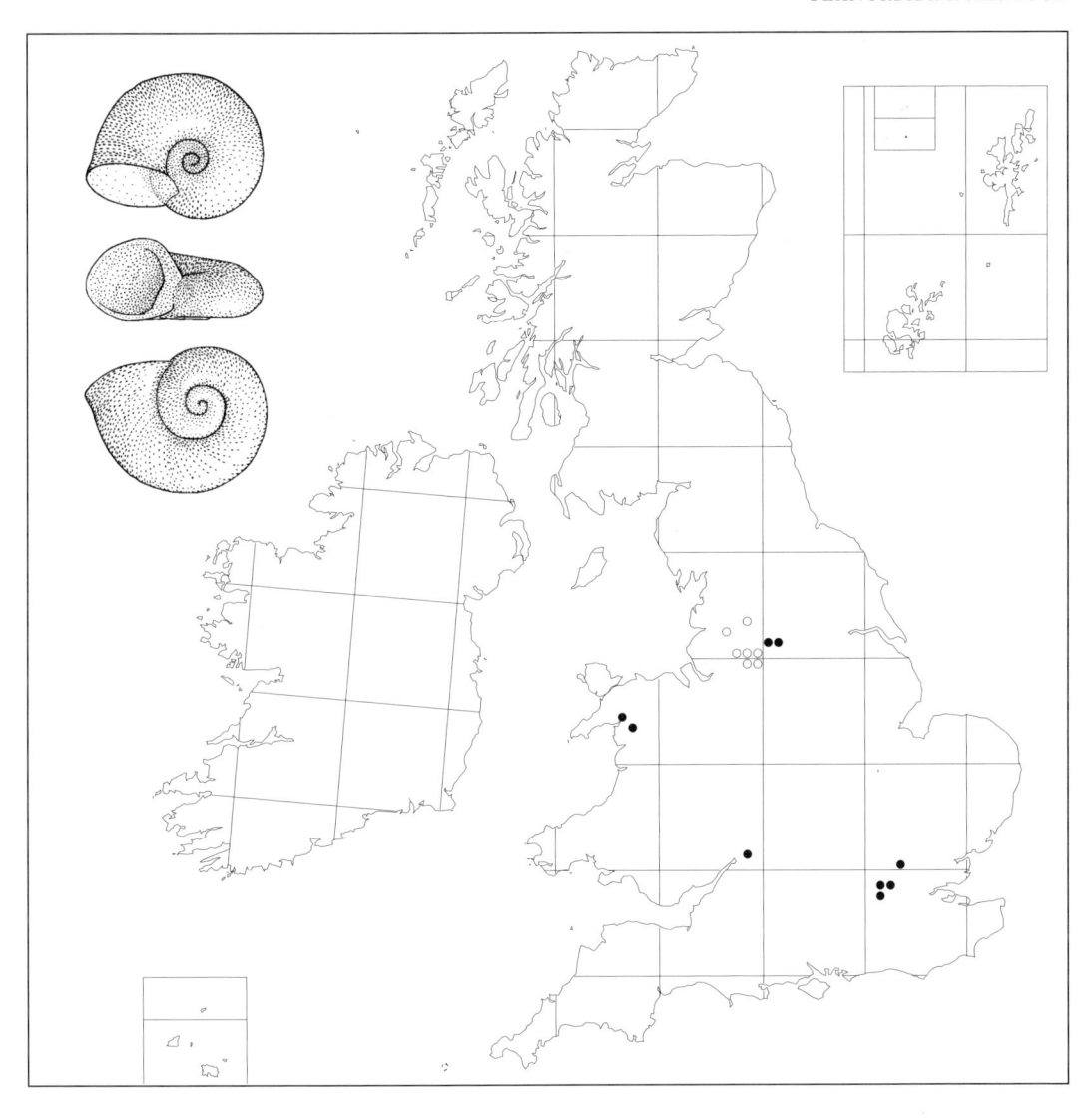

**Menetus dilatatus** (Gould, 1841) (*Planorbis dilatatus* Gould)

Trumpet ram's-horn   Size: 3–4mm

A rarity found in a few canals (around Manchester, near Gloucester and in London), in a canalized stream (Waltham Abbey, Essex), and in two lakes in Merioneth (Llyn Trawsfynydd, Llyn Mair). Several occurrences are associated with discharges of heated water: in the 19th century, characteristically, the condenser water from steam engines at cotton mills; in the 20th century, the cooling water from a now obsolete nuclear power station (Llyn Trawsfynydd). But it is not restricted to such places and can survive freezing temperatures. At Trawsfynydd, an artificial lake constructed in 1924–28, it lives attached to stones or among mosses and hornwort (*Ceratophyllum demersum*) (Dance, 1970a). In canals it appears to tolerate anaerobic bottom conditions.

Introduced. This American species was first recorded (at Pendleton, near Manchester) in 1869. It seems now to have disappeared from most of the canals in that area, with the exception of the Huddersfield Canal where it was discovered only in 1972. It was first found in North Wales in 1969, in the Grand Union Canal in London in 1974, and at Waltham Abbey in 1986.

Eastern U.S.A. (New England to Texas and Florida). Naturalized in Britain, Germany and Poland.

(Ellis: 128; Macan; 35)

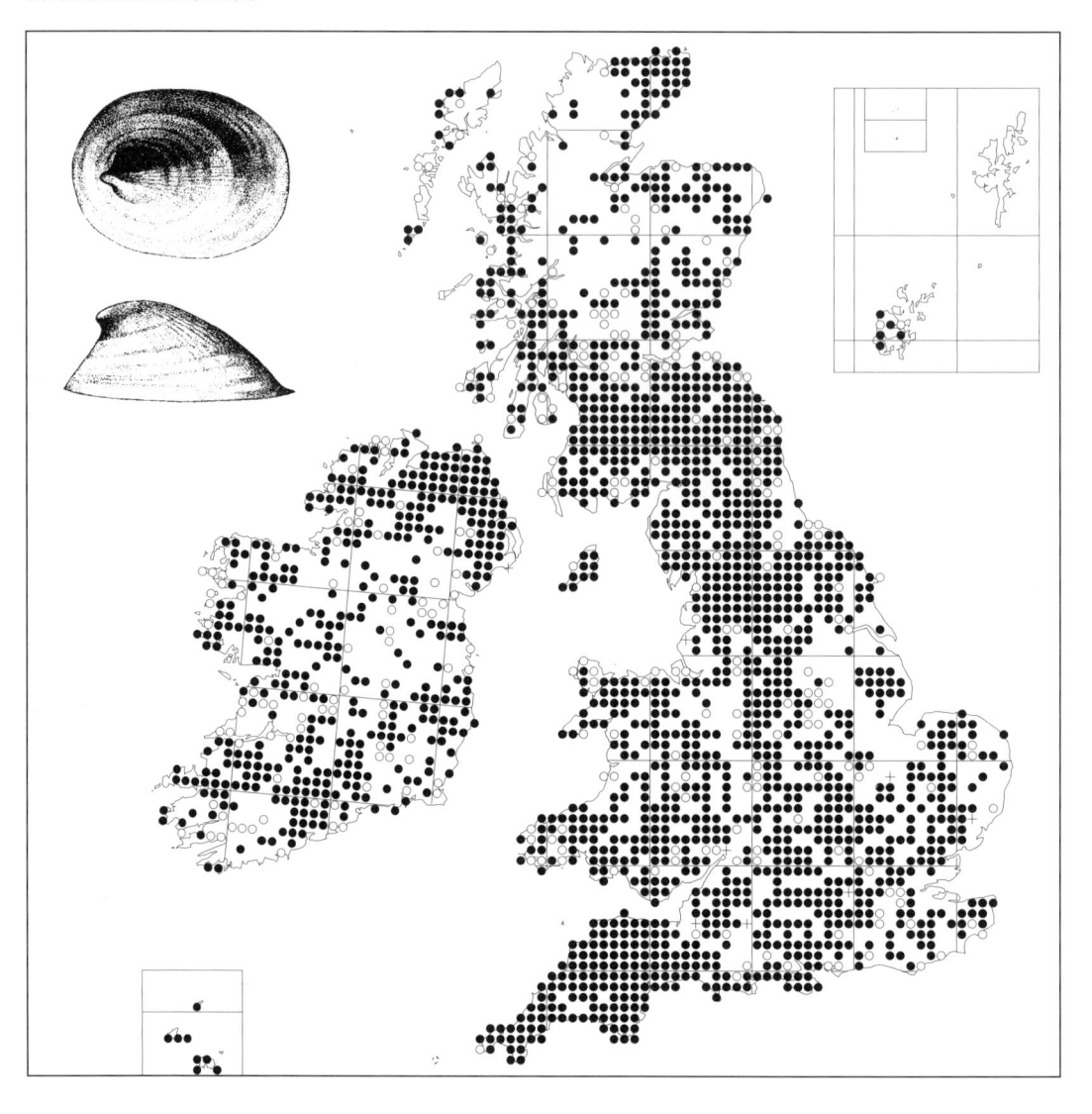

**Ancylus fluviatilis** Müller, 1774 (*Ancylastrum fluviatile* (Müller))

River limpet   Size: 5–8mm

The river limpet inhabits quick-flowing, sometimes turbulent water, hard or soft, adhering firmly to stones on which it grazes. It is found in rivers, streams, and even small mountain trickles provided they are not seasonal. Occasionally it may be found in lakes, on rocks in the marginal wash-zone. It requires clean water free from suspended matter and avoids muddy substrates or stones coated with mud or thick algae.

Native (Lgl). *A. fluviatilis* remains common in suitable habitats throughout the British Isles, though locally it may be affected by pollution. A frequent cause of its disappearance from lowland streams is eutrophication promoting excessive algal growth on the pebbles.

W. Palaearctic; in Europe widespread though absent from most of C. and N. Scandinavia.

(Ellis: 113; Macan: 19)

**Ferrissia wautieri** (Mirolli, 1960) (*F. clessiniana* (Jickeli))

Size: 4–6mm

This limpet attaches itself to firm surfaces (usually the leaves and stems of aquatic plants) in well-vegetated, stagnant or slowly moving water. Most of the British records are from shallow ponds (Holyoak, 1978b) but it has also been found in canals (Norris, 1982), in a large river (the Great Ouse) (Preece & Wilmot, 1979), and in greenhouse tanks in botanic gardens in London (Kew), Edinburgh and Glasgow.

It is said to tolerate seasonal desiccation, though this has not yet been confirmed in Britain.

Introduced. First detected in Britain in an open habitat in 1976 (Brown, 1977), though present in hothouse tanks at least as early as 1931 (Royal Botanic Gardens, Glasgow). Due to its superficial resemblance to *Acroloxus lacustris* it is likely to be underrecorded. Undoubtedly it is spreading.

Original home probably N. Africa; now naturalized in many places in Europe.

(Brown, 1977; van der Velde & Roelofs, 1977)

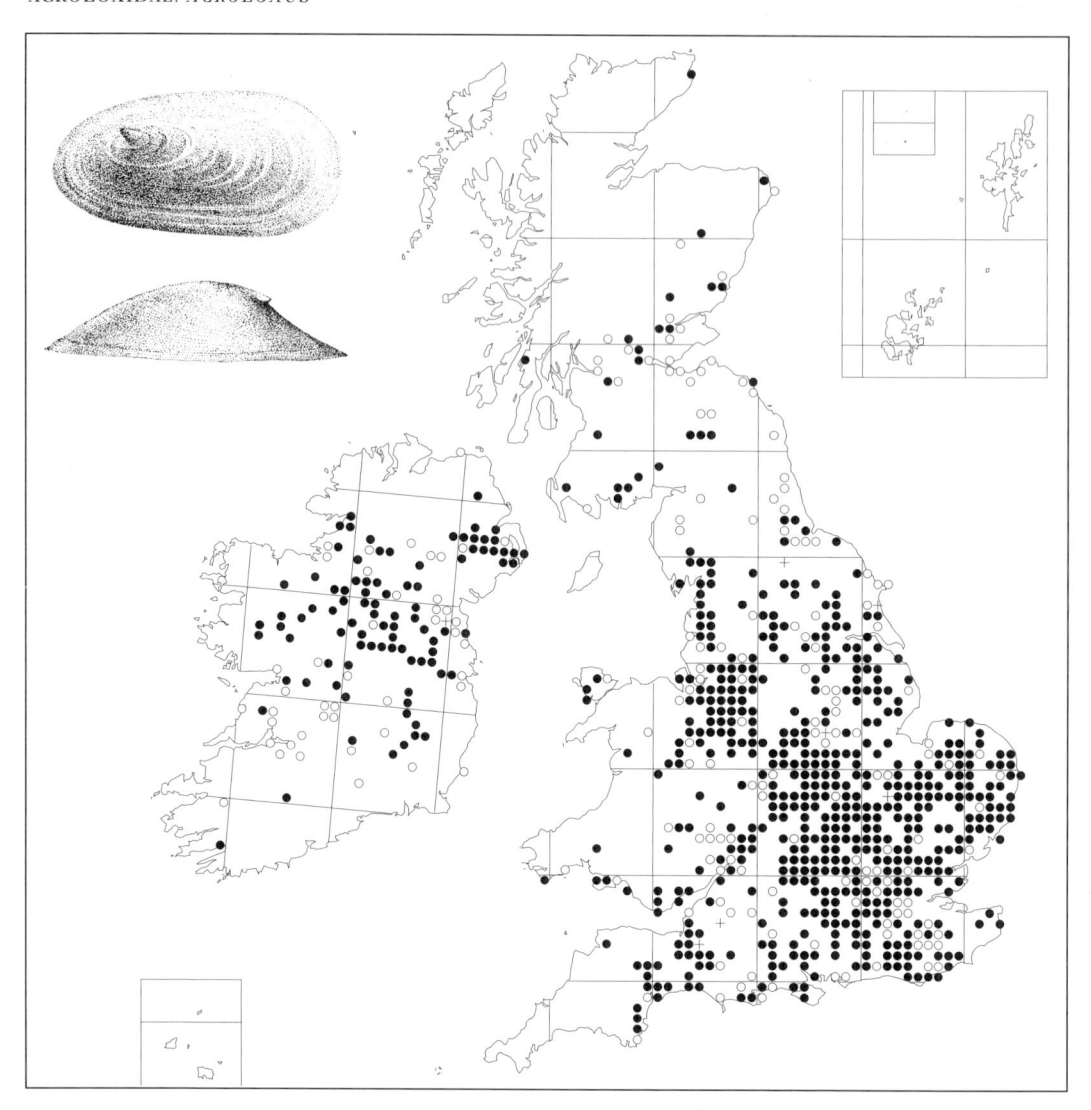

**Acroloxus lacustris** (L., 1758) (*Ancylus lacustris* (L.))

Lake limpet   Size: 4–7mm

A limpet inhabiting clean, quiet water in canals, slow lowland rivers, lakes and drainage ditches; it is quite frequent also in small closed ponds. It attaches itself to the stiff leaves and stems of aquatic plants, or to submerged timber or the shells of other molluscs, especially large bivalves. Though calciphile, it is tolerant of soft water.

Native (Lgl). Although there is little evidence of any significant regional decline, this species is rarer than formerly in some areas as a result of pollution and habitat destruction.

European; widespread in lowland areas north to about 61° in Scandinavia.

(Ellis: 114; Macan: 19)

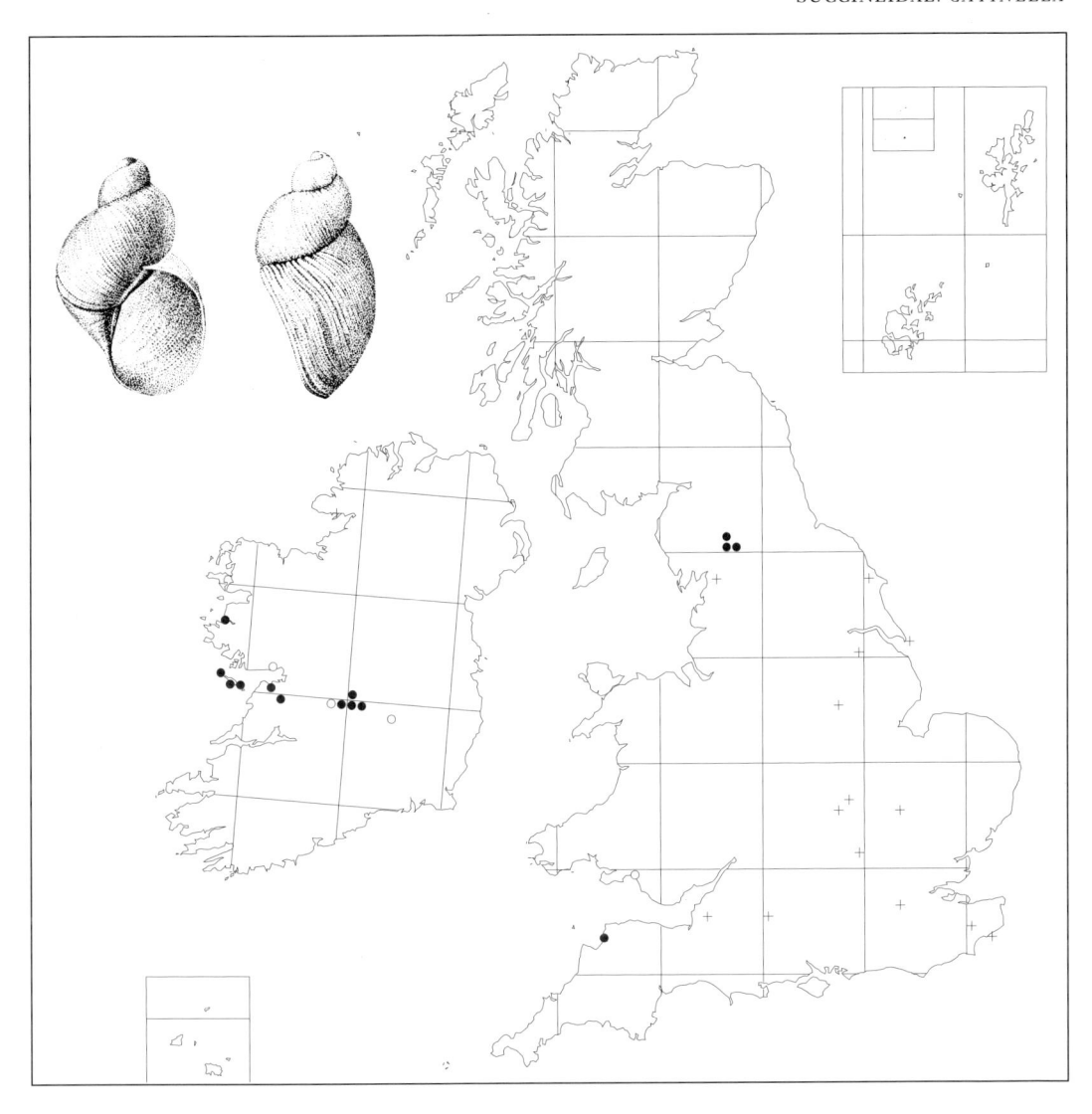

**Catinella arenaria** (Bouchard-Chantereaux, 1837)
(*Succinea arenaria* Bouchard-Chantereaux)

Sand amber snail    Size: 5–8mm

This rare snail lives in moist, grassy hollows in stabilized dunes (Braunton Burrows, N. Devon; Dooaghtry, Co. Mayo; also formerly near Swansea), or in base-rich fens and flushes (Orton and Crosby Gill, Westmorland; central Ireland; the Burren; the Aran Islands (Inishmore, Inishmaan)). It likes sparsely vegetated ground and is found mostly on the sand or mud at the base of sedges and other low vegetation, always in places with a stable water-level unaffected by deep flooding (Boycott, 1921; Quick, 1933; Coles & Colville, 1979). At several sites the open habitat is maintained by grazing.

Native (Lgl). This is a relict species which, at the end of the glacial period, was common in marshes in the lowlands of southern England; these habitats were largely destroyed by forest growth. The survival of the species must now depend on protection of the few remaining sites. Those in central Ireland are mostly threatened by bog drainage, but the remainder seem for the moment to be reasonably secure. RDB category: Endangered.

Isolated sites in W. Europe from the Alps to the Arctic Circle; mainly coastal and montane.

*C. arenaria* cannot always be distinguished from *Succinea oblonga* on shell characters. Most of the records shown here are based on dissections or, in the case of fossils, on well-characterized shells.

(K. & C.: 59; Quick, 1933)

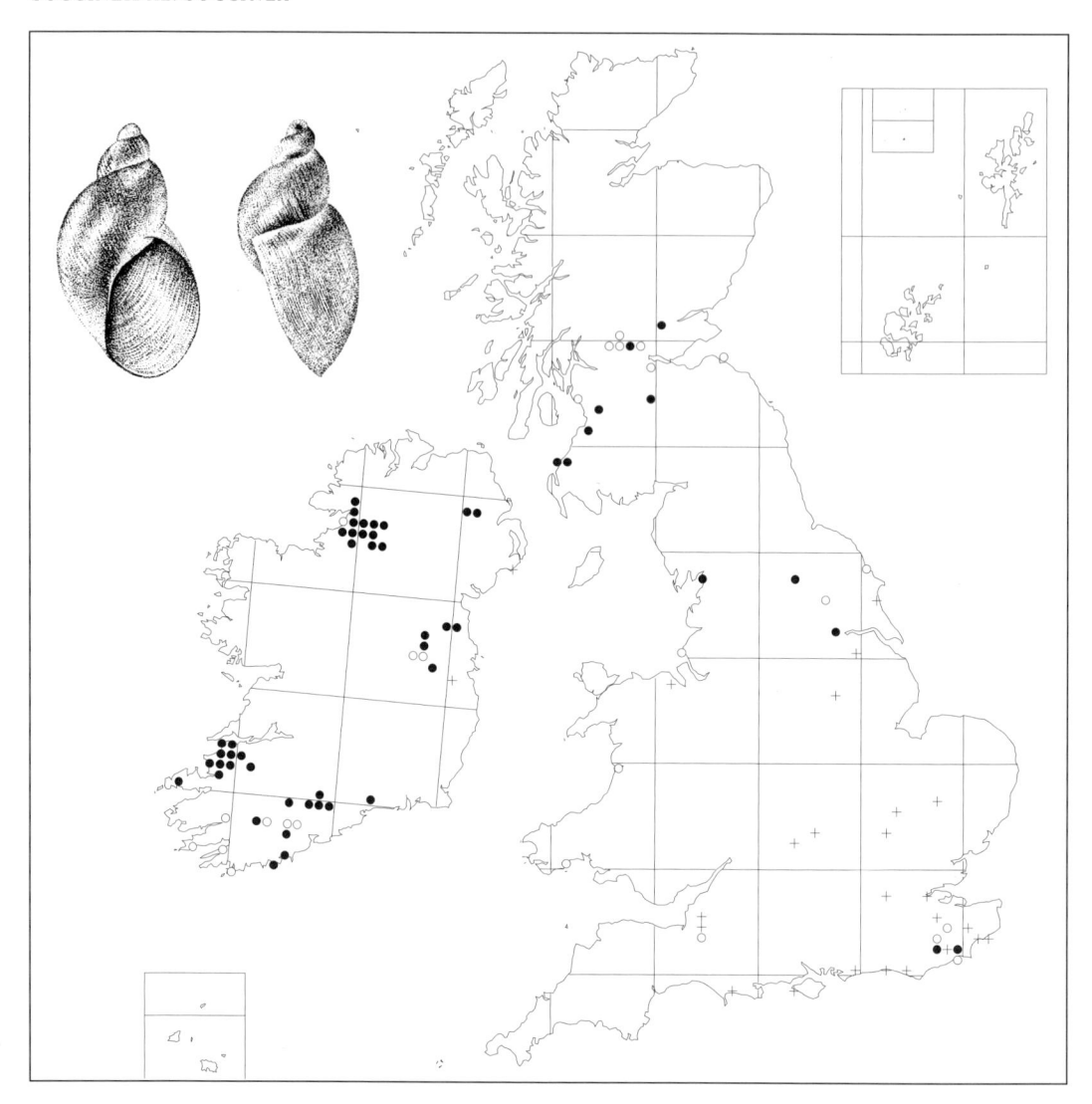

SUCCINEIDAE: *SUCCINEA*

**Succinea oblonga** Draparnaud, 1801

Small amber snail   Size: 6–8mm

A species of damp, unshaded, sparsely vegetated places: stony lake margins, the floodplains of rivers, damp cattle pastures. It likes bare, muddy surfaces and, in S.W. Ireland, is typical of damp mud in roadside ditches or among debris on the floor of abandoned stone quarries.

Native (Lgl). This is a receding species, common in lowland England in cold-climate deposits (especially windblown sediments) dating from the glacial periods. In Kent and Sussex it temporarily returned in the wake of prehistoric and Roman deforestation but has now largely receded once more from this area. There is no clear evidence of more recent decline, but in Britain the species must be considered potentially vulnerable because of its rarity and rather specialized requirements. Many of the Scottish sites are on floodplains (Kevan, 1931) which could be affected by river-management schemes. RDB category (Britain only): Rare.

European and W. Asiatic; in Europe to about 61°N in Scandinavia.

(K. & C.: 59; Quick, 1933)

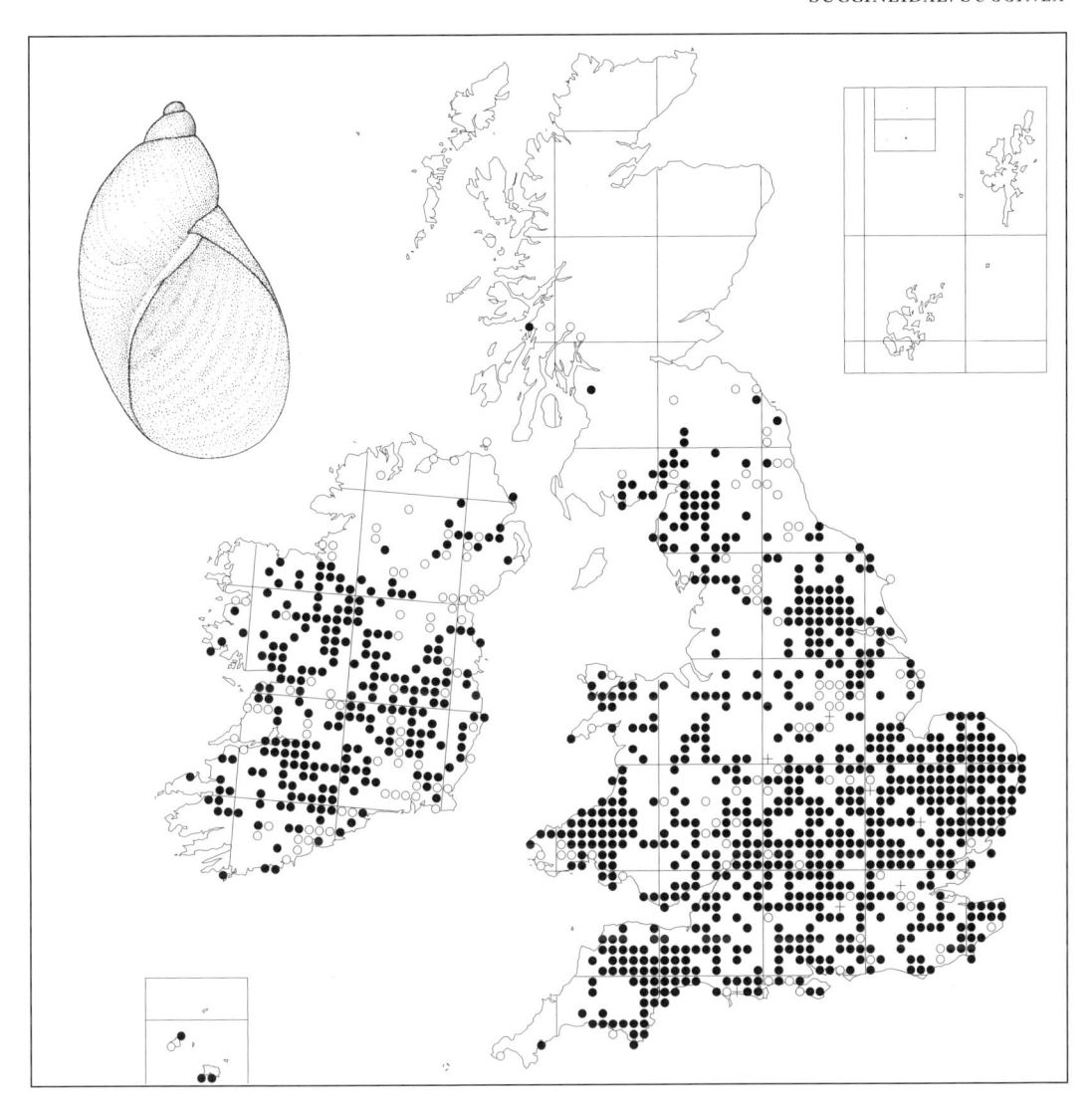

## Succinea putris (L., 1758)

Large amber snail    Size: 10–17 (occ.–22)mm

Inhabits wetlands, mainly of lowland type: fens, marshes, water meadows. It often climbs erect vegetation, such as *Phragmites*, *Carex* or *Iris*, growing in drainage ditches or at the margins of lakes and rivers. It can survive long periods in moist ground litter, sometimes overwintering a considerable distance from its normal feeding grounds (e.g. in woodland at the edges of river floodplains).

Native (E.Pgl). This species is still frequent over most of its range, though showing some local decline through habitat destruction.

European and Siberian; widespread in Europe, becoming sparser in the northern highland zones of Britain and Scandinavia.

*S. putris* cannot always safely be distinguished from *Oxyloma pfeifferi* on external criteria. Only records confirmed by dissection or based on well-characterized material have therefore been mapped. Many 19th-century records are unreliable, including some from northern Scotland.

(K. & C.: 60; Quick, 1933)

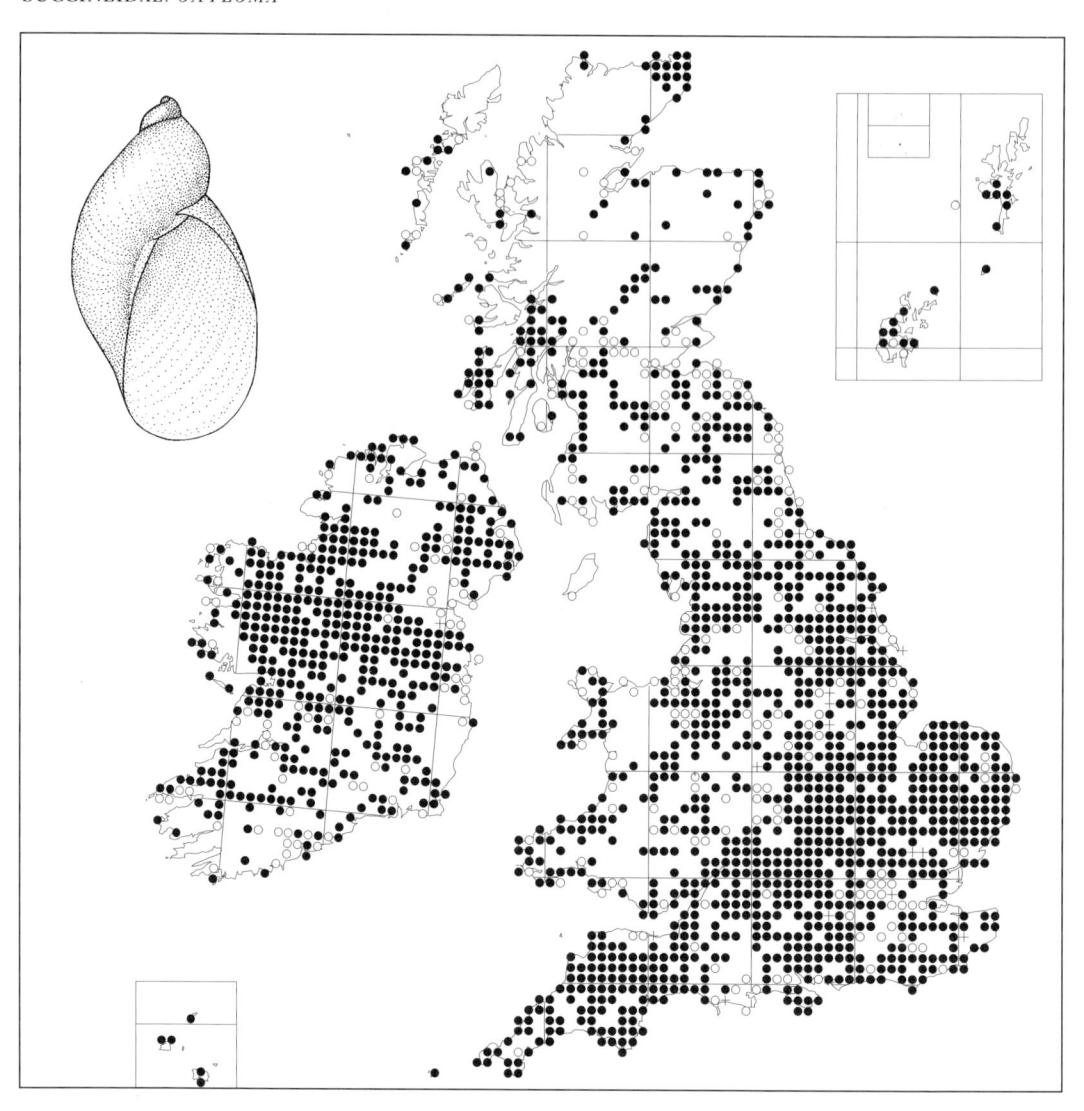

**Oxyloma pfeifferi** (Rossmässler, 1835) (*Succinea pfeifferi* Rossmässler, *Oxyloma elegans* of many Continental authors)

Pfeiffer's amber snail   Size: 9–12 (occ.–17)mm

This common succineid lives in wetlands of all kinds: fens, marshes, hillside flushes, dune slacks. It is more catholic than *Succinea putris* and tolerates poorly vegetated places, as on sea cliffs and lake margins. It is not found in woods.

Native (Lgl). No evidence of significant national decline, though locally adversely affected by land drainage.

Holarctic; widespread in Europe except in northern Scandinavia.

(K. & C.: 60; Quick, 1933)

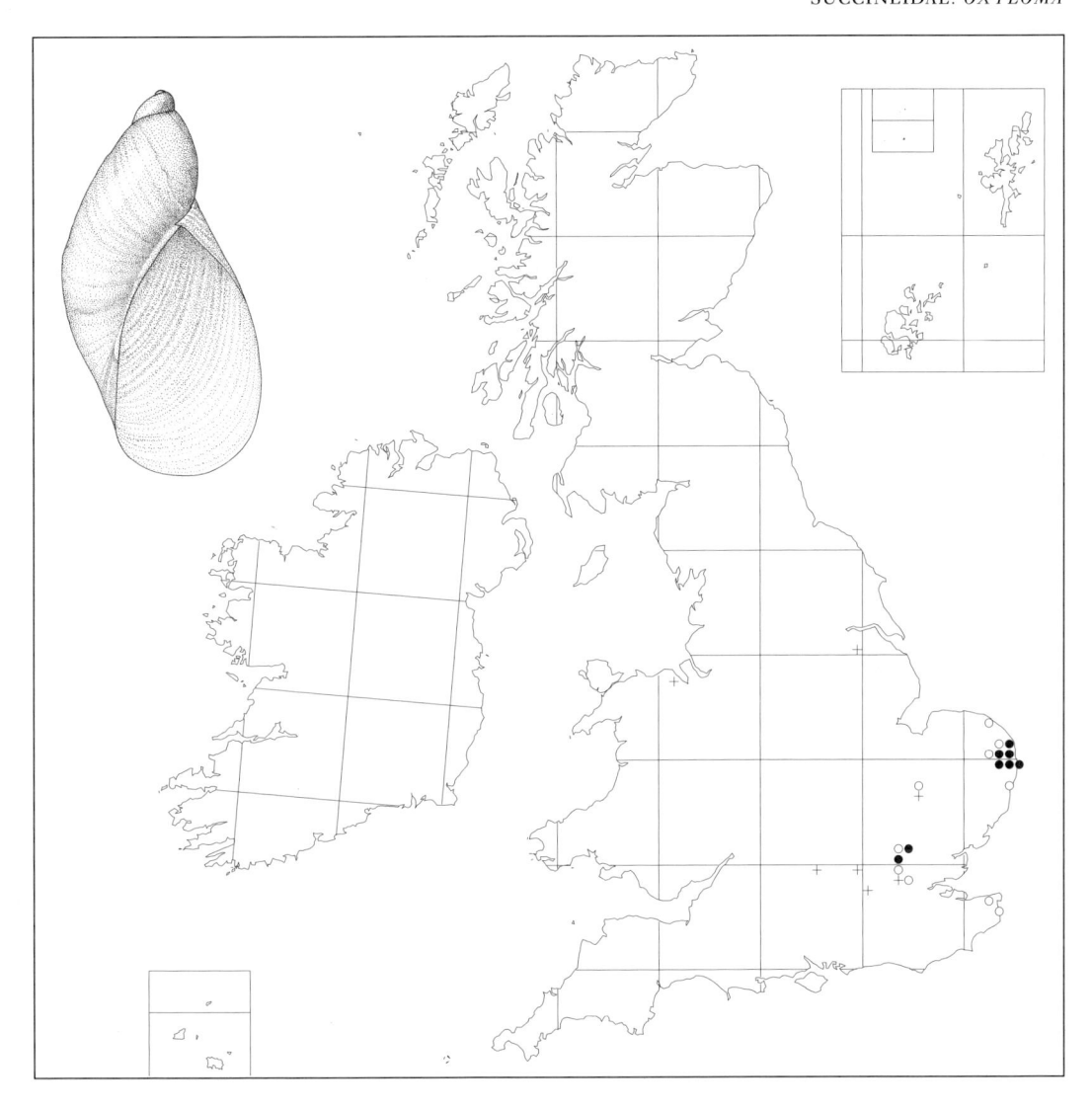

**Oxyloma sarsi** (Esmark, 1886) (*Succinea elegans* of Ellis, 1926 and Quick, 1933)

Slender amber snail   Size: 12–15 (occ.–18)mm

Very local in richly vegetated fens of lowland type at the margins of rivers and lakes and in drainage dykes in levels. It is more aquatic than *O. pfeifferi*, never straying far from standing water and usually found either on emergent vegetation such as *Phragmites* or crawling on *Glyceria* and other water plants. Other molluscan rarities, notably *Vertigo moulinsiana*, may be associated. Unlike *O. pfeifferi*, it appears to be calcicole.

Native (E.Pgl). This is a declining species, more widely distributed in the earlier part of the Postglacial period, when it occurred in Lincs and N. Wales. It is now known only at a few places in the Lea valley in Herts and Essex, in the lower Waveney valley and in the East Anglian Broadland (notably at Oulton Broad). Surviving populations are all at risk from pollution, dredging or river-management schemes. RDB category: Vulnerable.

N. European: known from scattered areas between the Alps and northern Scandinavia.

*O. sarsi* is often not easy to distinguish from *O. pfeifferi* externally. Dissection records and well-characterized shells only are plotted here.

(K. & C.: 61; Quick, 1933)

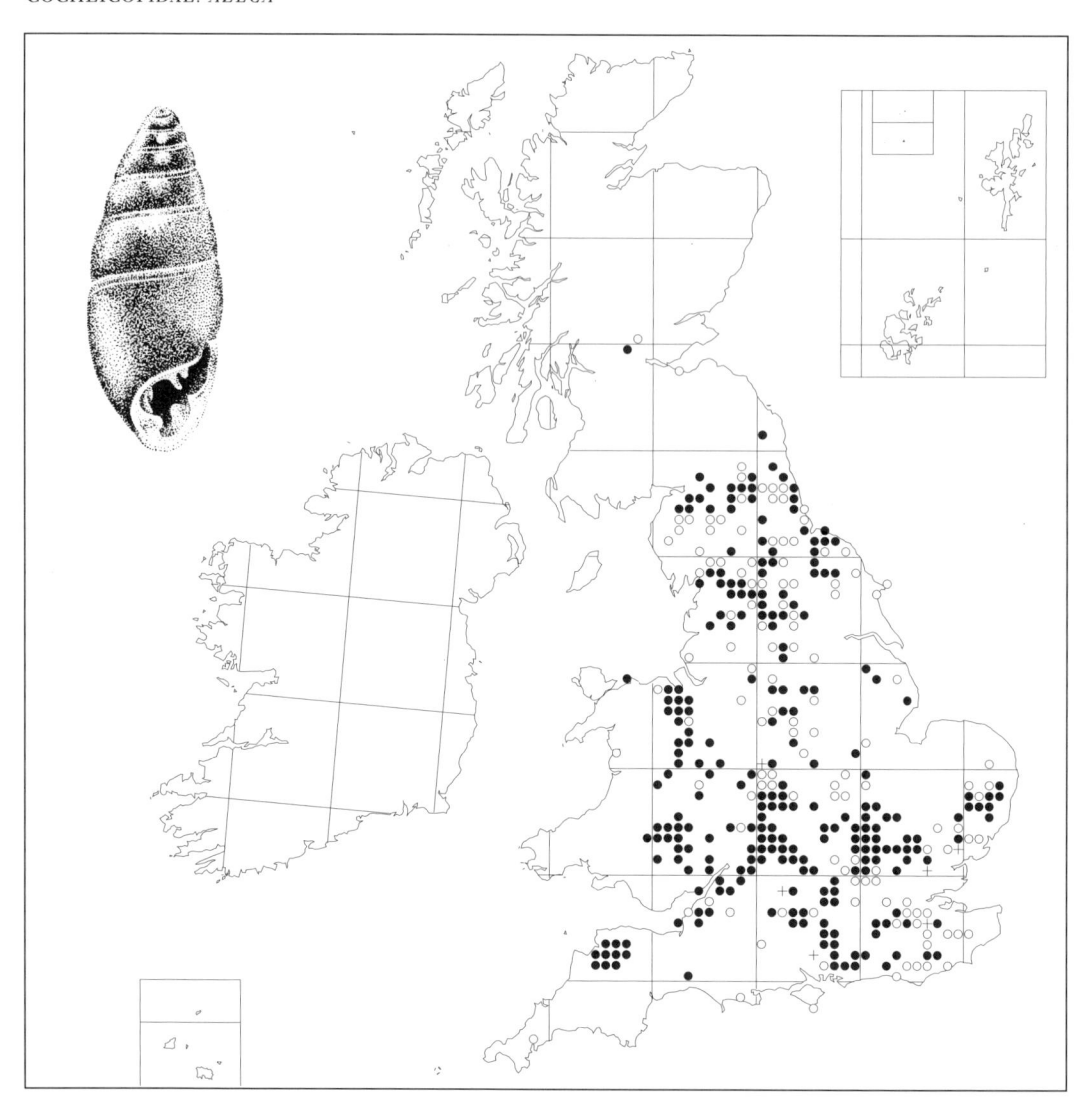

**Azeca goodalli** (Férussac, 1821) (*A. tridens* (Pulteney), *A. menkeana* (Pfeiffer))

Three-toothed moss snail   Size: 5.5–7.0mm

Local in moss, herbage and ground litter in deciduous woods, on hedge banks, and in undisturbed, scrubby places, usually though not always on calcareous soils. Colonies are typically small in extent and sharply defined, even within areas of apparently uniform character. It prefers light shade, avoiding both the extremes of open ground and dense woodland; the mossy edges of woodland rides are often favoured (Paul, 1974).

Native (E.Pgl). Though everywhere local, this species remains widespread, showing no evidence of significant recent change.

W. European: central Germany to the Pyrenees.

(Ellis: 167; K. & C.: 62)

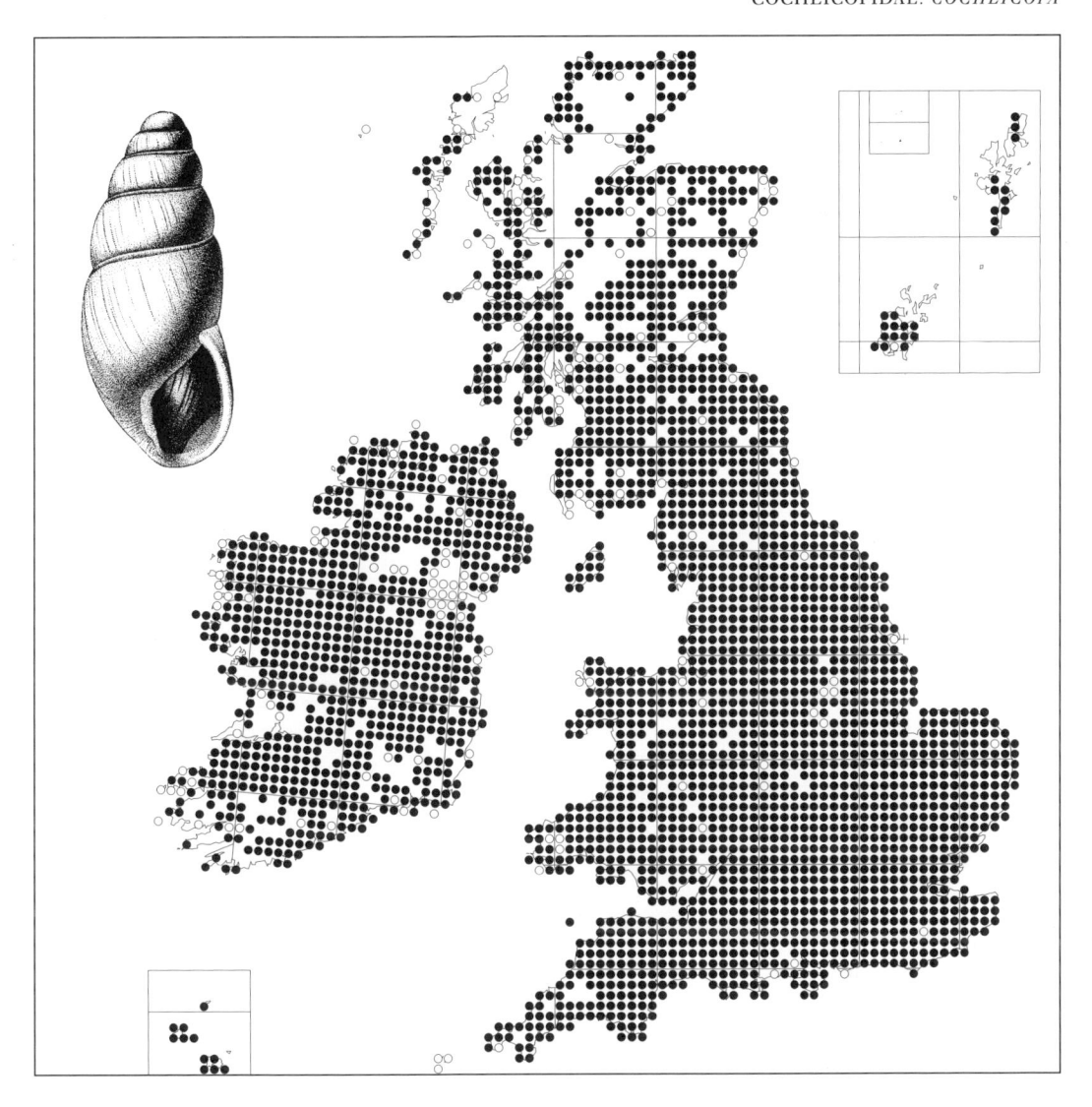

**Cochlicopa lubrica** (Müller, 1774)

Slippery moss snail   Size: 5–7mm

A catholic species common in all kinds of moderately damp, sheltered places: under ground litter in woods, grassy fields, hedge banks and marshes. It tolerates non-calcareous soils.

Native (Lgl). This species remains common everywhere.

Holarctic; throughout Europe.

It should be noted that the taxonomy of the genus *Cochlicopa* is not yet properly resolved. The distinction between the two existing British species is not always sharp and some populations and individuals are hard to name. Some Continental authors recognize a further species (*C. repentina* Hudec) with shell characters intermediate between those of *C. lubrica* and *C. lubricella*.

(K. & C.: 62; Quick, 1954)

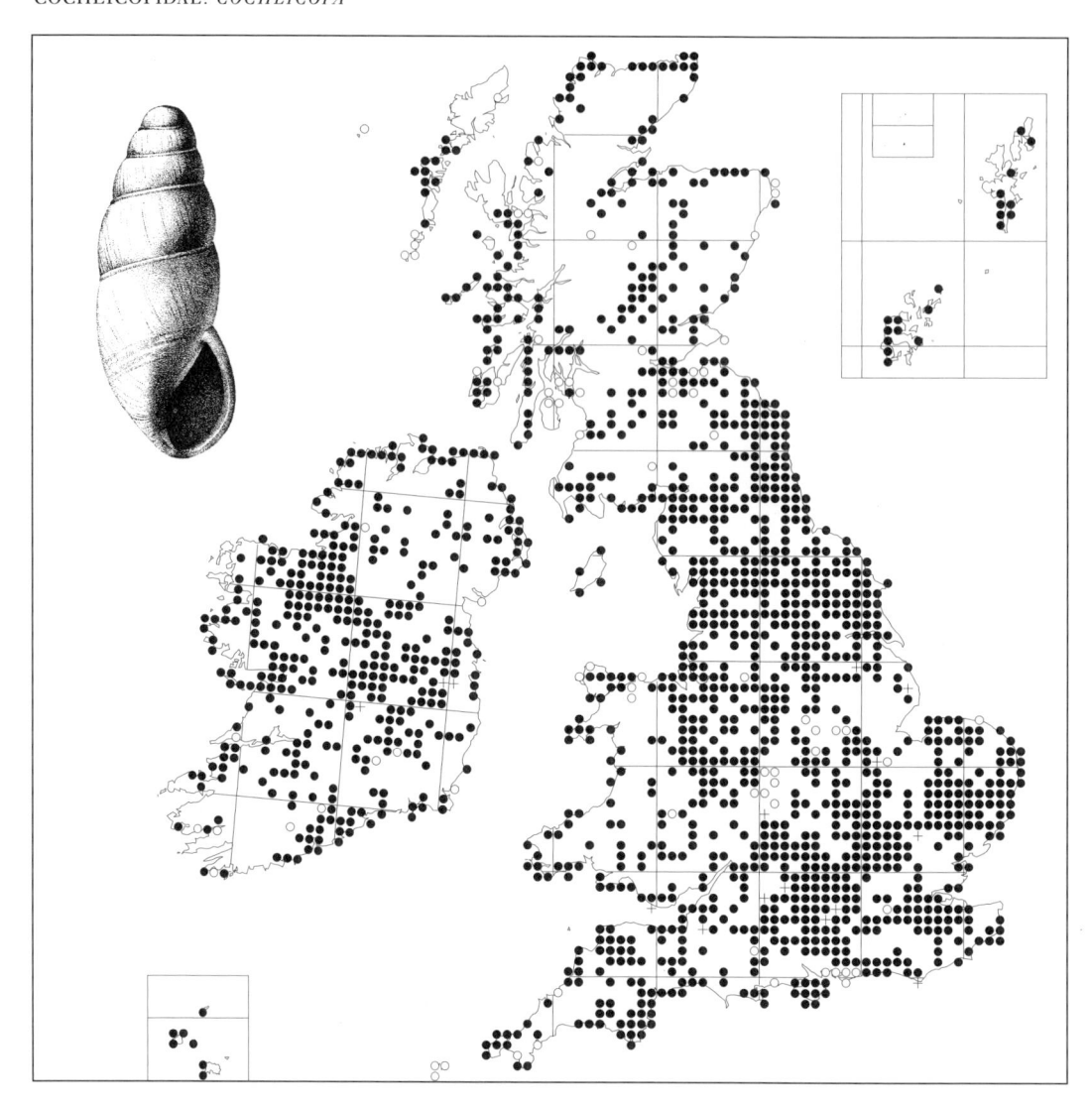

**Cochlicopa lubricella** (Porro, 1838) (*C. lubrica* var. *lubricella* Porro, *C. minima* (Siemaschko))

Size: 4.5–6.5mm

This species favours rather drier and more exposed places than *C. lubrica*, though avoiding poor, acidic soils. It is not normally found in marshes and is characteristic of dry grassland, sandhills, cliffs, old quarries and stone walls. The two species are however quite often associated, especially in woods.

Native (Lgl). No evidence of regional change or decline.

Holarctic; throughout Europe, but becoming rare in northern Scandinavia.

(K. & C.: 62; Quick, 1954)

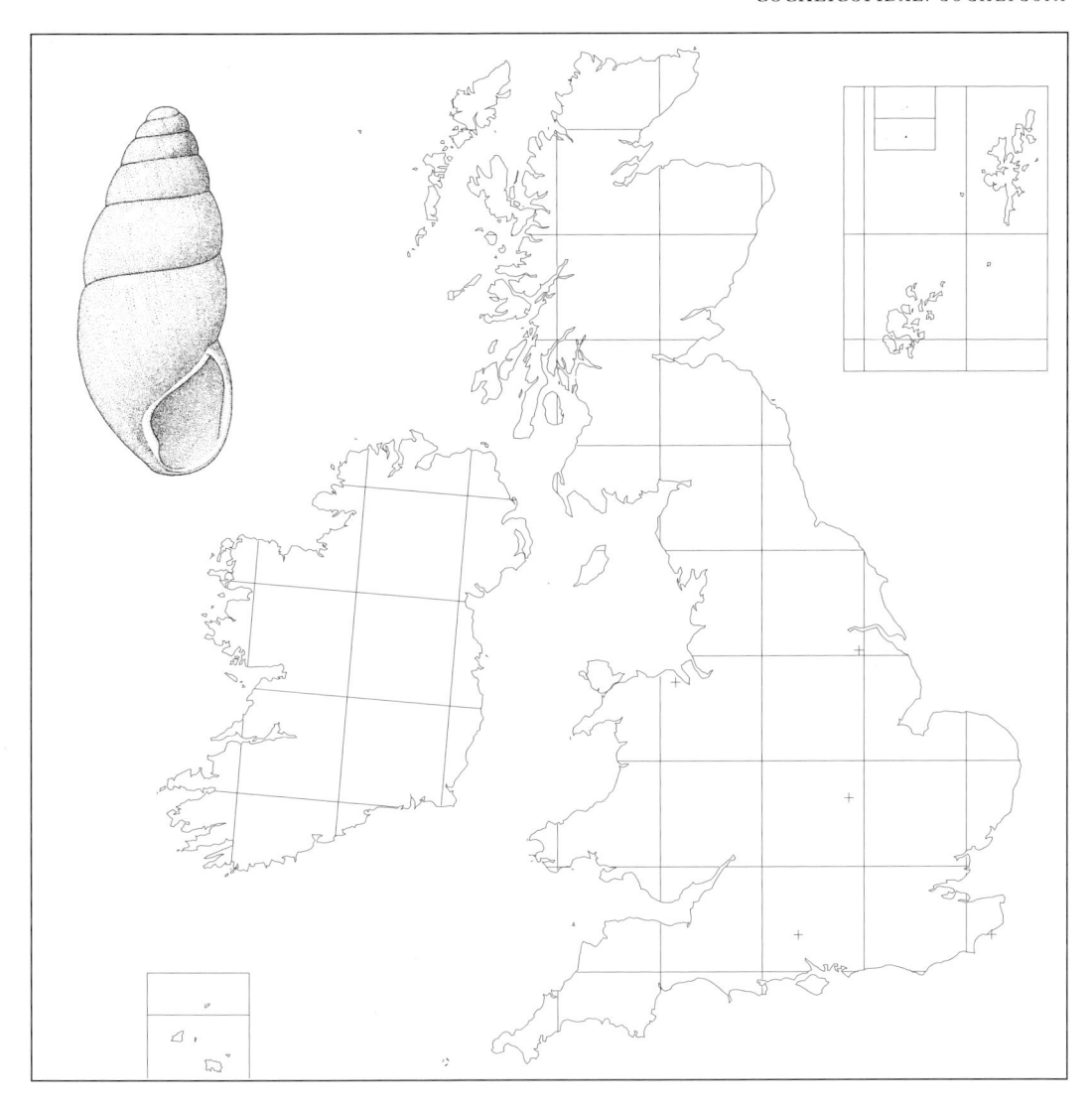

**Cochlicopa nitens** (Gallenstein, 1848)

Size: 6.2–7.5mm

On the Continent this is an uncommon species inhabiting calcareous swamps, often associated with *Vertigo moulinsiana*. Occasionally it is found in marshy, calcareous woodland.

Extinct (Lgl & E.Pgl fossils only). This species has been found in small numbers as a fossil in marsh deposits in Kent (Folkestone), Hants (Bossington), Northants (Great Harrowden), Flint (Caerwys) and Lincs. (Castlethorpe, Scawby). It became extinct very early in the Postglacial period, before the forest optimum (Preece, 1992). It lived also in Britain during the last interglacial period.

C. and E. European.

(K. & C.: 62; Preece, 1992)

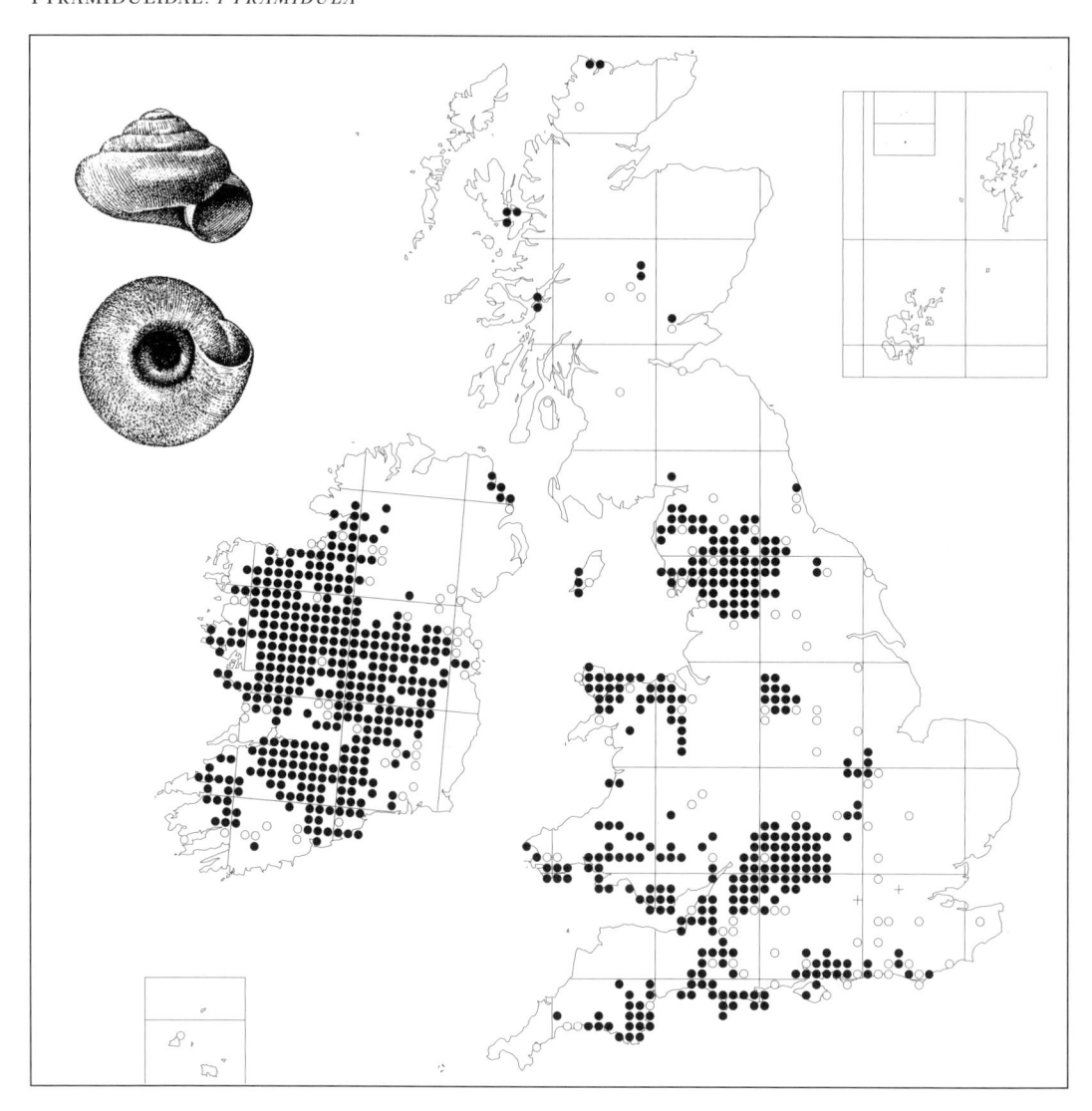

**Pyramidula rupestris** (Draparnaud, 1801) (*Helix rupestris* Draparnaud)

Rock snail   Size: 2.5–3.0mm

A species characteristic of limestone rocks and walls fully exposed to the sun and unshaded by trees. Occasionally it may be found on the calcareous mortar of walls built of non-calcareous rock, for example on flint walls in Sussex or on slates in Wales. It shelters in depressions and crevices, emerging at night or in wet weather to feed on lichens and algae. *Pyramidula* is ovoviviparous, and this has probably facilitated passive dispersal and colonization of remote and isolated areas of limestone separated by unfavourable terrain, notably in western Scotland (Lismore, Skye, Durness).

Native (E.Pgl). *P. rupestris* has been spread beyond areas of natural rock outcrop by the construction of stone walls. Though still common over most of its range, in some areas it is now at risk because of the removal of walls or their repair with Portland cement rather than lime mortar. Its food plants have also been affected by $SO_2$ pollution, possibly accounting for its disappearance from some sites in eastern England. The last records from certain counties are: Cambs., 1880; Herts., 1883; Notts., *c.*1890; Kent, 1896; N. Lincs., 1914; Surrey, 1916.

Mediterranean and W. European, extending north to British Isles (but see note below).

The taxonomy of European *Pyramidula* requires further elucidation. The British species (*Helix umbilicata* Montagu) may possibly be distinct from the true French *rupestris* (Gittenberger & Bank, 1996).

(Ellis: 160; K. & C.: 63)

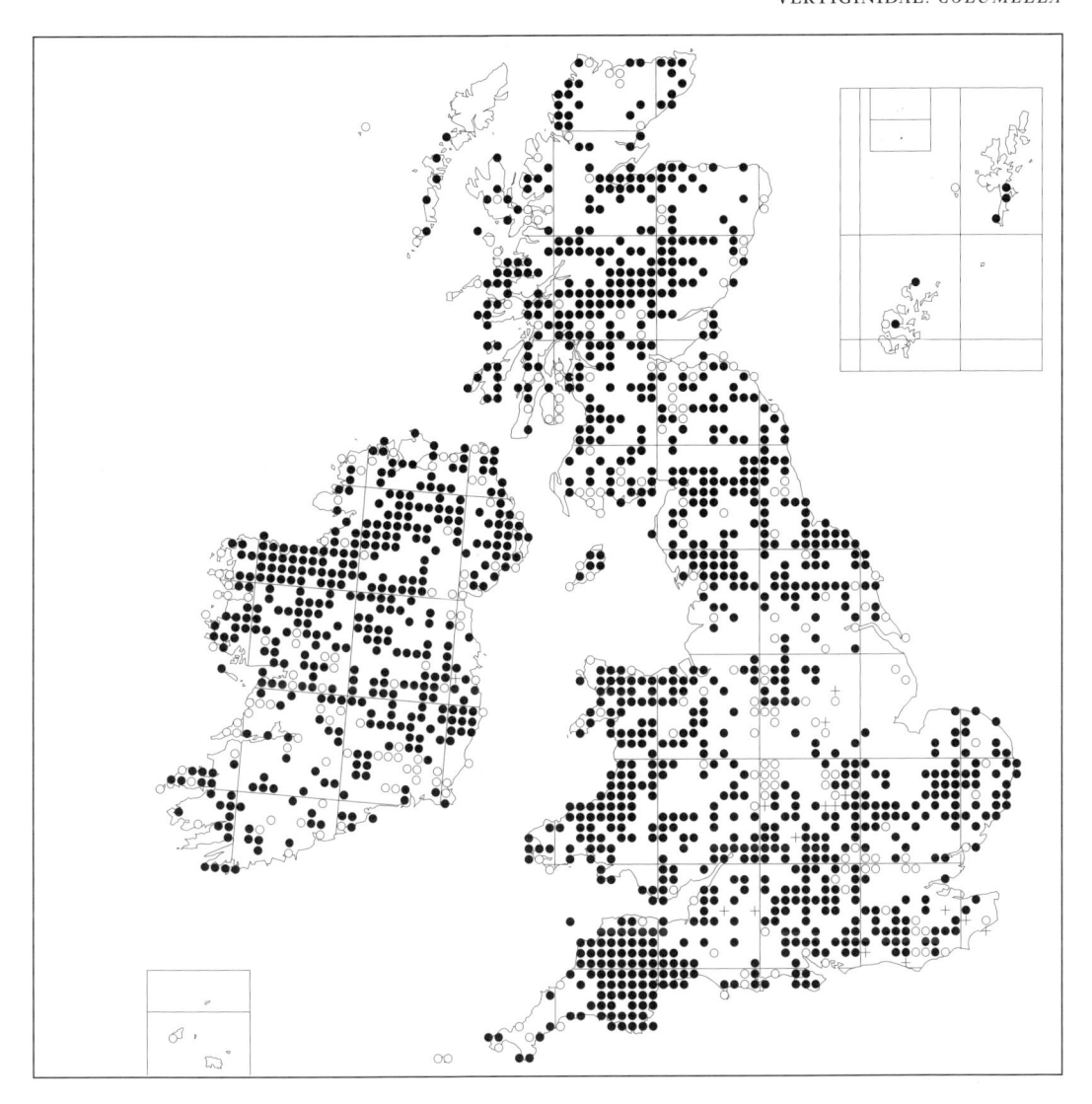

**Columella edentula** aggregate (*Vertigo edentula* of older British authors)

Toothless chrysalis snail

This map show all records of living *Columella* spp. (*C. edentula* (Draparnaud) + *C. aspera* Waldén) in the British Isles. The two segregate species, distinguished only in the 1960s, remain somewhat underrecorded.

In general the *Columella* species avoid intensive agriculture and human disturbance and are commonest in areas of marginal and upland farming.

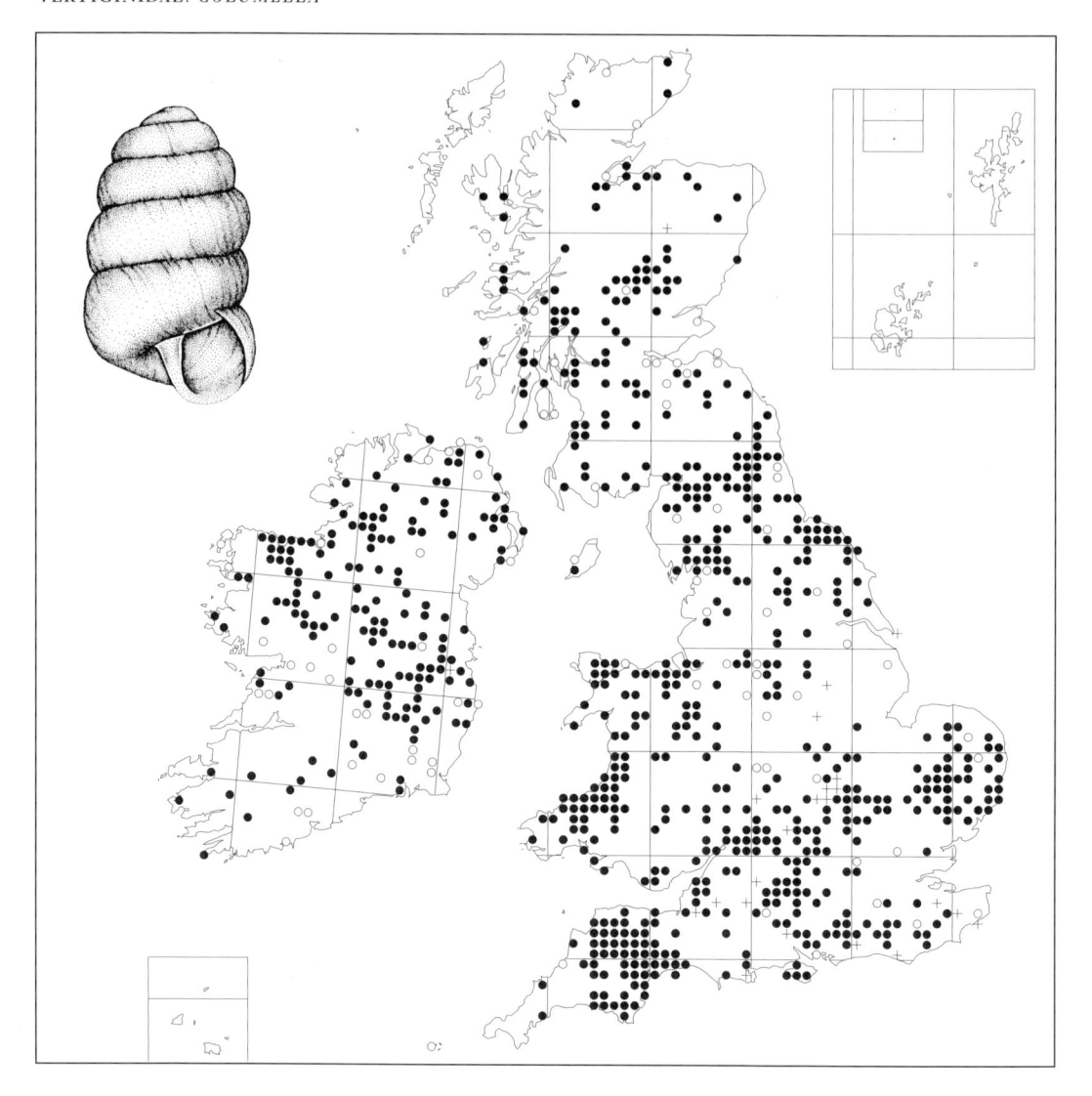

**Columella edentula** (Draparnaud, 1805)

Size: 2.5–3.0mm

A catholic species, inhabiting a variety of well-vegetated and relatively undisturbed places, in wetlands, woods and grasslands. In summer it tends to climb away from the ground and attach itself to dead twigs or to the stems and undersides of leaves of herbaceous plants. Unlike *C. aspera*, with which it is only rather rarely associated, it is typically a lowland snail and infrequent in harsh, non-calcareous mountain terrain (Paul, 1975).

Native (Lgl). No evidence of significant regional change. This species is somewhat underrecorded owing to confusion with *C. aspera*, from which it was distinguished only in 1966.

Palaearctic; in Europe imperfectly known, but occurring probably throughout.

(K. & C.: 66; Paul, 1975)

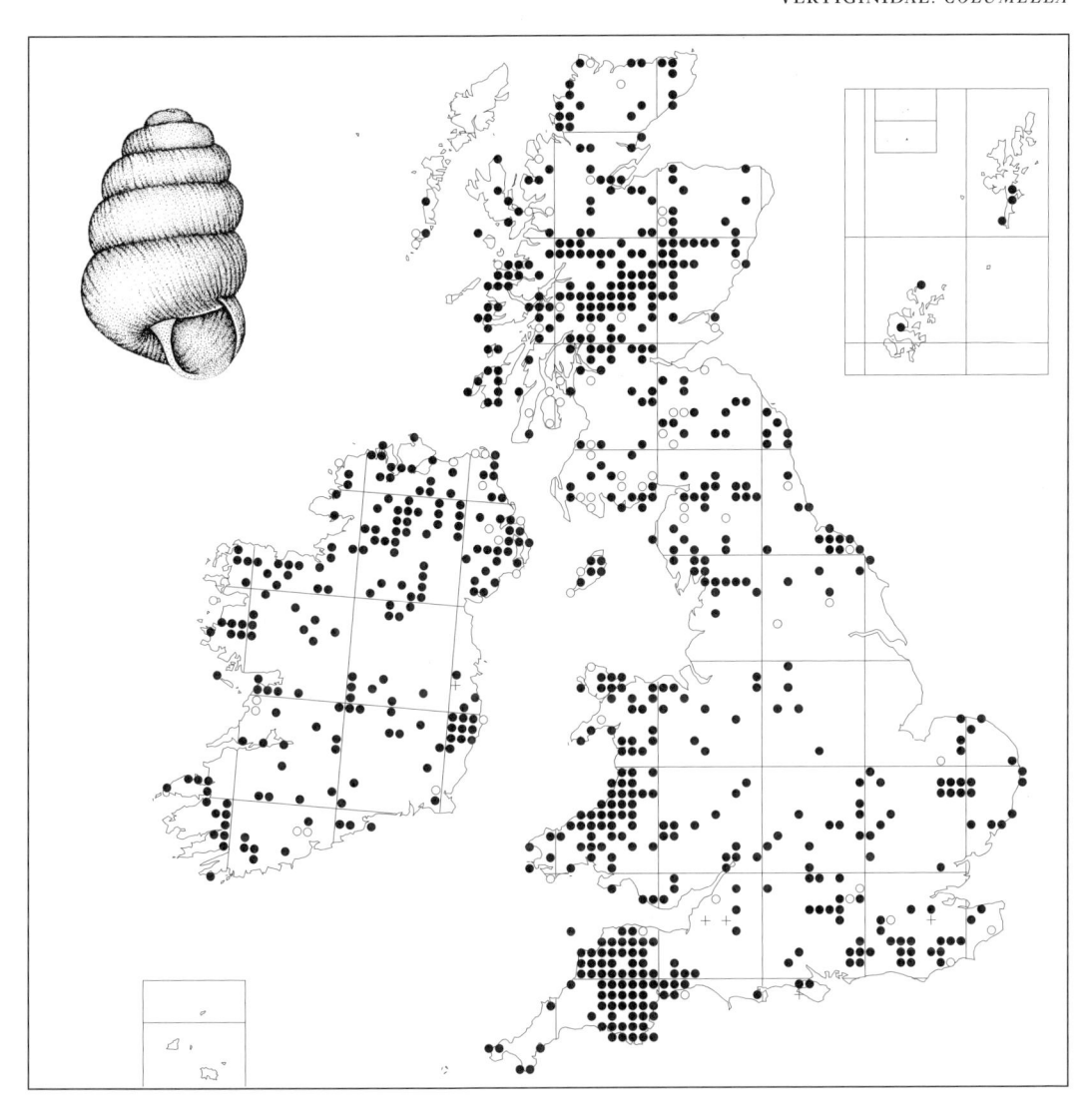

**Columella aspera** Waldén, 1966

Size: 2.2–2.5mm

Locally common in coniferous and deciduous woods, poor, uncultivated grassland, among rocks and in marshes. It is characteristic of drier and less base-rich places than *C. edentula* and can live even on bracken in acid moorland – a habitat usually devoid of all shelled molluscs. In general it is a highland rather than a lowland species (Paul, 1975).

Native (E.Pgl). Though as yet underrecorded, there is no evidence of significant national decline. In England a few sites have probably been lost through agricultural encroachment on lowland heath.

Probably Palaearctic, but as yet poorly known owing to confusion with *C. edentula*.

(K. & C.: 67; Paul, 1975)

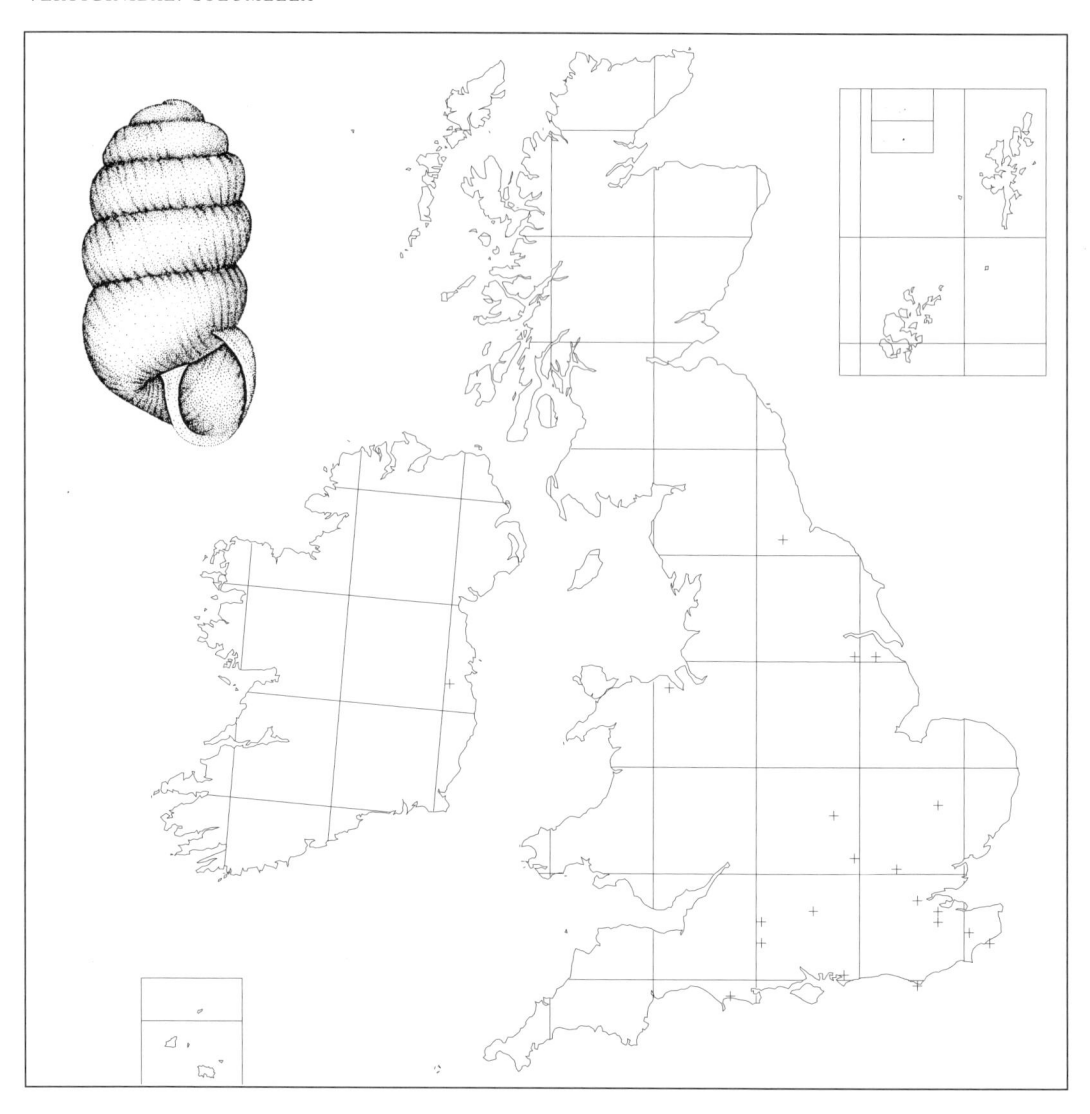

**Columella columella** (Martens, 1830) (*C. edentula* var. *columella* Martens)

Size: 2.7–3.5 (occ.–4.0)mm

On the Continent, *C. columella* lives in grassland or open woodland, often in stony places at high elevations (up to 2,000m in the Alps). It favours base-rich soils. In Scandinavia it is characteristic of marshy ground, either around flushes on open hillsides or in subarctic birch woods.

Extinct (Lgl & E.Pgl fossils only). *C. columella* was common in the British Isles at the close of the last glacial period, disappearing in the early Postglacial as a result of thermal improvement and habitat change. It is not impossible, however, that relict populations may yet be discovered in northern Britain (cf. *Vertigo modesta*).

Arctic and Alpine: Alps, Pyrenees and Carpathians, and mountains of northern Scandinavia, to beyond the Arctic Circle.

(K. & C.: 66)

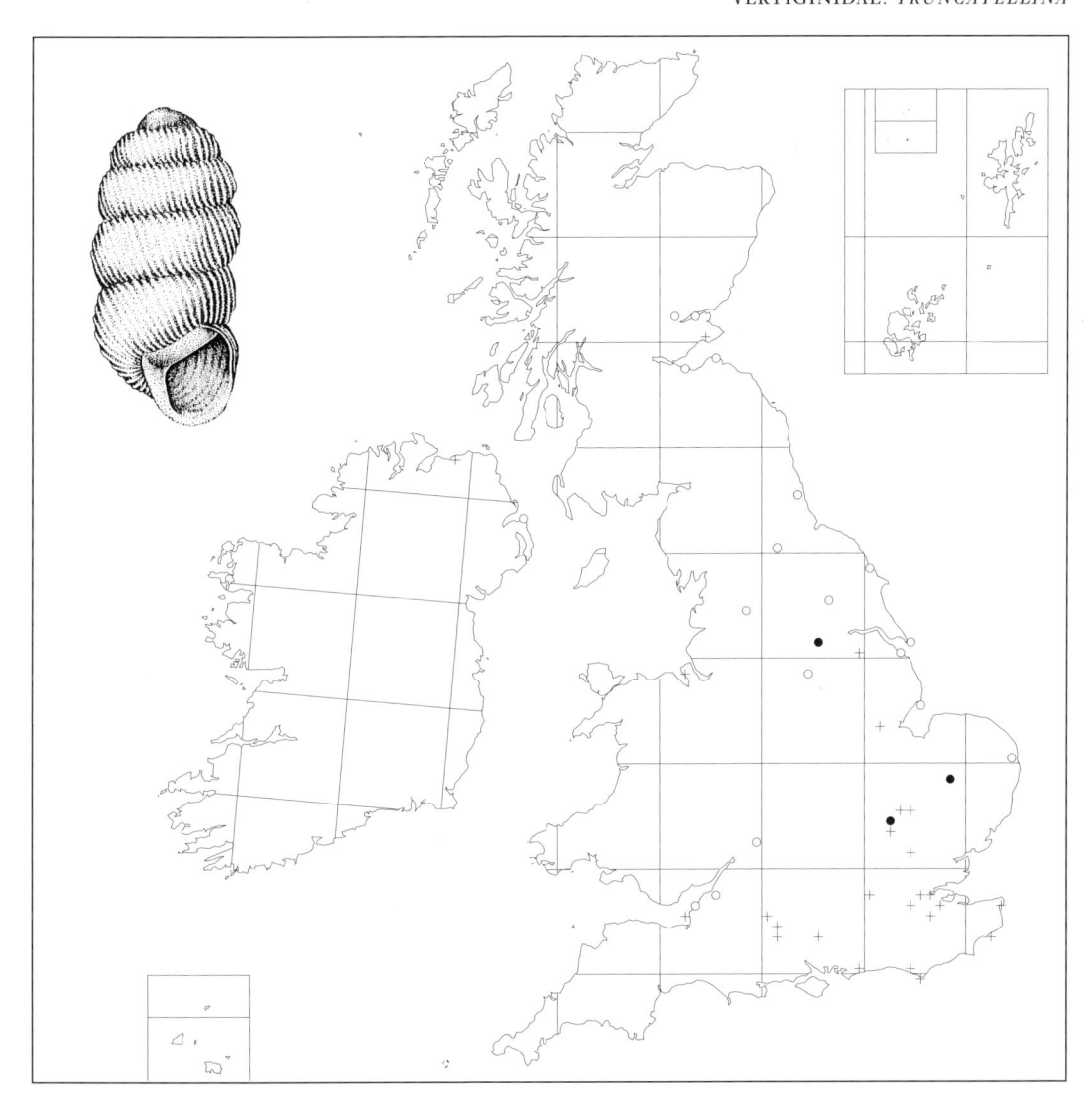

**Truncatellina cylindrica** (Férussac, 1807) (*Vertigo cylindrica* (Férussac))

Cylindrical whorl snail   Size: 1.7–1.9mm

A rare species found in short, dry, calcareous grassland in sandy or stony ground, or at the base of rocks or old walls, among *Sedum*, *Thymus*, *Artemisia* or similar plants. It is also recorded from coastal sandhills in E. Yorks. and Lincs. but was last seen in both counties in 1894.

Native (E.Pgl). Though this species is possibly underrecorded owing to its minute size, undoubtedly it is declining. Fossil records show it to have been more widely distributed in southern Britain in the Neolithic and Bronze Age periods, following primary clearance. Its subsequent decline may be connected with the increasing effect of arable farming, but climatic change may also have played a part (cf. *Pomatias elegans*). In the past fifty years it has been found living only at Potton (Beds.); Thetford (Norfolk) (Davis, 1952); and Went Vale (Yorks.) (Norris, 1976). Attempts to refind it at several old sites have failed, for example at Easby Abbey (Yorks.) (recorded 1883), Burgh Castle (Suffolk) (1909) and Arthur's Seat, Edinburgh (1836–1930s). At Burgh Castle the cause of its disappearance is almost certainly the stripping of the Roman walls, beginning in 1929. The Irish record (Groomsport, Co. Down) dates from about 1840. RDB category: Vulnerable (in Ireland: Extinct).

S. and W. European, widespread but local north to southernmost Scandinavia.

(Ellis: 152; K. & C.: 68)

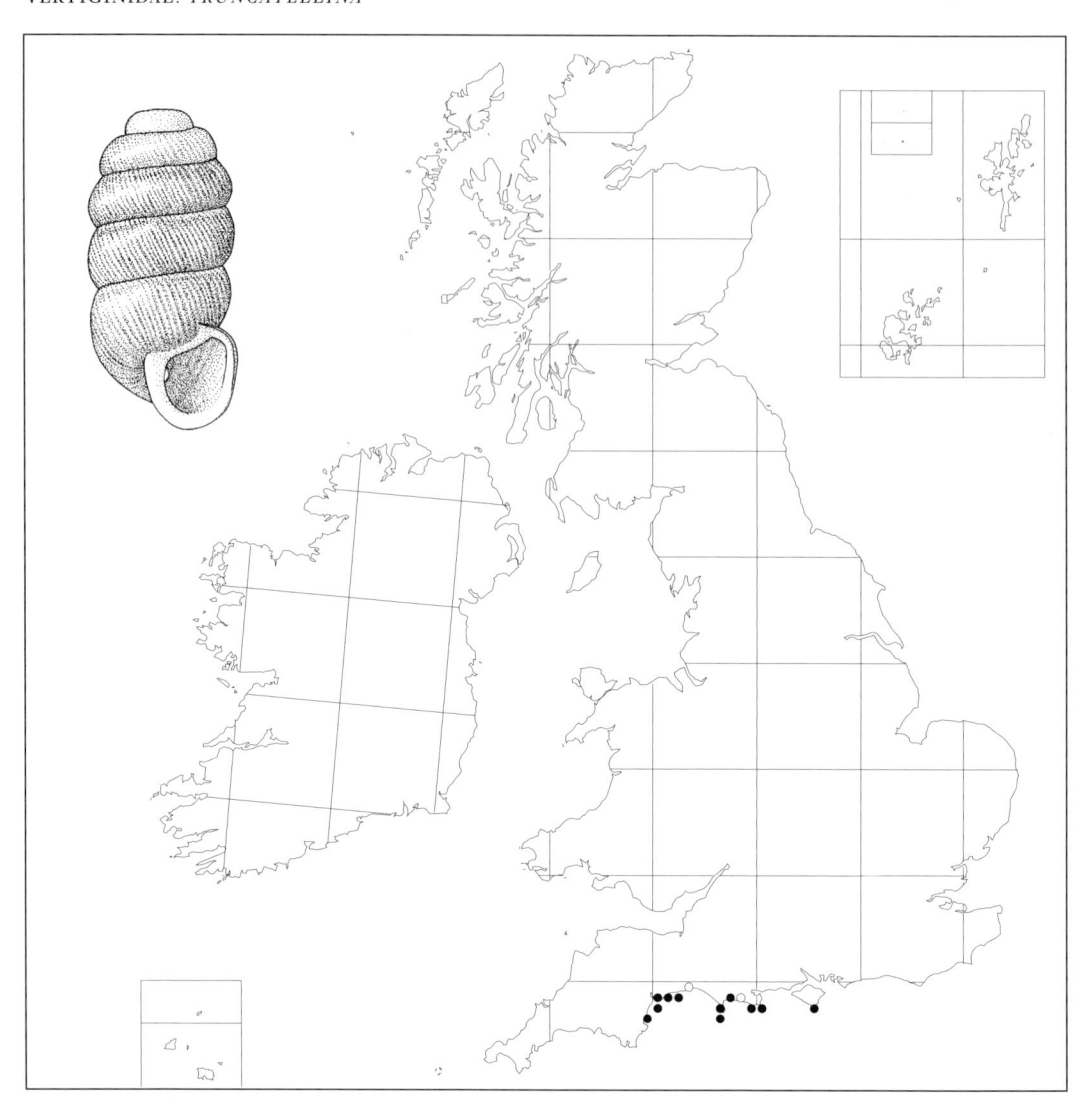

**Truncatellina callicratis** (Scacchi, 1833) (*T. britannica* Pilsbry, *Vertigo britannica* (Pilsbry))

British whorl snail   Size: 1.7–1.9mm

Inhabits short, dry, maritime grassland, on cliffs and screes, sometimes in abundance. Several occurrences are among old, stabilized quarry debris.

Probably native (L.Pgl fossils only). Though very local, *T. callicratis* appears to be secure within its range, especially at several sites on the Jurassic limestone cliffs between the Isle of Portland (Church Ope Cove) and Durlston Head, Dorset (Davis, 1955).

There is no evidence of overall decline (fossil records from Kent and Norfolk are erroneous). RDB category: Rare.

Mainly Mediterranean, Pyrenean & Alpine.

British specimens were originally considered to be a variety of *T. cylindrica*, and later an endemic British species (*T. britannica*); they differ from typical *T. callicratis* from southern Europe in their much weaker apertural dentition.

(Ellis: 153; K. & C.: 68)

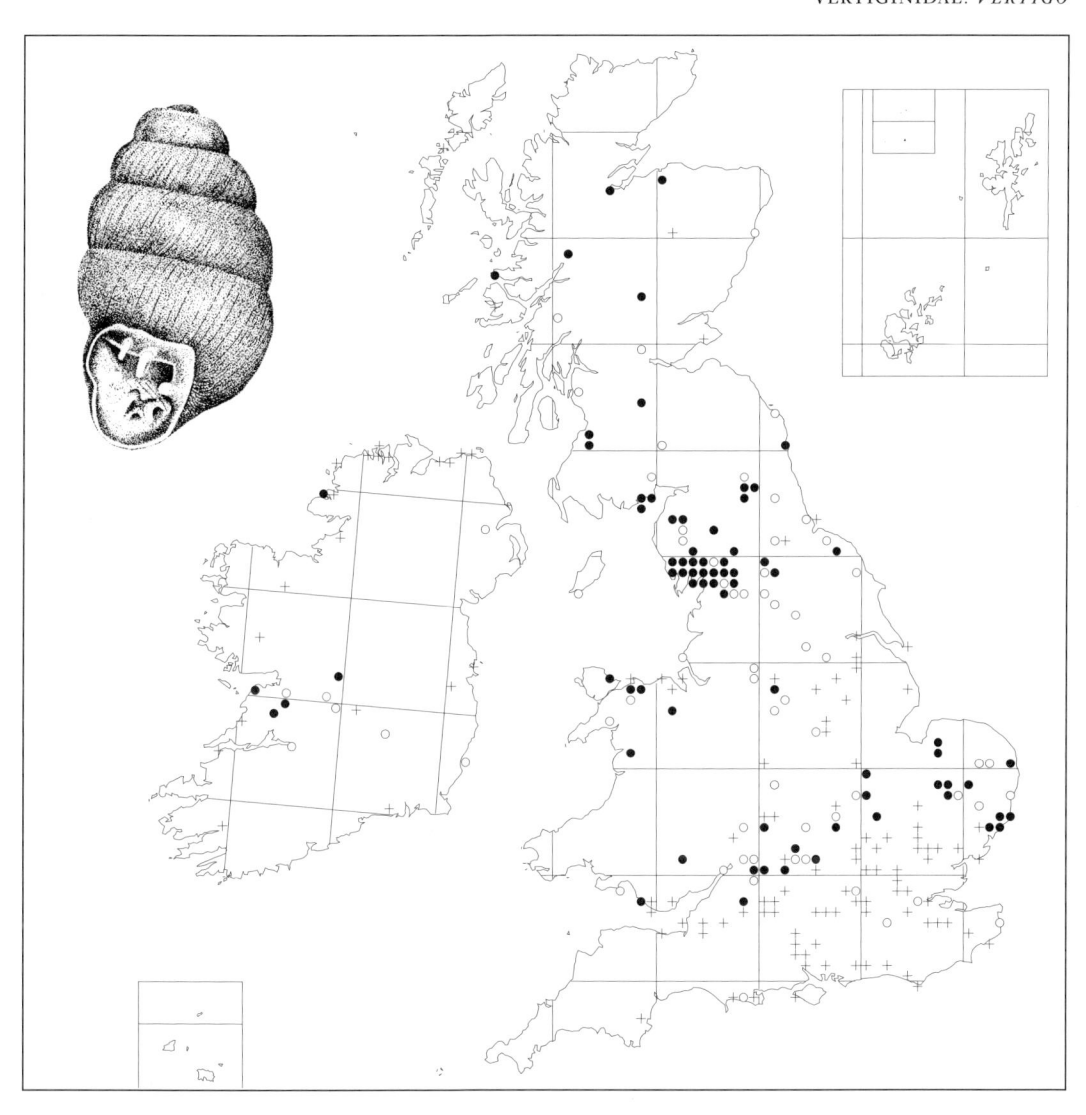

**Vertigo pusilla** Müller, 1774

Wall (or wry-necked) whorl snail   Size: 1.9–2.1mm

A species found most typically on or at the base of ivy-covered stone walls, usually shaded by trees. Most of the sites in the Lake District and the Cotswolds are of this type. It has also been reported from a wide range of rather dry and relatively undisturbed places: open woods, hedges, grassy banks, natural cliffs and occasionally fixed sand-dunes (Dance, 1969). It is rare in marshes and does not live in dense woodland.

Native (E.Pgl). In the earlier part of the Postglacial (Boreal and Atlantic periods) *V. pusilla* was a common species with a much wider distribution than today. Its decline roughly coincided with the beginnings of prehistoric farming, though whether the two are linked is unclear, and climatic change may also be involved (cf. *V. alpestris*). In N.W. England, where most of the habitats of *V. pusilla* are man-made, there is little evidence of recent decline. Elsewhere, however, isolated colonies may be at risk, and in some counties the species has not been seen for many years (e.g. Kent, 1803; Surrey, 1864; Warwicks., 1892; Dorset, 1898; Essex, 1935). RDB category (Ireland only): Rare.

European: widespread but local, reaching the Arctic Circle in Scandinavia.

(Ellis: 149; K. & C.: 70)

VERTIGINIDAE: *VERTIGO*

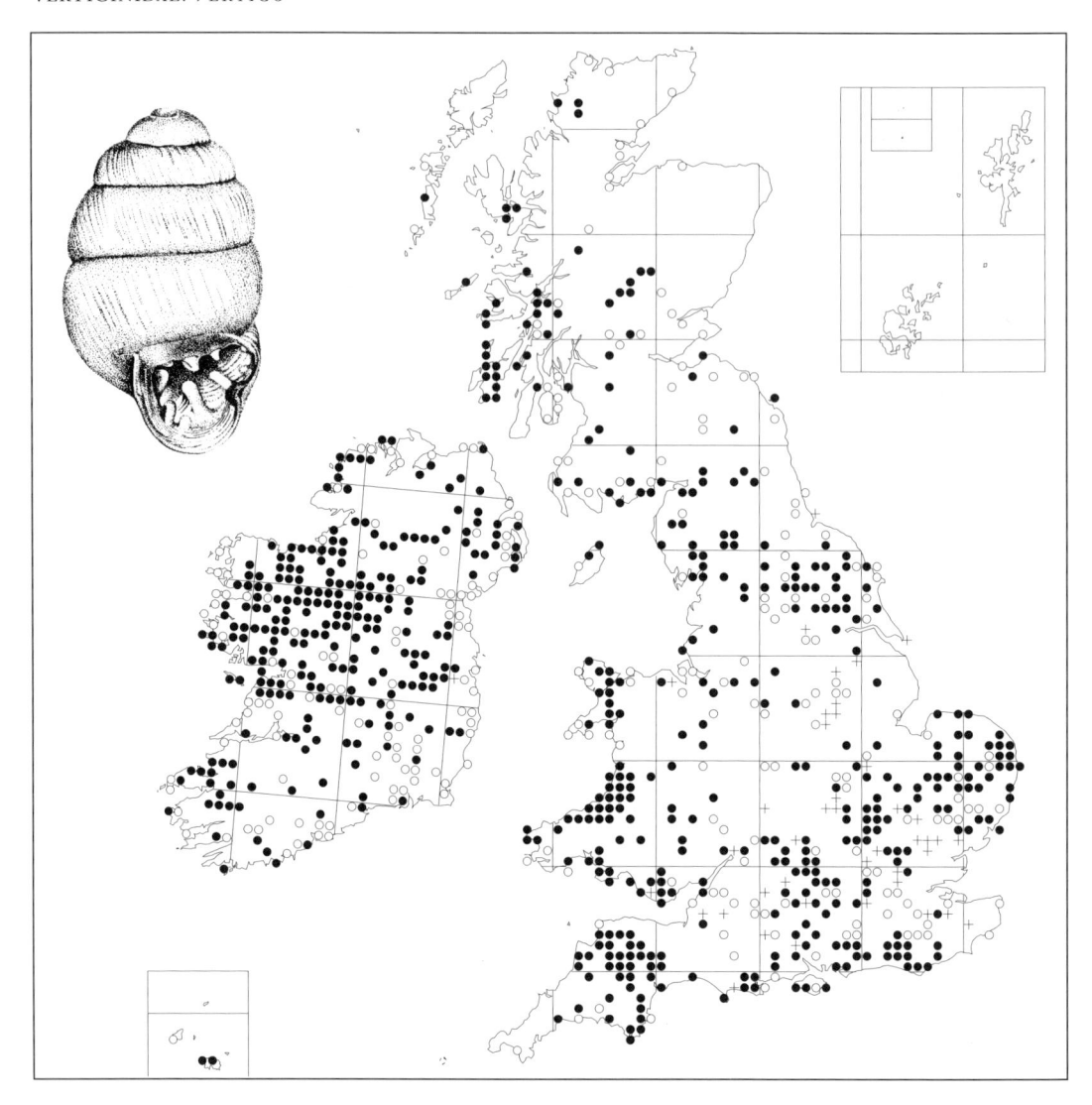

**Vertigo antivertigo** (Draparnaud, 1801)

Marsh whorl snail   Size: 1.9–2.1mm

A species restricted to wetlands, mainly, though not exclusively, of lowland type: marshes, fens, reed swamps, hillside flushes. It avoids places where the water-level fluctuates markedly. In sedge fens bordering open water it can often be found among flood rubbish or crawling on saturated ground litter. Occasionally it lives in dune slacks.

Native (Lgl). There is no evidence of major national decline, but undoubtedly this species is becoming rarer in parts of lowland England through agricultural drainage and river-management schemes, and has already become extinct in some 10km squares. In Ireland it is less affected.

Palaearctic; in Europe to about 63°N.

(Ellis: 141; K. & C.: 70)

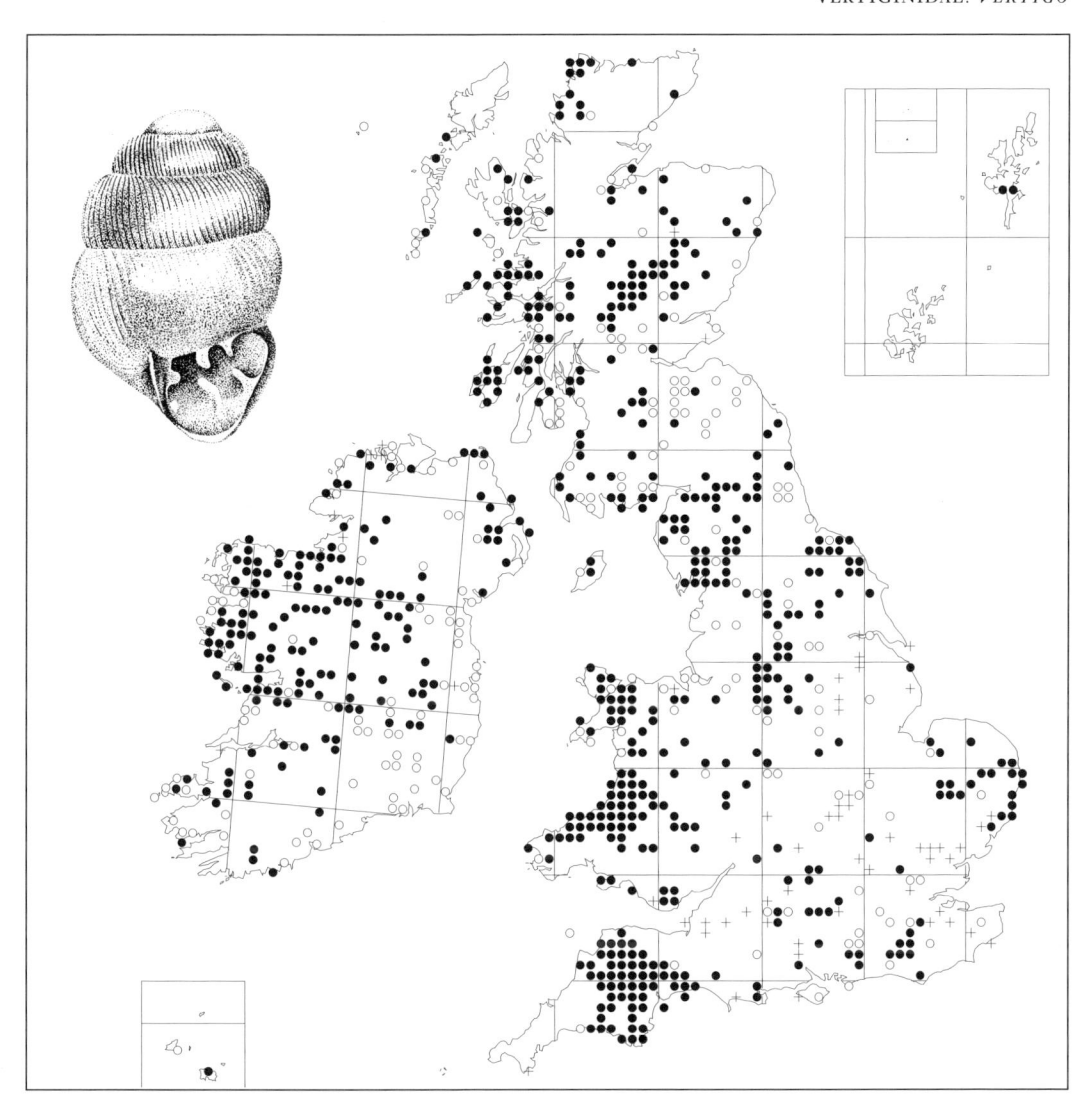

**Vertigo substriata** (Jeffreys, 1833)

Striated whorl snail    Size: 1.7–1.9mm

Restricted generally to moist habitats: marshes, damp woods, hillside flushes, dune slacks. In the highland zone it is not uncommon in boggy ground in areas of rough grazing. More rarely, it is found in ground litter in ordinary woodland, alongside such species as *Columella aspera*. It tolerates acid soils and is infrequent in base-rich habitats with a high diversity of species. In general, it is a snail intolerant of human disturbance.

Native (E.Pgl). This species was widespread during the earlier Postglacial (particularly the Boreal and Atlantic periods) but is now somewhat in decline. Around London and in the Midlands it must be accounted rare, with surviving populations at risk from intensive farming. In the more oceanic areas of the British Isles, however, it is certainly underrecorded.

C. and N. European, reaching to the Arctic Circle in Scandinavia.

(Ellis: 142; K. & C.: 71)

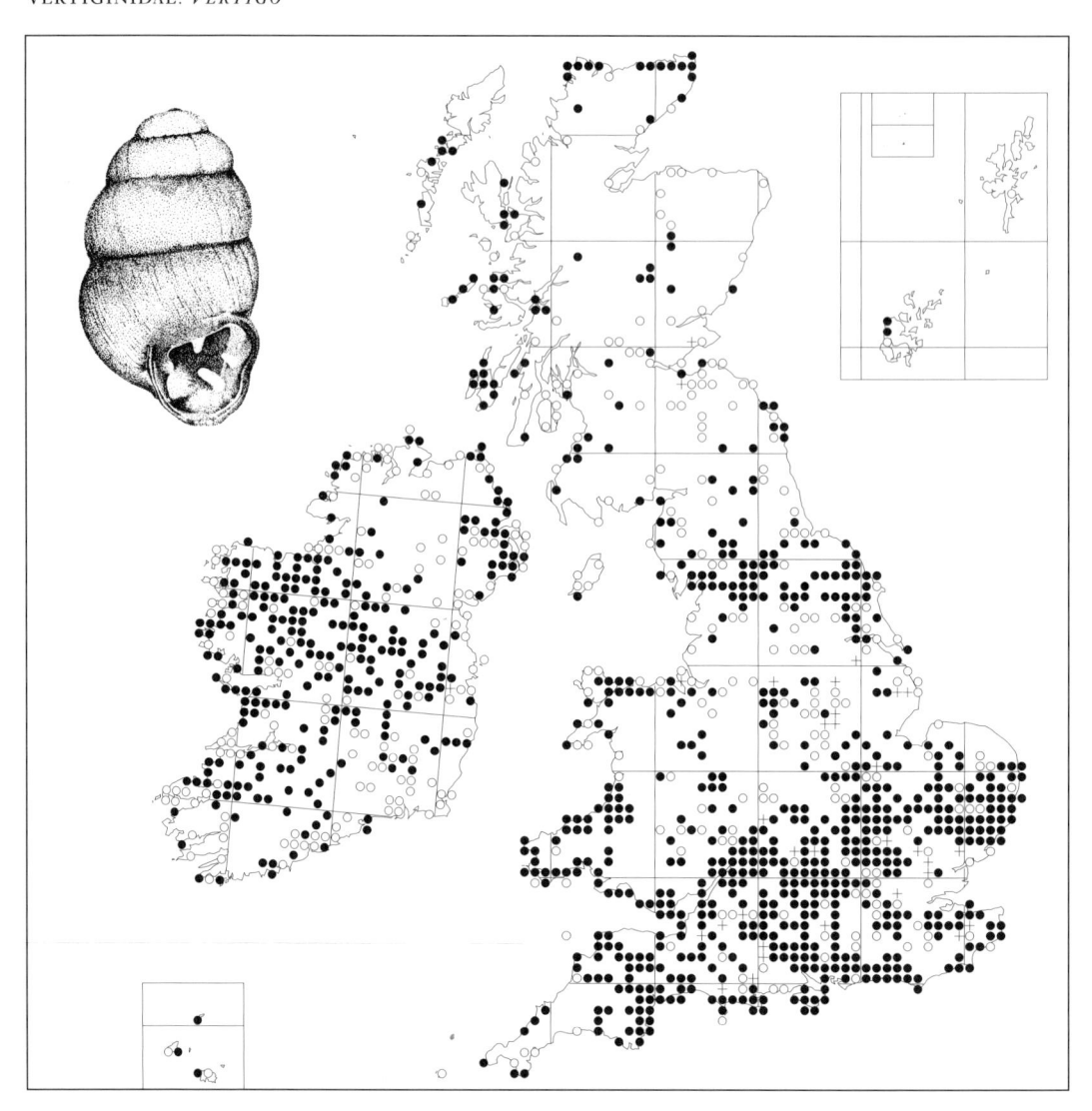

**Vertigo pygmaea** (Draparnaud, 1801)

Common whorl snail   Size: 1.7–2.2mm

A frequent species at the roots of grass in dry base-rich places, mostly in the lowland zone: roadside banks, chalk and limestone hillsides, cliffs and screes, maritime turf, the hollows of stabilized dunes. Sometimes (especially in western areas) it lives also in marshes, together with such species as *V. antivertigo*. It is intolerant of shade and not found in woods.

Native (Lgl). *V. pygmaea* remains widespread in suitable base-rich habitats. It is underrecorded owing to its minute size.

Holarctic; throughout Europe, to about 64°N in Scandinavia.

(Ellis: 144; K. & C.: 71)

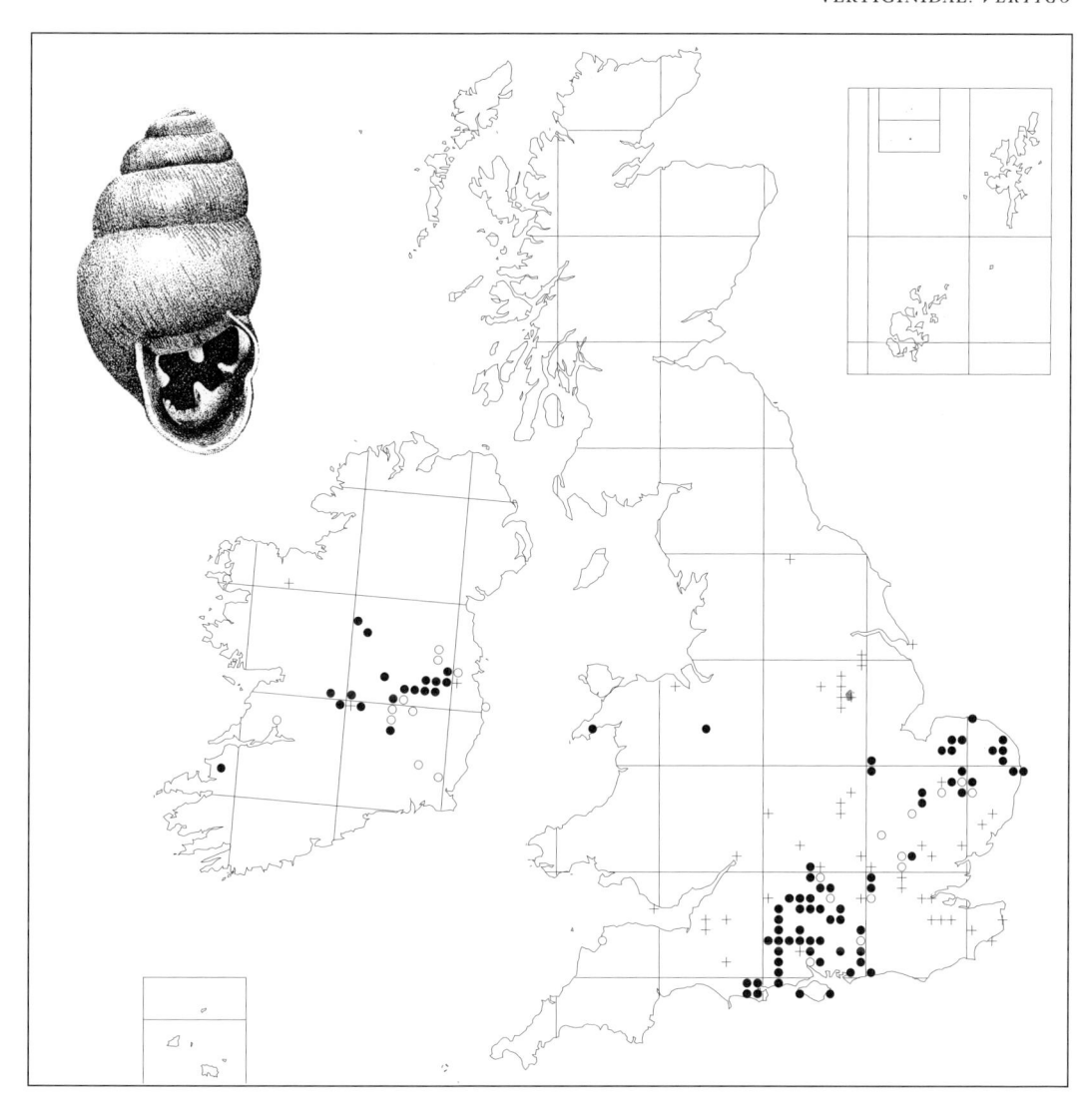

**Vertigo moulinsiana** (Dupuy, 1849)

Des Moulins' whorl snail   Size: 2.2–2.7mm

This species is restricted to old calcareous wetlands, usually adjacent to lowland rivers and lakes. Normally it lives on sedges and grasses (especially *Glyceria maxima*) close to the ground, but in autumn it may ascend taller marsh vegetation such as *Phragmites* and *Alnus* (Phillips, 1908). It is usually associated with *V. antivertigo* but is less tolerant of human disturbance.

Native (E.Pgl). In the earlier part of the Postglacial (Boreal and Atlantic periods) *V. moulinsiana* was more widely distributed than today, in England reaching as far north as Yorkshire. Its retreat may partly result from lower temperatures (the species is at the northern limit of its European range) but undoubtedly the main cause has been the progressive destruction of its habitats for agriculture from late prehistoric times onwards. Some sites have been lost in recent decades. The most extensive surviving populations are in central southern England and in Norfolk; elswhere they are very scattered and mostly at risk (Drake, 1997). The isolated colony in N. Devon (Braunton) was last seen in 1933. RDB category: Rare.

Probably Holarctic; in Europe very sparsely distributed, north to Denmark and southernmost Sweden.

(Ellis: 147; K. & C.: 72)

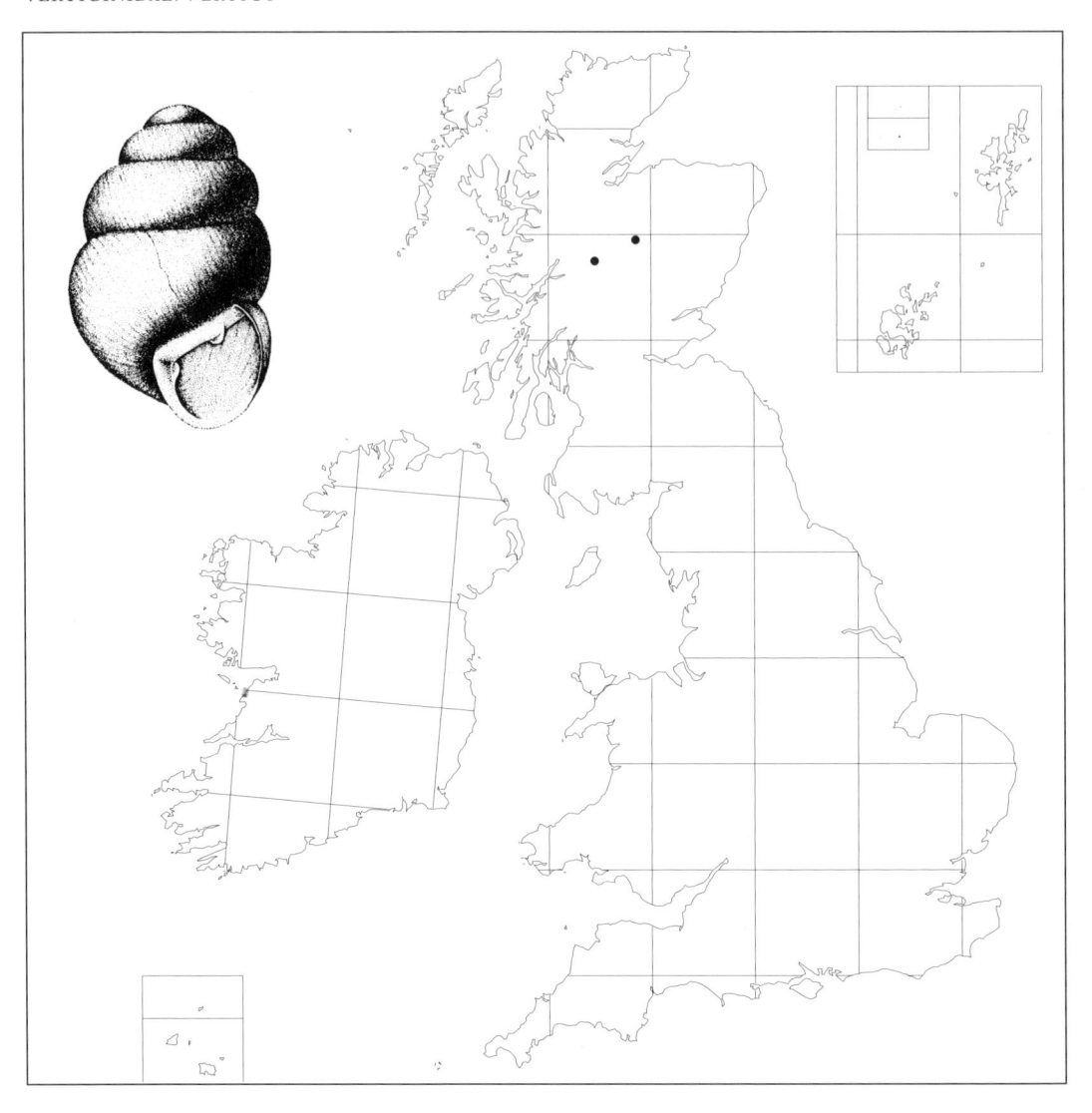

**Vertigo modesta** (Say, 1824) (*V. arctica* Wallenberg)

Size: 2.2–2.5mm

At its two known British localities in the Grampian Mountains (Geal Charn; Coire Garbhlach, Inverness-shire) *V. modesta* lives in north-facing corries between 800 and 980m. The vegetation is short turf and low heath/scrub rich in Arctic-alpines, including such species as *Salix lanata*, *S. reticulata* and *Dryas octopetala*. The substrate at Geal Charn is hard Dalradian limestone. The habitats are similar to those in the Scandinavian mountains.

Probably native (no fossil records). *V. modesta* was discovered on Geal Charn in 1987 (Marriott & Marriott, 1988) and at Coire Garbhlach in 1993. There can be little doubt that both populations are Lateglacial relicts. Other colonies may perhaps be found by careful search of calcareous ground elsewhere at high elevations in the Scottish mountains; the habitat type is nevertheless uncommon in Britain and for biogeographical reasons the species must be considered an endangered rarity. RDB category: Endangered.

Circumpolar Holarctic; in Europe mainly Arctic and high Alpine.

(K. & C.: 73)

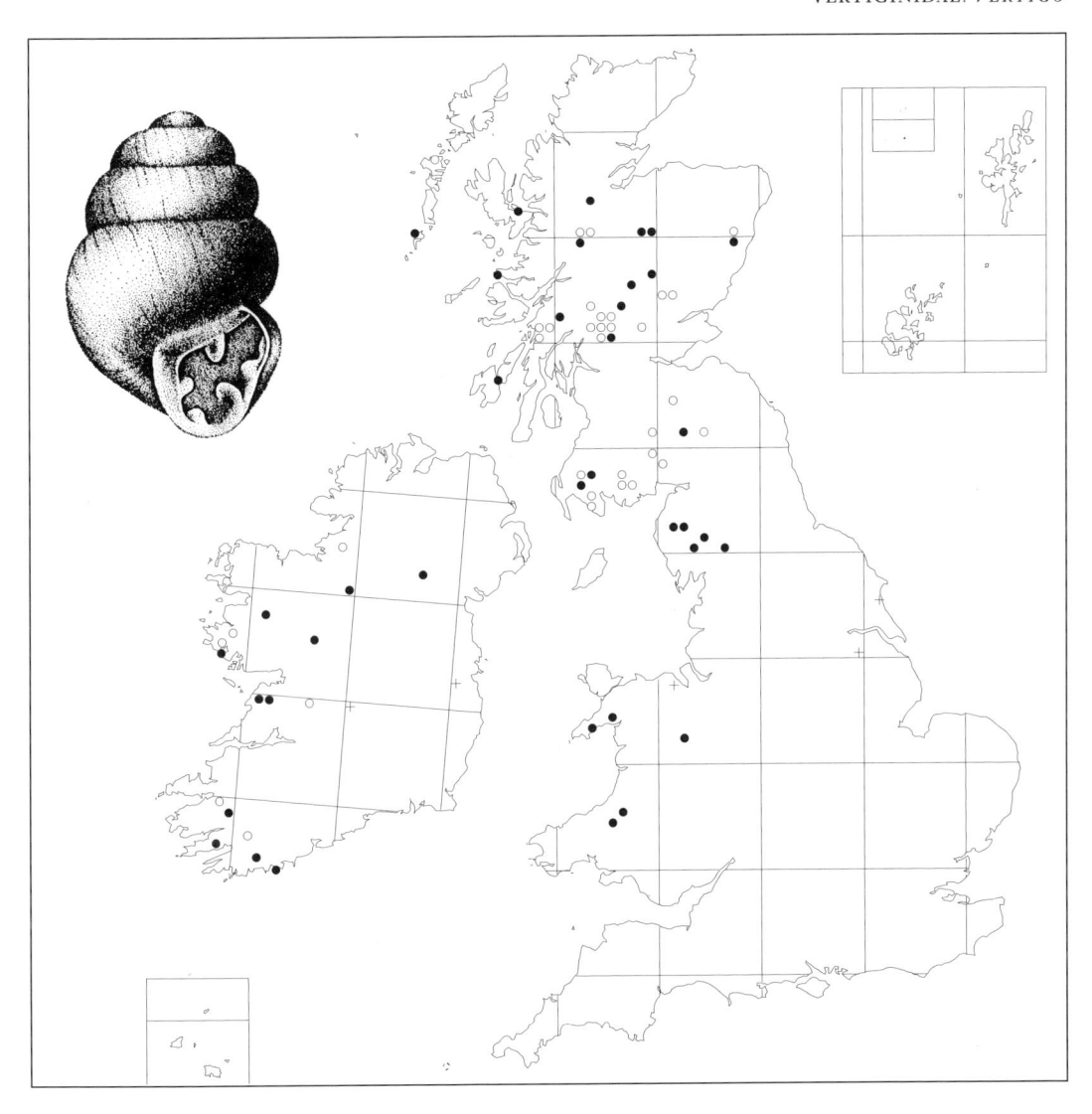

**Vertigo lilljeborgi** (Westerlund, 1871)

Lilljeborg's whorl snail   Size: 1.9–2.2mm

Restricted to saturated decaying vegetation in *Carex* and *Juncus* swamps, sometimes shaded by alders. Most sites are at the margins of highland lakes and rivers, in places subject to deep natural flooding; occasionally it is found in small, isolated mires. *V. lilljeborgi* tolerates acidic conditions. It is only rarely immediately associated with other species of *Vertigo* (Kevan & Waterston, 1933; Dance, 1972; Long, 1974; Chater, 1985; Cameron, 1992).

Native (Lgl). *V. lilljeborgi* is a boreo-arctic snail which is probably relict in the British Isles. A few Lateglacial or very early Postglacial fossils (Flint, Lincs, E. Yorks, Co. Dublin, Co. Tipperary) show that it was formerly more widely distributed. Today the main threat to the species comes from agricultural drainage or from the raising or controlling of water-levels in lakes and rivers for reservoirs and hydro-electric schemes. Several colonies have been destroyed in this way since the 1930s. RDB category: Rare.

N. European: mainly Scandinavian, British and Irish.

(Ellis: 146; K. & C.: 74)

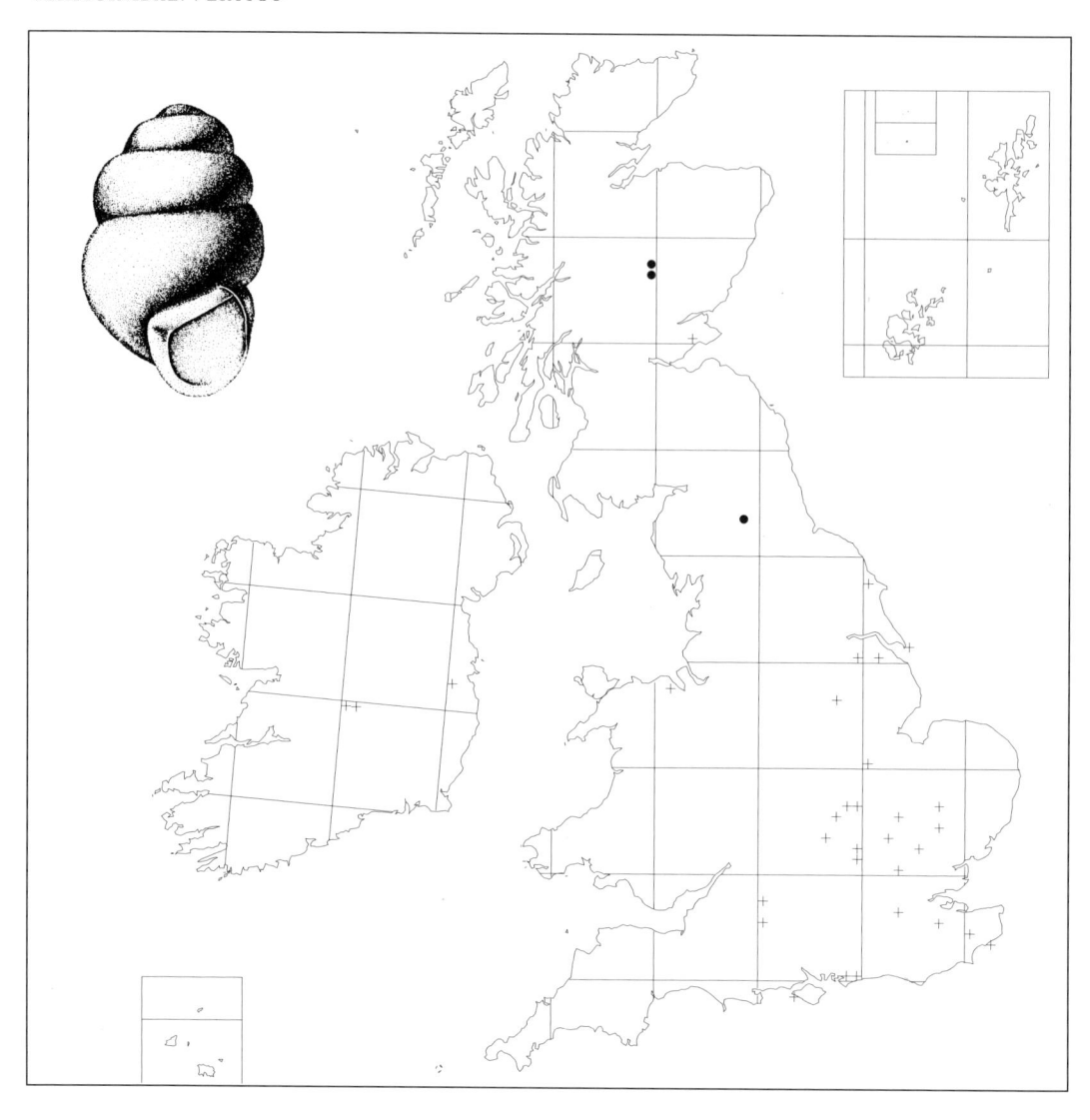

**Vertigo genesii** (Gredler, 1856)

Size: 1.7–2.1mm

At its only English locality in Teesdale *V. genesii* lives among mosses and low-growing sedges (mainly *Carex demissa*) in an alkaline flush fed by springs within a north-facing hillside depression at 495m (Coles & Colville, 1980). In winter the hollow may fill with several metres of snow. Arctic-alpine plants, most notably bog sandwort (*Minuartia stricta*), occur close by, though not in direct association. The site is lightly grazed by sheep. In Scotland, *V. genesii* is found in similar calcareous flushes at somewhat lower elevations. The habitats are typical of the occurrence of the species in the Scandinavian mountains.

Native (Lgl). The Teesdale colony (Widdybank Fell) and those more recently found in Perthshire (Glen Tilt) are doubtless Lateglacial survivals. Fossil occurrences show that the species was common in lowland England (and probably also in Ireland) during the Lateglacial period. It rapidly became extinct in the early Postglacial as a result of increasing warmth and the suppression of its habitats by the spread of forests. The sites are under no immediate threat, though their very small size renders them vulnerable to accidental damage. RDB category: Endangered.

Boreo-Alpine; in Europe virtually only in the mountains of C. Scandinavia and in the Alps.

(K. & C.: 74)

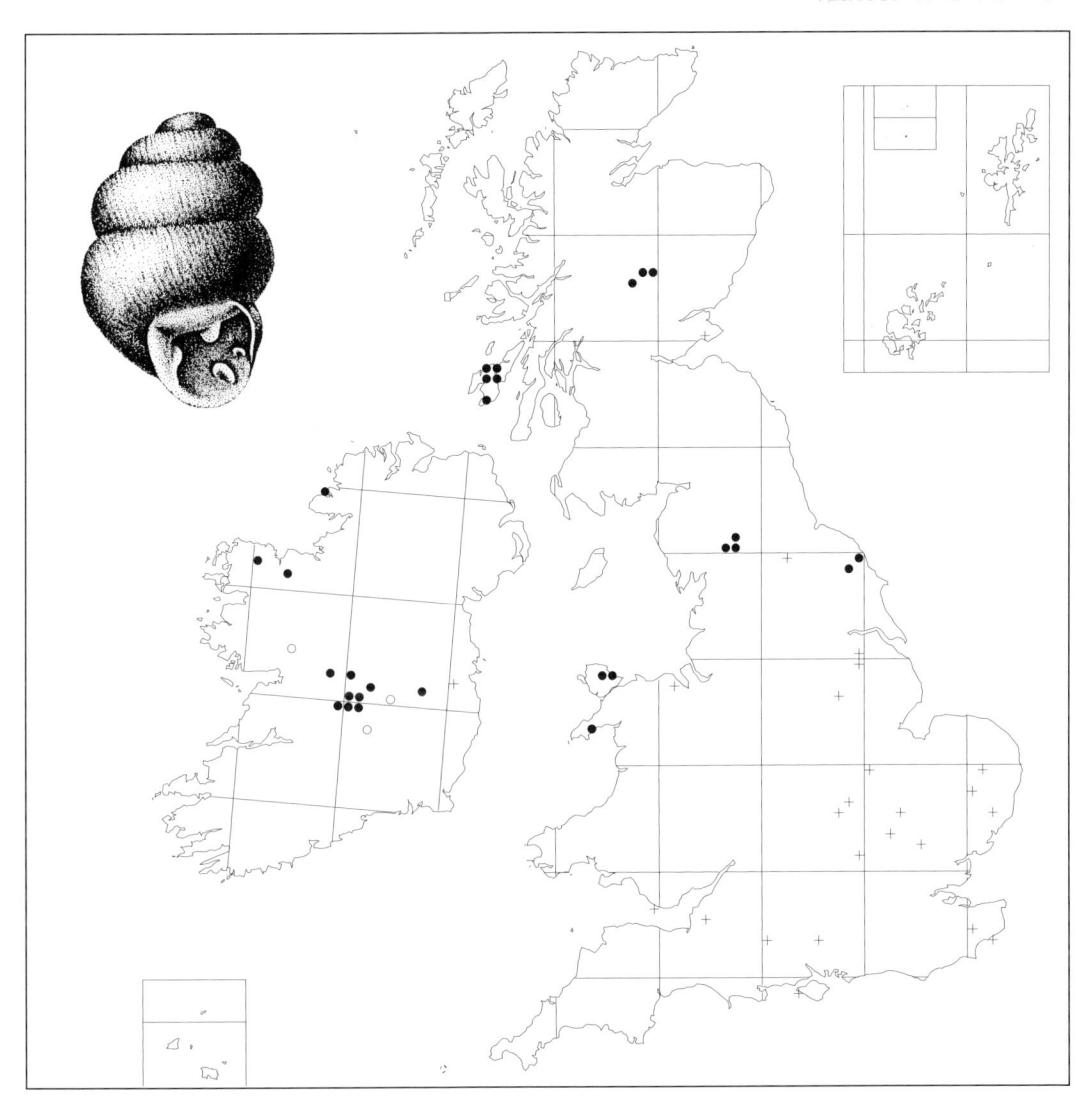

**Vertigo geyeri** Lindholm, 1925 (*V. genesii* of Ellis, 1926, *non* Gredler)

Size: 1.7–1.9mm

A rare snail found in neutral to base-rich bogs and fens, in very wet, swampy places not subject either to periodic drying or to flooding. It requires a low-growing but luxuriant vegetation dominated by the finer-leaved grasses and sedges (e.g. *Schoenus nigricans*, *Eleocharis quinqueflora*, *Carex*, *Juncus*) and relatively free from *Sphagnum* and other mosses. The Irish sites include flat, lowland fens (Stelfox & Phillips, 1925; Phillips, 1935; Norris & Pickrell, 1972; Cawley, 1996a). The British sites (Anglesey, Caernarvon, Westmorland, N. Yorks., Perthshire, Islay) are mostly alkaline flushes on gently sloping ground (Coles & Colville, 1979; Colville, 1994b). The species is sometimes associated with *Catinella arenaria*.

Native (Lgl). This relict species was common in lowland Britain immediately after the last glaciation. Surviving populations have escaped both mid-Postglacial forest growth and subsequent drainage by man. All are vulnerable, but more especially those at the flatter sites which are easily dried out by a lowering of the water-table. Several Irish sites have been lost by drainage in recent years, as at Cloonascragh Bog (Co. Galway) and Fiagh Bog (Co. Tipperary). Pollboy Bog (Co. Galway) where the species was common in 1934, is now poor, gorse-covered grazing land following drainage in the 1950s. RDB category: Endangered.

Boreo-Alpine; in Europe mainly in Scandinavia and in the Swiss and Austrian Alps.

(Ellis: 145; K. & C.: 75)

VERTIGINIDAE: *VERTIGO*

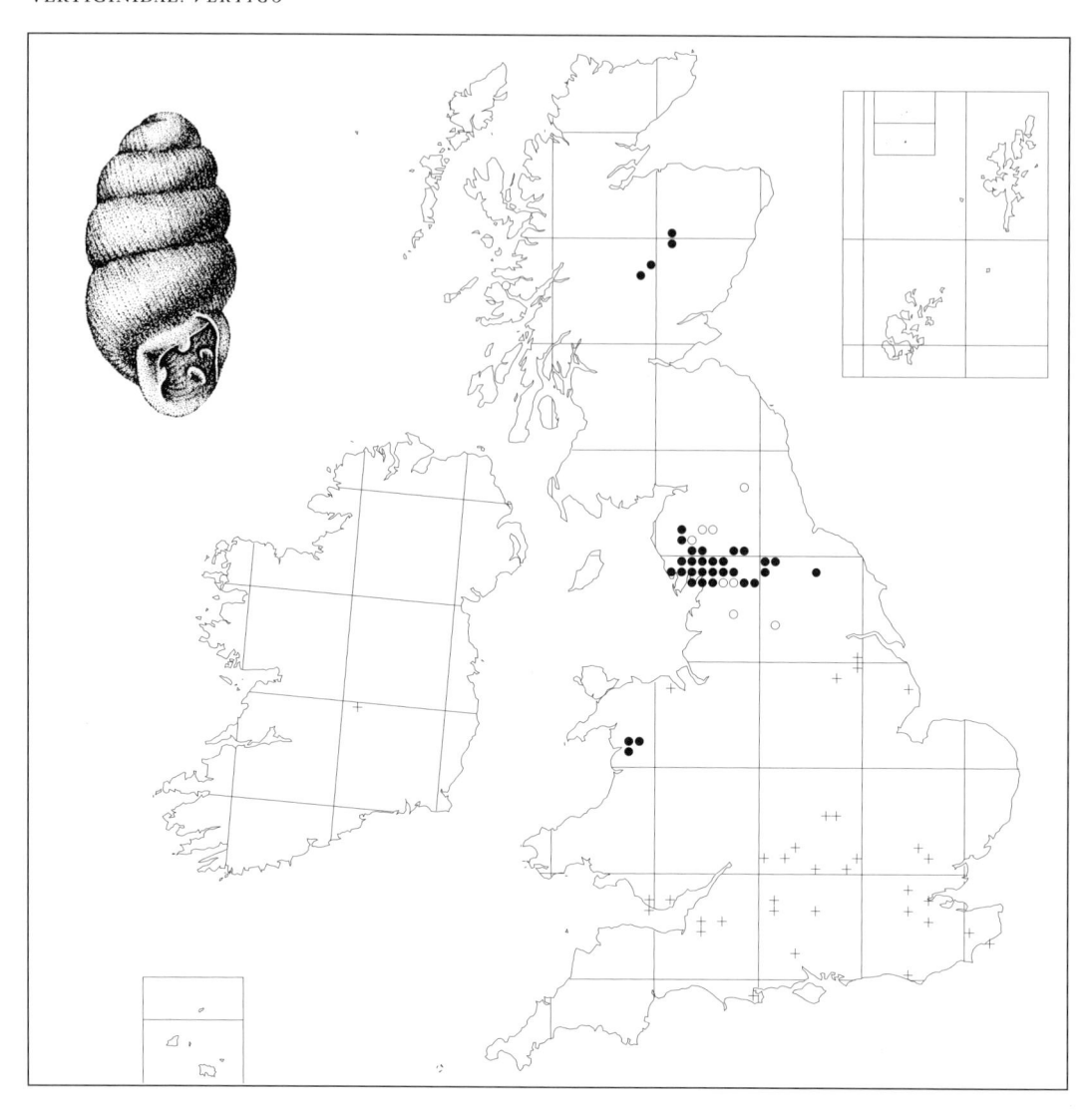

**Vertigo alpestris** Alder, 1838

Mountain whorl snail   Size: 1.8–2.0mm

In England and Wales this species is found almost exclusively on old stone walls or in leaf litter at their foot. The walls are usually covered with patches of ivy and lightly shaded by trees. In Scotland it lives among mosses on natural limestone screes in Perthshire, Banffshire and Aberdeenshire (Marriott & Marriott, 1982a; 1984). On walls it is generally associated with *Vertigo pusilla*.

Native (E.Pgl). *V. alpestris* is frequent within its principal area of distribution in N.W. England (Dean & Kendall, 1909) and shows no evidence there of recent decline. Fossils however show that it was much more widely distributed in the earlier Postglacial period, occurring in southern England and also in Ireland (where it is now apparently extinct). In fossil contexts, *V. alpestris* is associated with purely woodland mollusc faunas. Forest clearance may therefore have played some part in its retreat but, as with *V. pusilla*, it is likely that long-term climatic change has also been important in bringing about the decline of this essentially boreo-alpine snail. The association with walls must be secondary, and of fairly recent origin.

N. European and Alpine: widespread in Scandinavia, elsewhere mainly montane.

(Ellis: 149; K. & C.: 75)

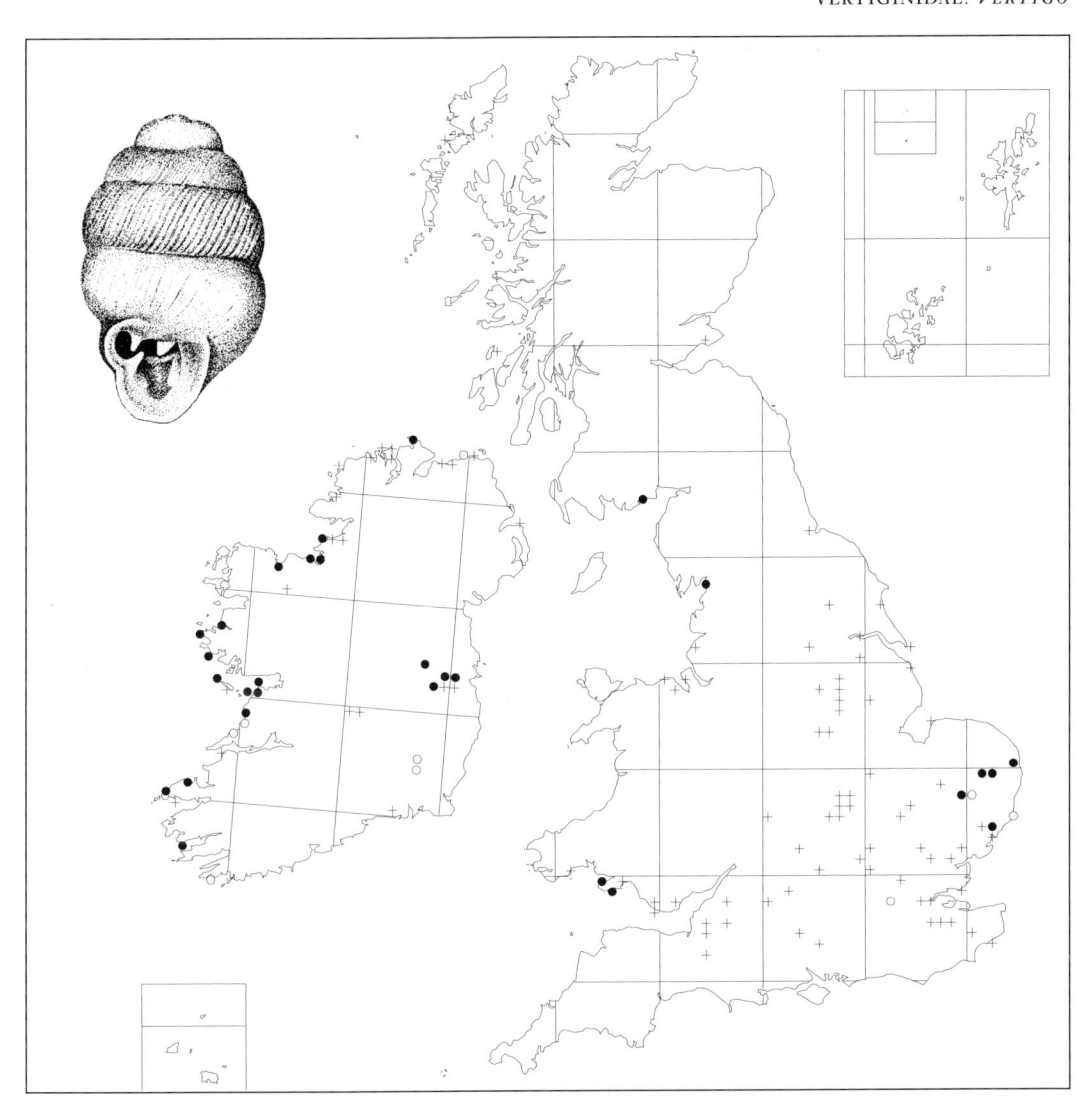

**Vertigo angustior** Jeffreys, 1830

Narrow-mouthed whorl snail    Size: 1.7–1.9mm

Restricted to moist places which are affected neither by periodic desiccation nor by flooding. It requires open conditions quickly warmed by the sun, inhabiting short vegetation of grasses, mosses or low herbs. In the British Isles it has been found in wet, base-rich meadows (sometimes grazed), in coastal marshes and dune slacks, in maritime turf (chiefly in Ireland) and in the depressions of a limestone pavement (Gait Barrows, Lancs.) (Norris & Colville, 1974; Marriott & Marriott, 1982b; Killeen, 1983, 1997; Preece & Willing, 1984; Colville, 1994a).

Native (E.Pgl). *V. angustior* is a declining species. It was common in the open landscapes of lowland Britain at the beginning of the Postglacial (Preboreal and Boreal periods) but its habitats were largely suppressed by mid-Postglacial forests. Some of the existing sites may represent chance survivals from that early period in places where the hydrological balance remained unchanged and which escaped a dense tree cover; this is likely to be true, particularly, of colonies in or near dunes and on hard limestones where deep-rooted trees could not establish themselves. The sites may easily be destroyed by drainage, afforestation or other changes in land use. RDB category: Endangered (in Ireland: Vulnerable).

European: throughout to southernmost Scandinavia but everywhere very local.

(Ellis: 15: K. & C.: 75)

CHONDRINIDAE: *ABIDA*

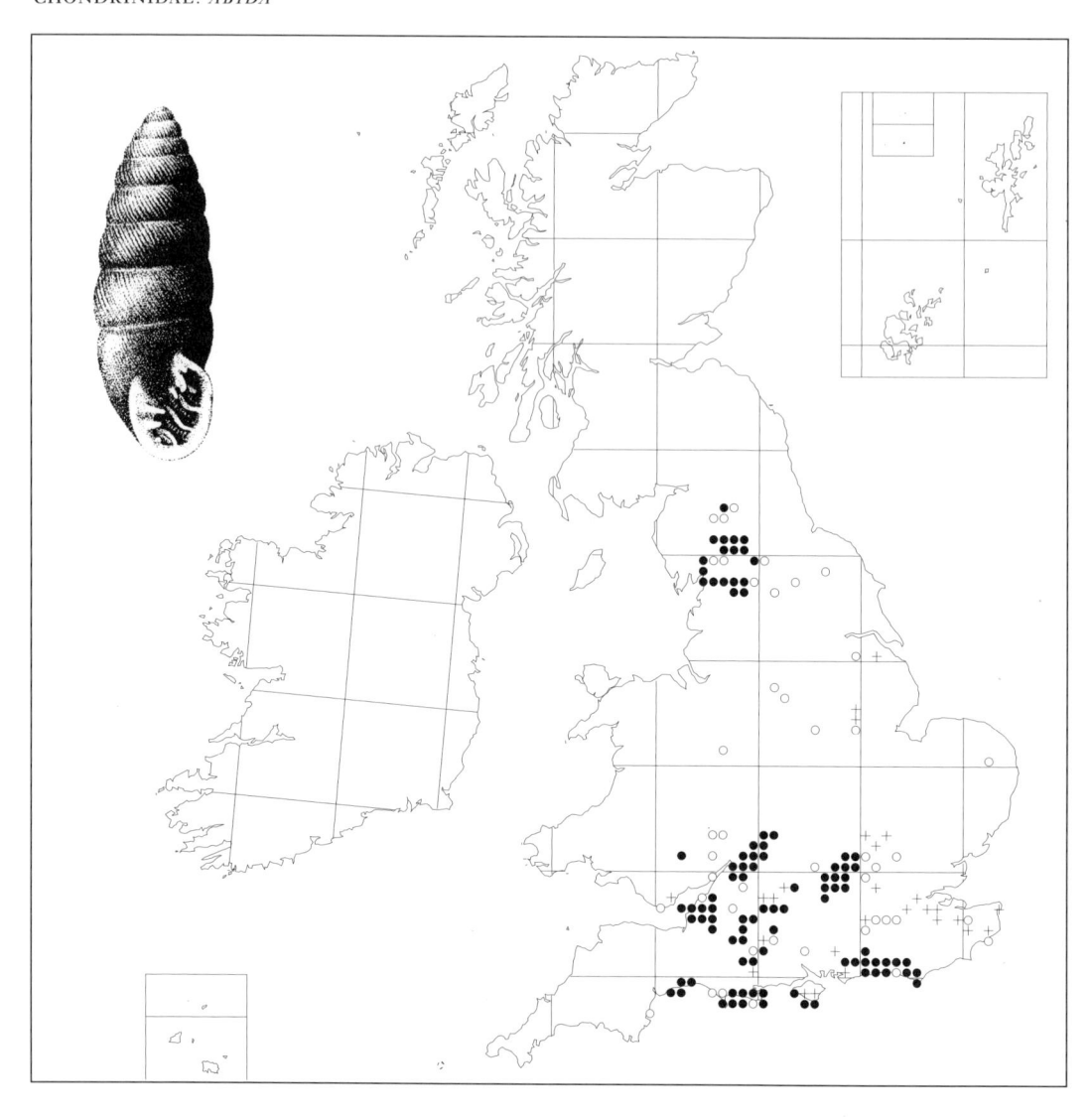

**Abida secale** (Draparnaud, 1801) (*Pupa secale* Draparnaud)

Large chrysalis snail    Size: 6.0–8.5mm

Locally common in dry, stony situations mainly on chalk or limestone: short-turfed grassland, natural screes and cliffs, disused quarries, occasionally at the base of old walls. It likes places where the soil cover is broken and some bare rock shows through. Most of its habitats are unshaded and rather exposed, but sometimes it may be found under a light woodland canopy.

Native (Lgl). This was a common snail on the chalk of southern England in the Lateglacial and earliest Postglacial periods, prior to the spread of forests and the development of deep soils. It has made only a poor recovery in the wake of human clearance. A few sites, notably in the Lake District, may indeed be relicts from early Postglacial time. It is maintaining its ground in its strongholds, especially in Dorset, the Cotswolds (Long, 1970) and in N.W. England (Cameron & Letanka, 1976), but elsewhere many isolated populations seem to have become extinct over the past century, probably through changes in land use. The last records for certain counties are: Shropshire, 1840; Notts., *c*.1860; Herefs., 1898; Derbys., 1902; Lincs., 1911; Surrey, 1912; Kent, 1920.

W. European and Alpine, extending north to Britain and central Germany.

(Ellis: 159; K. & C.: 84)

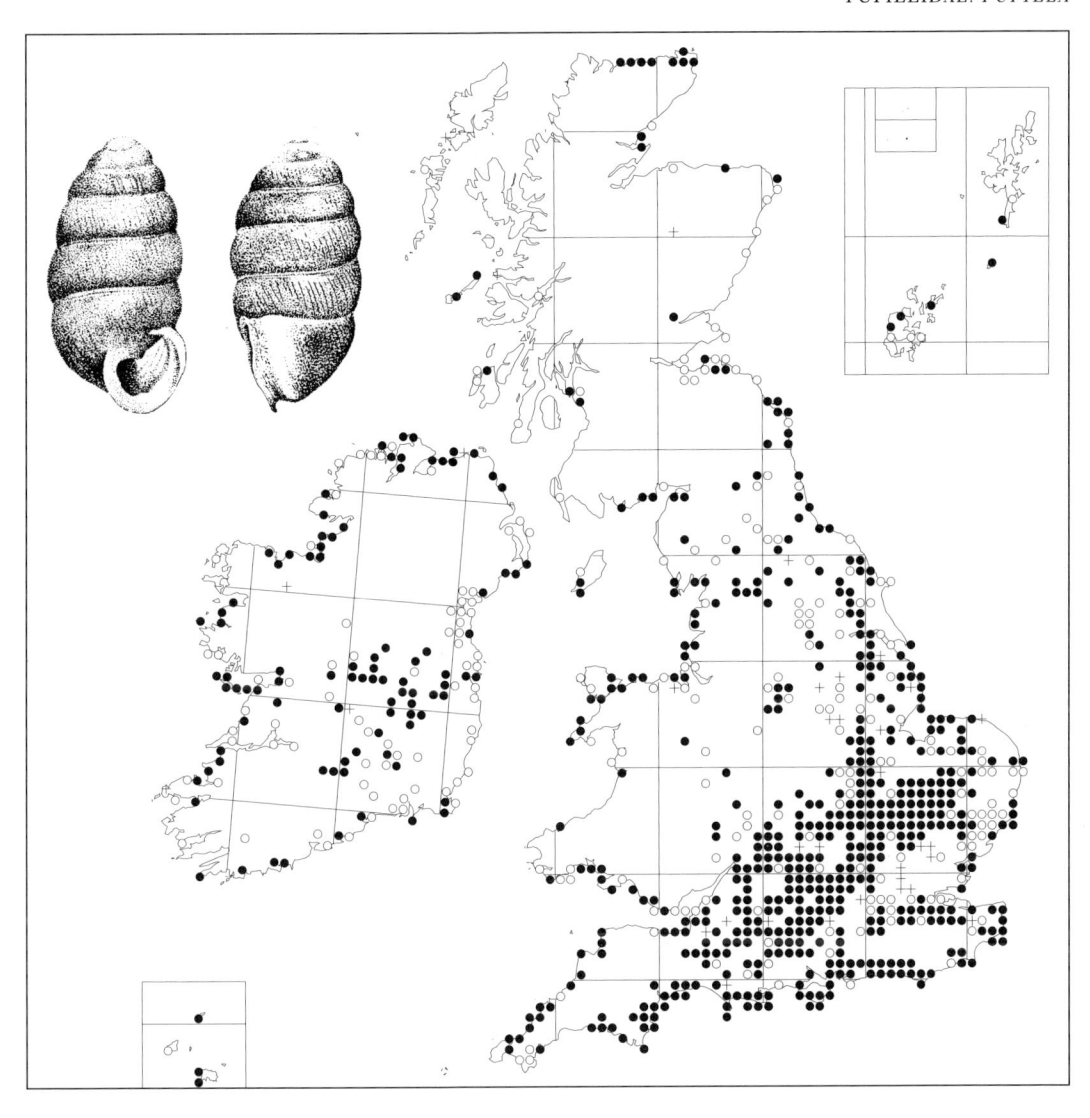

**Pupilla muscorum** (L., 1758) (*Pupa marginata* Draparnaud)

Moss chrysalis snail    Size: 3–4mm

Locally common in dry, exposed calcareous places, especially in stony or sandy ground. It is typical of sheep-grazed chalk or limestone grassland and of maritime turf in areas of shell-sand. It also inhabits cliffs and screes, abandoned quarries, and the sunny tops of old limestone walls bearing stonecrops (*Sedum* spp.) and similar plants. Unlike *Lauria cylindracea*, it rarely climbs vertical surfaces.

Native (Lgl). This species is still fairly frequent over most of its range, though it shows local decline and has become extinct in a number of 10km squares (e.g. in Suffolk; Killeen, 1992) because of the ploughing up or reversion to scrub of old calcareous grassland.

Holarctic; in Europe nearly throughout, but rare in the northern highland zone.

(Ellis: 156; K. & C.: 90)

PUPILLIDAE: *LEIOSTYLA*

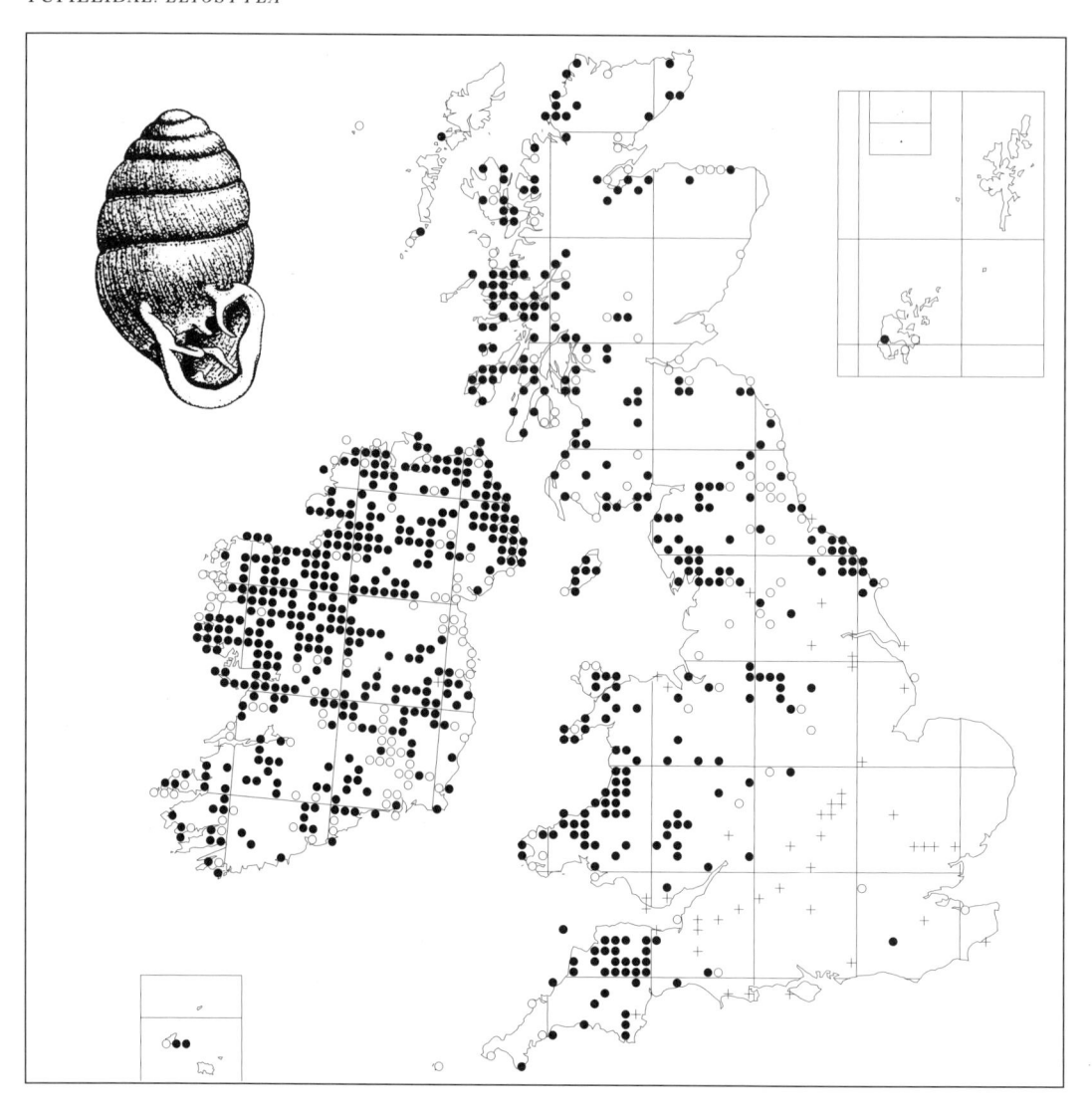

**Leiostyla anglica** (Wood, 1828) (*Pupa anglica* Wood, *Lauria anglica* (Wood))

English chrysalis snail   Size: 3.0–3.7mm

In western oceanic areas of Ireland and Scotland this species lives in humid places of all kinds, whether shaded or open: coniferous and deciduous woods, marshes, dune slacks, sea cliffs. Further east most sites are in undisturbed deciduous woodland, under moist ground litter or in boggy hollows. In central and southern England its presence is a useful indicator of ancient woodland that has never been clear felled.

Native (E. Pgl). This is a declining species. Fossils show that it was more widespread during the forest optimum of the Postglacial (Boreal and Atlantic periods) and began to recede from the English lowlands by the time of the first prehistoric clearances. This retreat doubtless continues.

Mainly British Isles; known also from a few places in western France, Spain, Portugal and Algeria.

(Ellis: 158; K. & C.: 92)

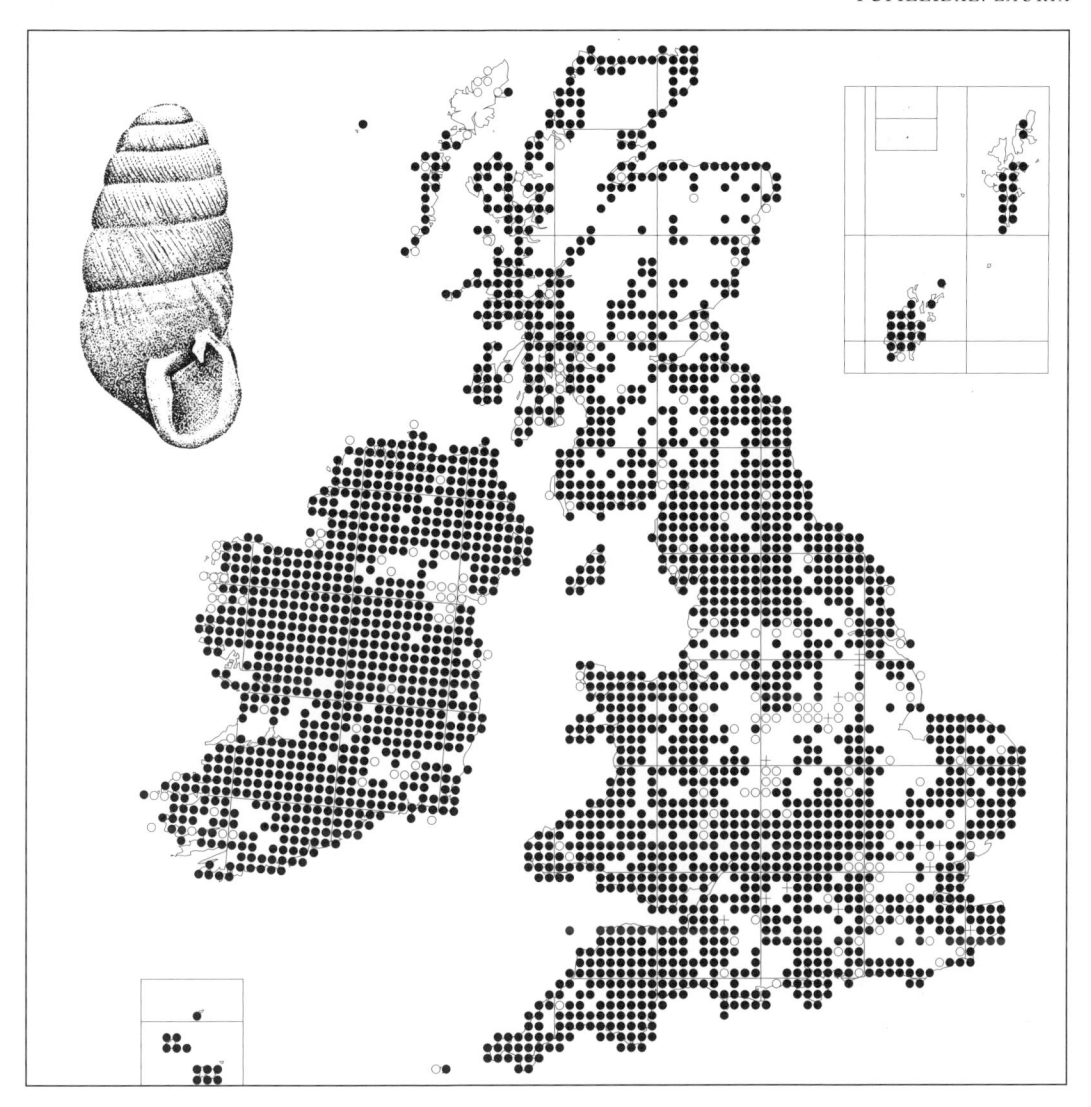

**Lauria cylindracea** (da Costa, 1778) (*Pupa umbilicata* Draparnaud)

Common chrysalis snail   Size: 3.0–4.4mm

In western oceanic areas this is a catholic snail, inhabiting rocks and screes, quarries, woods, hedgerows, grasslands and waste ground, avoiding only marshes and other very wet places. A characteristic habitat in which it may occur in great abundance is on stone walls, especially at the top under loose mats of ivy. In the climatically more continental east of the British Isles it becomes increasingly restricted to rocks, walls and woods, and is scarce or absent in some of the highly agricultural lowlands of central and eastern England like the Fenland basin where suitable habitats are few.

Native (E.Pgl). No clear evidence of regional change, though possibly receding in the English midlands.

Mediterranean and W. European Atlantic (inland not east of the Rhine valley). In Scandinavia coastal only.

(Ellis: 157; K. & C.: 92)

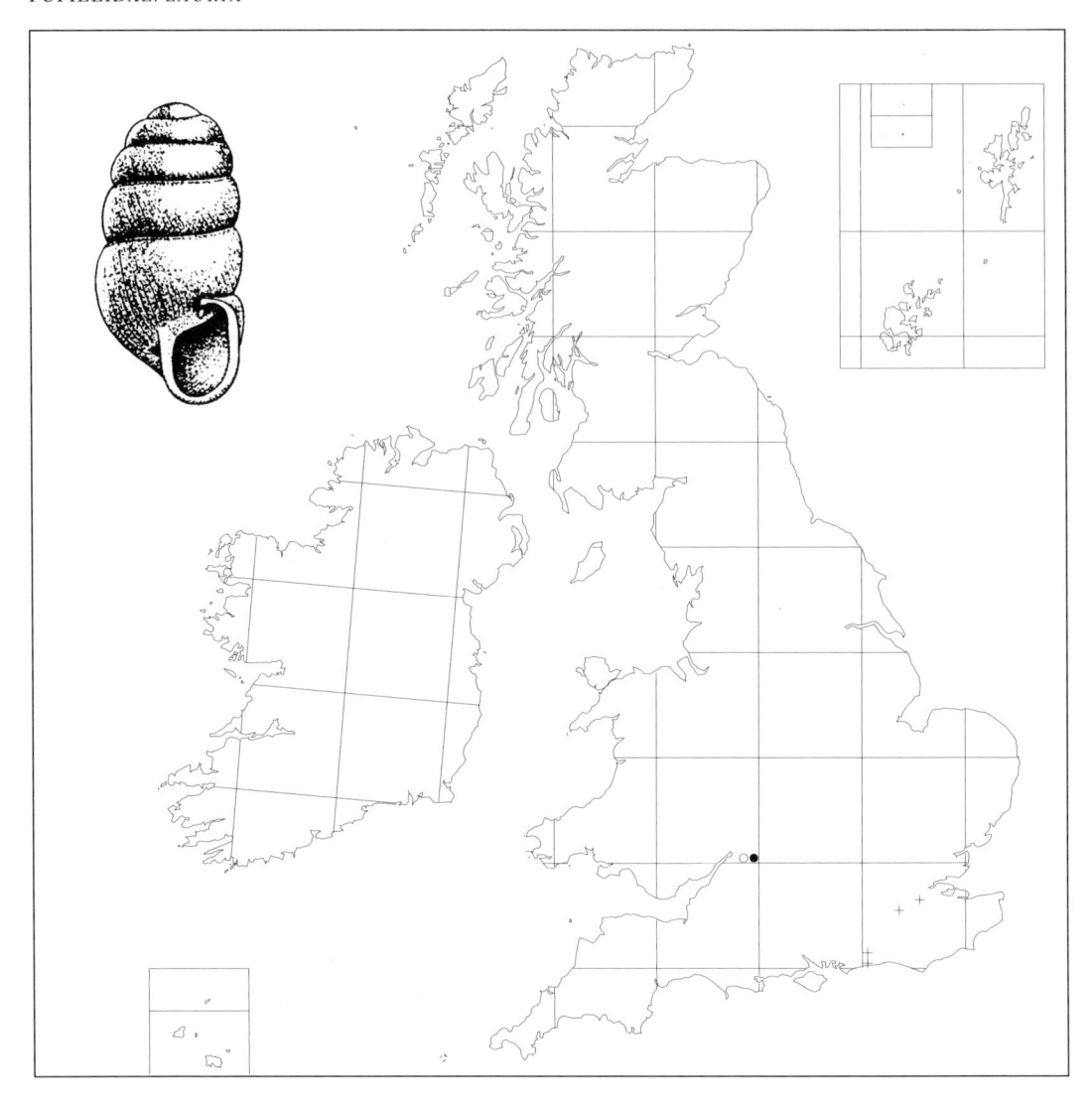

**Lauria sempronii** (Charpentier, 1837)

Size: 3.0–3.2mm

On the Continent, this species inhabits similar places to *Lauria cylindracea*: woods, rocks and stone walls. It is mainly montane. At the single known extant British site, it lives on an unmortared limestone wall among *Sedum*, *Hedera* and *Geranium robertianum*. Though unshaded, the wall is in a relatively moist, sheltered situation on the side of a valley.

Probably native (L. Pgl fossils only). *L. sempronii* was first found living in Britain in 1894 at Haresfield Beacon (Glos.), though the specimens were not recognized until 1971 (Kerney & Norris, 1972).

Recent searches of the area – a mixture of deciduous woodland, calcareous grassland, limestone rocks and walls – have failed to relocate the species. In 1985 a second colony was discovered at Edgeworth (Glos.), about 13km east of Haresfield (Long, 1986). The colony is of limited extent. *L. sempronii* occurs also as a fossil in a few deposits of Neolithic and later date on the Chalk in Kent, Surrey and Sussex, in all cases associated with woodland faunas (Kerney, 1983).
RDB category: Endangered.

S. European: mainly Pyrenean and Alpine.

(K. & C.: 93)

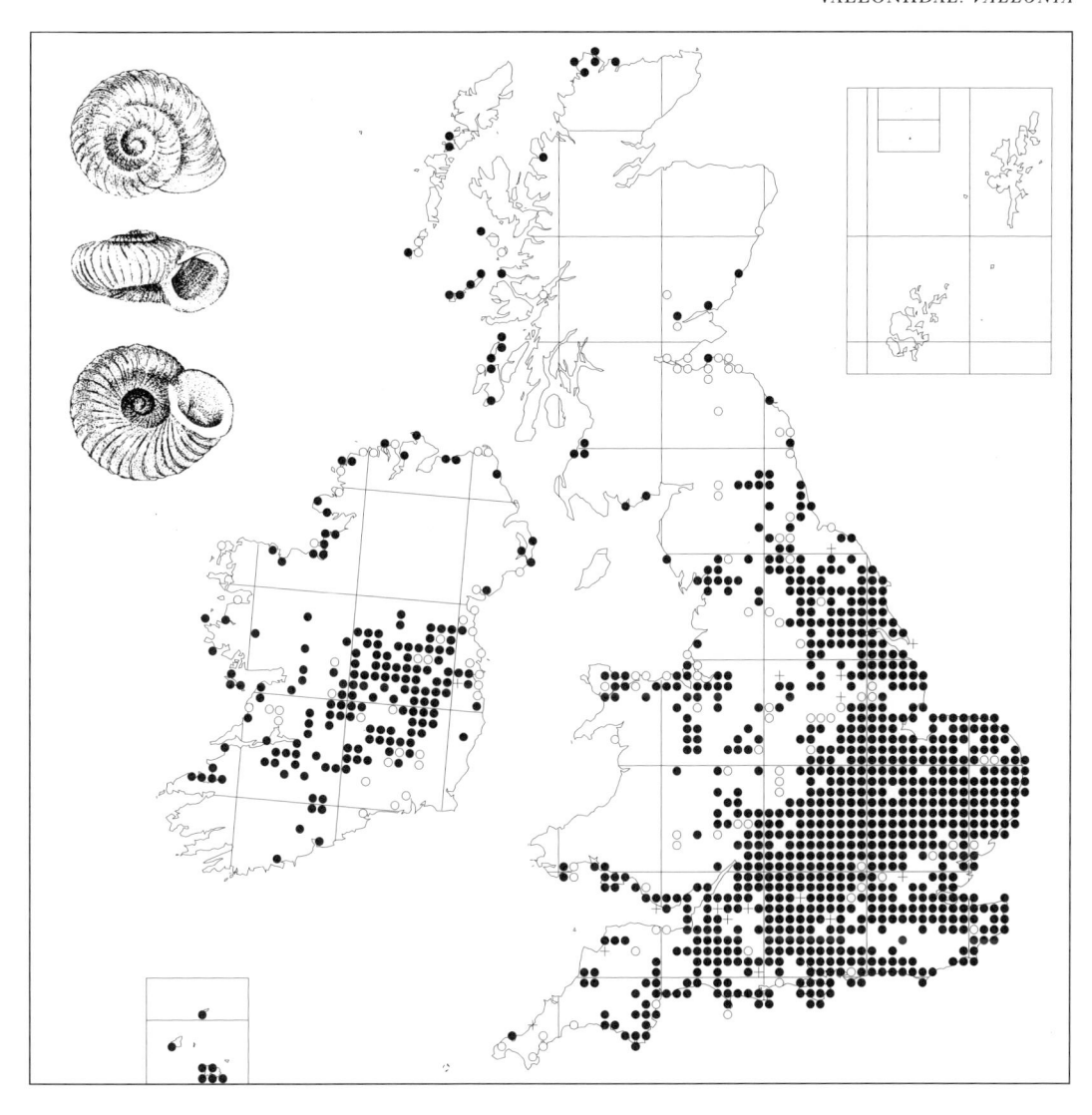

**Vallonia costata** (Müller, 1774)

Ribbed grass snail    Size: 2.0–2.7mm

This species inhabits dry, open places, mainly on chalk or limestone but also on shell-sands and other base-rich soils such as the alluvial silts of the East Anglian fenland. It is typical of short-turfed grassland, natural cliffs and screes, disused quarries and old walls. It is also not uncommon under stones and human litter in waste ground and, unlike the other *Vallonia* species, can tolerate light shading in dry, scrubby places. Occasionally it lives in marshes, where, however, it is usually replaced by *V. pulchella*.

Native (Pgl). *V. costata* shows no evidence of significant recent distributional change, though it is perhaps not as tolerant of cultivation as *V. excentrica*.

Holarctic; widespread in Europe, but becoming mainly coastal in the northern highland zone.

(Ellis: 162; K. & C.: 95)

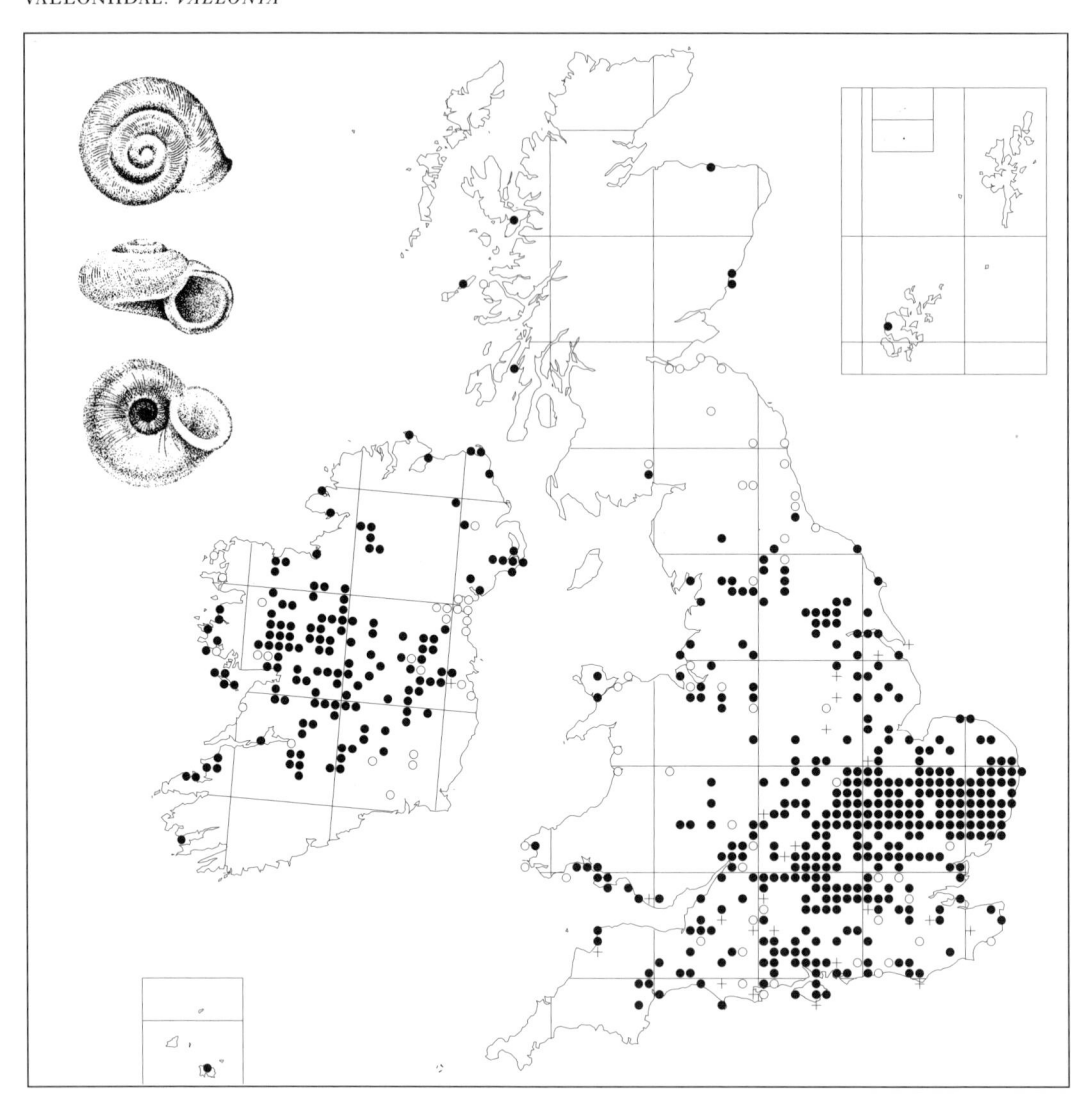

VALLONIIDAE: *VALLONIA*

**Vallonia pulchella** (Müller, 1774)

Smooth (or beautiful) grass snail   Size: 2.0–2.5mm

This is a lowland species found in grassy, base-rich places, nearly always under much wetter conditions than *V. costata* or *V. excentrica*. It is typical of water meadows on river floodplains, moist pastures, marshes, and dune slacks. Occasionally it may occur sparsely in fairly dry grassland, associated with *V. excentrica*. It is not found in woods.

Native (Lgl). This species gives little evidence of serious national decline, though certainly scarcer than formerly in many lowland areas due to the drainage of water meadows and other suitable grassy wetlands.

Holarctic; widespread in Europe, becoming rare in the highland zone of Britain and Scandinavia.

(Ellis: 162; K. & C.: 96)

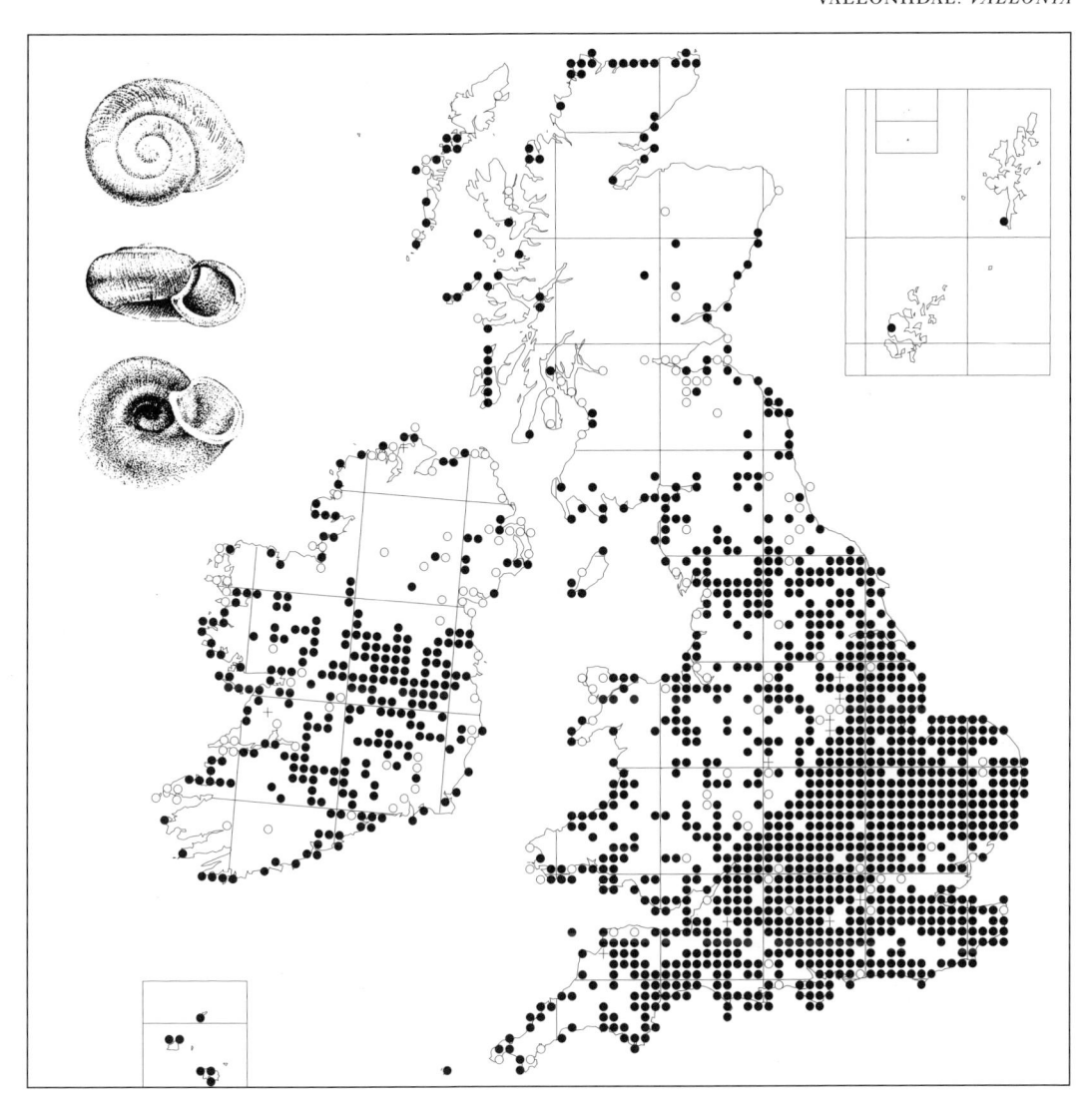

**Vallonia excentrica** Sterki, 1892 (*V. pulchella* var. *excentrica* Sterki)

Eccentric grass snail   Size: 2.0–2.2mm

A common snail of dry, open grassy places, such as roadside verges, short-turfed chalk or limestone grassland, maritime pastures and stabilized dunes. It is also found among rocks on natural cliffs and screes, in disused quarries and on walls, though in these stony places it is usually less common than *V. costata* and tends to be replaced by it. It is tolerant of less calcareous soils than *V. costata*. Unlike that species, it does not occur in shaded habitats or in bare waste ground among rubbish. Occasionally it may live in marshes, where, however, it is usually replaced by *V. pulchella*.

Native (Lgl). *V. excentrica* remains frequent everywhere within its range.

Holarctic; common throughout most of Europe but scarce beyond 60°N in Scandinavia.

(Ellis: 163; K. & C.: 96)

VALLONIIDAE: *ACANTHINULA*

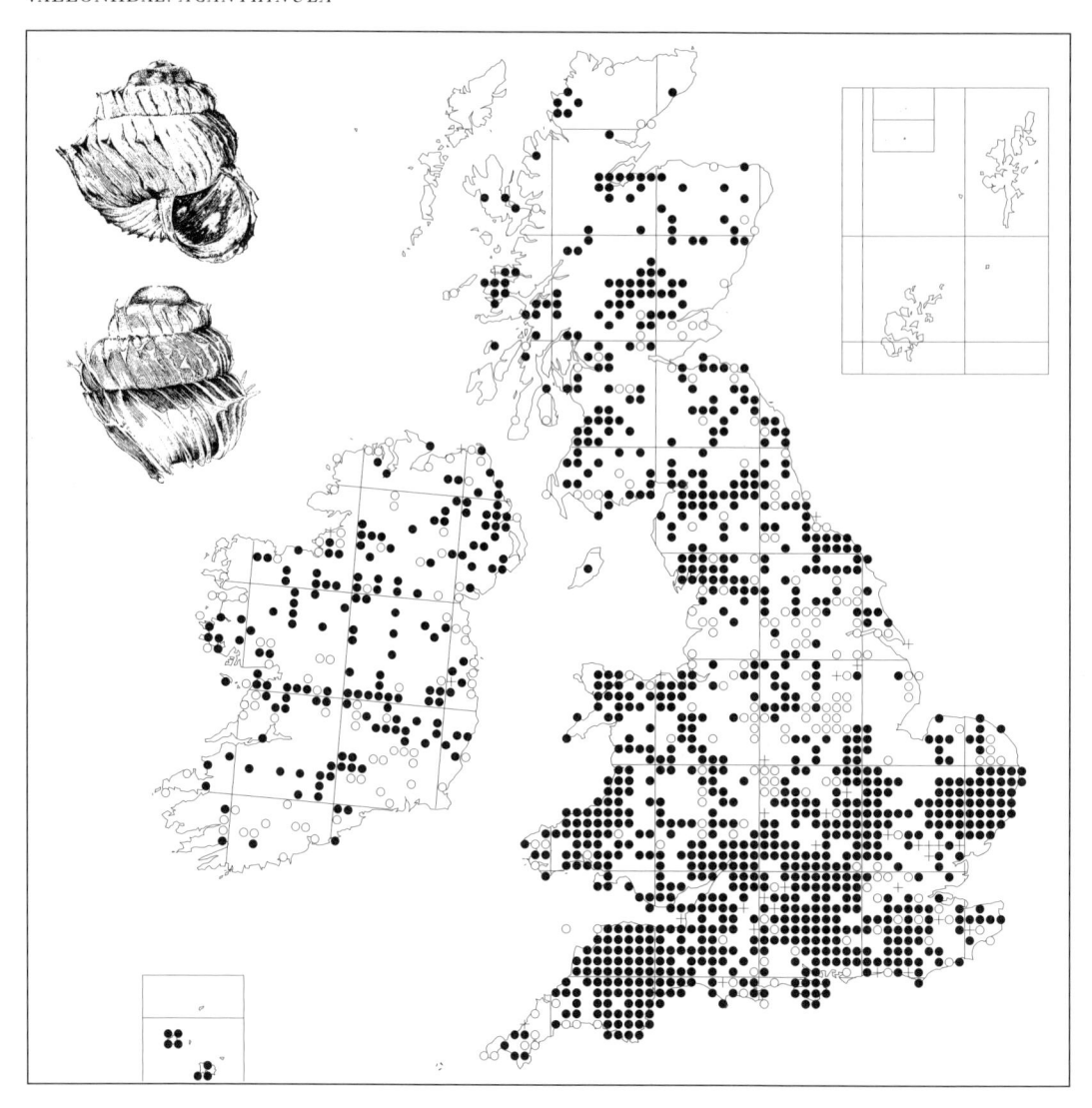

**Acanthinula aculeata** (Müller, 1774) (*Helix aculeata* Müller)

Prickly snail   Size: 2.0–2.2mm

This is a snail mainly of deciduous woodland, commonest on base-rich soils and typical of leaf litter and fallen timber in shady places. Occasionally it may live in more open situations, on hedge banks, in undisturbed, scrubby waste ground, or at the base of moist vegetation in stable grassland or at the edges of marshes.

Native (E.Pgl). *A. aculeata* shows some local recession through habitat loss, especially in intensively farmed counties like Lincolnshire where there may be a few 10km-square extinctions. In general, however, it remains frequent throughout its range. It is easily overlooked, and hence somewhat underrecorded.

W. Palaearctic; widespread in Europe, but scarcely beyond 62°N in Scandinavia.

(Ellis: 164; K. & C.: 97)

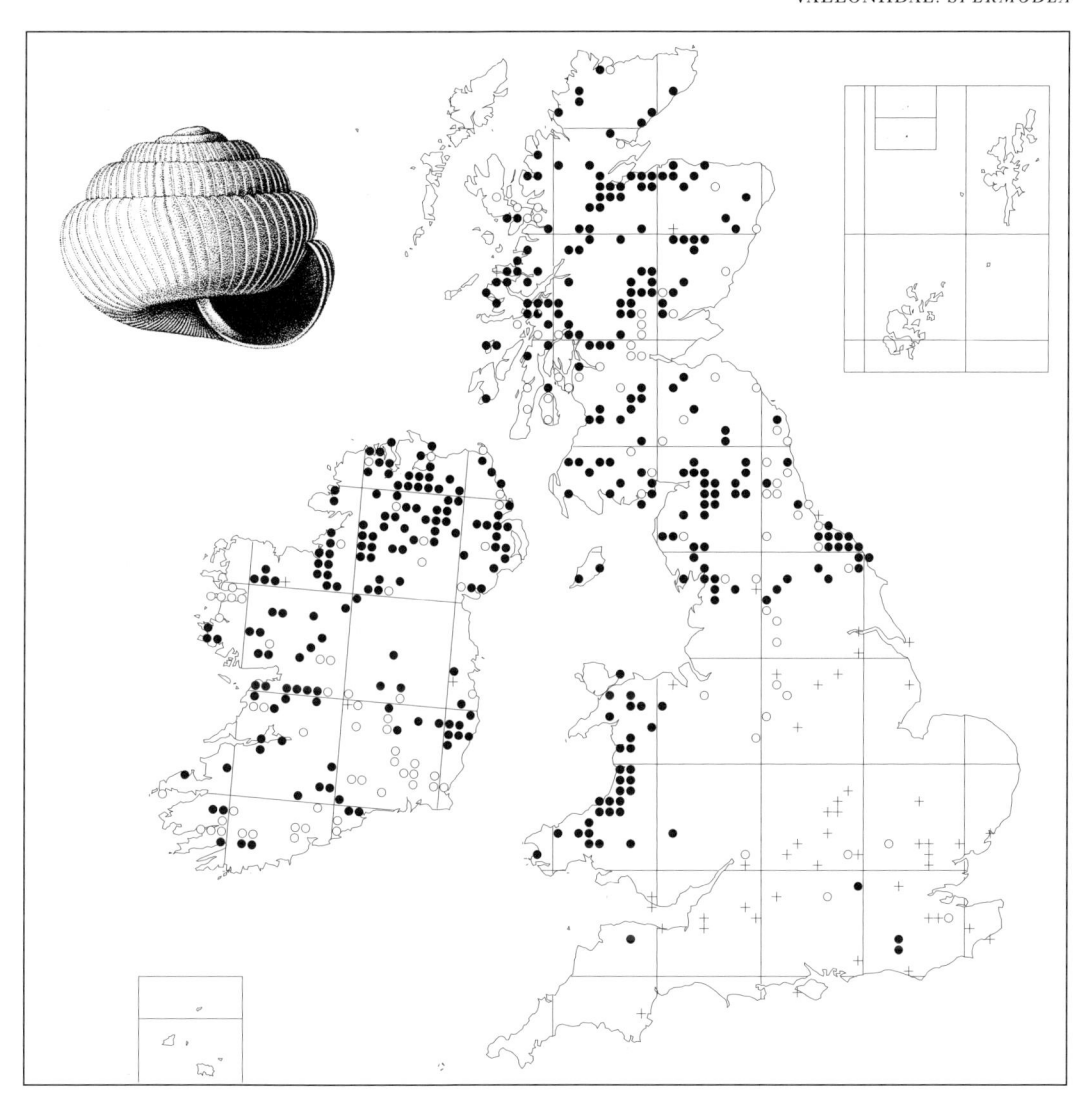

**Spermodea lamellata** (Jeffreys, 1830) (*Helix lamellata* Jeffreys, *Acanthinula lamellata* (Jeffreys))

Plated snail   Size: 2.0–2.2mm

A very local snail of old deciduous woodland, found in moist leaf litter or on fallen twigs and branches. It is a useful indicator of ancient woodland that has never been clear felled. It tolerates non-calcareous soils and is often associated with great wood-rush (*Luzula sylvatica*). Sometimes it lives in boggy ground overhung with trees, especially on the margins of floodplains. It avoids pure conifer plantations, but will otherwise live in a variety of mixed, partially replanted, or managed woods such as hazel coppices. In Scotland it has occasionally been reported under rhododendron and other exotics.

Native (E.Pgl). This is a declining species, both through human pressures and long-term climatic change. During the forest optimum of the Postglacial it was common in many parts of southern and eastern England where it is now extinct. There are relict populations in Bucks. (Burnham Beeches) and Sussex (between Balcombe and Ardingly). Other finds in central and southern England have not been confirmed in recent years (Cheshire, 1886; Berks., 1888; Derbys., 1890; Kent, 1918; Glos., 1920; Herts., 1925; Staffs., 1930). In northern Britain and Ireland numerous populations survive, though many are vulnerable, being in small woodland remnants easily destroyed by clearance or replanting by conifers.

N.W. European Atlantic: mainly British Isles and coasts of S. Scandinavia, to about 64°N in Norway.

(Ellis: 164; K. & C.: 98)

ENIDAE: *ENA*

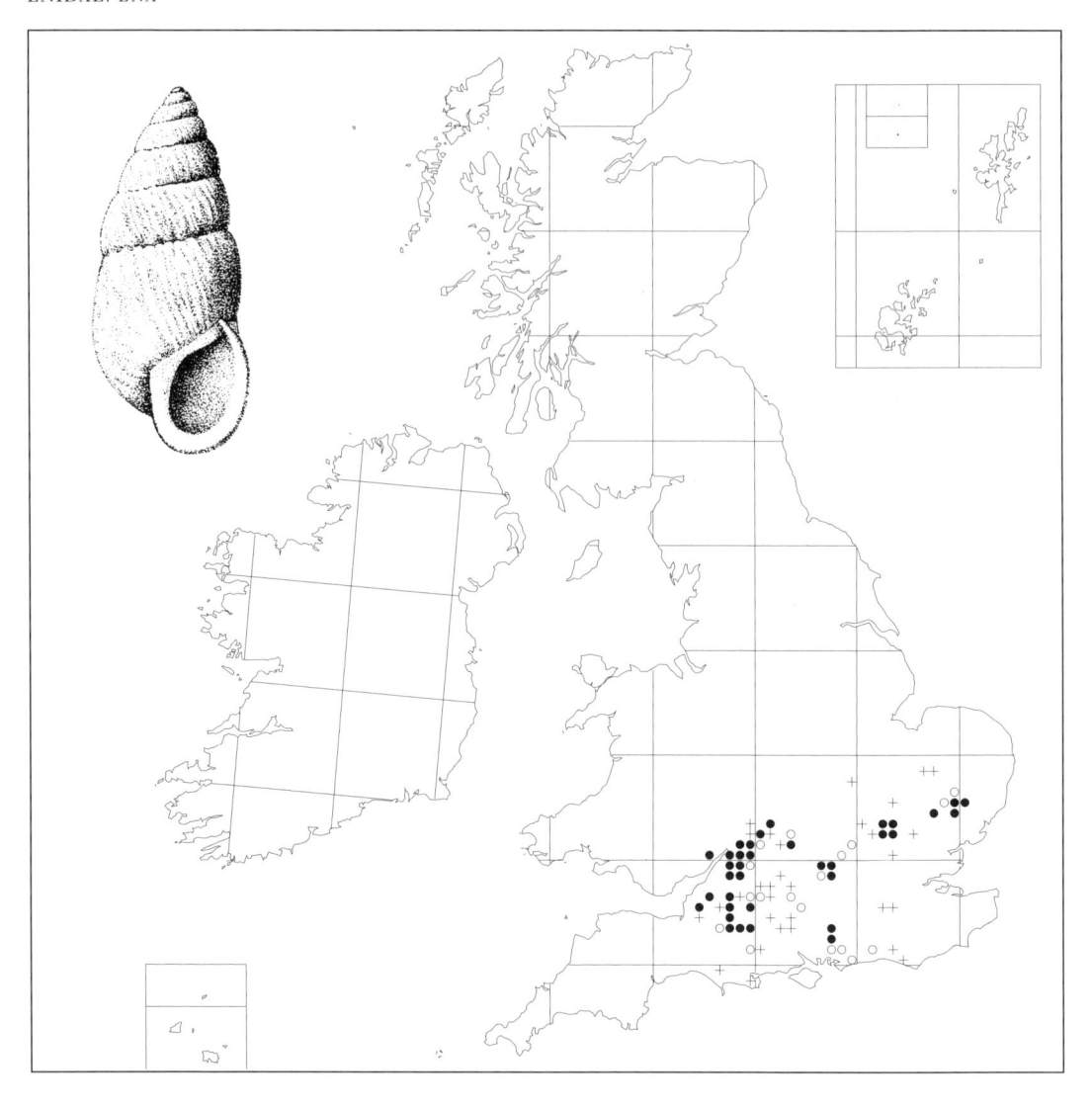

**Ena montana** (Draparnaud, 1801)

Mountain bulin   Size: 14–17mm

Principally a species of old deciduous woods on well-drained calcareous soils. It lives among ground litter and fallen timber, but in wet weather will climb a few feet up smooth tree trunks, especially of beech and ash. It also lives at the base of ancient hedgerows, and in the Mendips can be found in hazel or ash scrub among limestone rocks (Boycott, 1939). It is rarely present in any numbers, though at one anomalous site in Herts. (Buntingford) it abounds in a roadside ditch among nettles and human rubbish.

Native (E.Pgl). This species attained its widest distribution in southern England during the Neolithic and Bronze Age periods, possibly helped by higher summer temperatures. Since then it has progressively declined. Populations are now very scattered, with the greatest concentration in the Cotswolds. Many sites are reasonably secure, but some, especially in hedgerows as in East Anglia, are at risk from intensive farming (Killeen, 1992). RDB category: Rare.

Mainly C. European and Alpine, with scattered populations west to Britain and north to S. Sweden.

(Ellis: 168; K. & C.: 100)

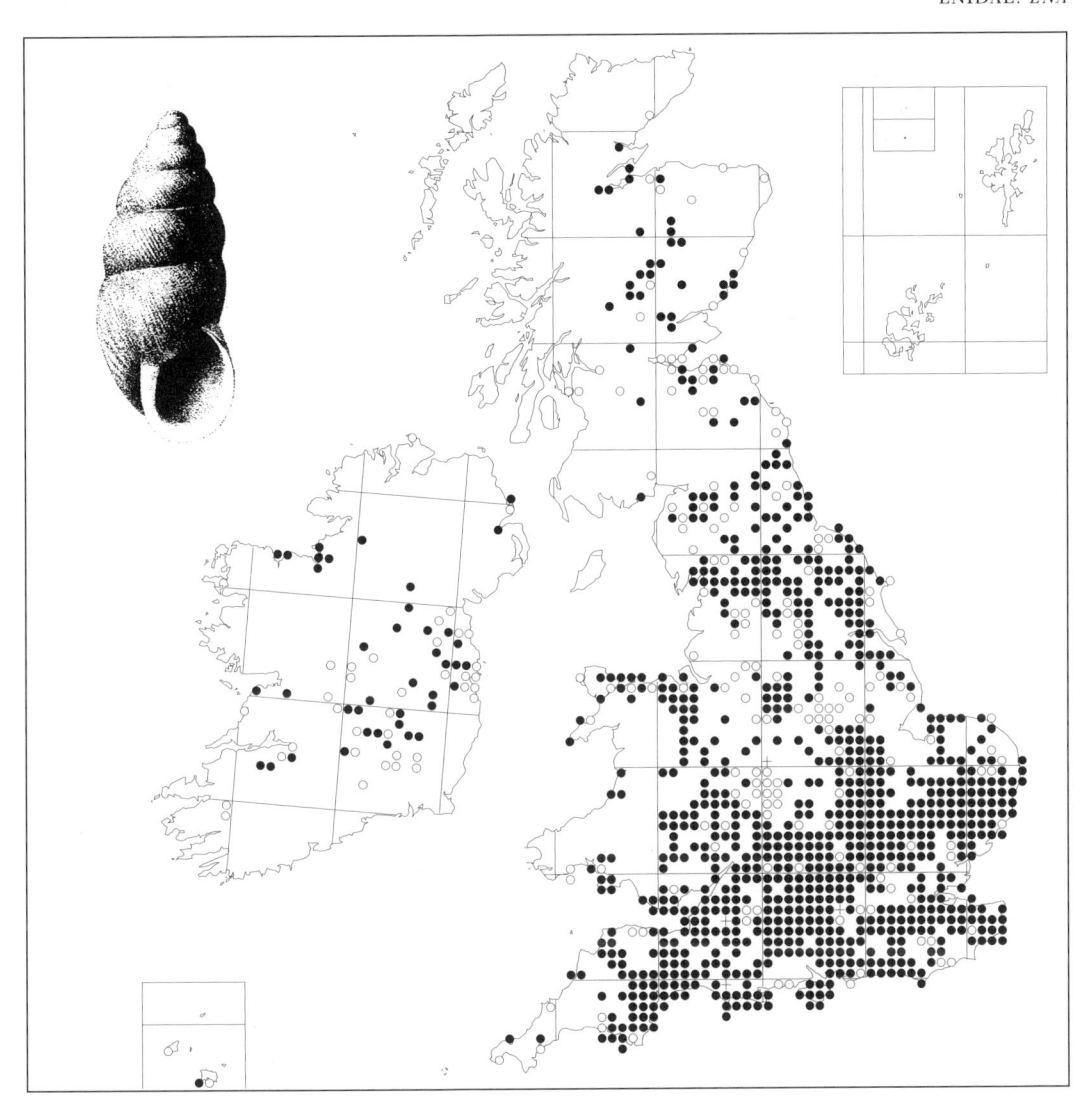

**Ena obscura** (Müller, 1774) (*Merdigera obscura* (Müller))

Lesser bulin   Size: 8.5–9.0mm

This species lives in all kinds of relatively undisturbed, shady places mainly on base-rich soils: deciduous woods, hedgerows, scrubland, the base of walls, among rocks. Normally it shelters among ground litter, but in wet weather the young especially will climb vertical surfaces on walls and tree trunks.

Native (E.Pgl). No evidence of significant national decline.

European: widespread, as far north as the British Isles and southernmost Scandinavia.

(Ellis: 169; K. & C.: 100)

PUNCTIDAE: *PUNCTUM*

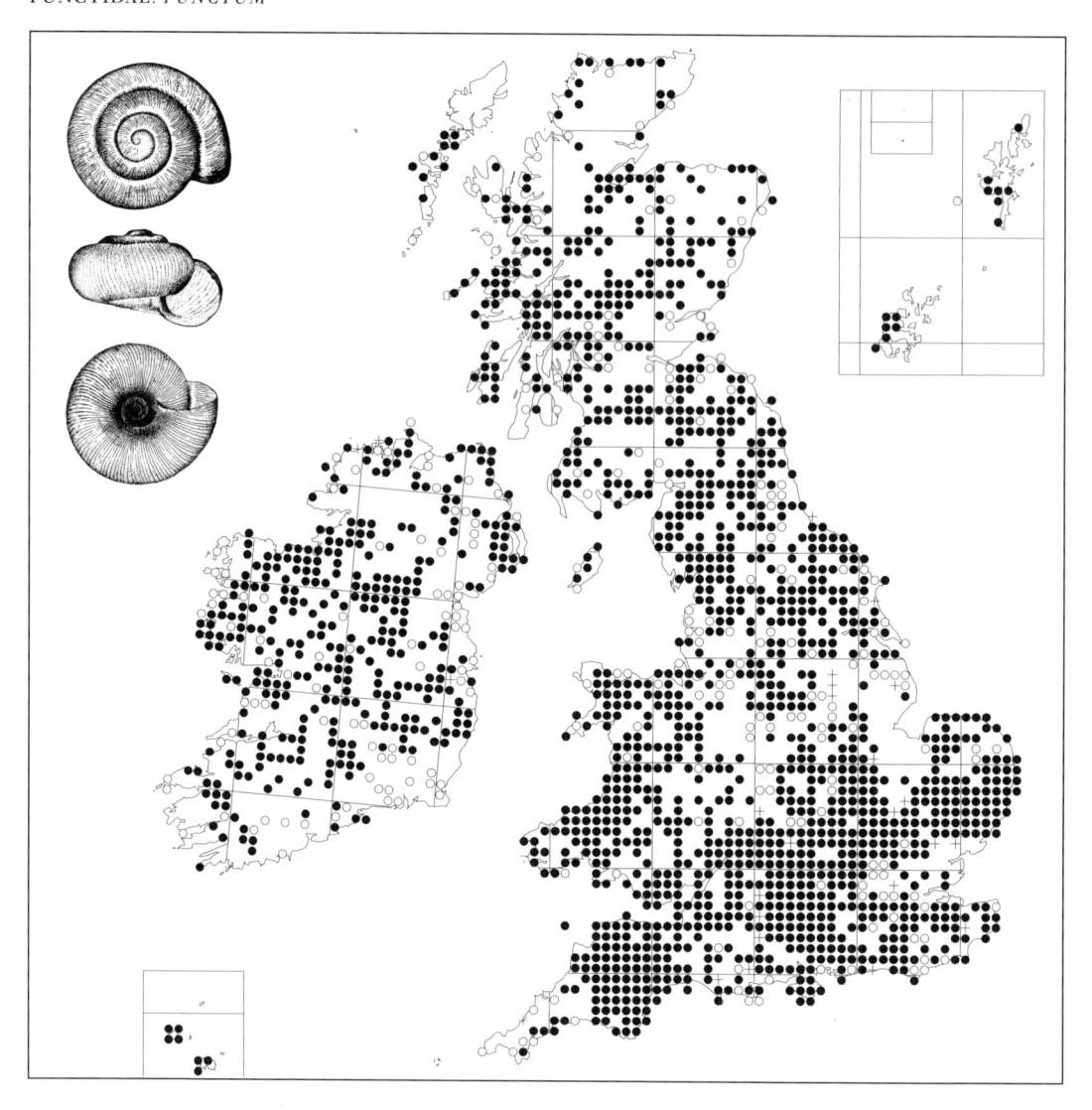

**Punctum pygmaeum** (Draparnaud, 1801)

Dwarf snail   Size: 1.2–1.5mm

A catholic species, found in a wide variety of well-vegetated places, avoiding only the driest and most exposed habitats and those heavily disturbed by man. It is particularly characteristic of leaf litter in deciduous woods, but lives also in marshes, hedge banks, stable grassland and dune slacks. It is commonest on base-rich soils, but will tolerate acid conditions.

Native (Lgl). There is no evidence of significant regional or local change. The species is under-recorded owing to its minute size. It appears genuinely absent from the Fenland basin.

Holarctic; common throughout Europe.

(Ellis: 170; K. & C.: 101)

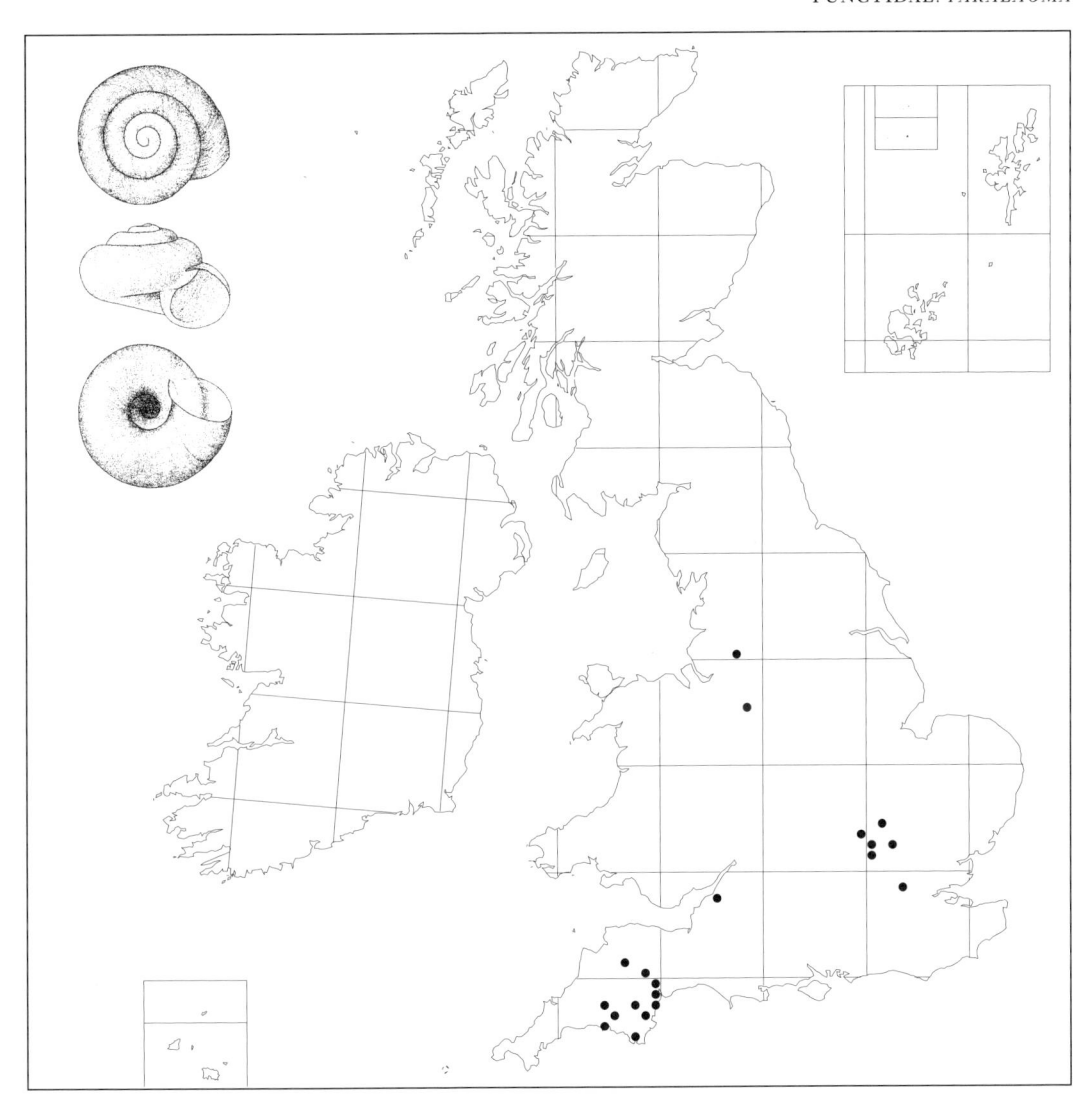

**Paralaoma caputspinulae** (Reeve, 1852) (*Punctum pusillum* (Lowe), *Toltecia pusilla* (Lowe), *Pleuropunctum micropleurum* (Paget))

Size: 1.8–2.0mm

Around the western Mediterranean this is a common snail in ground litter in moist, sheltered places. In Britain it is found in disturbed habitats such as roadside banks and gardens, and in *Salix* litter in a sandy, coastal warren (Dawlish, Devon).

Introduced. First noted in 1985 in a nursery at Luton (Beds.), living in the open in moist gravel between plant pots (Guntrip, 1986). Subsequently it was detected in nurseries, garden centres and urban waste ground as far north as Manchester. In S.W. England (Devon and Cornwall) it is now widespread and fully naturalized.

Australasian; widely distributed by man in the Mediterranean region and the Atlantic islands.

(Kerney, Cameron & Jungbluth, 1983: 136)

DISCIDAE: *HELICODISCUS*

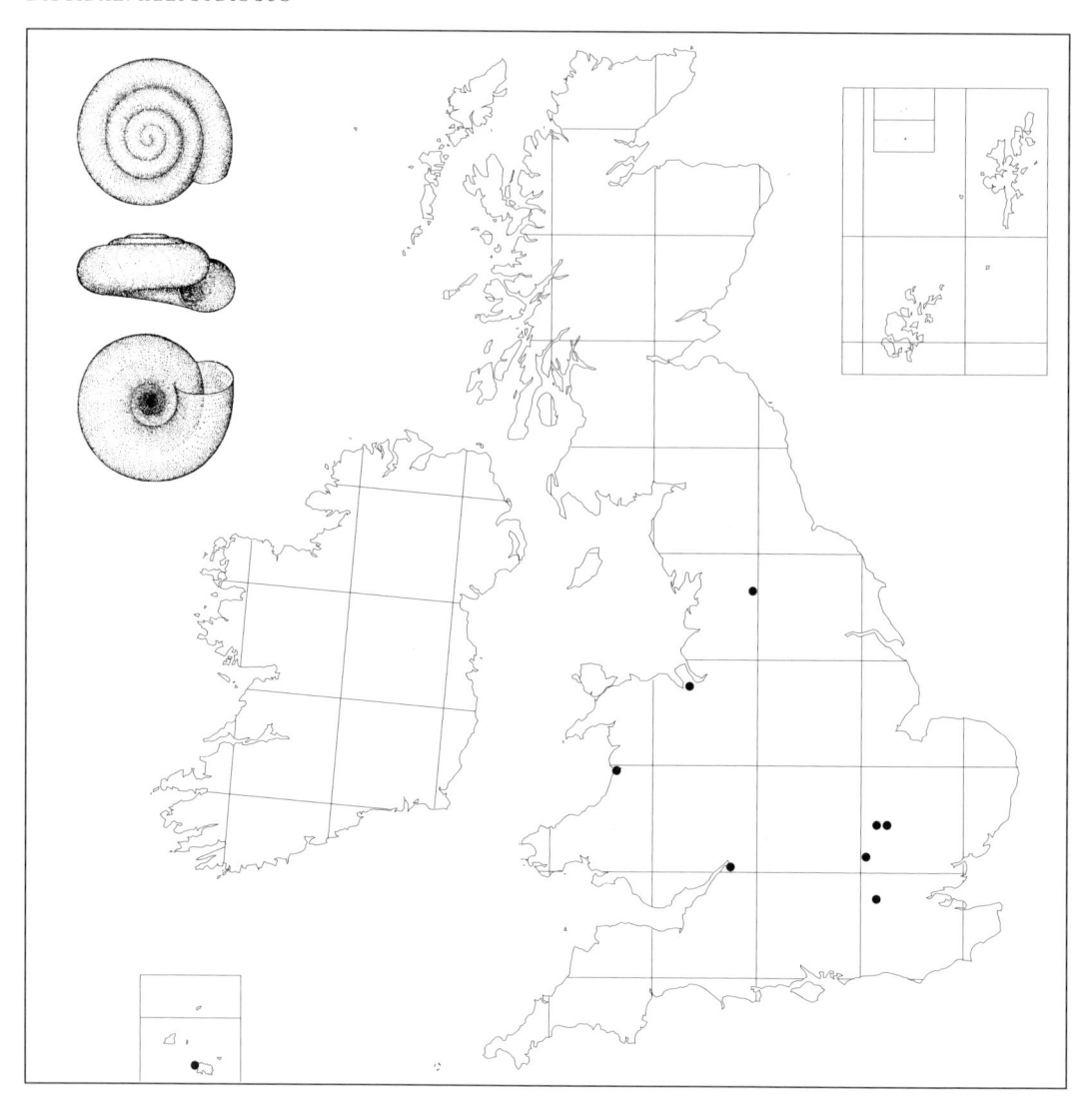

**Helicodiscus singleyanus** Pilsbry, 1890

Size: 1.8–2.5mm

This is a blind, subterranean species, living in rootlet holes and shrinkage cracks down to depths of a metre from the surface. Most British finds have been of empty shells (sometimes quite fresh) sieved from loose friable soil and ground litter at the base of walls, dry grassy banks, and similar places (Chatfield, 1977). The Gloucestershire (Fretherne) and Yorkshire (Grassington) finds are from the flood debris of rivers. It has also been found in greenhouses.

Introduced, probably from N. America. The history of this species in the British Isles remains unclear. It was first reported in 1975, and there are no fossil records. It is certainly uncommon, though doubtless underrecorded owing to its mode of life, small size, and undistinctive appearance.

Mainly N. American; in Europe reported sporadically from most central and West European countries, north to Sweden. Note that the British finds belong to the form *inermis* Baker, which some authorities regard as distinct from *singleyanus*.

(K. & C.: 101)

DISCIDAE: *DISCUS*

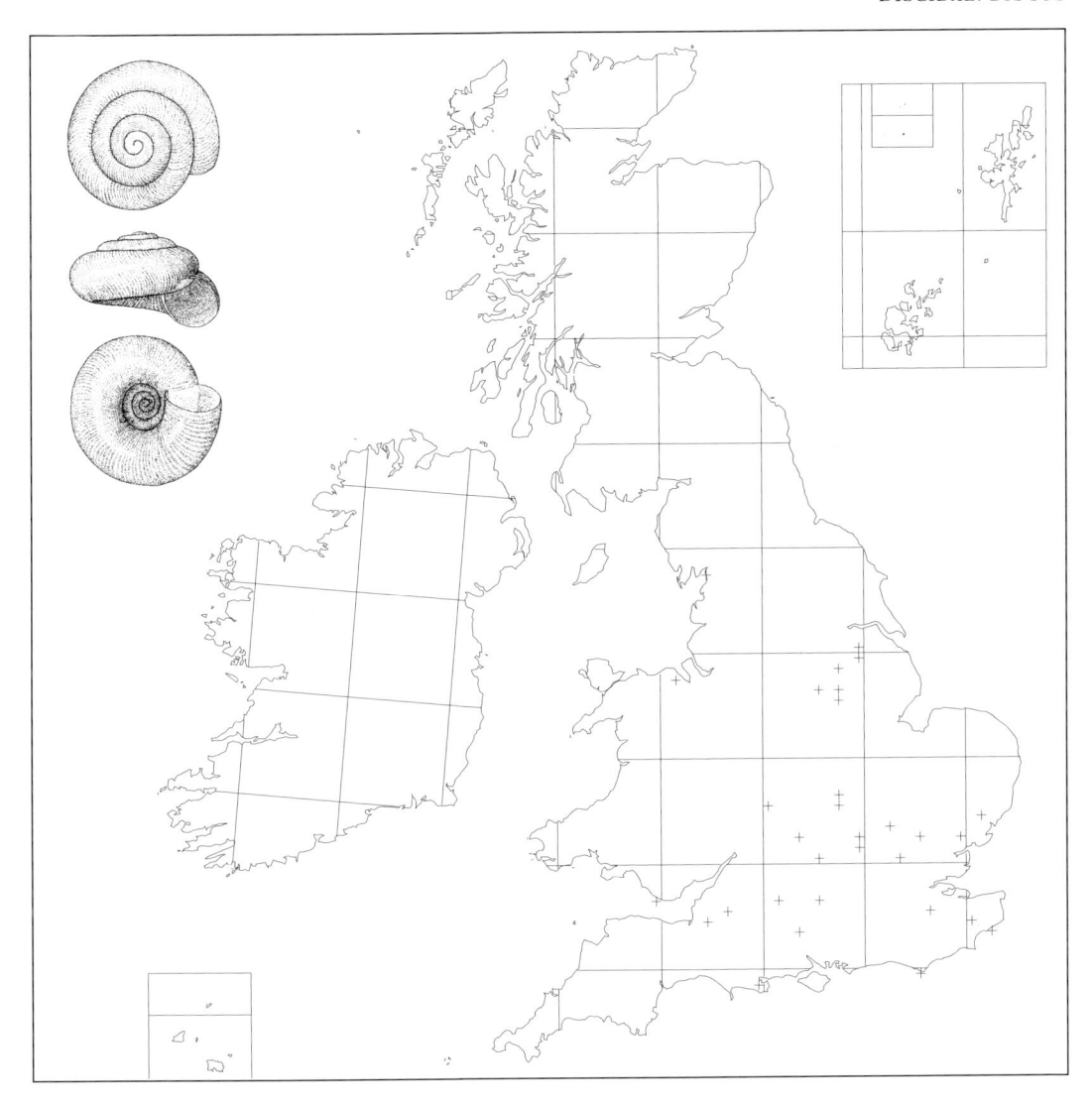

**Discus ruderatus** (Férussac, 1821)

Size: 5.5–7.0mm

Within its main northern range in Scandinavia this is a common species, living in woods (mainly birch and conifer) under ground litter and fallen timber. It is also found in more open places in marshes or in moist grassland. It tolerates poor, acidic soils.

Extinct (E.Pgl fossils only). *D. ruderatus* lived in the earliest Postglacial forests which covered southern Britain soon after the retreat of the ice (Preboreal and early Boreal periods). It apparently became extinct about 8,000 years ago, during the mid-Postglacial forest optimum, and was replaced by *D. rotundatus*.

Holarctic; in western Europe principally Scandinavian and montane (Alps, Pyrenees, Carpathians), with a few relict populations elsewhere.

(K. & C.: 102)

DISCIDAE: *DISCUS*

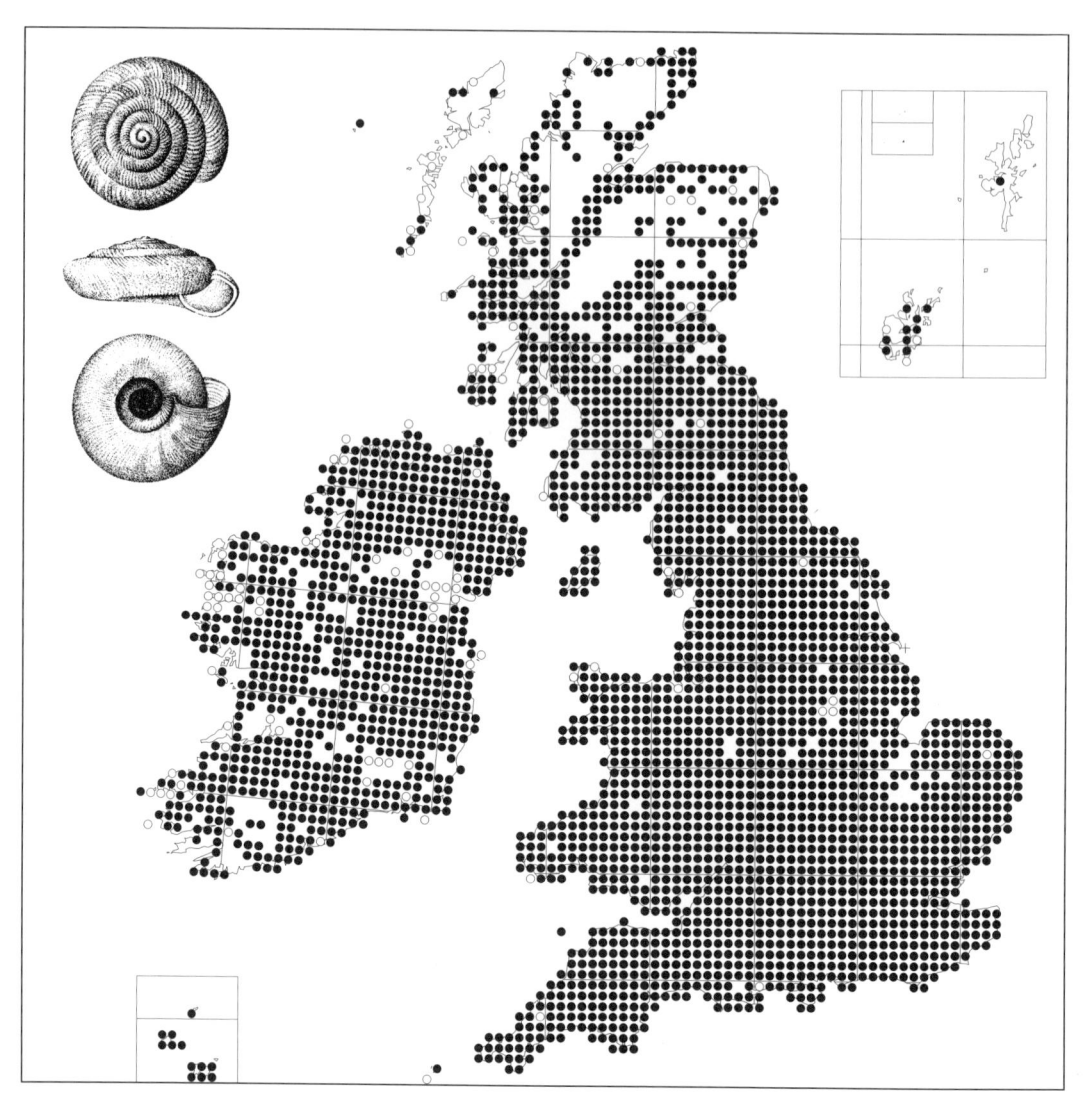

**Discus rotundatus** (Müller, 1774) (*Helix rotundata* Müller, *Pyramidula rotundata* (Müller), *Goniodiscus rotundatus* (Müller))

Rounded snail; radiated snail   Size: 5.5–7.0mm

Nearly ubiquitous in all kinds of moderately moist and sheltered places: ground litter in coniferous and deciduous woods, damp herbage, under logs and stones in waste ground, and in gardens. Though commonest on base-rich soils, it tolerates poor, acidic conditions. At the edges of its distributional (and altitudinal) range in Scotland, an association with human settlements becomes marked. It is surprisingly uncommon in parts of the East Anglian fenland, and here tends to be restricted to uncultivated refugia such as churchyards.

Native (E.Pgl). No evidence of recent significant geographical change, though probably being spread by man in parts of the highland zone.

W. and C. European, northwards to S. Scandinavia.

(Ellis: 171; K. & C.: 102)

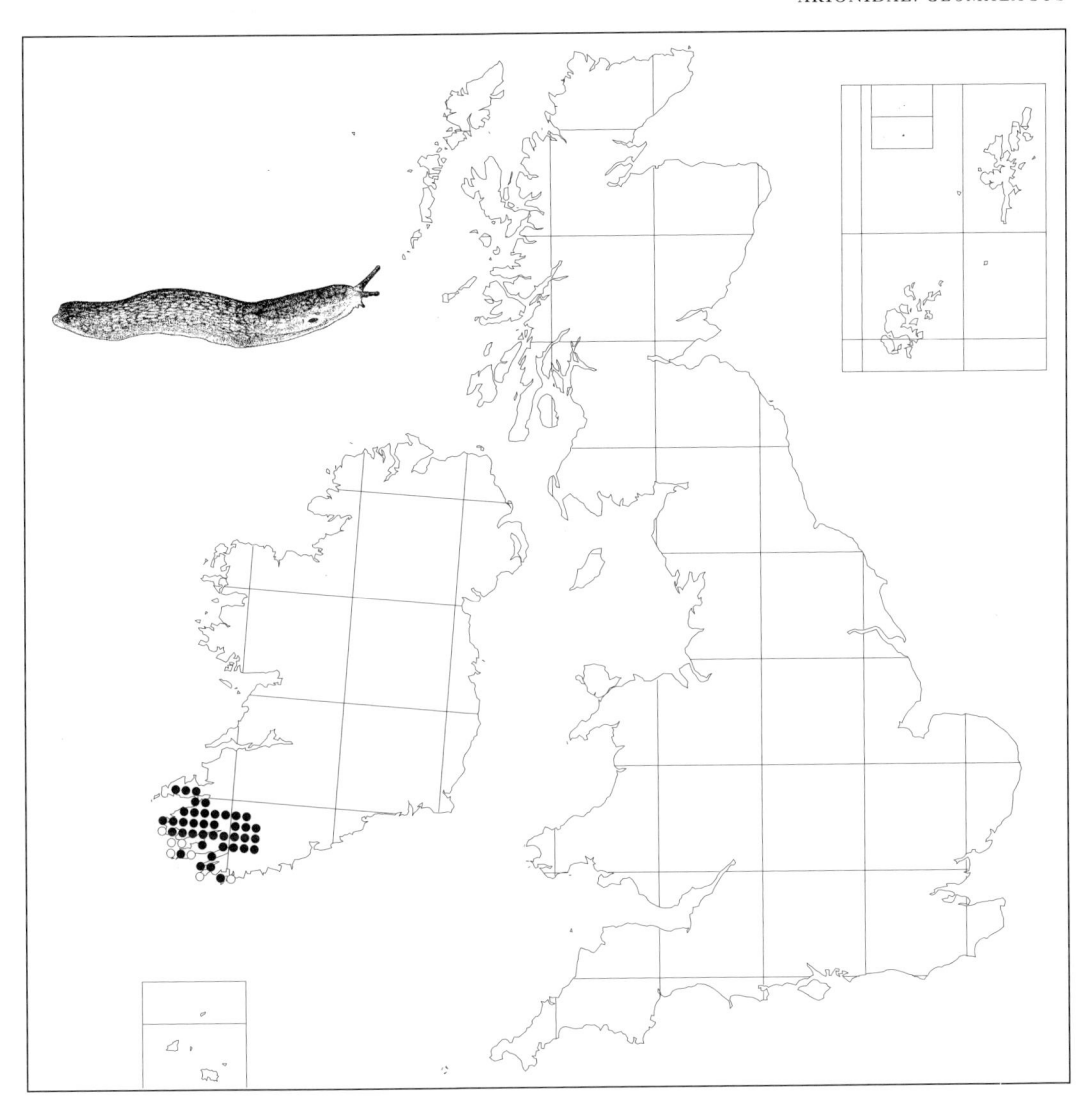

**Geomalacus maculosus** Allman, 1843

Kerry slug   Size: 60–90mm

*Geomalacus* lives among rocks in heather moorland or rough pasture, from sea-level to 300m. It feeds on lichens and algae (Boycott & Oldham, 1930). In dry weather it hides in crevices or under carpets of moss and may be difficult to find, even in places where it is common. More rarely it lives in oak woods: on the trunks of trees, under fallen timber covered in moss and lichens, or among boulders (Zoer & Visser, 1972; Platts & Speight, 1988). Most of its habitats are on acid Devonian sandstones (Old Red Sandstone) on which it is associated with few other molluscs apart from slugs.

Introduced? The presence of this Lusitanian species in Ireland is an old biogeographical puzzle: is *Geomalacus* a preglacial survival in an ice-free refugium, a natural postglacial immigrant, or a relatively recent, accidental introduction? Opinion is now veering towards the last of these hypotheses. The species is common over a considerable area and there is no evidence that it is declining.

Lusitanian: S.W. Ireland and Iberian peninsula only.

(Ellis: 174; K. & C.: 103)

ARIONIDAE: *ARION*

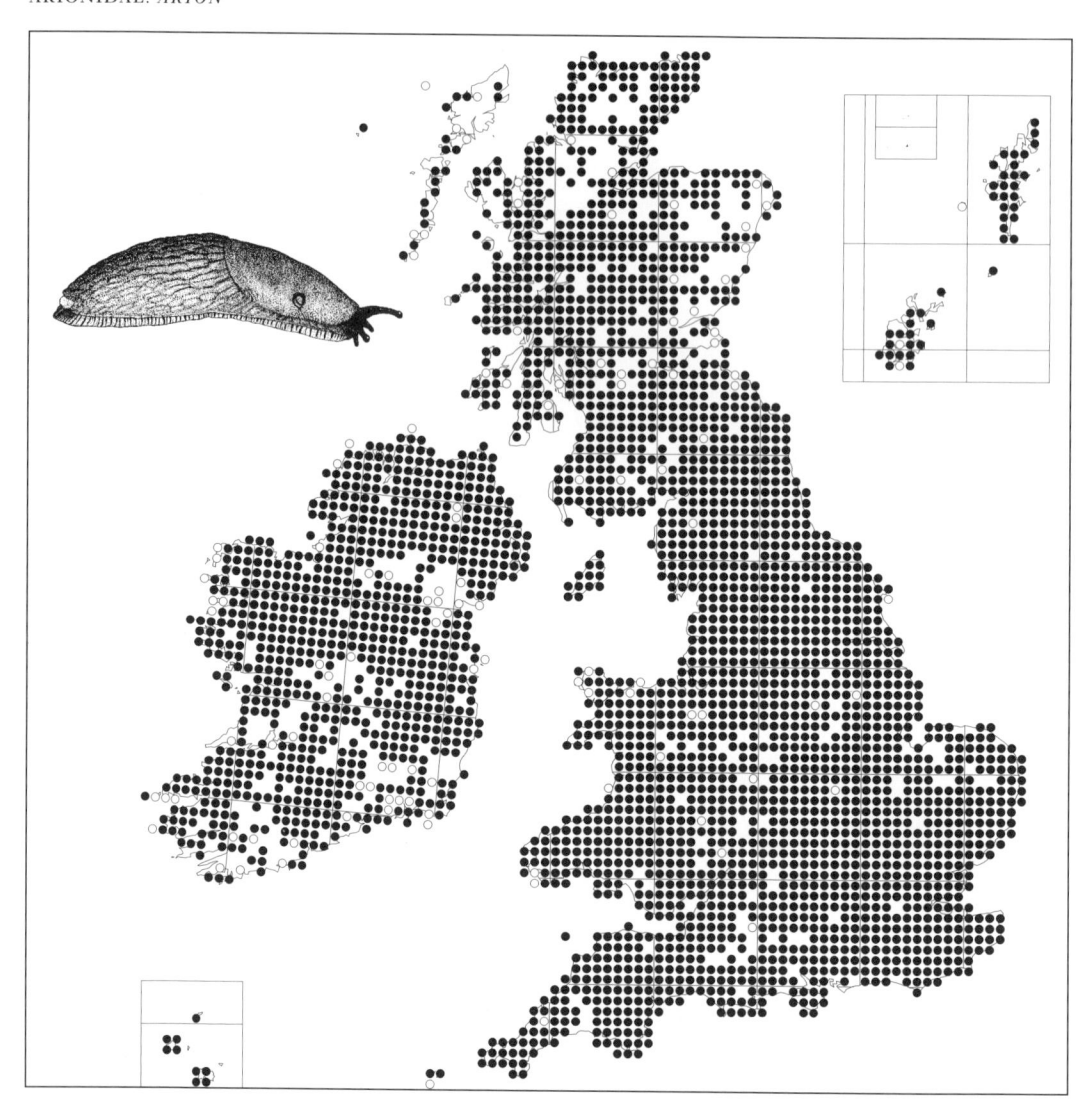

### Arion ater (L., 1758)

Large black slug; black snail (in Scotland)   Size: 60–130mm

This is a catholic species, found in all kinds of moderately damp places: fields, roadsides, woods, waste gound and gardens (though not usually in small urban gardens). It is frequent in acid moorland devoid of all other molluscs, and extends to high altitudes in the Scottish and Welsh mountains.

Probably native. Common nearly everywhere in suitable places; no evidence of geographical change.

N.W. European, in Scandinavia becoming coastal beyond about 61°N. Similar forms in C. Europe and Iberia may be separate species or subspecies.

Included here are the *A. ater* (L.) and *A. rufus* (L.) of Quick (1947). Though normally separate and often regarded as distinct taxa, these appear to have interbred in Britain. The British *rufus* form (which is often black with an orange fringe and a creamy sole) is commonest in lowland and man-made habitats mainly in southern Britain and Ireland, while typical *ater* is characteristic of wilder, harsher places mainly in the north and west. The map probably also includes some records of *A. lusitanicus*.

(Ellis: 181; K. & C.: 104)

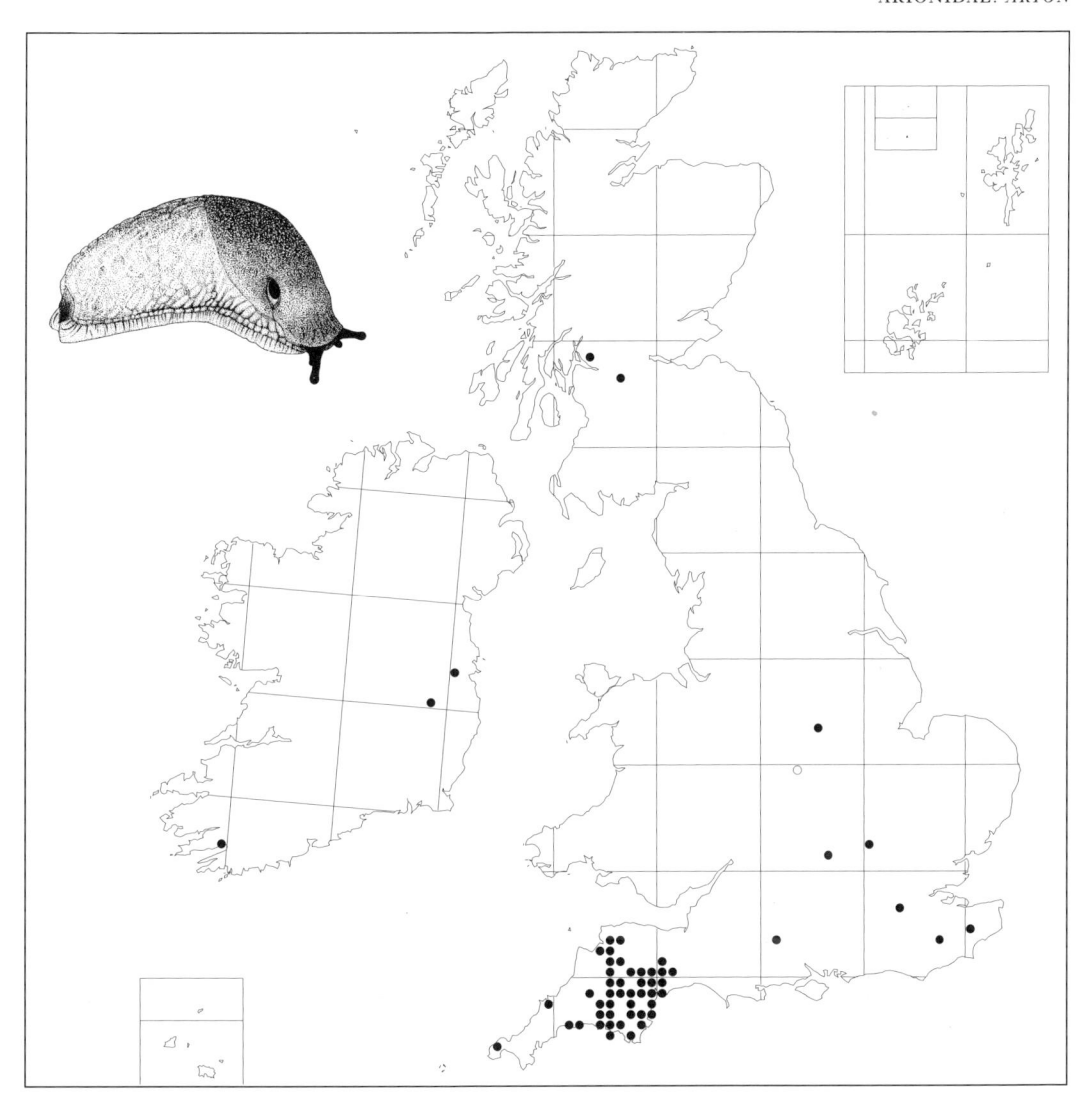

**Arion lusitanicus** Mabille, 1868

'Lusitanian' slug   Size: 60–130mm

Known occurrences include damp, sheltered places in gardens, roadsides, waste ground, and a cemetery (Glasgow Necropolis). Most habitats are more or less disturbed.

Introduced? This species shows signs of association with disturbance and is possibly a relatively modern introduction into Britain. Though seriously underrecorded it is undoubtedly local. The earliest certain record dates from 1954 (garden at Attleborough, Warwicks.). Rapidly increasing (colonizing) populations have been recognized at some sites in recent years and the species is probably spreading.

Distribution as yet poorly known (seemingly not Lusitanian). France; also introduced in many places in C. Europe and in Sweden.

It should be noted that older British records of '*A. lusitanicus*' (e.g. Quick, 1952, 1960) refer mainly to *A. flagellus* (Davies, 1987). The name of this species may need to be changed, as the British slug seemingly differs from that present at Mabille's type locality for *lusitanicus* in Portugal (Castillejo, 1998).

(Davies, 1987)

ARIONIDAE: *ARION*

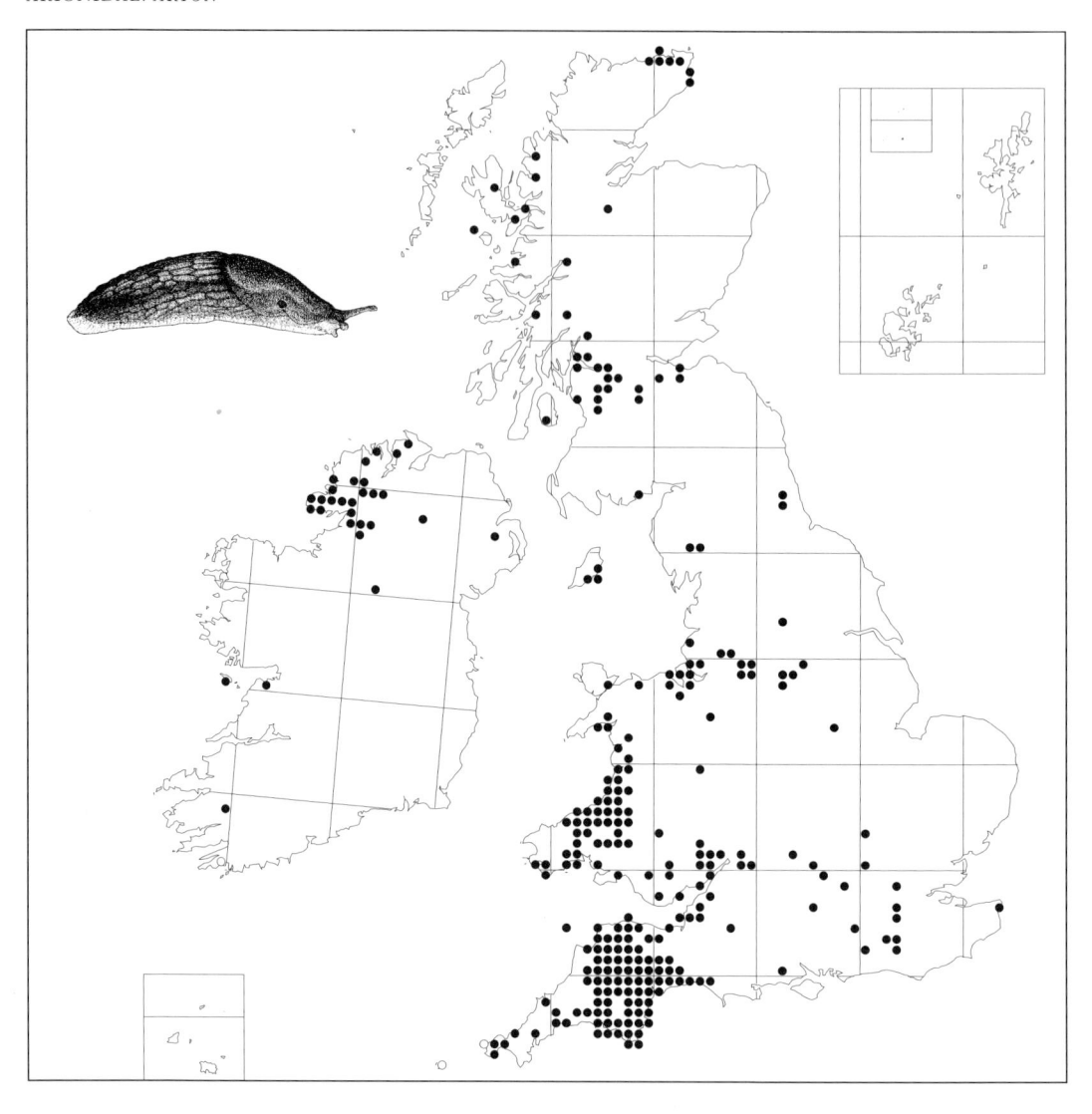

**Arion flagellus** Collinge, 1893 (*A. lusitanicus* of Quick, 1952, *non* Mabille)

Durham slug   Size: 60–100mm

This slug has been found in a wide variety of moderately damp, sheltered places, both wild and disturbed: gardens, waste ground, sea cliffs, rocky moorland and deciduous woods.

Possibly native. Though it remains underrecorded, it is certainly widespread in many areas, especially in western Britain, and appears either native or long naturalized. Some conspicuous, isolated populations (especially in gardens) suggest recent colonization.

As yet, known with certainty only from the British Isles.

*A. flagellus* was described as a new species from Schull, Co. Cork, in 1893. Subsequently it was synonymized first with *A. subfuscus*, and later with *A. lusitanicus*, from which it has only recently been distinguished (Davies, 1987).

(K. & C.: 104 (as '*lusitanicus*'); Davies, 1987)

ARIONIDAE: *ARION*

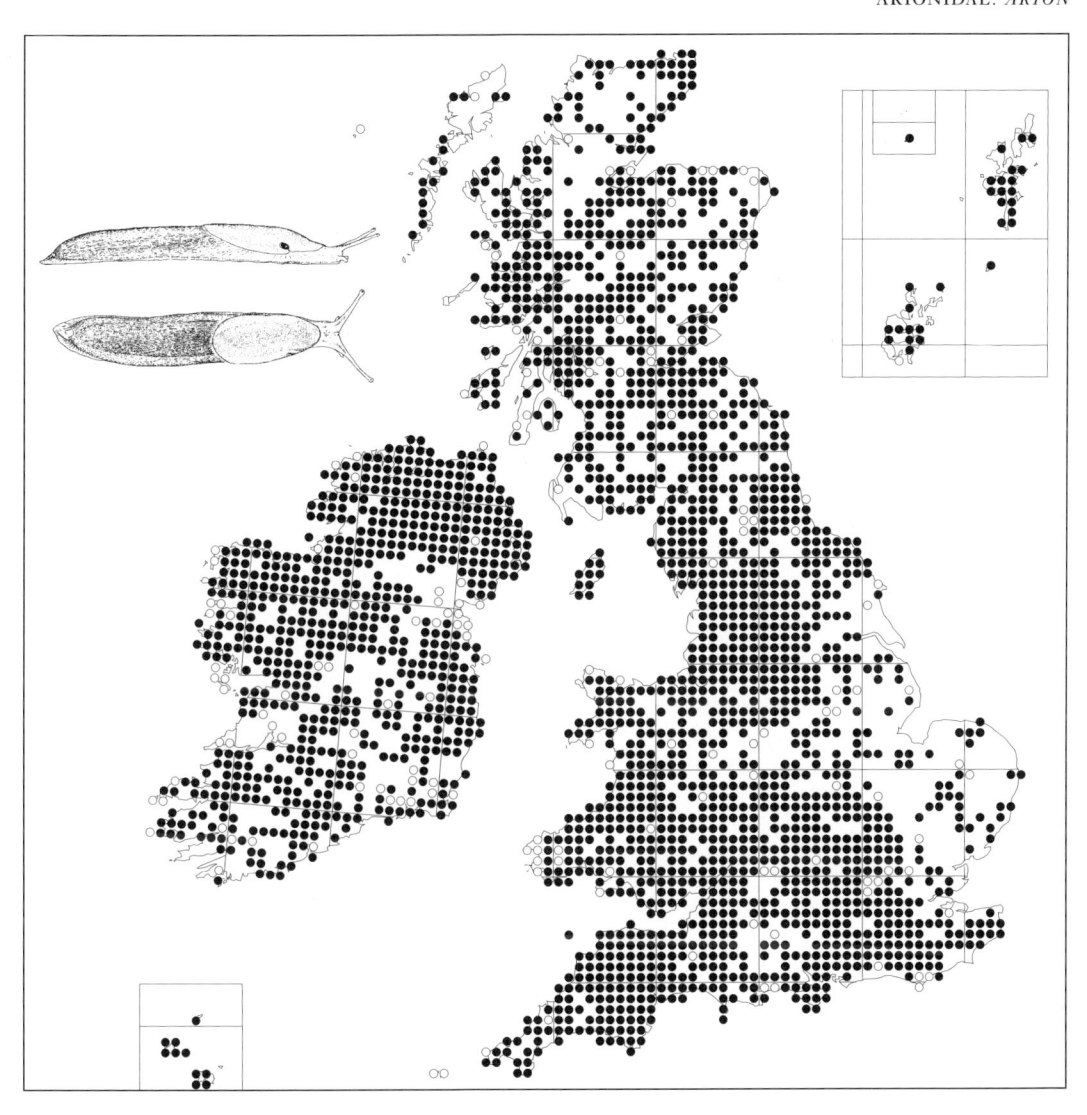

**Arion subfuscus** Draparnaud, 1805

Dusky slug   Size: 50–70mm

In northern and western areas this is a species common among herbage and moist ground litter in all kinds of sheltered places: deciduous and coniferous woods, roadsides, waste ground, gardens. In eastern England, and especially in East Anglia, it is much scarcer and restricted mainly to old woods.

Probably native. No evidence of recent distributional change. Its rarity in East Anglia was commented on nearly a century ago.

European: common nearly throughout. Well established in N. America, and perhaps native there.

(Ellis: 180; K. & C.: 104)

123

ARIONIDAE: *ARION*

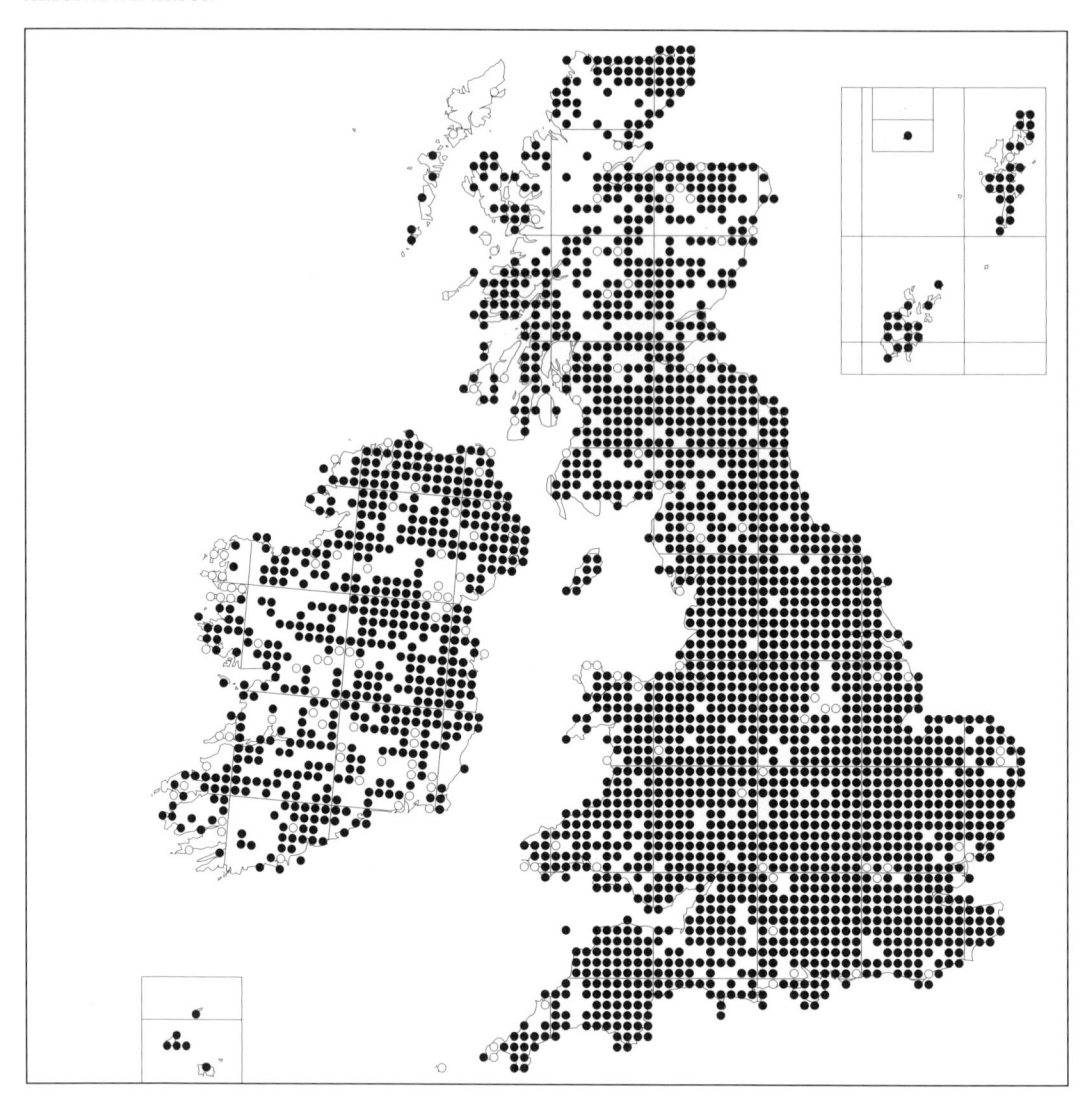

**Arion circumscriptus** aggregate (*A. fasciatus* of Quick, 1960, *non* Nilsson)

Bourguignat's slug

This map shows all records of the two closely allied species *A. circumscriptus* Johnston and *A. silvaticus* Lohmander. These were not distinguished in Britain until the mid 1960s and remain imperfectly recorded. A few records of the true *A. fasciatus* (Nilsson) may also be included.

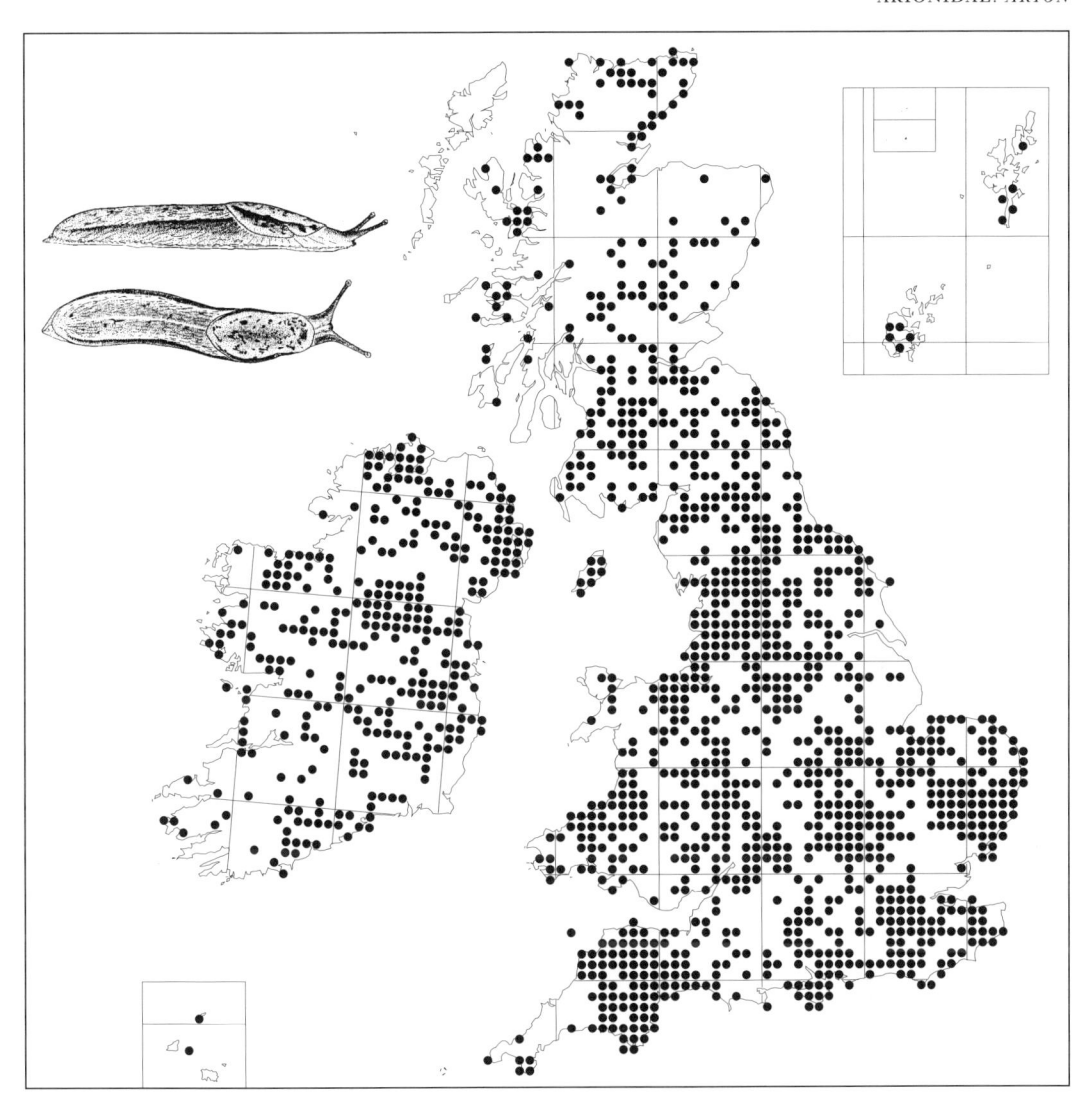

**Arion circumscriptus** Johnston, 1828

Size: 30–40mm

Found in moist, sheltered places of all kinds, under litter in woods, by roadsides, in waste ground, occasionally in gardens. It is mainly a lowland species and less common than the closely related *A. silvaticus* in exposed, acid, upland terrain.

Probably native. This species is likely to be common throughout the British Isles, though still underrecorded owing to confusion with related species. No evidence of distributional change.

European: imperfectly known, but possibly throughout except in C. and N. Scandinavia.

(K. & C.: 105)

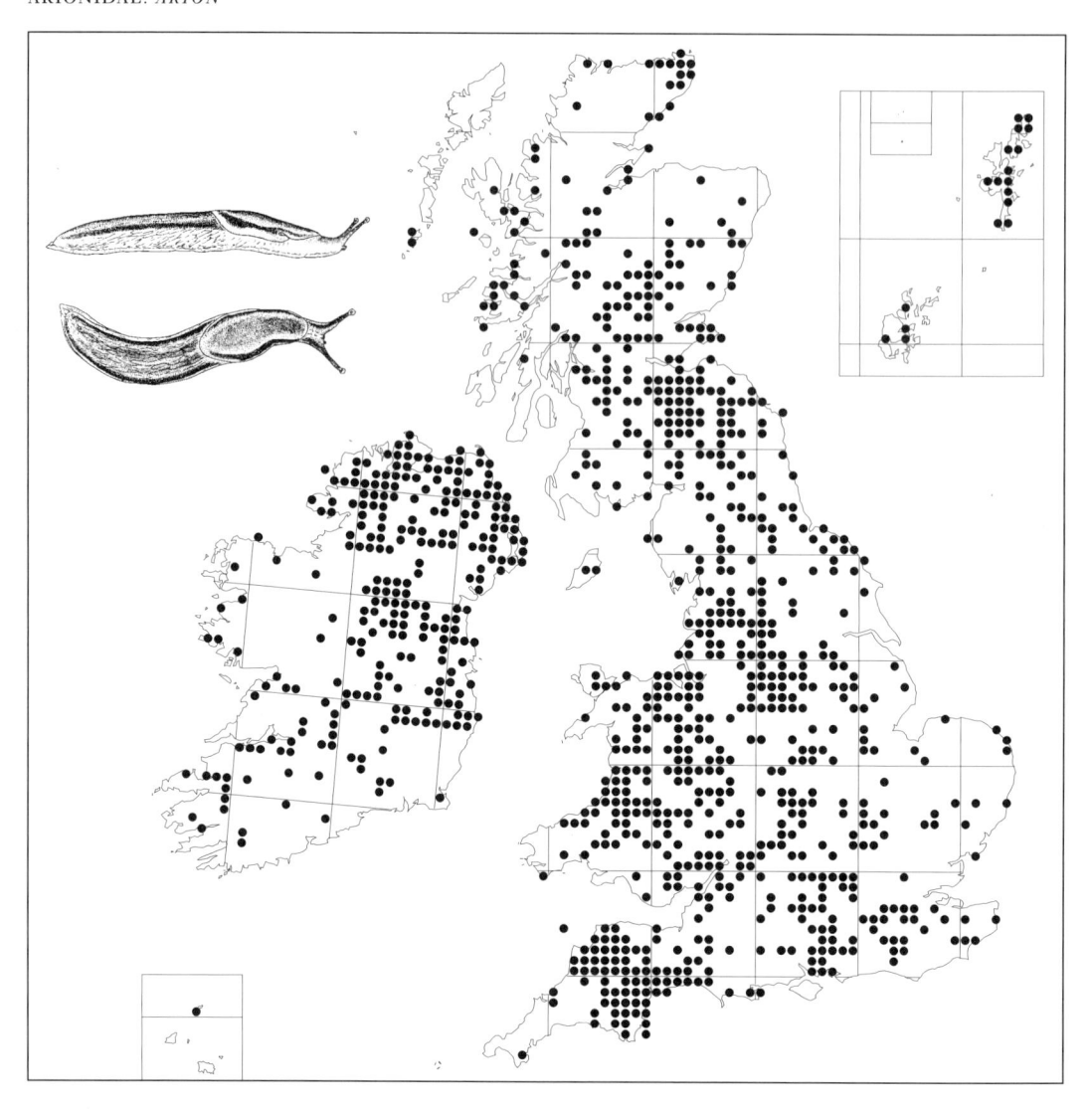

## Arion silvaticus Lohmander, 1937

Size: 30–40mm

This species lives in moist, sheltered places of all kinds, in herbage and under ground litter. It is a slug tolerant of poor, non-calcareous habitats, such as acid woodland, heathland, sea cliffs and mountains. In some lowland areas, notably in eastern England, it is much scarcer than *A. circumscriptus* but the two species are not infrequently associated.

Probably native. Still somewhat underrecorded due to confusion with *A. circumscriptus*, from which it was first distinguished in the British Isles only in the 1960s.

European: poorly known, but probably widespread, except in N. Scandinavia. Also Iceland and the Faroe Islands.

K. & C.: 105

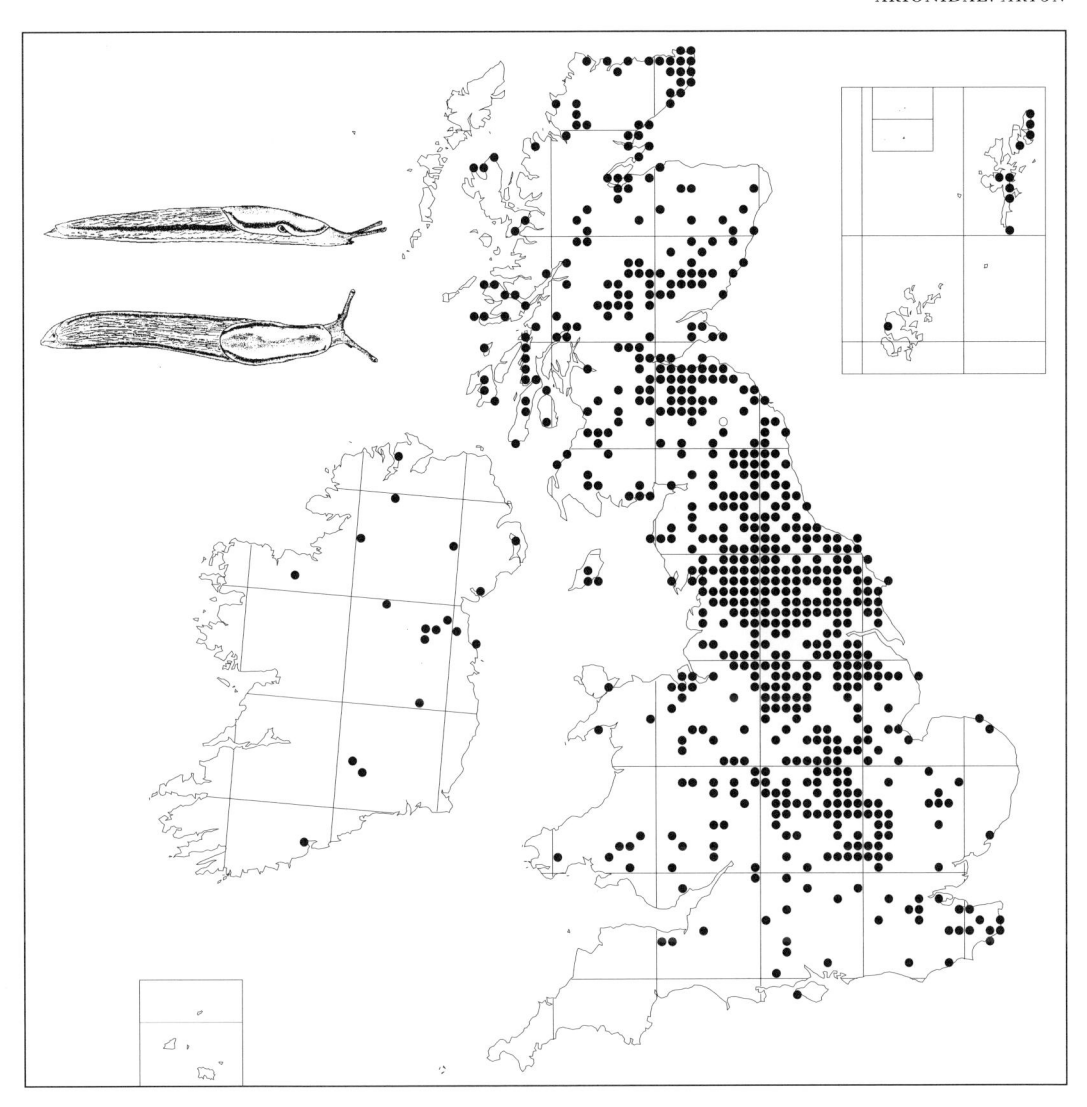

## Arion fasciatus (Nilsson, 1823)

Size: 35–45mm

Occurs among herbage and ground litter in moist, sheltered places of all kinds. It is most typical perhaps of disturbed habitats, as under rubbish in waste ground, but is also to be found in wild places, especially within its main area of distribution in the north of England.

Probably native. Though first recognized in the British Isles only in 1962 and still imperfectly recorded, *A. fasciatus* may be spreading, particularly in southern Britain and in Ireland where an association with human disturbance is most marked.

N. European: exact distribution still uncertain.

Note that old records of '*A. fasciatus*' (e.g. Quick, 1960) refer mostly to *A. circumscriptus* or to *A. silvaticus*, and not to the present species.

(K. & C.: 105)

ARIONIDAE: *ARION*

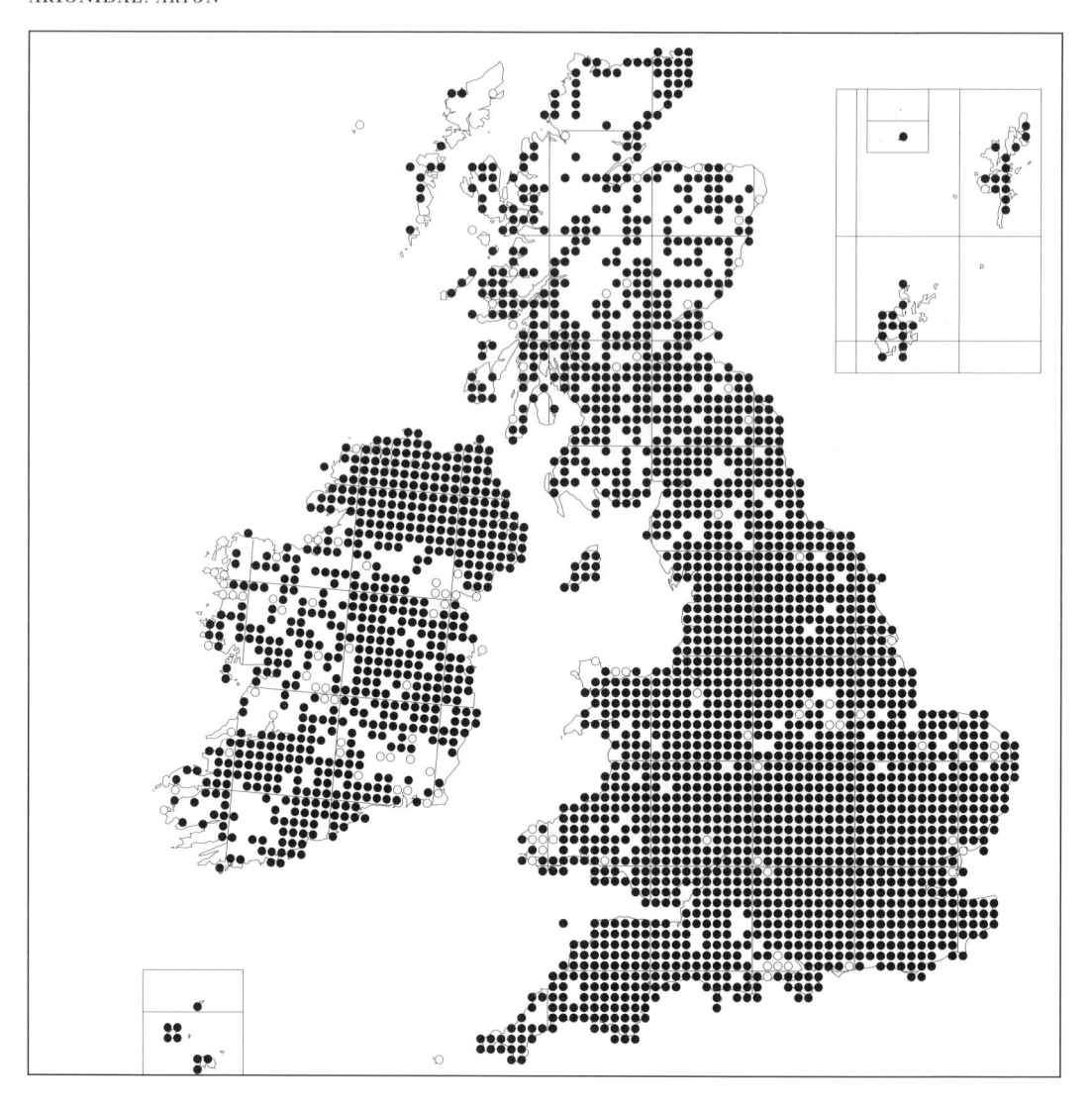

**Arion hortensis** aggregate

Garden slug

This map shows all records of the *Arion hortensis* group (*A. hortensis* Férussac + *A. distinctus* Mabille + *A. owenii* Davies). The separate species were not distinguished prior to the 1970s and remain inadequately recorded.

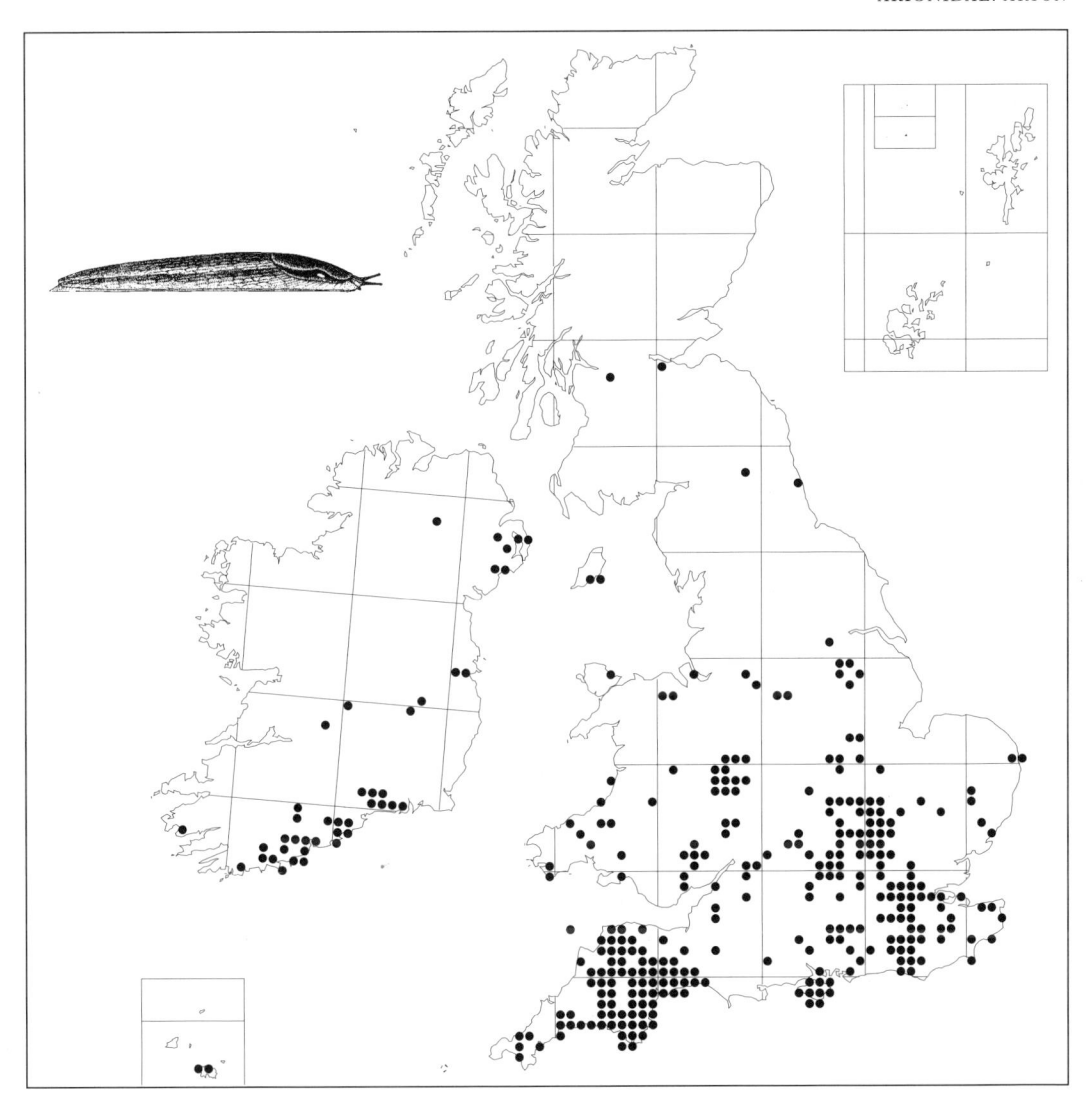

**Arion hortensis** Férussac, 1819 ('*A. hortensis* species R' of Davies, 1977)

Size: 25–35mm

*A. hortensis*, sensu stricto, lives among herbage and ground litter and in soil crevices, both in natural and humanly disturbed habitats. It can be a garden and agricultural pest. In northern Britain the few known sites are all in gardens or waste ground. It is often associated with *A. distinctus* (though, typically, maturing at different seasons).

Probably native. Though underrecorded, this species is evidently locally common in wild habitats in the south of England and Ireland. Further north there is evidence that it is being spread by man.

W. and S. European, but exact Continental distribution as yet uncertain. Rare or absent in Scandinavia.

Note that the majority of old records of '*Arion hortensis*' refer to *A. distinctus*, the following species.

(Davies, 1977; 1979)

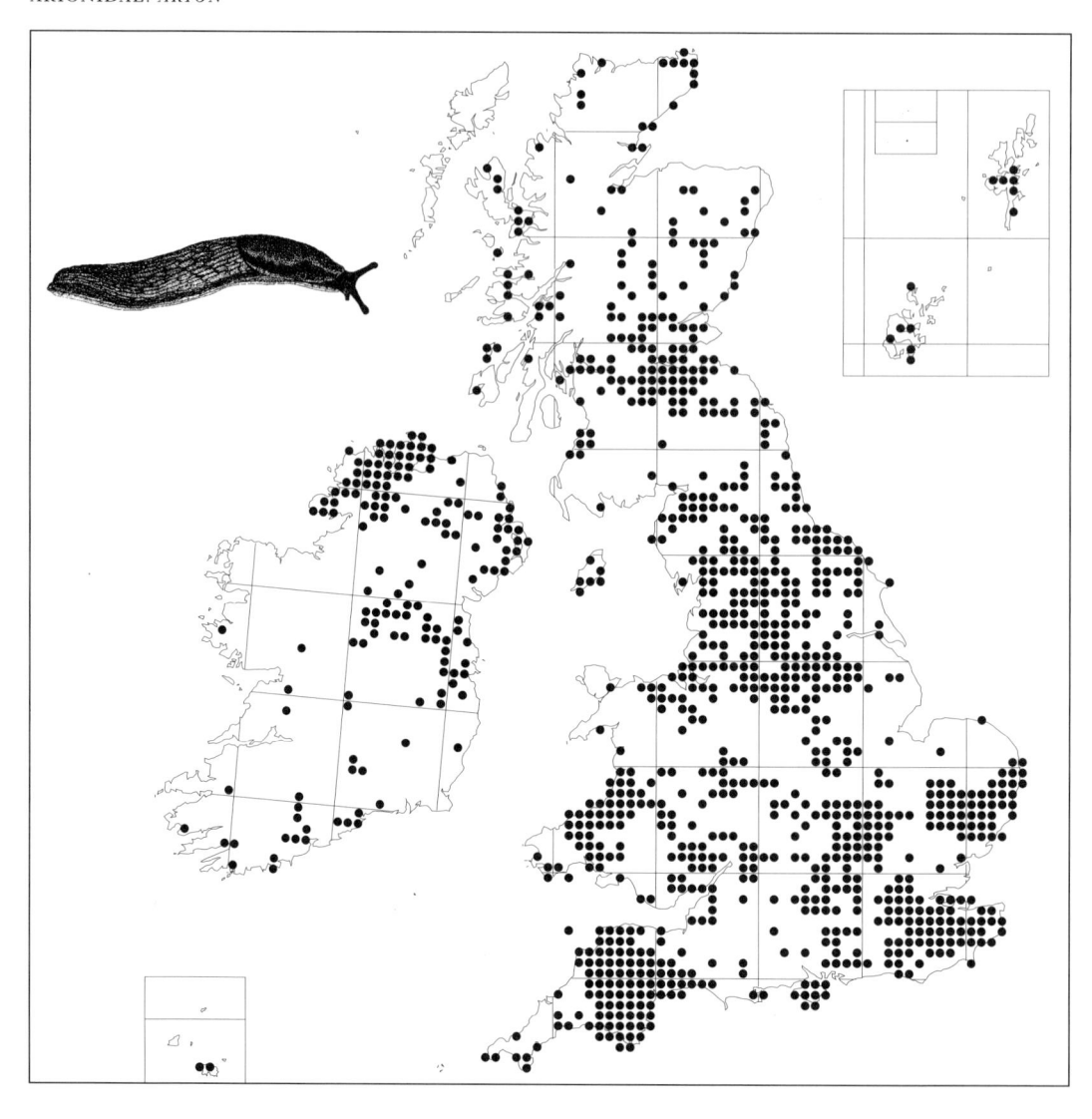

**Arion distinctus** Mabille, 1868 ('*A. hortensis* species A' of Davies, 1977)

Size: 25–35mm

A common slug of moist, sheltered places, among herbage and ground litter, often burrowing into the soil. It is typical of gardens, urban waste ground and crops, where it can be a serious agricultural pest, but is found equally in many natural or semi-natural habitats. It has a slight preference for base-rich soils and is absent from much harsh, upland terrain tolerated by *A. ater* or *A. intermedius*.

Probably native. Though still underrecorded, this species is likely to occur throughout most of the British Isles. It is a rapid colonist of waste ground and widely spread by man.

W. and S. European, extending to the Faroes and to S. Scandinavia (to 63°N in Norway)

It should be noted that the majority of old records of '*Arion hortensis*' refer to this species.

(Davies, 1977; 1979)

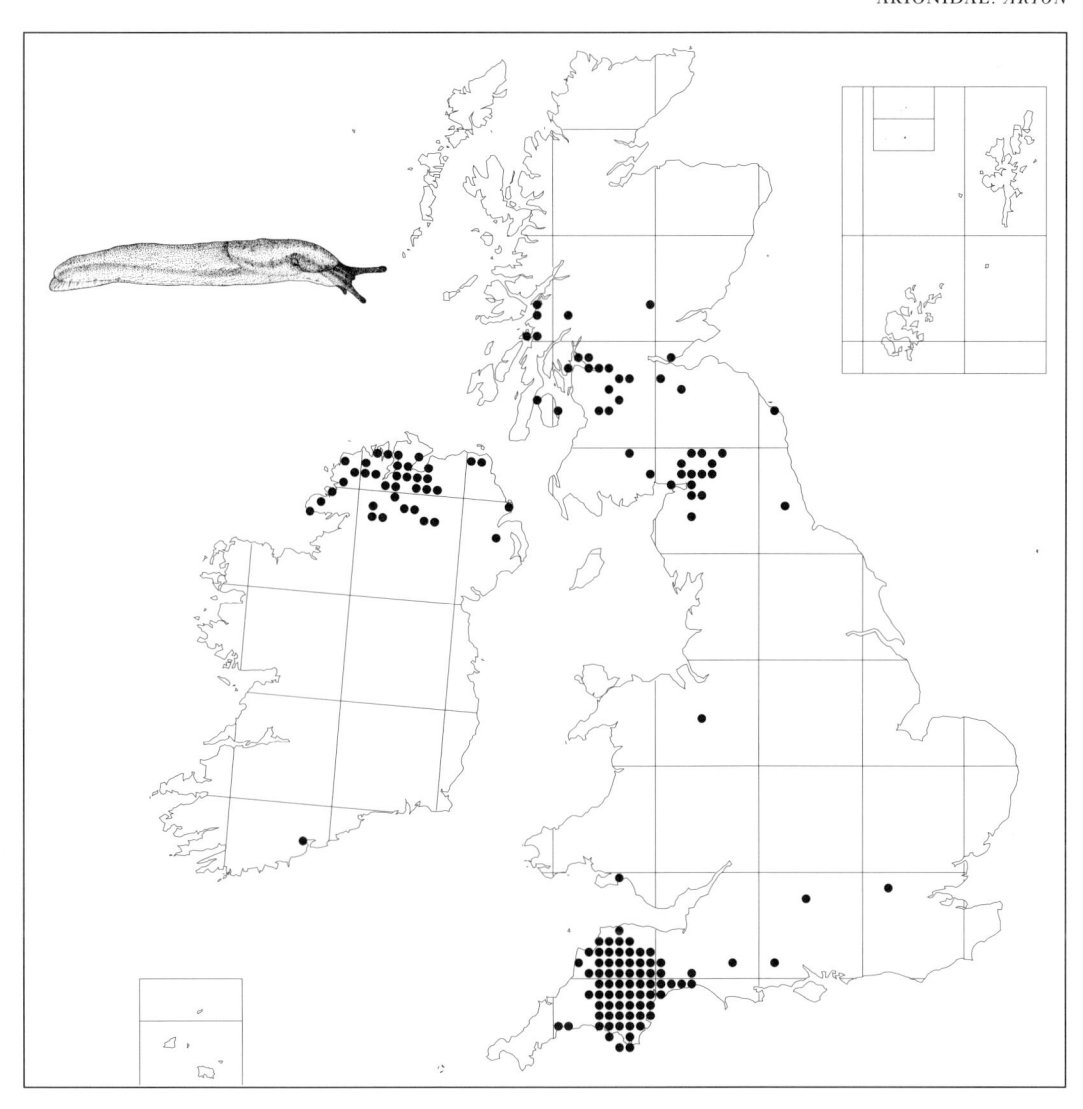

**Arion owenii** Davies, 1979 ('*A. hortensis* species B' of Davies, 1977)

Size: 25–35mm

A species found in moist, sheltered places, under ground litter and in the soil. It occurs in both wild and disturbed habitats.

Probably native. Though first distinguished only in the 1970s and remaining seriously underrecorded, nevertheless clearly a local species, common in certain areas (e.g. Devon) but not detected by intensive mapping elsewhere (e.g. in Cardigan, Beds. or Suffolk).

Very few records yet available from outside the British Isles. Reported from Brittany and from Norway.

(Davies, 1977; 1979)

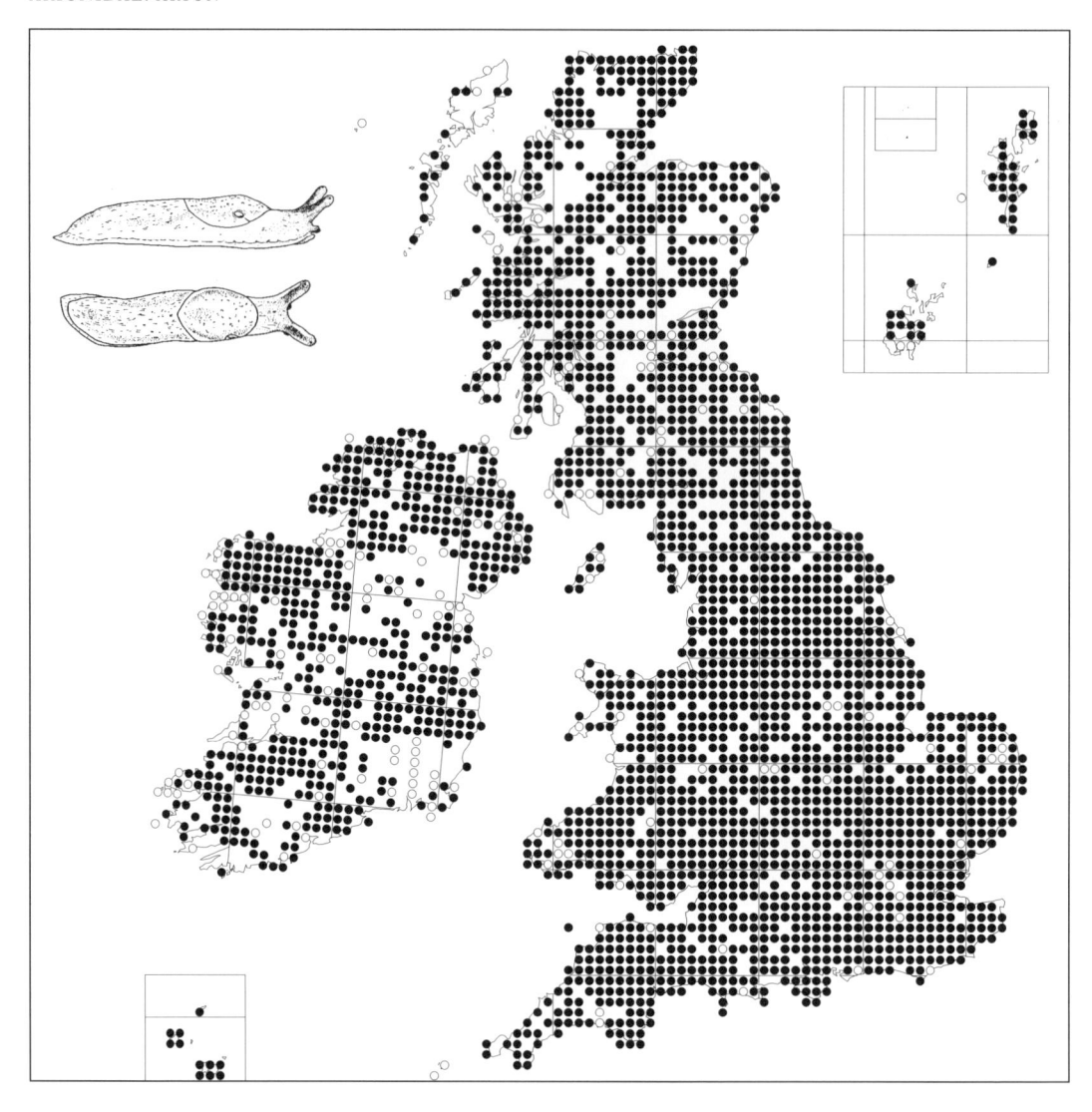

**Arion intermedius** Normand, 1852 (*A. minimus* Simroth)

Hedgehog slug   Size: 15–20mm

This is a very hardy, ecologically catholic species, found in herbage and under ground litter in coniferous and deciduous woods, heathland, pastures, roadside banks and waste ground (but not usually in gardens). It tolerates poor, acidic soils on which it is associated with few other molluscs. It ascends to high altitudes in the Scottish highlands but may be absent from base-rich lowland habitats of apparently suitable character.

Probably native. No evidence of distributional change.

W. European: mainly British Isles, France, the Low Countries, N. Germany and S. Scandinavia (to 65°N on the coast of Norway). Also Iceland and the Faroes.

(Ellis: 179; K. & C.: 106)

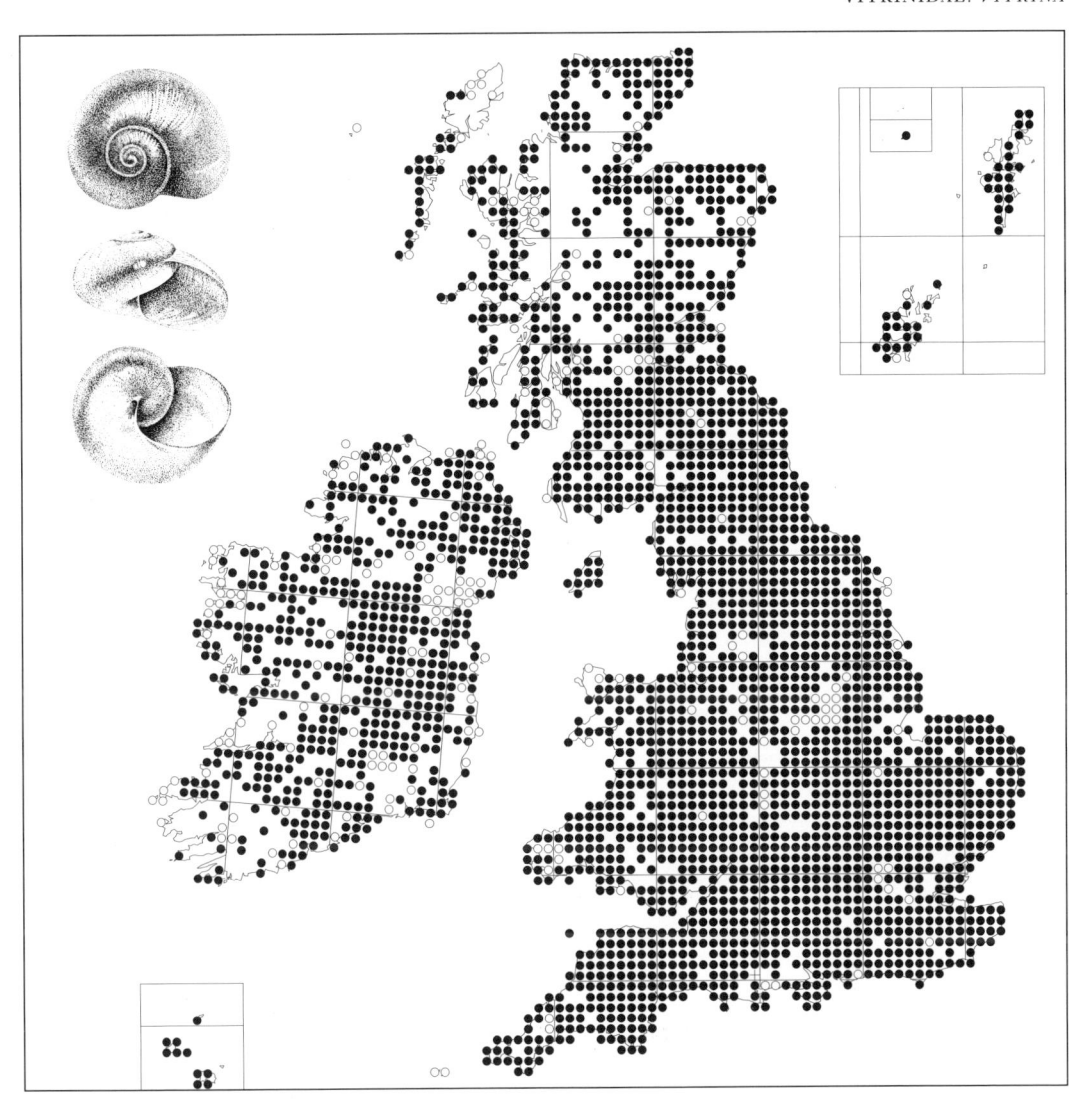

**Vitrina pellucida** (Müller, 1774)

Pellucid glass snail   Size: 4.5–6.0mm

A catholic species found in a wide variety of habitats, moist or dry, in coniferous and deciduous woods, grasslands and waste ground. It sometimes abounds in sparsely or incompletely vegetated places such as natural screes, old quarries, or the grassy hollows of stabilized dunes. It tolerates poor, non-calcareous soils, and ascends to great altitudes.

Native (Lgl). No evidence of distributional change.

Holarctic; throughout Europe to beyond the Arctic Circle.

(Ellis: 248; K. & C.: 109)

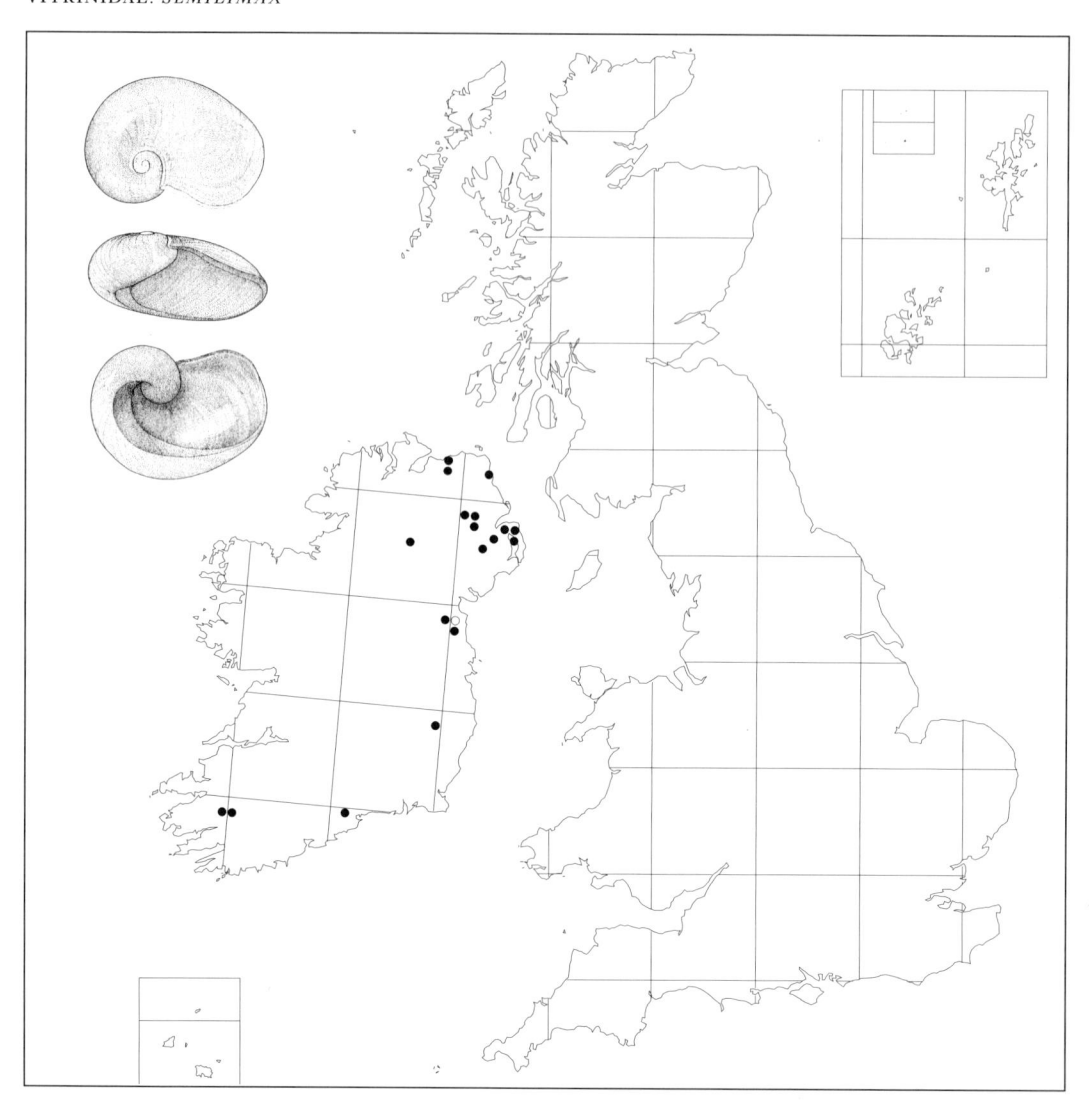

**Semilimax pyrenaicus** (Férussac, 1821) (*Vitrina pyrenaica* (Férussac), *V. hibernica* Taylor)

Pyrenean glass snail    Size: 5–6mm

Locally common in damp ground litter in woodland. The woods are mostly on non-calcareous soils and very mixed in character, wholly or partly replanted with a variety of conifers or deciduous trees such as spruce, larch, beech, birch, ash or hazel. Sometimes *S. pyrenaicus* occurs in more open, marshy places, either in alder carr or in sedgy grassland. A common feature of all the sites is lack of trampling by grazing animals (Fogan, 1969; Anderson, 1974).

Introduced? (No fossil records.) The status of *S. pyrenaicus* in Ireland is enigmatic. Nearly all of its known habitats have been altered by man, and some clearly result from 18th- or 19th-century estate planting which included exotic, ornamental trees and shrubs, as at Humewood Castle, Co. Wicklow (Kerney, 1978a) and Temple Demesne, Collon, Co. Louth. The species may have been accidentally introduced to Ireland in this way. Certainly there is some evidence of very recent spread (Anderson, 1991; 1992). Alternatively, it may be a native, surviving in areas that are vegetationally much modified but nevertheless in historical continuity with the early Postglacial forests of Ireland.

Pyrenees and Ireland only.

(Ellis: 250; K. & C.: 111)

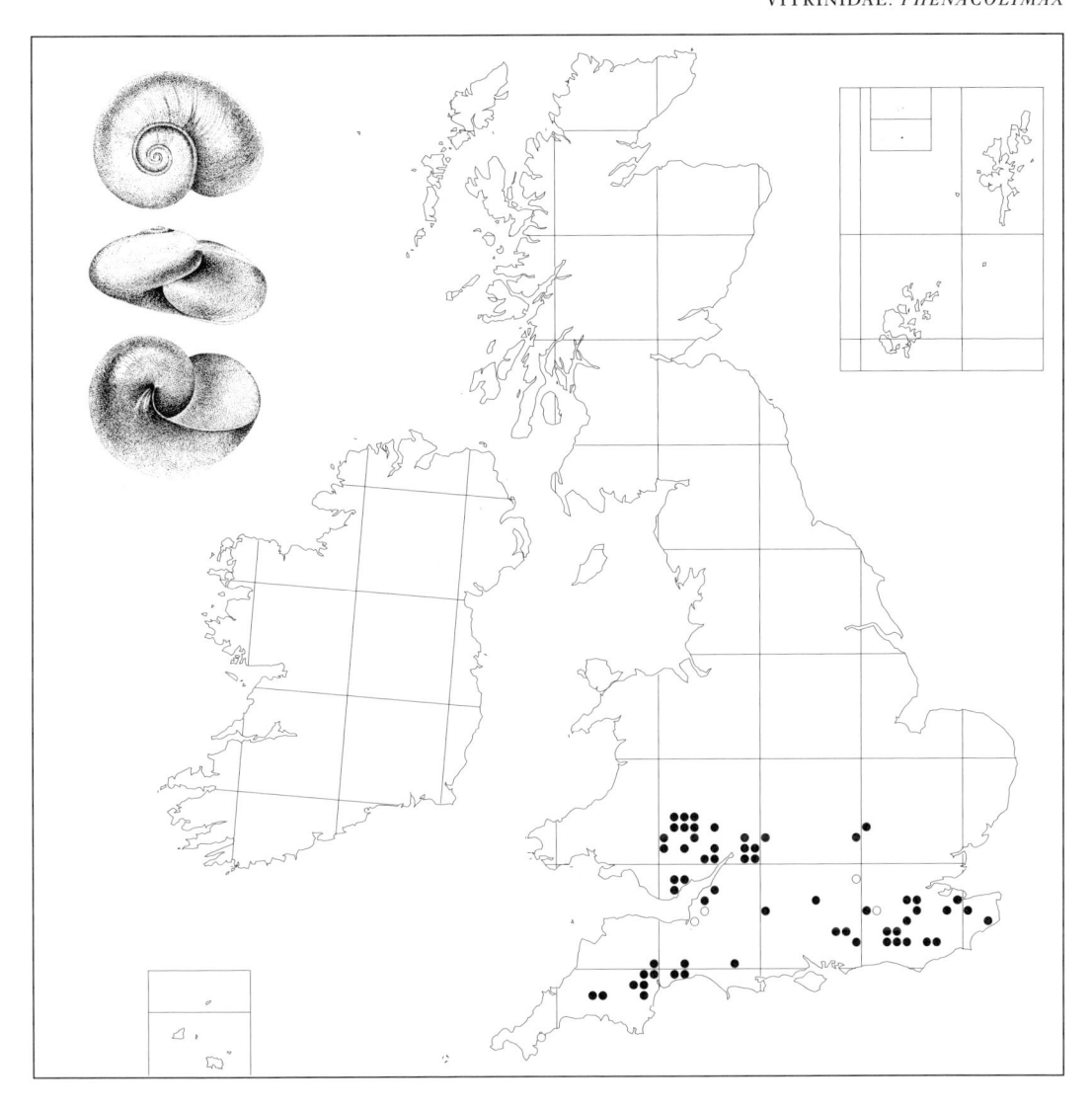

**Phenacolimax major** (Férussac, 1807) (*Vitrina major* (Férussac))

Greater pellucid glass snail   Size: 5–6mm

A very local species found under moist leaf litter and fallen branches in old deciduous woodland, both on calcareous and non-calcareous soils (Boycott, 1927; Quick, 1957). It occurs in a variety of woods, ranging from those of mature beech or oak to traditionally managed coppices of hazel, hornbeam or chestnut. Sometimes it inhabits dry-stone walls or crevices in natural rocks overhung with trees.

Probably native (no certain Postglacial fossils). In Britain, though not further south in Europe, *P. major* is a useful indicator of primary woodland. Though still secure in many places and doubtless somewhat underrecorded (shells alone can be difficult to name with certainty), it is probably declining overall. Isolated colonies are particularly at risk from clear felling or replanting with conifers.

W. European: mainly France, W. Germany and southern England.

(Ellis: 249; K. & C.: 115)

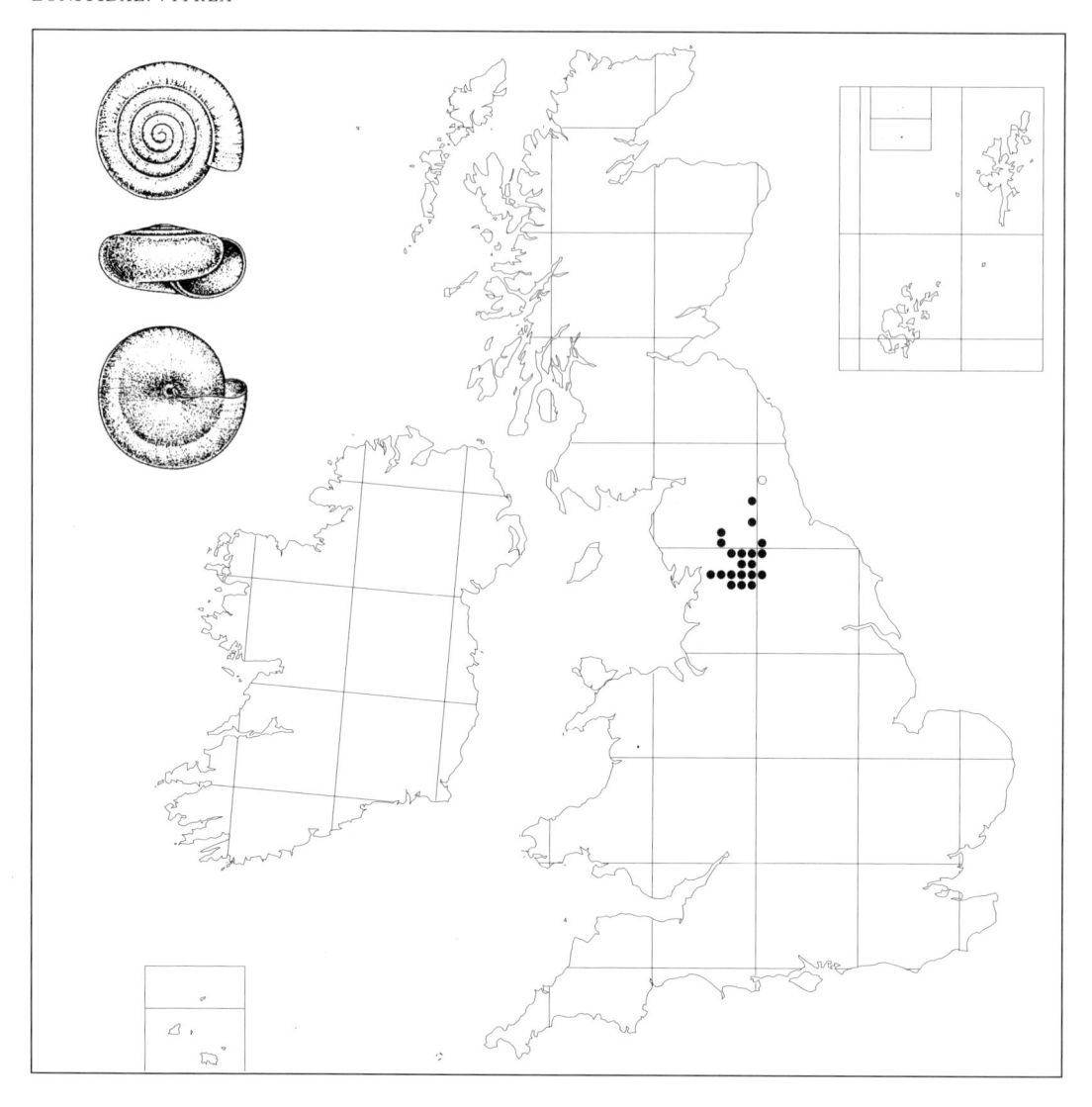

**Vitrea subrimata** (Reinhardt, 1871) (*V. diaphana* var. *subrimata* Reinhardt)

Size: 2.5–3.0mm

A species restricted to limestone screes and rocky hillsides where the soil cover is broken, either in open situations or under thin woodland of hazel, rowan or hawthorn. It inhabits leaf litter and relatively moist sheltered places under stones or in crevices. It is nearly always associated with *V. contracta*, and, more rarely, with *V. crystallina*. In open situations it may be found with *Abida secale*. The known sites lie mostly between 250m and 600m.

Native (E.Pgl). Detected in Britain only in 1966 (Kerney & Fogan, 1969). It is locally common in suitable habitats in the northern Pennines and is known as a fossil within this area. There is no evidence that it formerly possessed a wider distribution though this is likely on biogeographical grounds (cf. *Clausilia dubia*). Afforestation may be a threat to some populations.

Alpine, Carpathian and S. European (mainly montane).

(K. & C.: 118)

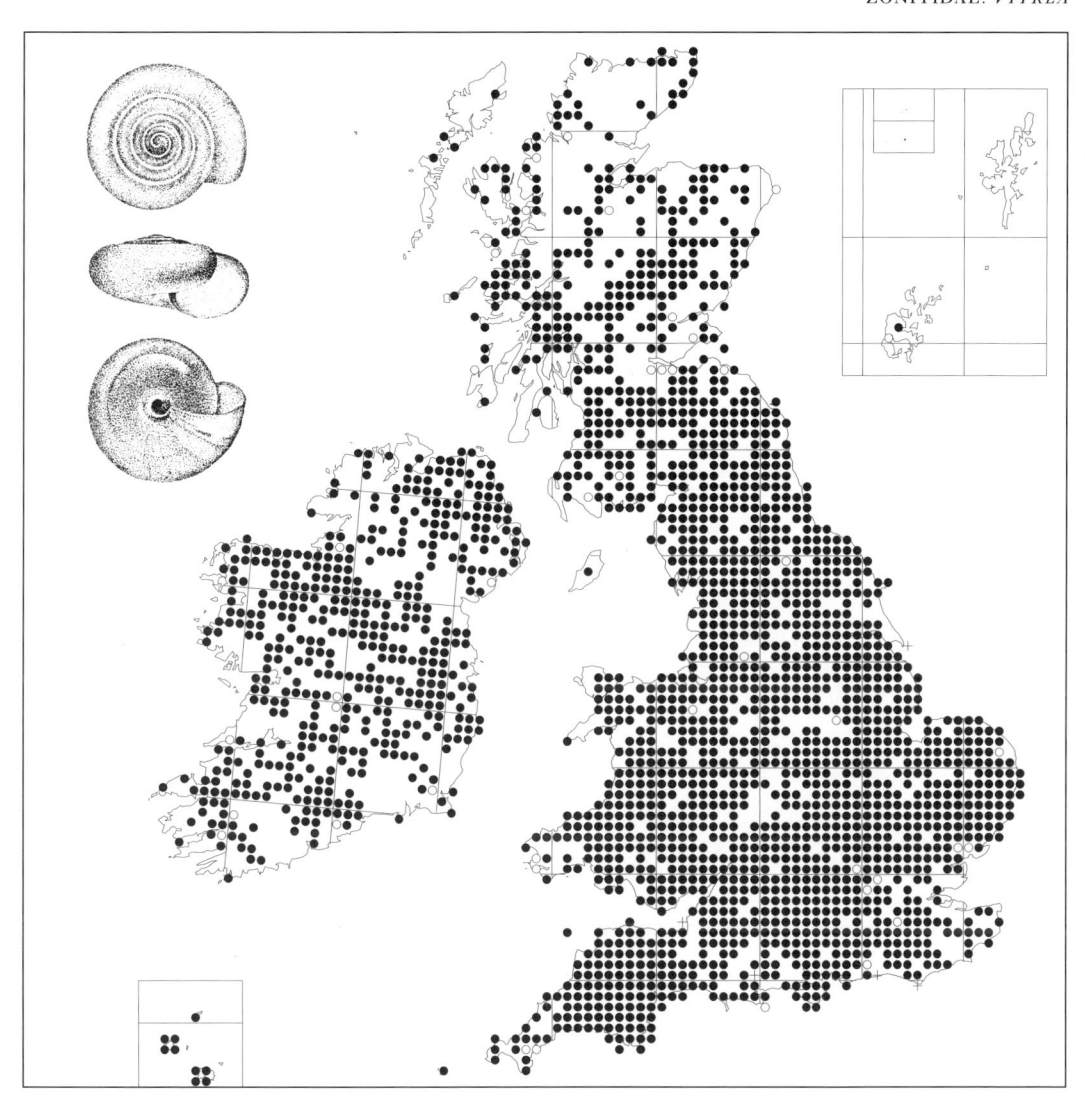

**Vitrea crystallina** (Müller, 1774) (*Hyalinia crystallina* (Müller))

Crystal snail   Size: 3–4mm

This ecologically catholic species lives in all kinds of moist, sheltered places, in coniferous and deciduous woods, grassy pastures, roadside banks and marshes. It has a slight preference for base-rich soils. Empty shells are often common in the flood rubbish of rivers.

Native (E. Pgl). No evidence of distributional change.

European: widespread nearly throughout, to southernmost Scandinavia (65°N in Norway).

(K. & C.: 118)

ZONITIDAE: *VITREA*

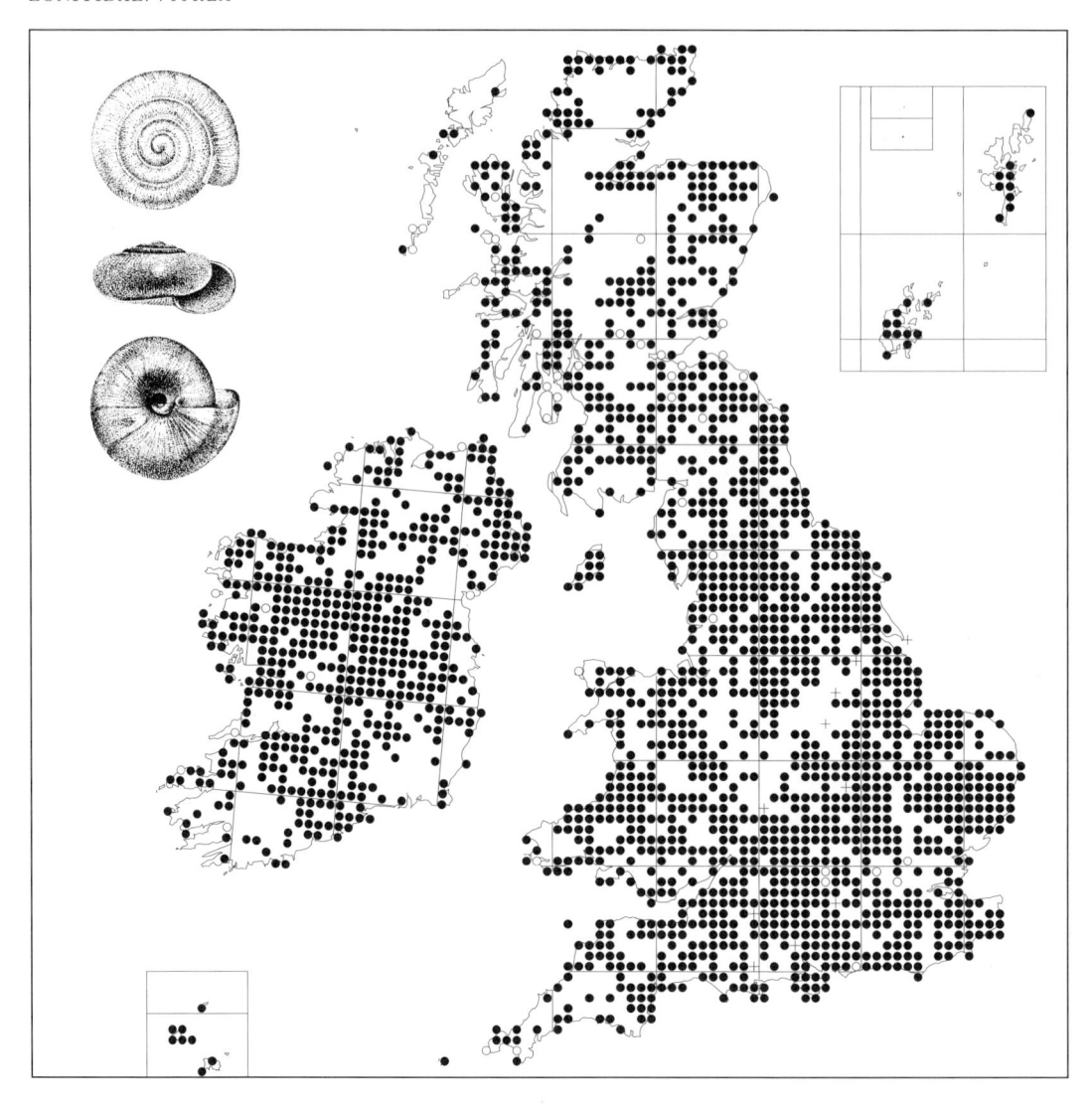

**Vitrea contracta** (Westerlund, 1871) (*V. crystallina* var. *contracta* Westerlund)

Milky crystal snail    Size: 2.2–2.5mm

A snail found in a wide variety of terrestrial habitats. It tends to favour drier places than *V. crystallina*, though the two species are often associated, as in the leaf litter of many deciduous woods. It is found in base-rich grasslands, maritime turf, natural cliffs and stone walls, but is uncommon in marshes. In rocky places it may penetrate deeply into crevices, and can sometimes be found in caves.

Native (E.Pgl). No evidence of recent distributional change.

W. Palaearctic; widespread throughout Europe, though absent from much of C. and N. Scandinavia.

(K. & C.: 119)

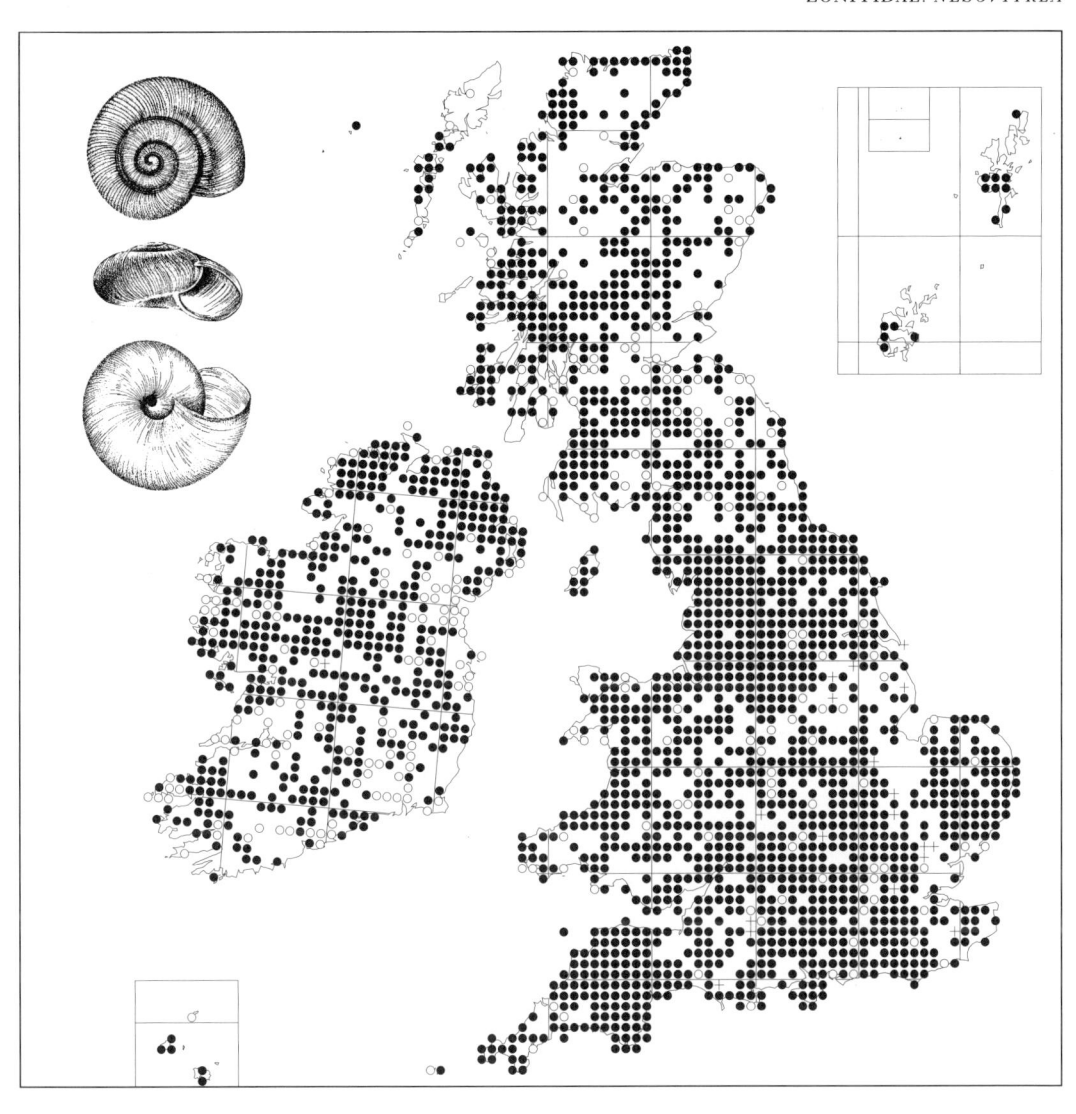

ZONITIDAE: *NESOVITREA*

**Nesovitrea hammonis** (Ström, 1765) (*Hyalinia radiatula* (Alder), *Retinella radiatula* (Alder))

Rayed glass snail   Size: 3.5–4.2mm

A catholic snail, found in vegetation and ground litter (though not usually in great numbers) in all kinds of terrestrial habitats ranging from wet to moderately dry: coniferous and deciduous woods, grassland, roadside banks, fens and marshes. It tolerates acid soils, ascending to considerable altitudes in the highland zone, but may be scarce or absent in some base-rich lowland habitats with a high diversity of species, such as beech woods on chalk.

Native (Lgl). No evidence of distributional change.

Palaearctic; throughout Europe to beyond the Arctic Circle.

(Ellis: 241; K. & C.: 120)

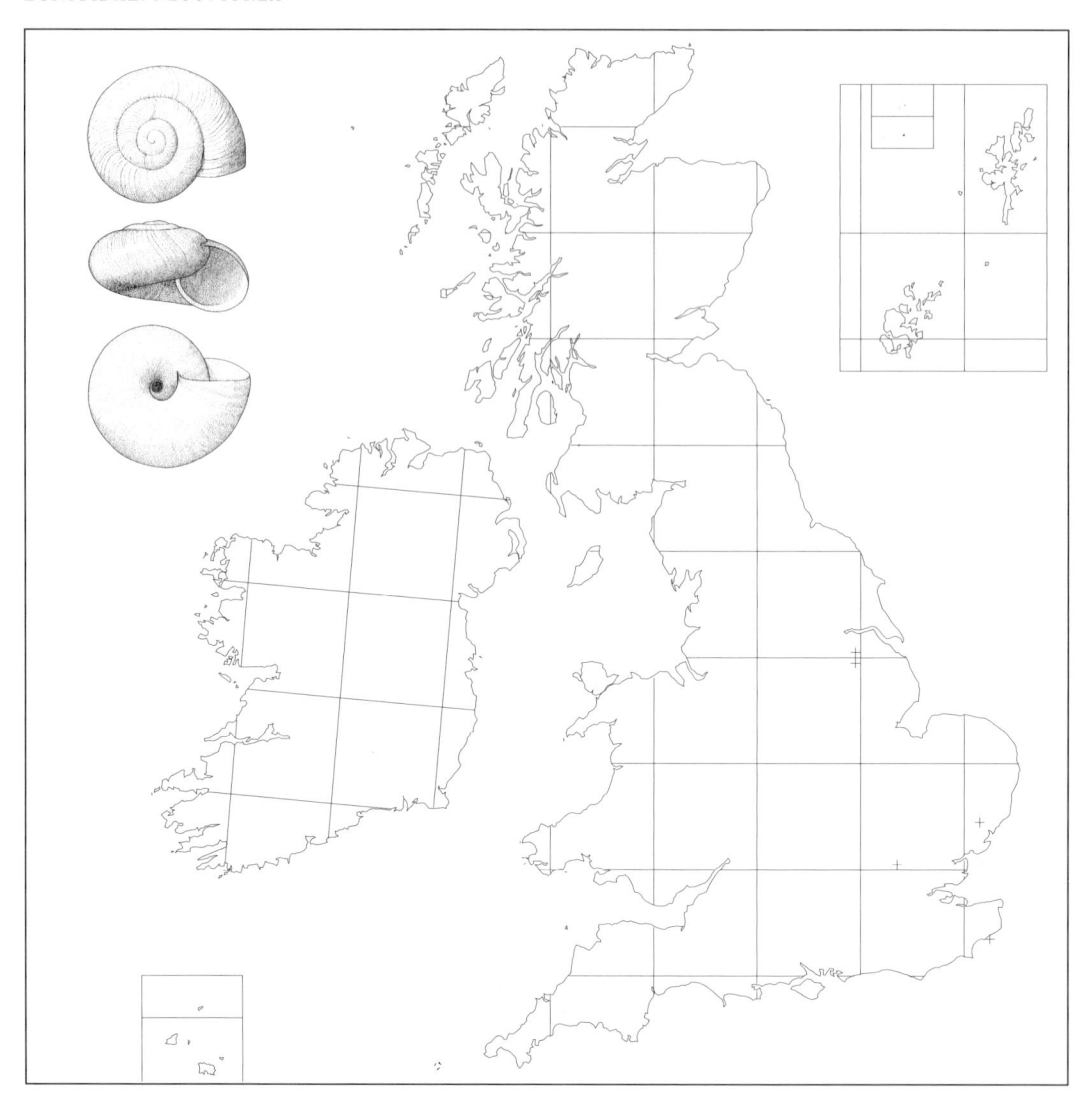

ZONITIDAE: *NESOVITREA*

**Nesovitrea petronella** (L. Pfeiffer, 1853) (*Retinella radiatula* var. *petronella* Pfeiffer)

Size: 4.2–5.0mm

On the Continent, *N. petronella* lives in rather similar places to *N. radiatula*, but is especially characteristic of open, montane woodland (to 2,500m in the Alps). It tolerates poor acidic soils.

Extinct (E.Pgl fossils only). *N. petronella* is known from a few deposits of early Postglacial (pre-Atlantic) date in Kent, Herts., Essex, Suffolk and Lincs., associated with *Discus ruderatus* (Preece & Robinson, 1984). It may perhaps yet be found living in Scotland.

Boreal and Alpine; in Europe common in Scandinavia and in the Swiss and French Alps, with scattered populations in the mountains of S. and C. Germany.

(K. & C.: 120)

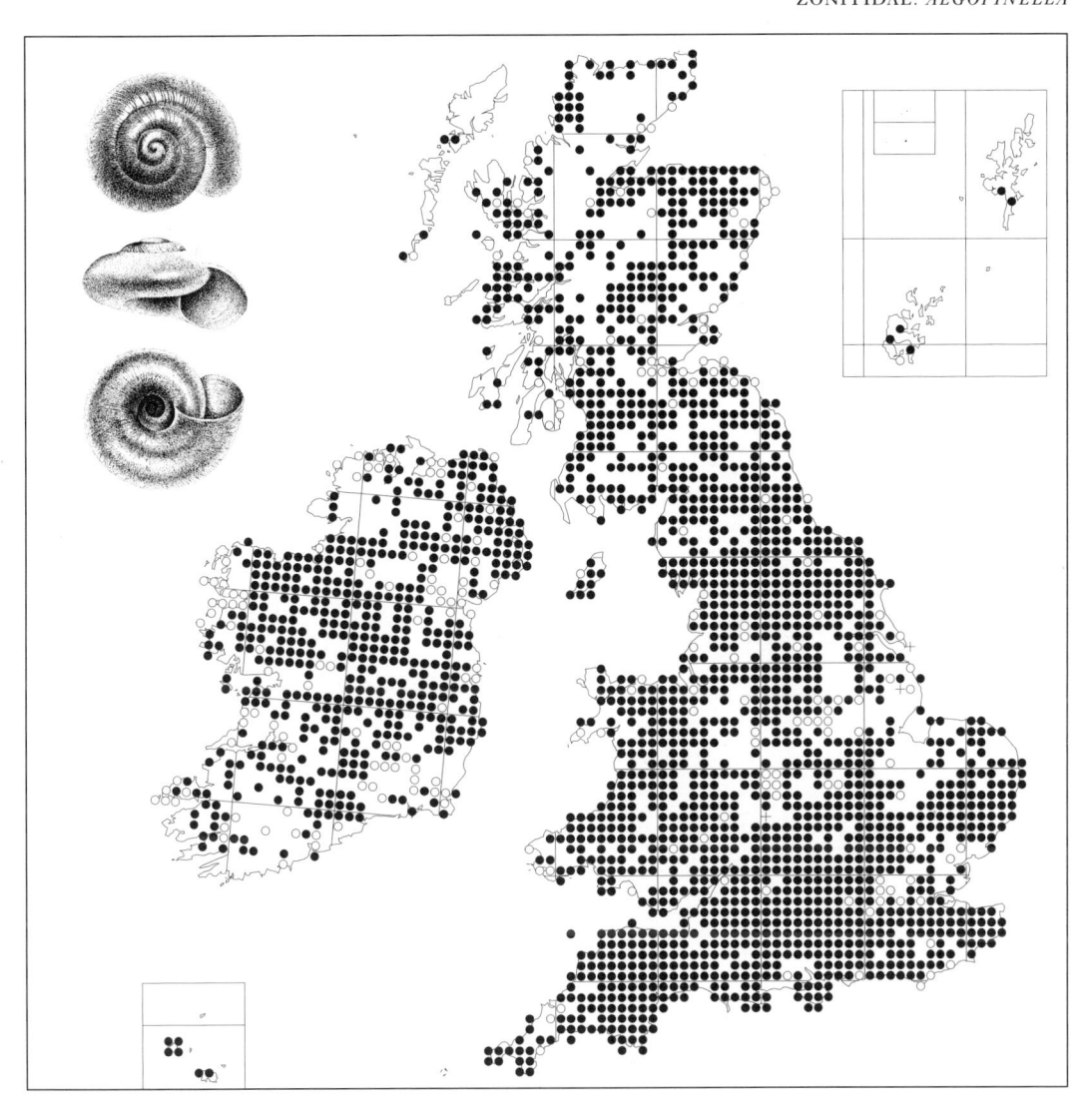

**Aegopinella pura** (Alder, 1830) (*Hyalinia pura* (Alder), *Retinella pura* (Alder))

Clear (or delicate) glass snail   Size: 3.5–4.2mm

A snail of moderately moist (but not wet) places, under ground litter or at the base of vegetation. It slightly favours base-rich soils. It is a species especially characteristic of leaf litter in deciduous woods, but can also be found in a wide variety of more open habitats of a reasonably sheltered kind, though not usually in waste ground or gardens.

Native (E.Pgl). No evidence of recent distributional change.

European: widespread, becoming mainly coastal in Scandinavia.

(Ellis: 242; K. & C.: 120)

ZONITIDAE: *AEGOPINELLA*

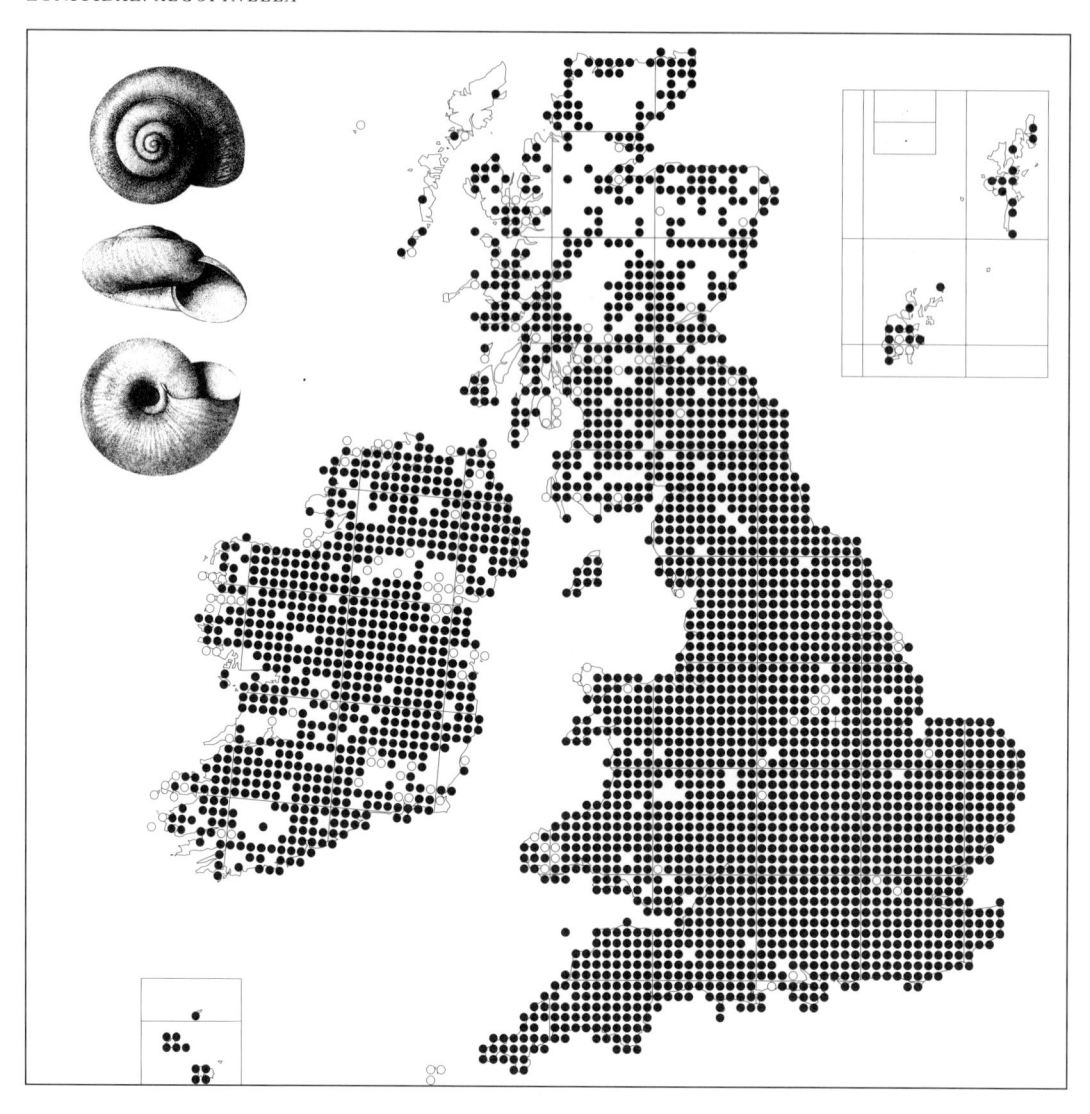

**Aegopinella nitidula** (Draparnaud, 1805) (*Hyalinia nitidula* (Draparnaud), *Retinella nitidula* (Draparnaud))

Smooth (or dull) glass snail   Size: 8–10mm

A common and catholic species, found under ground litter in a wide variety of sheltered places, such as coniferous and deciduous woods, the base of rocks and walls, moist grassland and roadside banks. It tolerates human disturbance and is not infrequent in waste ground and in gardens.

Native (E.Pgl). No evidence of recent distributional change.

N.W. European: mainly France, British Isles, Low Countries, N. Germany and S. Scandinavia.

The closely related, central European montane species *A. nitens* (Michaud) has occasionally been reported from Britain. The difference between the shells of these two species is, however, not reliable, and British *Aegopinella* of *nitens*-like shape have so far proved to be *nitidula* on dissection.

(Ellis: 243; K. & C.: 121)

ZONITIDAE: *OXYCHILUS*

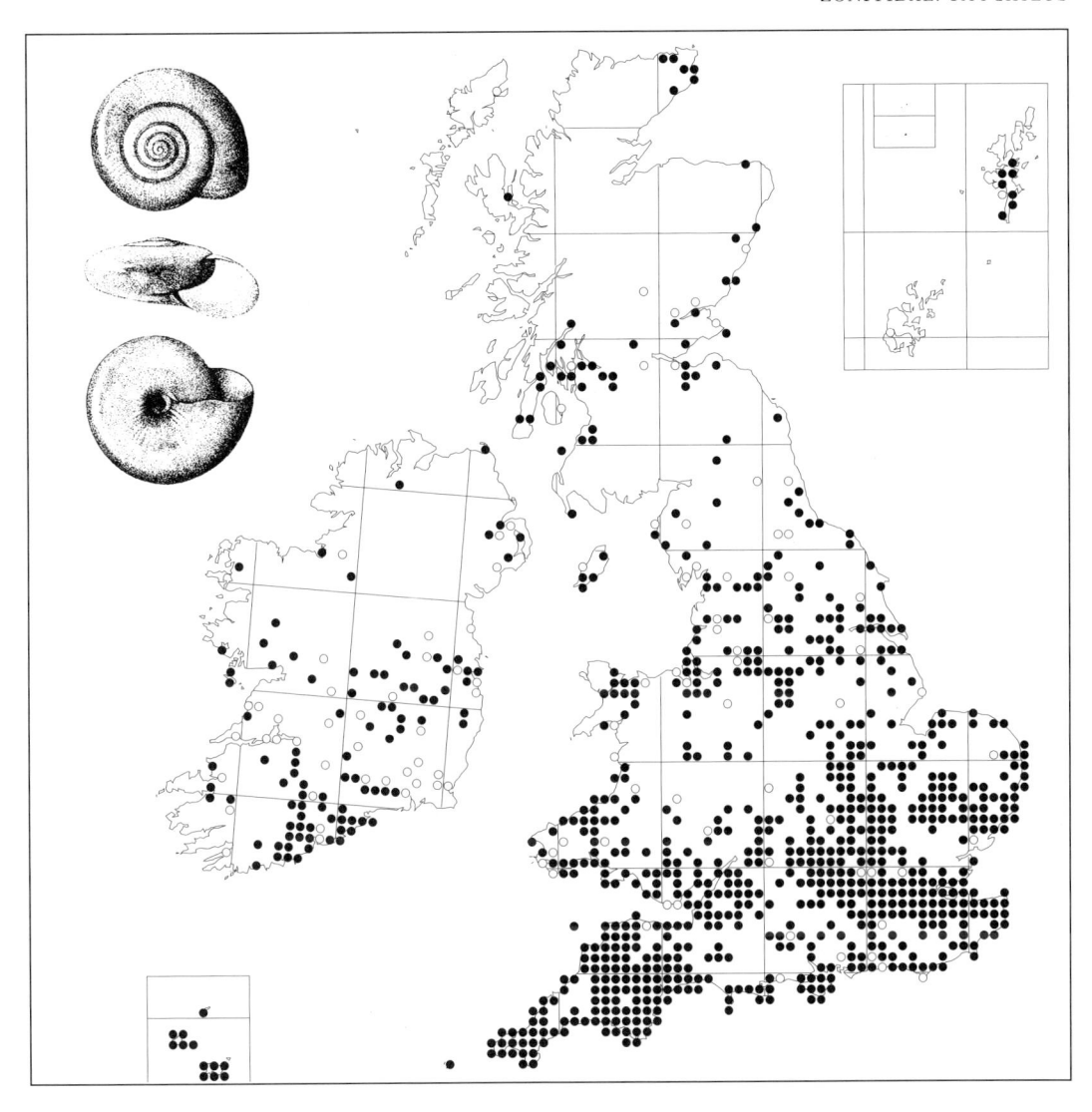

**Oxychilus draparnaudi** (Beck, 1837)   (*O. lucidus* (Draparnaud), *Hyalinia lucida* (Draparnaud))

Draparnaud's glass snail   Size: 11–15mm

This is a species of ground litter in moist sheltered places, characteristic especially of gardens, urban waste ground and roadside rubbish, and often associated with other synanthropic molluscs like *Tandonia sowerbyi*, *T. budapestensis* and *Deroceras panormitanum*. In the central and northern parts of the British Isles virtually all its habitats are of this type, but in the south (especially in S.W. England, southernmost Ireland and the Channel Islands) it lives also in wild places: in woods, among rocks and on sea cliffs.

Introduced. This is a Roman or post-Roman introduction. It has colonized many new areas during the 20th century and is probably still extending its range. Over most of the British Isles its presence is indicative of habitat disturbance.

Originally Mediterranean, but now widely spread by man in W. Europe, northwards to the British Isles and S. Scandinavia.

(Ellis: 246; K. & C.: 123)

ZONITIDAE: *OXYCHILUS*

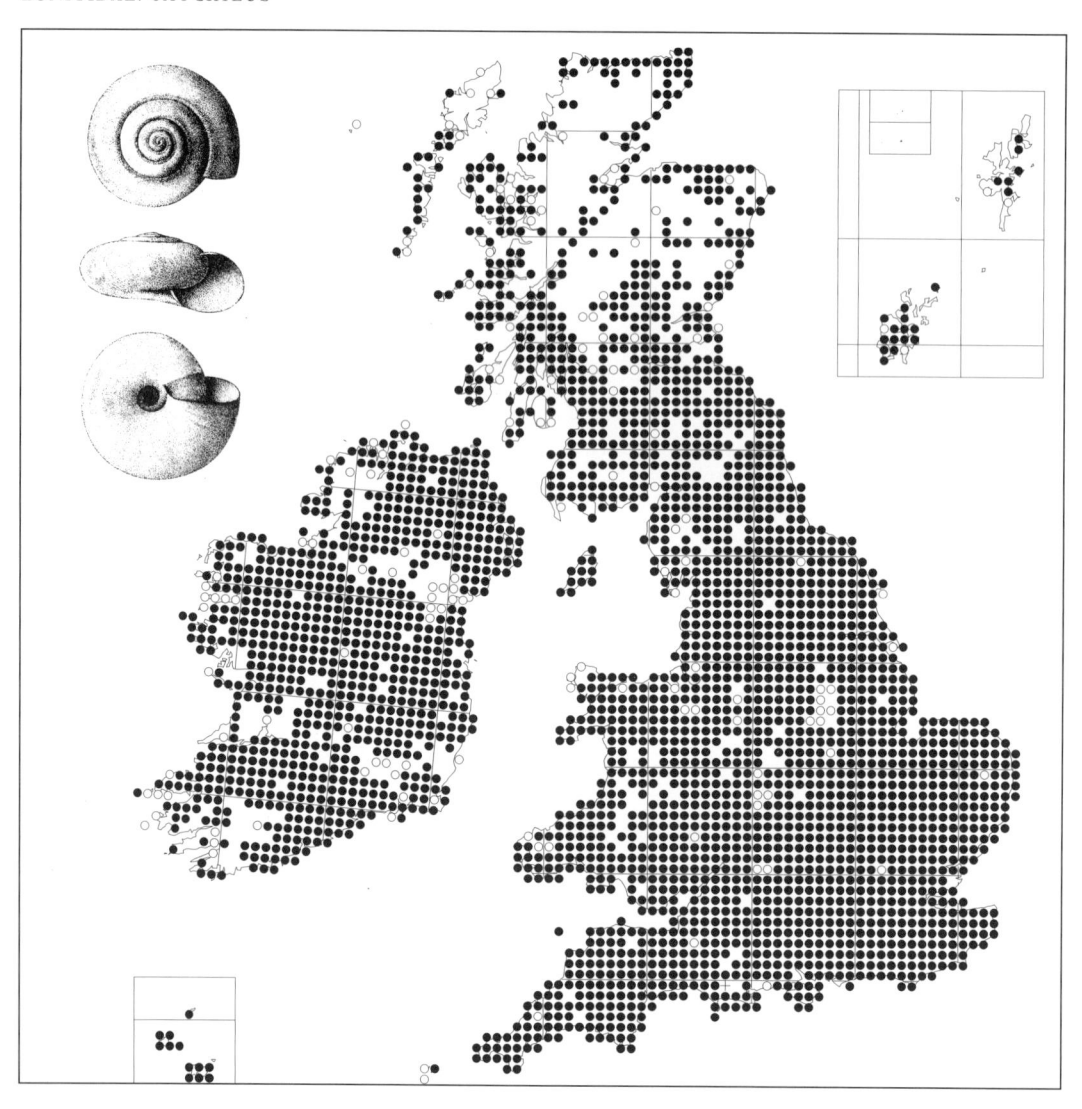

**Oxychilus cellarius** (Müller, 1774) (*Hyalinia cellaria* (Müller))

Cellar snail    Size: 9–12 (occ.–14)mm

A species of moist, sheltered places, under wood, stones or other ground litter. It prefers base-rich soils. In rocky habitats it will penetrate deeply into crevices, and can be found in caves, cellars and tombs. It is common in gardens and waste ground and in many highland areas is entirely restricted to disturbed places of this kind, tending to be replaced by *O. alliarius* in adjacent wild habitats.

Native (E.Pgl). This species shows no clear evidence of recent distributional change, though possibly it is still spreading in parts of Scotland. In harsh, non-calcareous or upland areas it is a useful indicator of habitat disturbance.

W. European: widespread, as far as southernmost Scandinavia (where it is mainly coastal and synanthropic).

(Ellis: 245; K. & C.: 124)

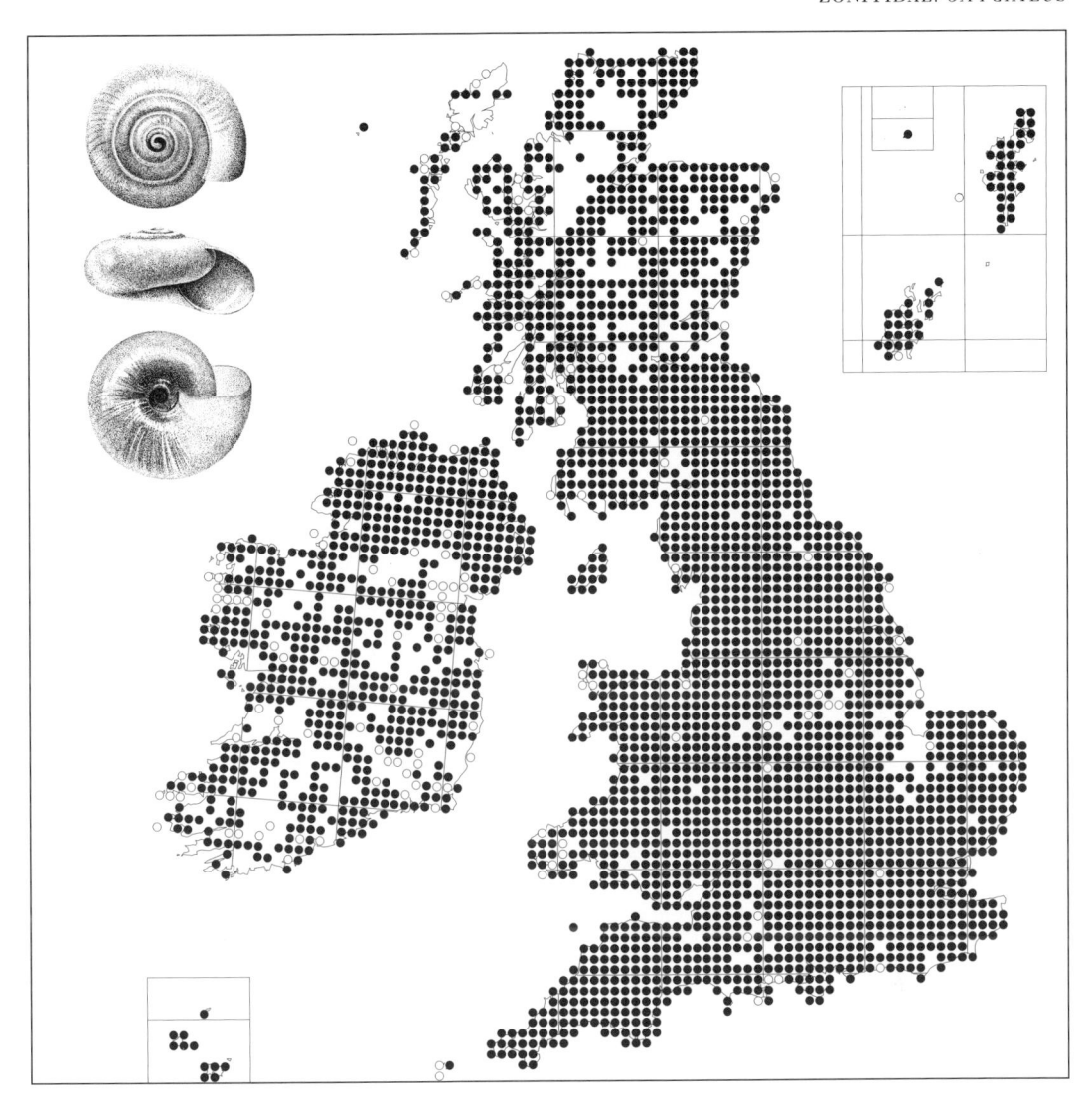

**Oxychilus alliarius** (Miller, 1822) (*Hyalinia alliaria* (Miller))

Garlic snail   Size: 5.5-7.0mm

An ecologically catholic species, living in moist, sheltered places in ground litter in woods, fields, cliffs, hedge banks and waste ground (occasionally in gardens and greenhouses). It avoids marshes. It tolerates acid soils and can occur in poor heathy upland country where other shelled molluscs are hard to find; it is relatively frequent in the Scottish mountains but scarce in some rich lowland habitats.

Native (E. Pgl). No evidence of distributional change.

W. European: mainly France, British Isles, the Low Countries, N. Germany and S. Scandinavia (to beyond the Arctic Circle on the coast of Norway).

(Ellis: 244; K. & C.: 125)

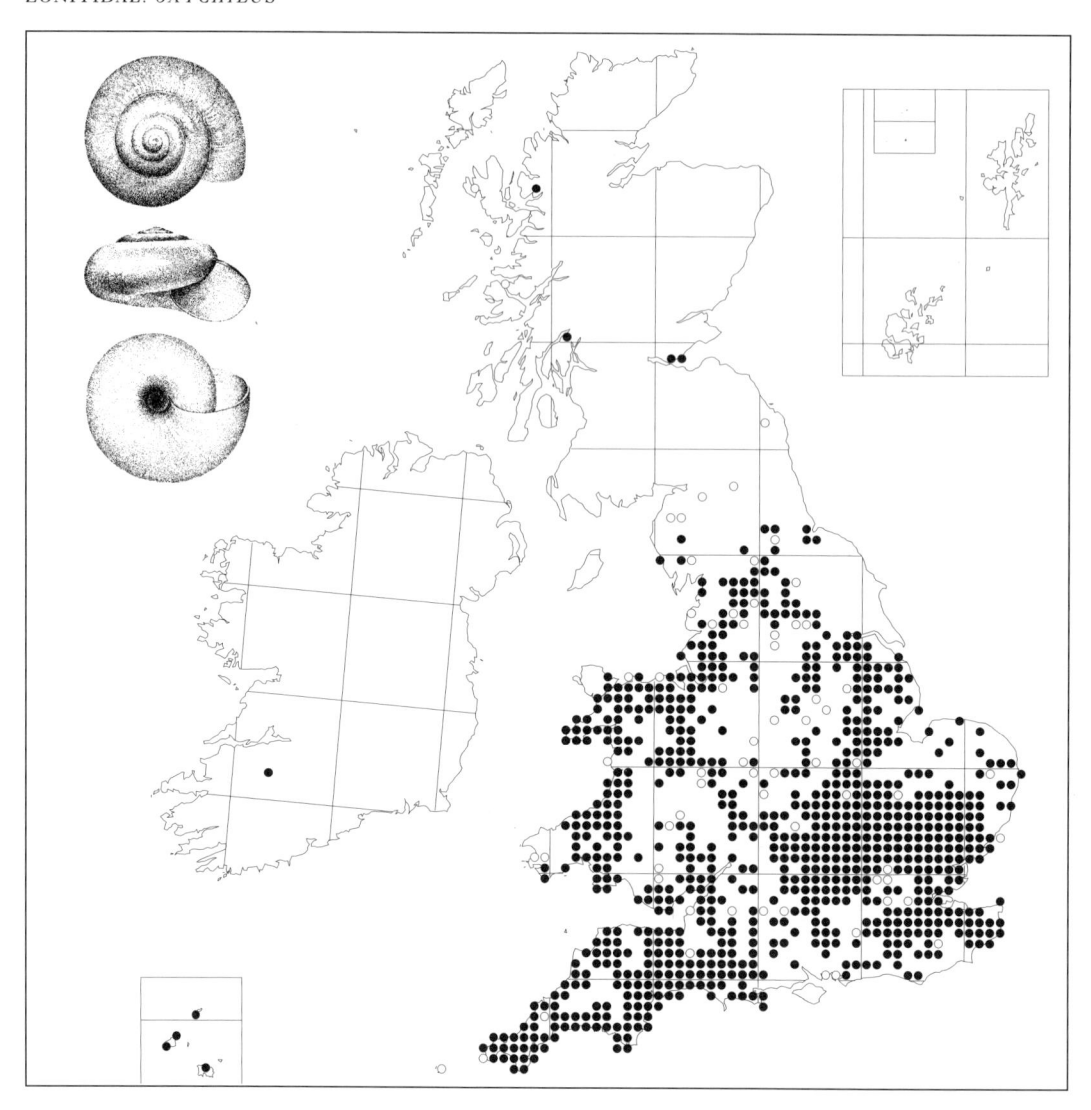

ZONITIDAE: *OXYCHILUS*

**Oxychilus helveticus** (Blum, 1881) (*Hyalinia rogersi* (Woodward), *H. helvetica* (Blum))

Glossy (or Swiss) glass snail   Size: 8–10mm

A species locally common in moist, sheltered places under ground litter. It is most frequent in woods but occurs also in more open situations, such as sea cliffs, old quarries and roadside banks. It tolerates poor soils. It is not uncommon under human rubbish and is occasionally found in gardens.

Introduced? (post-Roman fossils only). Though now completely naturalized in wild habitats in southern Britain, this is probably a relatively modern introduction. Its patchy distribution and absence from many apparently congenial areas of lowland Britain suggest that it is still spreading. It is often associated with rubbish and disturbance; the isolated colonies in Scotland and Ireland (Kerney, 1978b) are of this kind, making it likely that *O. helveticus* has only recently become established in those countries through human agency.

N.W. European: mainly British Isles, France, Belgium and Switzerland.

A number of published records of this species from northern England and Scotland are here omitted, being probably based on dark-bodied forms of *O. cellarius*. Old Irish records are similarly erroneous.

(Ellis: 244; K. & C.: 125)

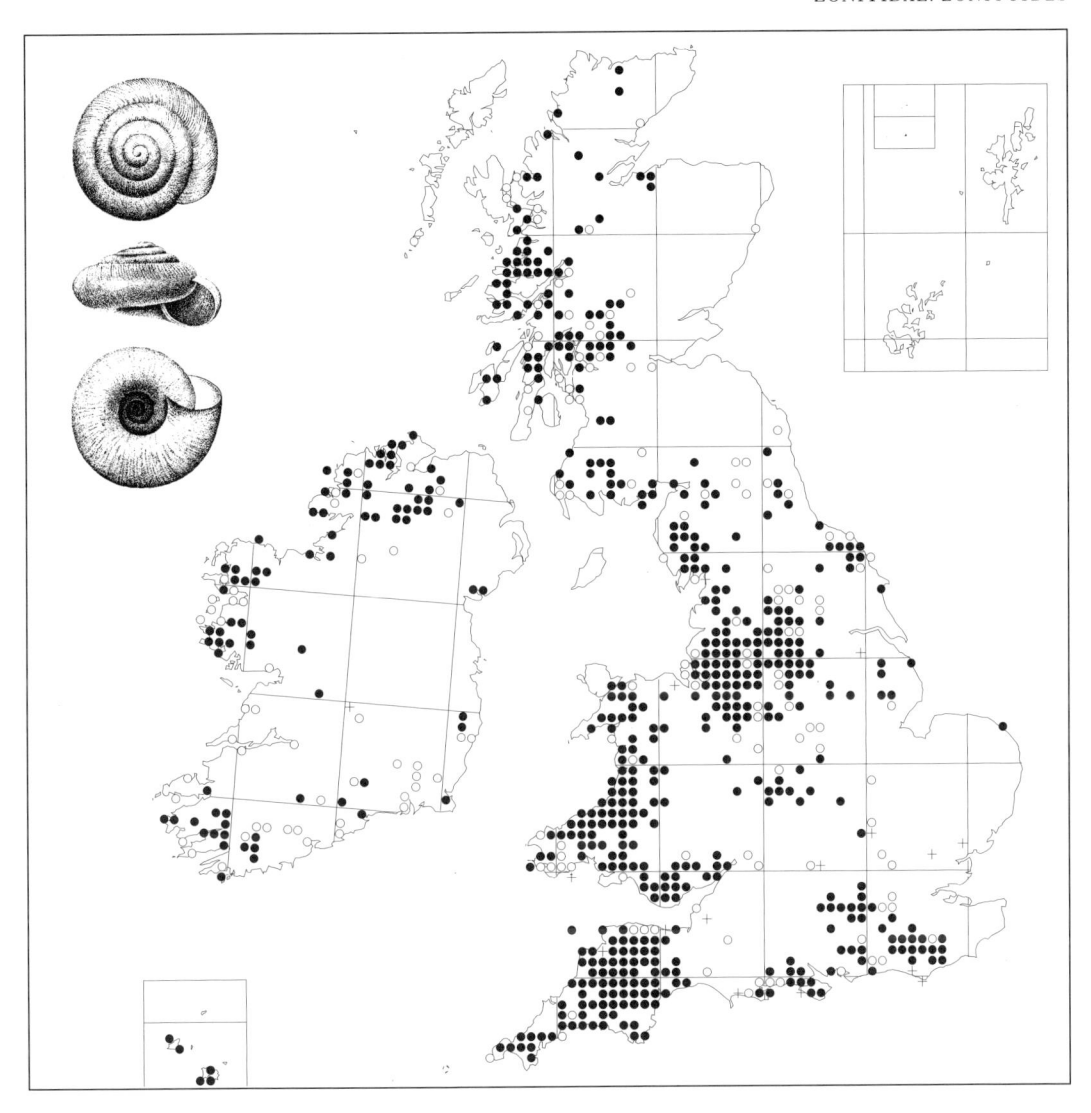

ZONITIDAE: *ZONITOIDES*

**Zonitoides excavatus** (Alder, 1830)

Hollowed glass snail   Size: 5.3–6.0mm

This is the only British terrestrial mollusc which is clearly calcifuge, avoiding base-rich soils and absent from areas of chalk or limestone. Its usual habitat is moist, old-established acid woodland of, for example, birch, oak, beech or larch where it lives under dead wood and ground litter, often in poorly drained places along the banks of streams. Though it tolerates some degree of human disturbance and replanting, it is not normally found in forestry plantations. In the oceanic west it occurs also in unshaded habitats such as marshes or coarse grassland.

Native (E.Pgl). *Z. excavatus* had a somewhat wider distribution in England in the mid-Postglacial (Boreal and Atlantic periods) than today. The temporary acidification of calcareous soils by deep leaching during the forest optimum was no doubt the principal reason. There is no clear evidence of more recent geographical decline.

N.W. European Atlantic: British Isles and coastal regions adjacent to the North Sea from Belgium to S. Denmark.

(Ellis: 239; K. & C.: 127)

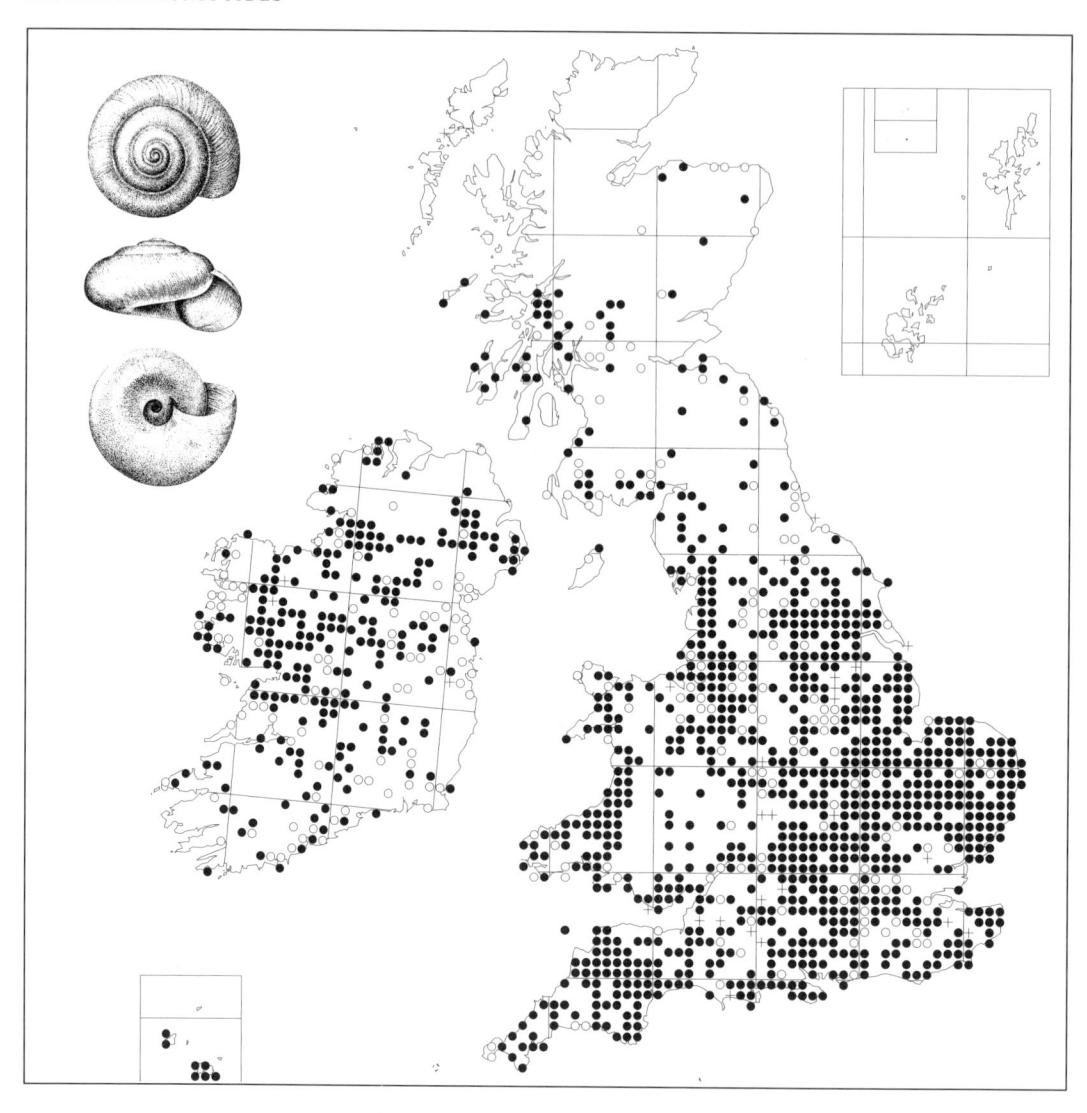

ZONITIDAE: *ZONITOIDES*

**Zonitoides nitidus** (Müller, 1774)

Shiny glass snail    Size: 6–7mm

This is a characteristic wetland species. A typical habitat is the zone of emergent vegetation at the edges of lakes and rivers, where it lives on decaying *Phragmites* or *Carex* litter, or on driftwood or other flotsam lying in muddy ground. Like the Succineidae (amber snails), it is virtually amphibious and will survive long periods of flooding. It is mainly a lowland species, commonest on base-rich soils.

Native (E.Pgl). *Z. nitidus* is receding in some areas through drainage and embanking but shows no serious overall decline. Man-made habitats, such as pools in old quarries, are sometimes colonized after a few years.

Holarctic; widespread in Europe, though rare in the northern highland zone.

(Ellis: 240; K. & C.: 127)

148

MILACIDAE: *MILAX*

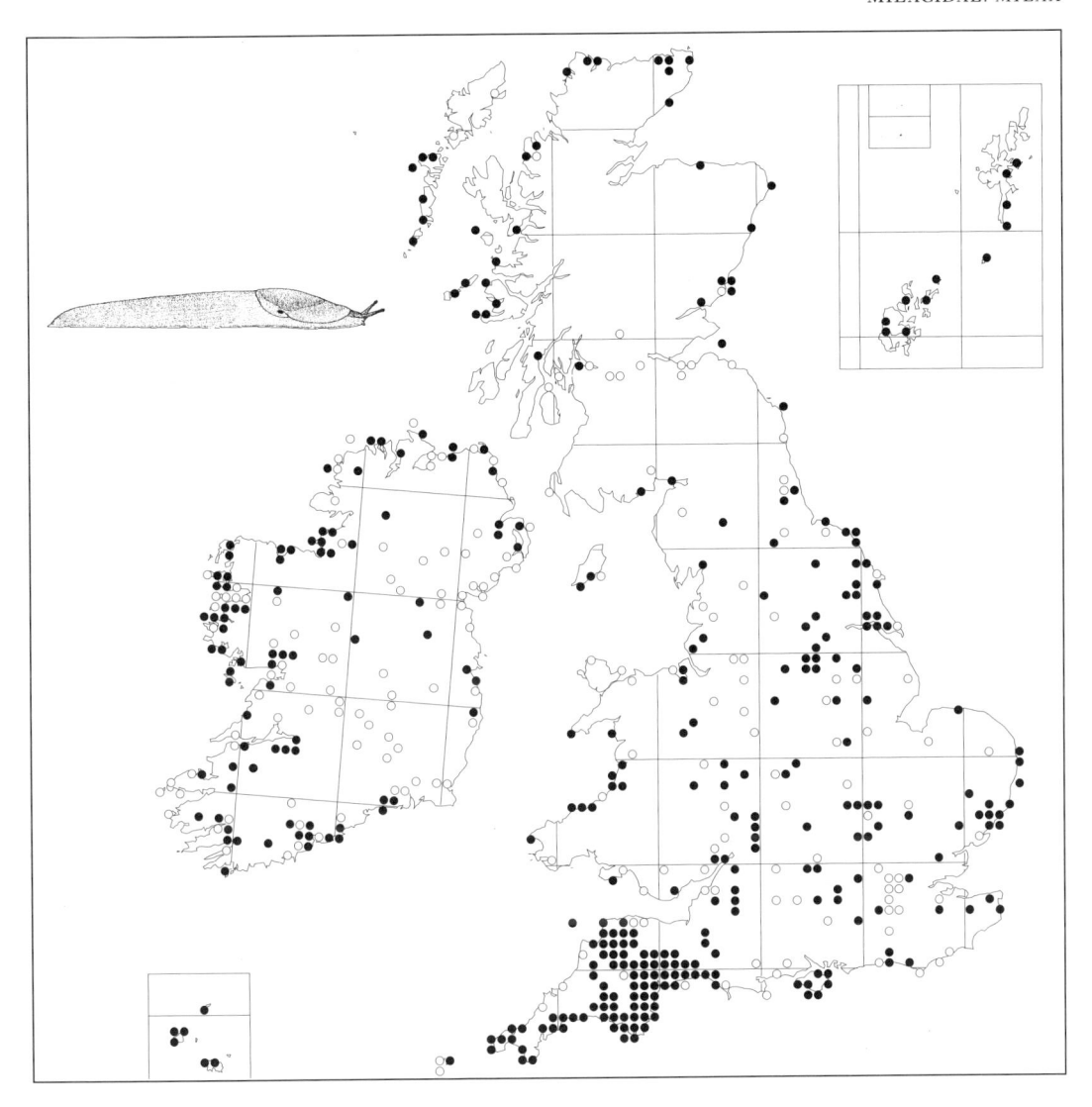

**Milax gagates** (Draparnaud, 1801)

Jet slug   Size: 45–55mm

Except in the south-west, this is generally a local slug, found under moist ground litter, in soil cavities, and among herbage. Like other British Milacidae (apart from *Tandonia rustica*), it is typical of gardens and waste ground but, unlike them, is not uncommon also in wild places, especially on cliffs, roadside banks and rough pastures adjacent to the sea. Occasionally in western areas it may be an agricultural pest.

Native? *M. gagates* shows a weaker association with man than *Tandonia sowerbyi* or *T. budapestensis* and may perhaps be indigenous. There is no clear evidence of significant geographical change.

W. Mediterranean and W. European Atlantic, with scattered populations as far east as the Rhine valley; introduced in Finland.

Old (pre-1930) records of this species are unreliable and many have been discarded. A related Mediterranean species, *M. nigricans* (Philippi) (=*M. insularis* (Lessona & Pollonera)), was found in a garden at Bexhill, Sussex, in 1948 (Quick, 1960) but there is no evidence that it is naturalized in Britain.

(Ellis: 253; K. & C.: 130)

149

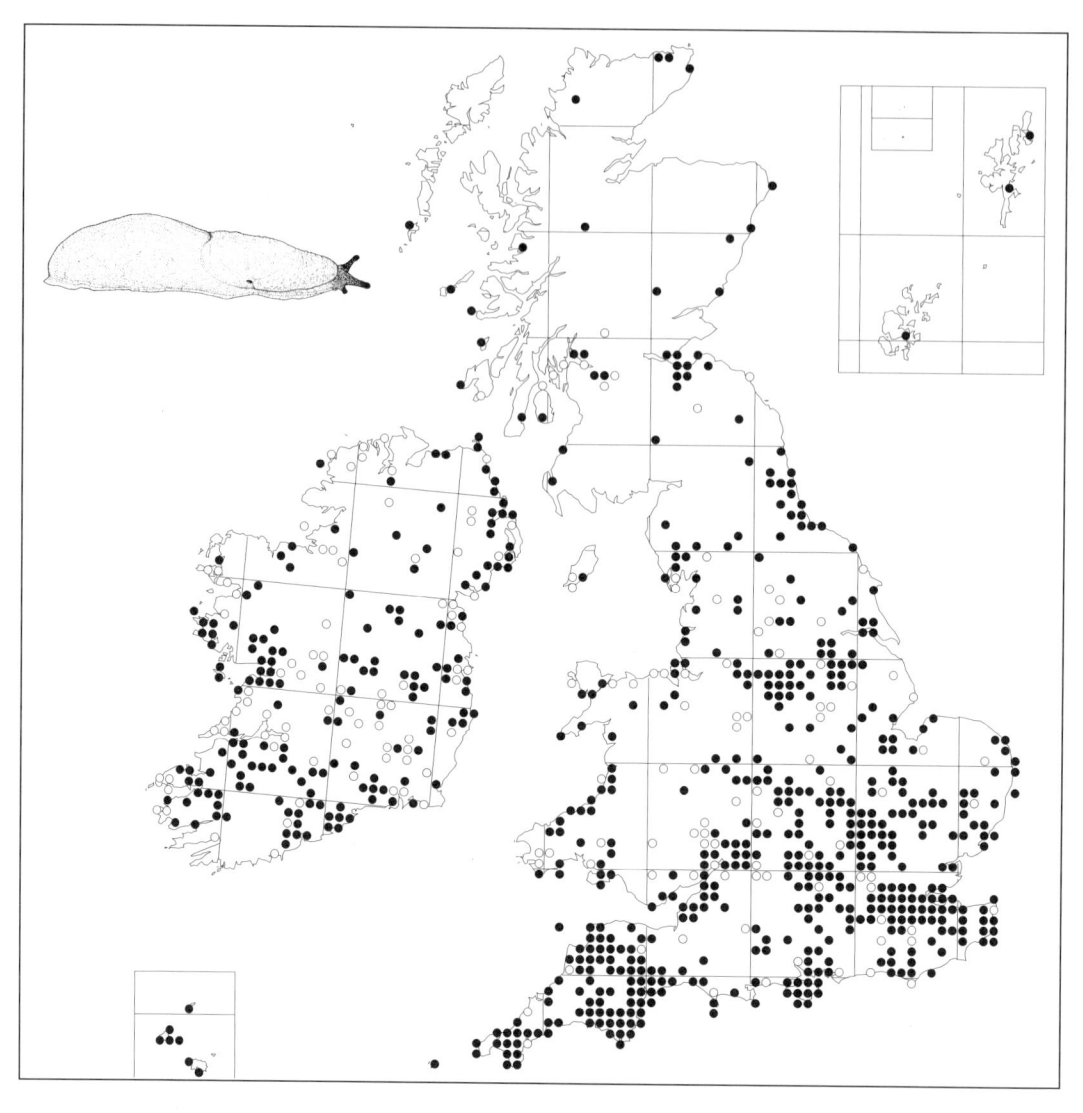

**Tandonia sowerbyi** (Férussac, 1823) (*Milax sowerbyi* (Férussac))

Keeled (or Sowerby's) slug   Size: 60–75mm

Locally common under moist ground litter, at the base of plants, or in soil cavities and crevices, nearly always in humanly disturbed places such as gardens, churchyards, rubbish tips and the margins of ploughed fields; it can be an agricultural pest, feeding on roots and tubers. Occasionally, especially on the south coast, it can also be found in relatively wild places by the sea, on cliffs and in maritime grassland.

Probably introduced. Though likely to be spreading, there is no clear evidence of recent distributional change.

Mediterranean and W. European Atlantic, northwards to the British Isles and eastwards to the Netherlands; not in Scandinavia.

It should be noted that old (pre-1930) records of this species made prior to the recognition of *Tandonia budapestensis* are unreliable and have mostly been discarded.

(Ellis: 254; K. & C.: 130)

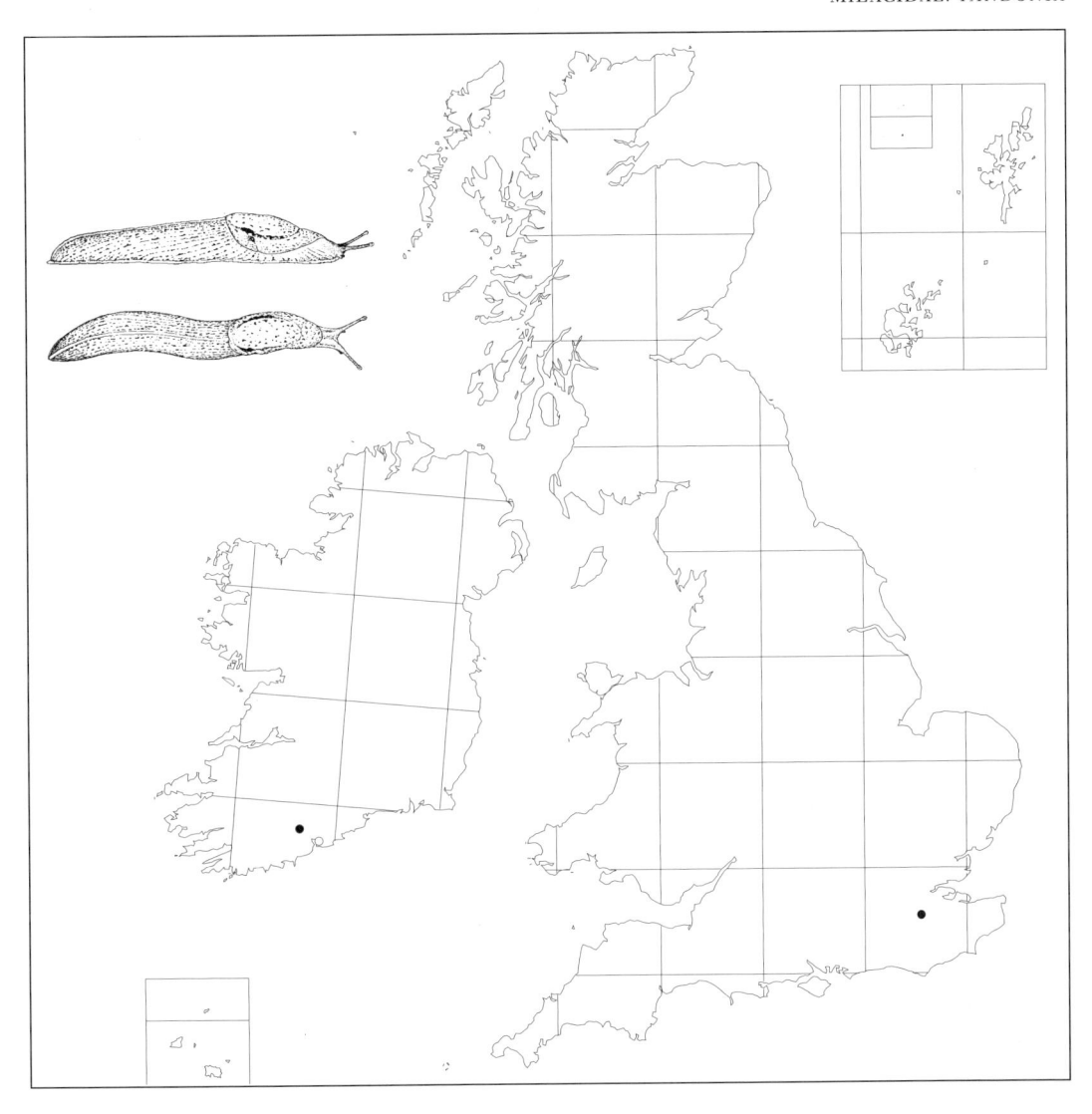

**Tandonia rustica** (Millet, 1843) (*Milax rusticus* (Millet), *M. marginatus* (Draparnaud))

Size: 50–70mm

On the Continent this species frequents woods and uncultivated places, especially on calcareous soils. Limestone screes are a favourite habitat. Unlike its close relative *T. sowerbyi*, it shows no obvious association with man. At the two known sites in the British Isles (One Tree Hill, near Sevenoaks (Kent); Blarney (Co. Cork)) it was discovered in old, mixed, deciduous woodland over limestone.

Native? *T. rustica* was discovered in England in 1986 (Philp, 1987) and in Ireland in 1995 (Cawley, 1996b). However, references to a 'spotted' milacid found at Blarney and at Aghada early this century (Stelfox, 1911) make it likely that the species has long been present in Co. Cork. Its restriction to wild habitats suggests that it is not an introduction, though this cannot be ruled out entirely. It is a slug of distinctive appearance and probably, therefore, genuinely rare in the British Isles rather than merely overlooked. RDB category: Insufficiently Known.

C. European, extending westwards to E. France, Belgium and S. Netherlands.

(K. & C.: 131)

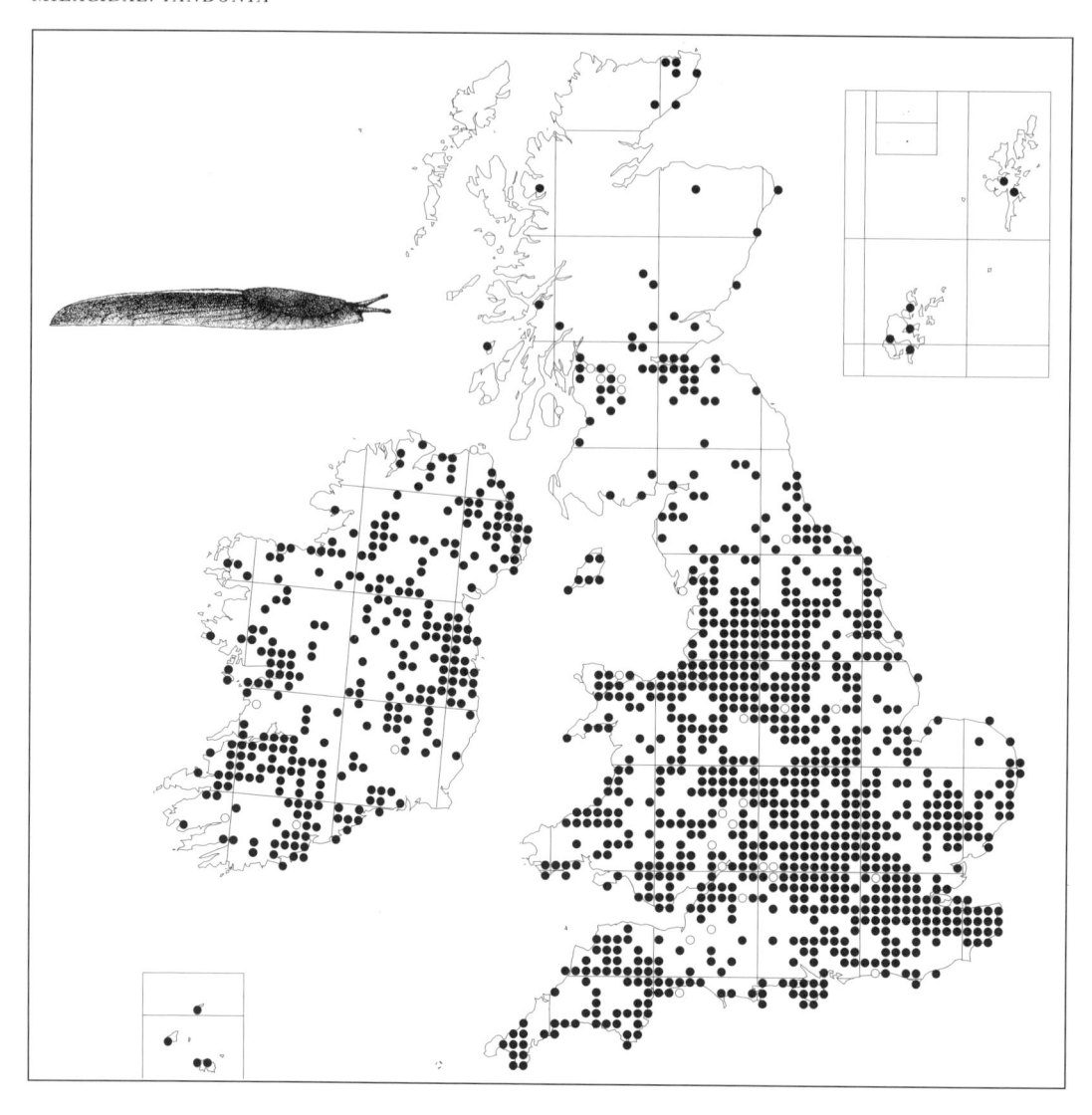

**Tandonia budapestensis** (Hazay, 1881) (*Milax budapestensis* (Hazay), *M. gracilis* (Leydig))

Budapest slug   Size: 50–70mm

A slug mainly associated with human disturbance. It is a characteristic species of urban gardens, roadsides, ploughed fields, and waste ground of all kinds. It lives in damp places under stones and ground litter, and in cavities beneath the surface, burrowing even into heavy soils. In lowland England it is a serious agricultural pest, especially of potatoes.

Probably introduced. First recognized in the British Isles only in 1921 (Phillips & Watson, 1930) but there is evidence that it was present here at least from the 1880s. It is still spreading. Sporadic records from semi-natural habitats such as old deciduous woodland can usually be traced to some degree of disturbance.

European: originally probably S.E. European only, but widely spread by man north to S. Germany, the Low Countries and the British Isles.

(K. & C.: 131)

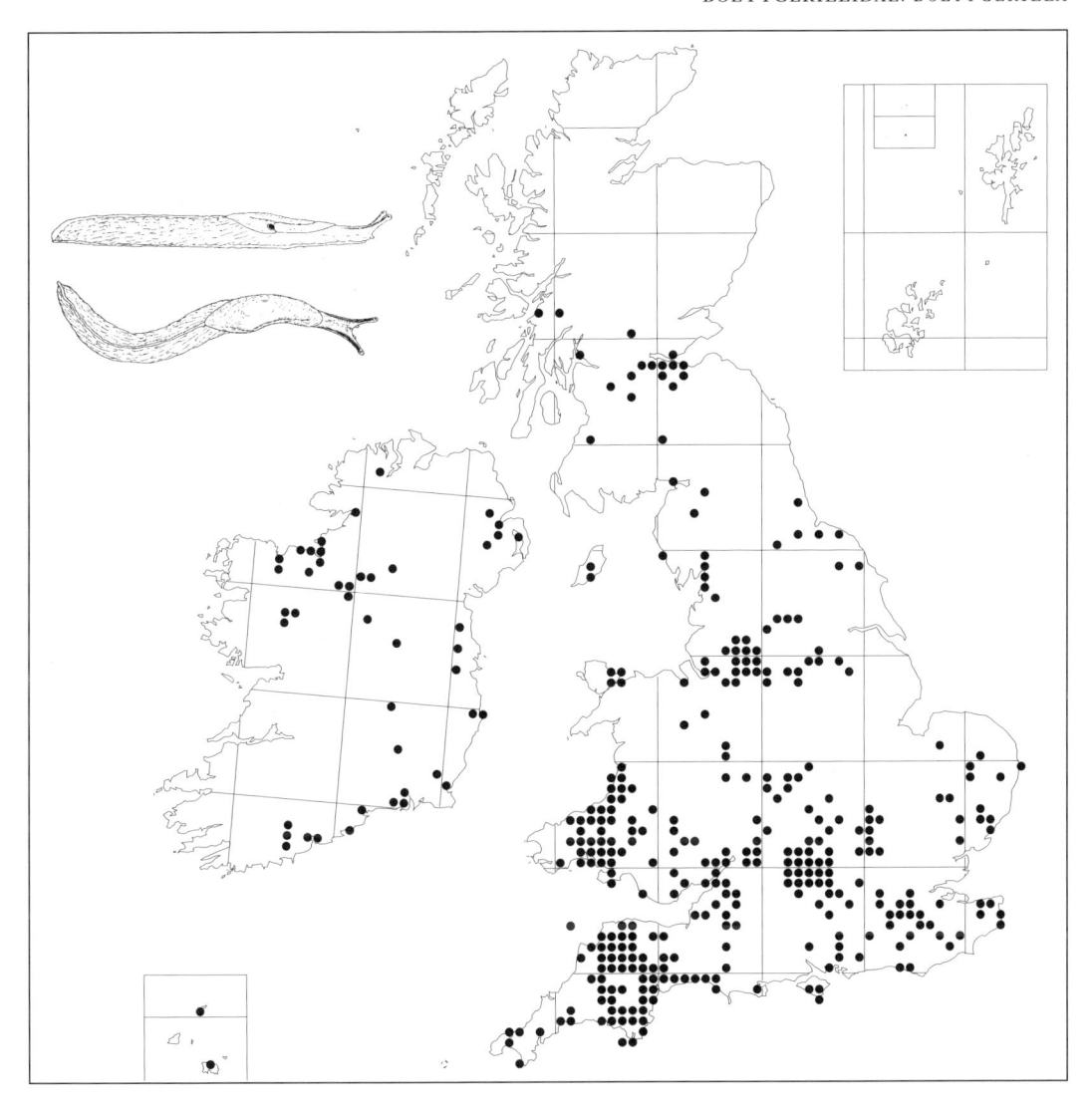

**Boettgerilla pallens** Simroth, 1912 (*B. vermiformis* Wiktor)

Worm slug   Size: 35–55mm

This is a slug of damp ground and subsurface crevices, usually in disturbed places such as gardens, old quarries, and unkempt, roadside verges. A favourite haunt is under stones or bricks lightly bedded into moist, clay soil. Its worm-like, extensible body adapts it to life below the surface in shrinkage cracks, root holes and worm burrows. *Boettgerilla* is often found with other 'rubbish' species like *Tandonia budapestensis* or *Oxychilus draparnaudi* but has also successfully invaded semi-natural woodland, especially in the west.

Introduced. *B. pallens* was first noted in England and Wales in 1972 (Windermere, Westmorland; Colville *et al.*, 1974), in Ireland in 1973 (Mountstewart, Co. Down; Anderson & Norris, 1974) and in Scotland in 1983 (Inverkeithing, Fife). It has excellent powers of dispersal and is evidently still spreading rapidly, though unevenly and unpredictably.

Originally Caucasian and S.E. European; recently spread by man into central and western Europe; also noted in British Columbia.

(K. & C.: 131)

LIMACIDAE: *LIMAX*

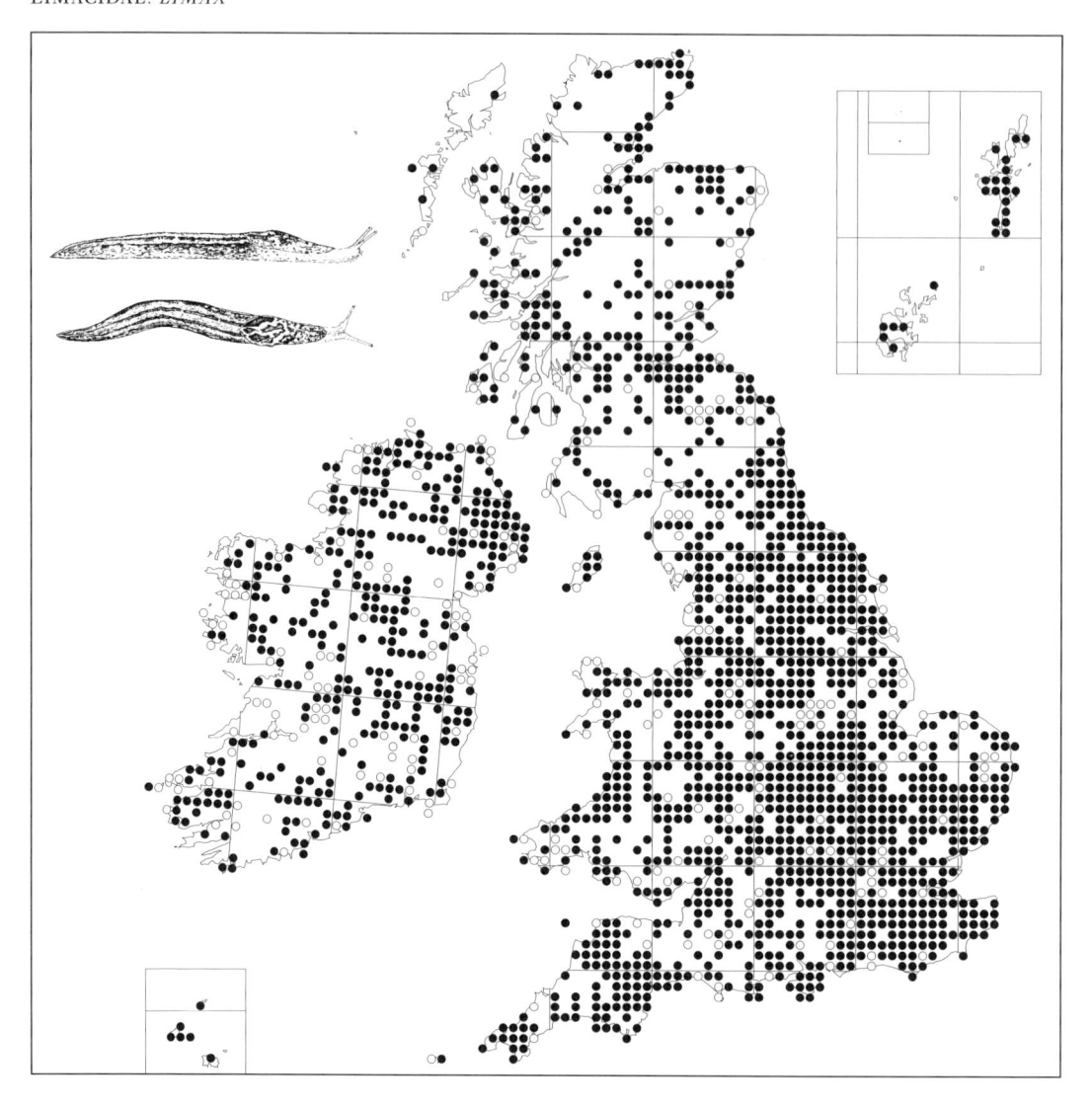

**Limax maximus** L., 1758

Great grey slug   Size: 100–160 (occ.–200)mm

This large species inhabits moist, sheltered places, hiding solitarily by day beneath logs and large pieces of ground litter, in the crevices of rocks, or under loose bark on tree trunks. It is a strong climber, at night ascending to considerable heights on walls and trees. Though common in some ancient deciduous woods, *L. maximus* is a species more characteristic of waste ground, farmyards, large gardens and roadside rubbish dumps, and in the highland areas of Scotland is largely restricted to such places.

Probably native. No evidence of distributional change, though doubtless locally spread by man.

W. and C. European, though rare in Scandinavia beyond 60°N.

(Ellis: 258; K. & C.: 136)

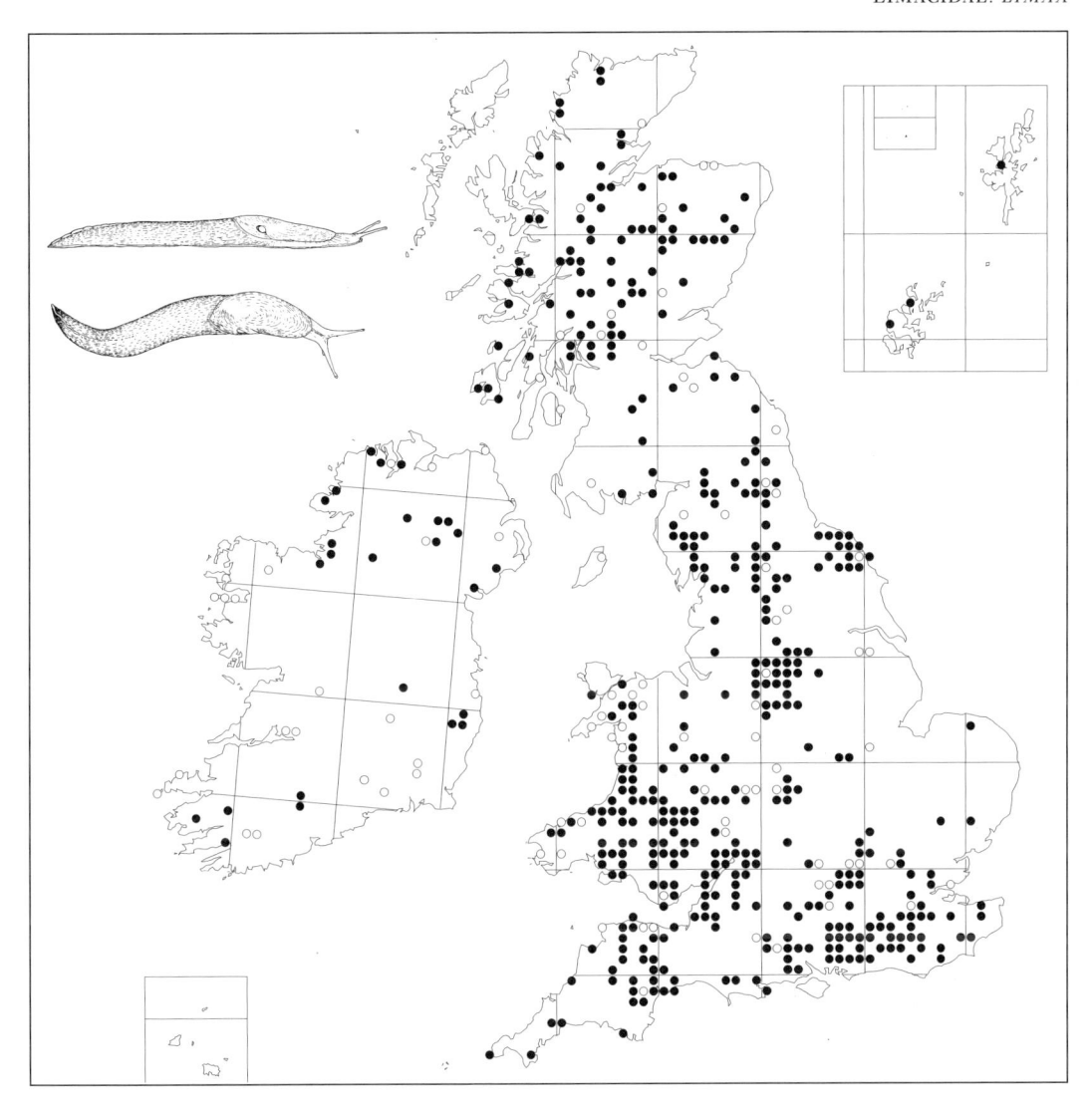

**Limax cinereoniger** Wolf, 1803

Ash-black slug   Size: 100–200mm

Britain's largest slug is typical of old deciduous or coniferous woods, sheltering by day under logs or loose bark. Occasionally it may be found on rocks or in rough grassland adjacent to woods. In the west of Ireland there are old records from sea cliffs. Most sites are on acid soils, though it is not averse to chalk or limestone. It is strongly nocturnal, emerging at dusk or in moist weather to feed and climb. It occurs in a variety of mixed and partially replanted woods, but is absent from historically modern plantations. It is only rarely associated with *L. maximus*.

Probably native. *L. cinereoniger* is a good indicator of ancient woodland, even if greatly modified by replanting or by traditional methods of management such as coppicing. Though still common at many sites it is likely to be retreating. In Ireland it is genuinely rare. RDB category (Ireland only): Rare.

European: widespread but local, and absent from northern Scandinavia.

(Ellis: 257; K. & C.: 136)

LIMACIDAE: *LIMAX*

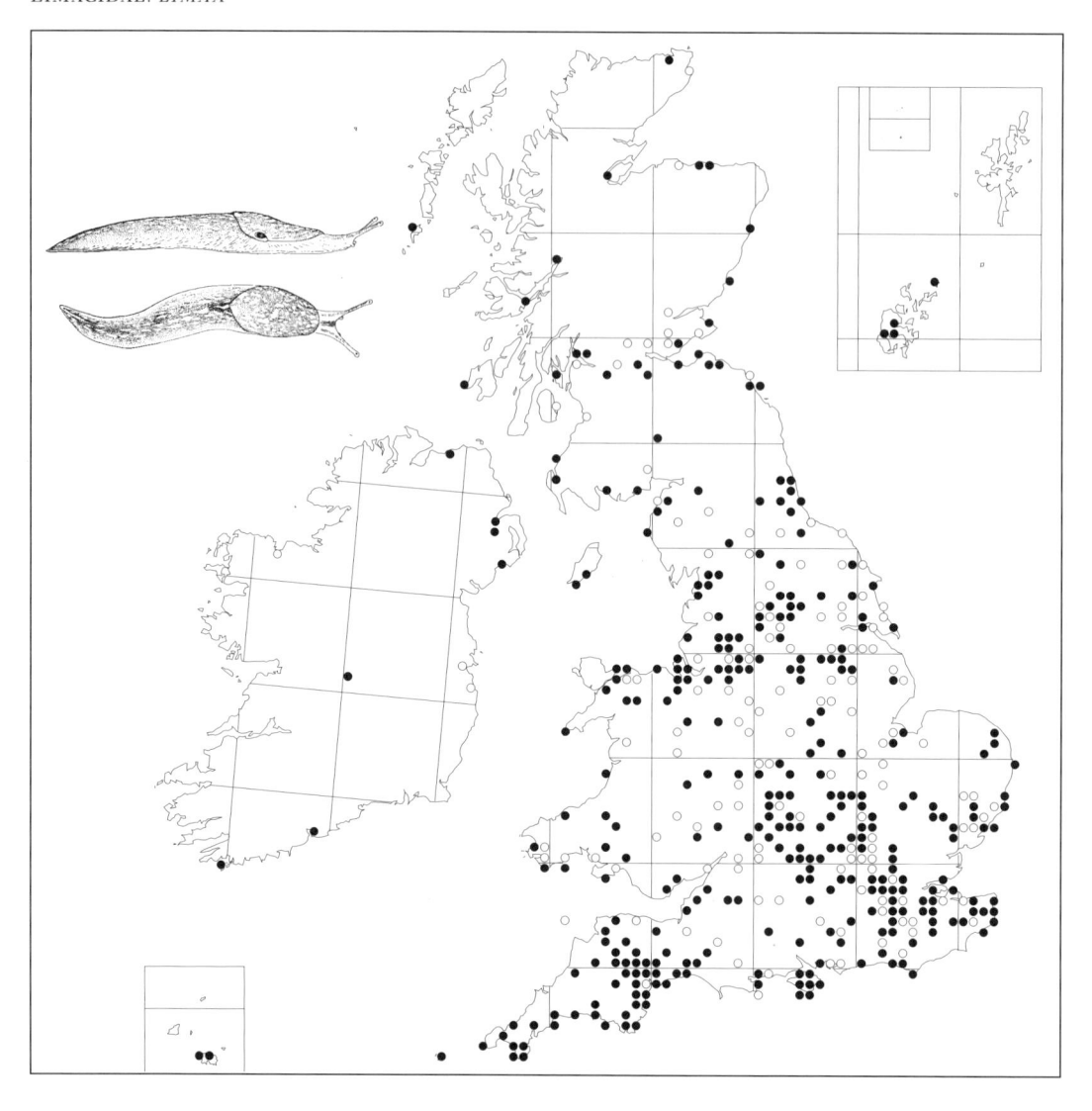

**Limax flavus** L., 1758 (*Limacus flavus* (L.))

Yellow slug   Size: 80–130mm

In the British Isles this is a slug closely associated with man. It lives in crevices and under rubbish in gardens, cellars, outhouses and farmyards; in cities, old stone-floored sculleries and damp basements were traditional haunts but it is not unknown in modern suburban houses. Occasionally it is found in woods. It is strongly nocturnal and a vigorous climber.

Probably introduced, though certainly present in England in the 17th century, when its anatomy was described by Martin Lister (1694). In spite of the high proportion of old records there is no evidence of recent geographical change; the species is underrecorded since its nocturnal habits allow it to escape detection even in places where it is common. In Ireland it appears to be genuinely local except in some urban areas.

Originally Mediterranean but widely spread by man, northwards as far as the British Isles, Denmark and southernmost Sweden.

Note that some old British records may be referable to *L. maculatus* though it is likely that the majority are correct. On the other hand, most old Irish records of '*flavus*' do probably refer to *maculatus*, and have been discarded.

(Ellis: 262; K. & C.: 137)

LIMACIDAE: *LIMAX*

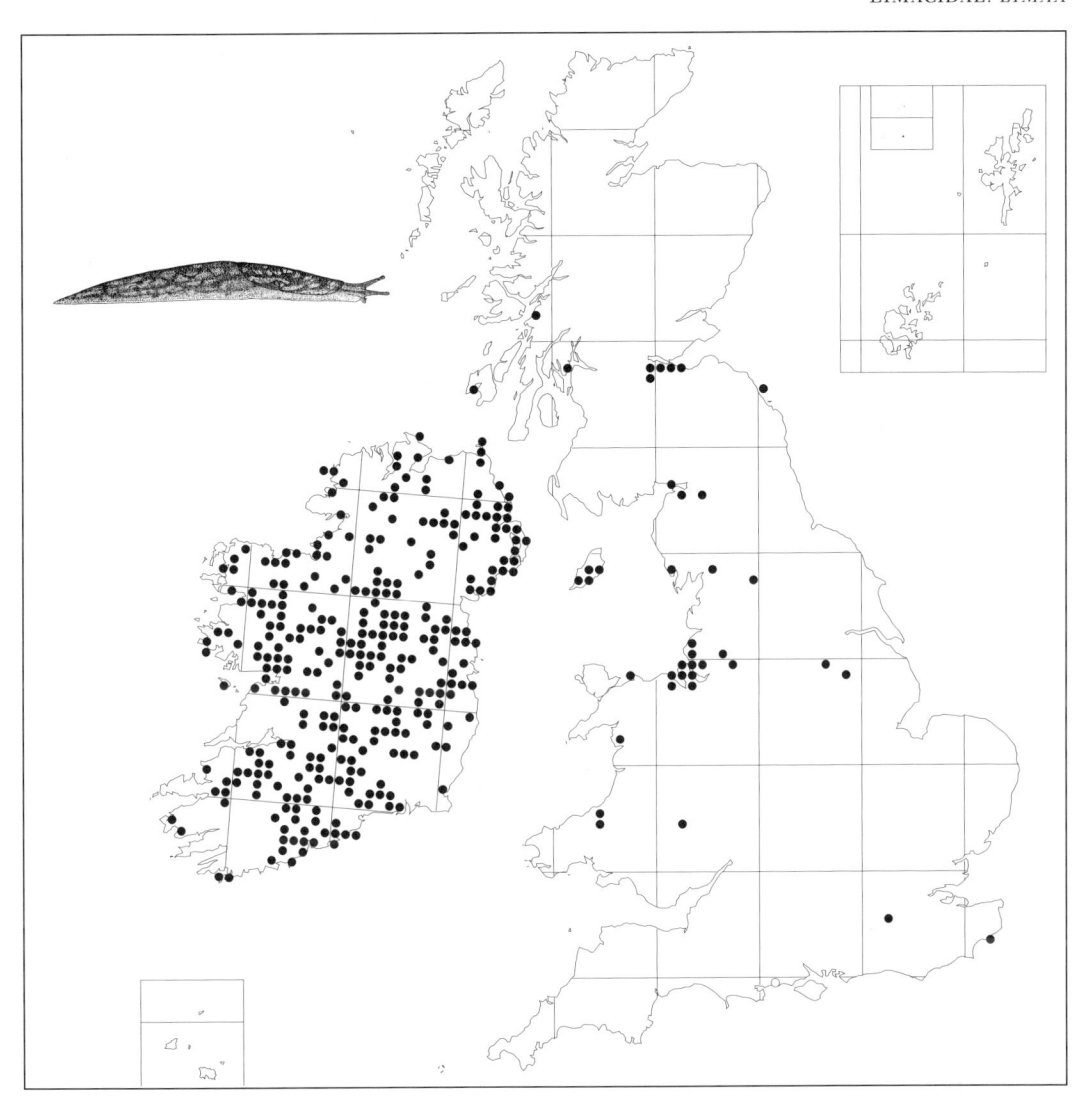

**Limax maculatus** (Kaleniczenko, 1851) (*Limacus maculatus* (Kaleniczenko), *Limax grossui* Lupu, *L. pseudoflavus* Evans)

Size: 80–130mm

In Ireland this species occurs freely in roadside spinneys and plantations, hiding under logs, loose bark, or in deep holes in tree trunks, emerging at night to feed and climb. It lives also in more open habitats around ruined buildings and in the crevices of stone walls (Evans, 1978). Like *L. flavus* it is fond of human rubbish, sheltering gregariously under large pieces of litter dumped by roadsides or in neglected farmyards and gardens. In Britain most finds are from such places.

Introduced. A native of the deciduous forests of the Crimea and the Caucasus, this species has been spread by man in Europe (Wiktor & Norris, 1982). Though now common in Ireland there is some evidence that it was much rarer there a century ago. In Britain (where it is probably underrecorded) it was first noted with certainty in 1884 at Christchurch, Hants (Kerney, 1986). It may still be extending its range.

Full range uncertain; recorded from the British Isles, France, Romania, Bulgaria, Turkey and Russia (Black Sea region, and introduced to vicinity of Moscow and St Petersburg).

(K. & C.: 137)

157

LIMACIDAE: *MALACOLIMAX*

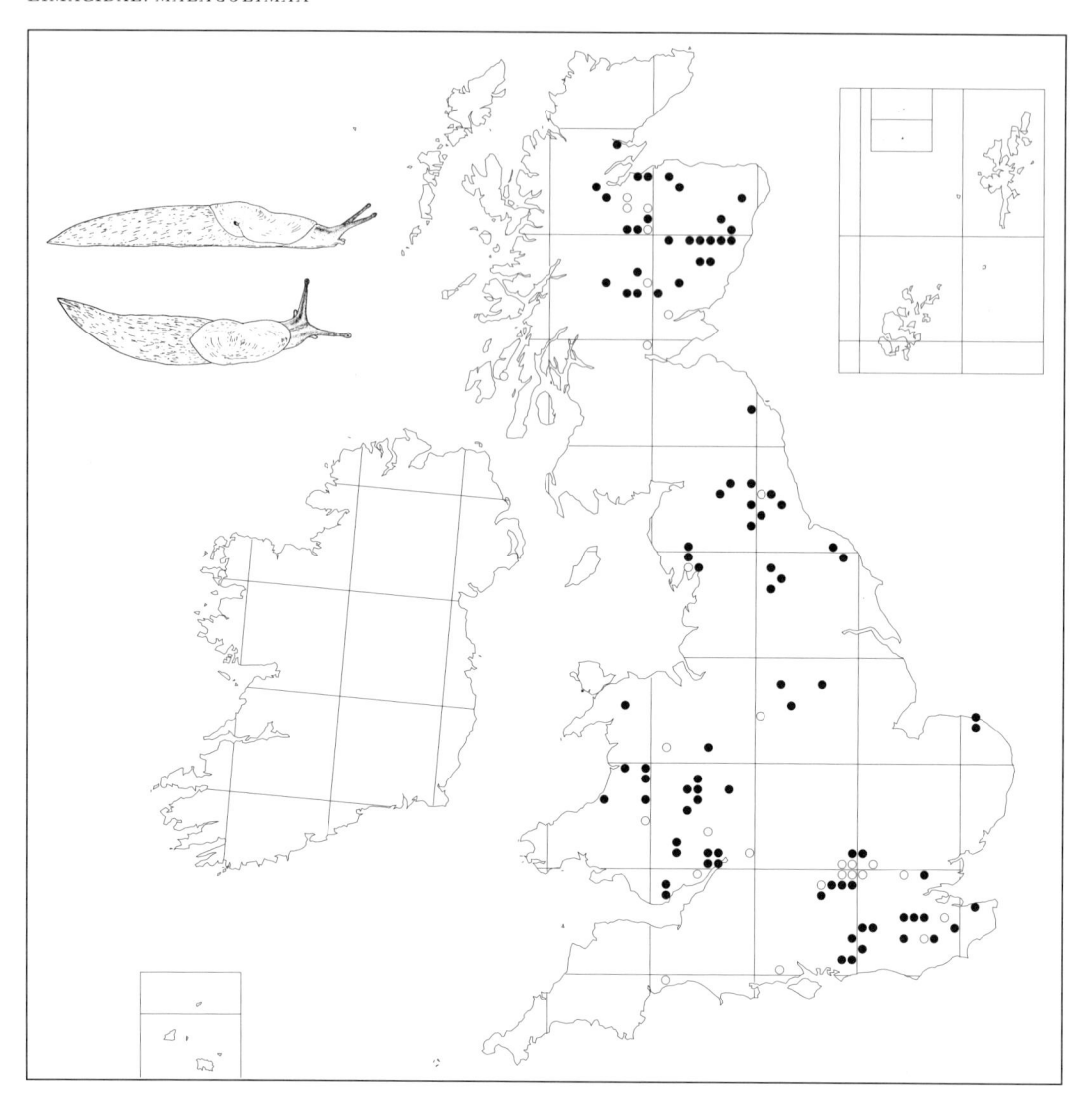

**Malacolimax tenellus** (Müller, 1774) (*Limax tenellus* Müller)

Slender (or tender) slug   Size: 35–50mm

A species restricted to ancient deciduous or coniferous woods, occasionally on chalk or limestone but more commonly on poor, acid soils, often in remnant woodland on steep slopes. It shelters under ground litter, logs and loose bark, emerging at night to feed, typically on fungi. Although it lives in a wide variety of mixed and partially replanted woods (for example, of beech, oak, chestnut, hornbeam, birch or pine) and tolerates traditional methods of management such as pollarding and coppicing, it is unknown in modern plantations. *Limax cinereoniger* is usually found with it (Oldham, 1909; 1922b).

Probably native. Undoubtedly receding, and at risk in many places from replanting and clear felling. Nevertheless it is not an easy species to find and the absence of modern records at certain sites (for example, in the New Forest and in Epping Forest) is not necessarily evidence of extinction.

N. and C. European: widespread but local, and absent from northern Scandinavia.

(Ellis: 259; K. & C.: 136)

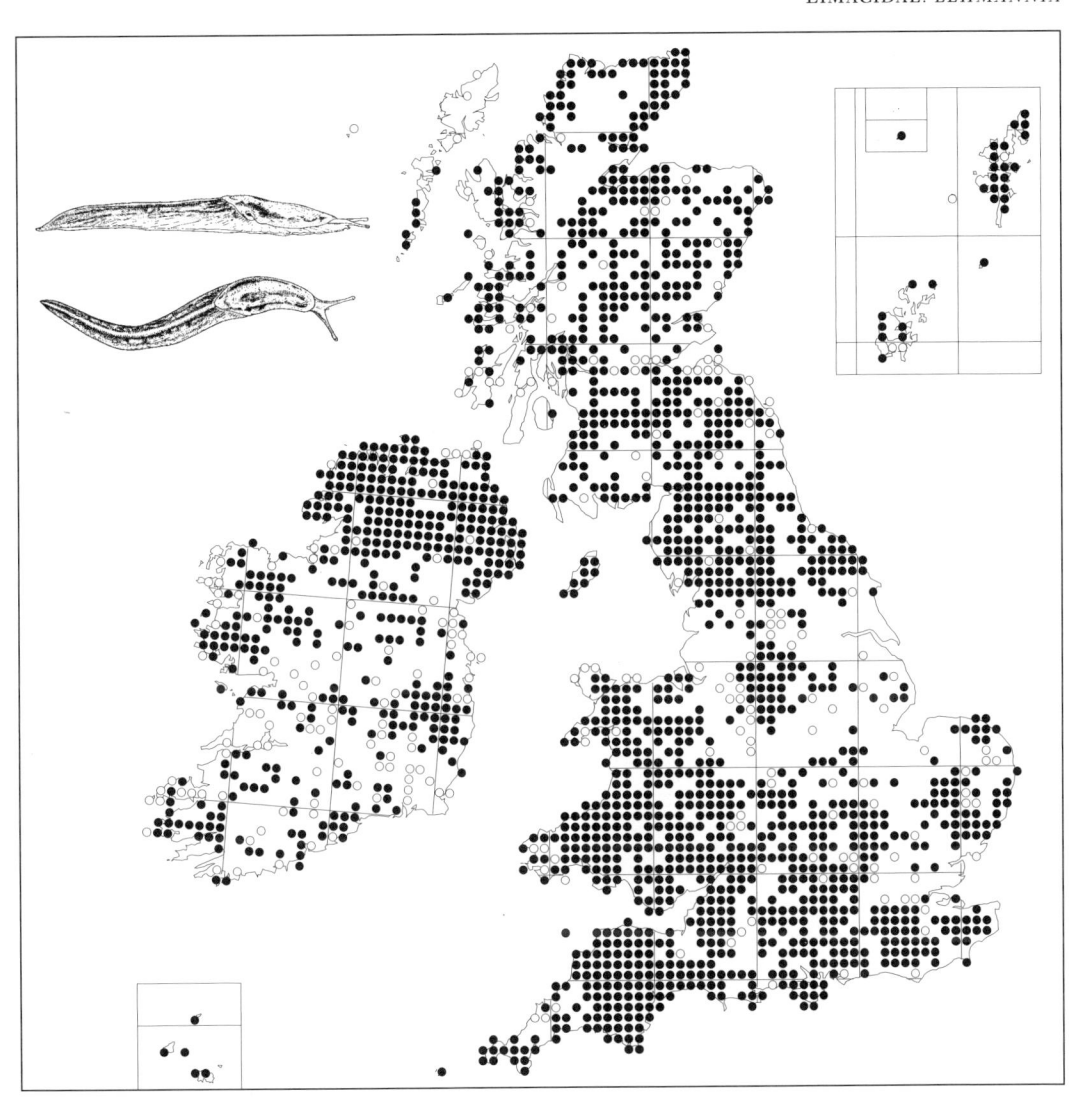

**Lehmannia marginata** (Müller, 1774) (*Limax marginatus* Müller, *L. arborum* Bouchard-Chantereaux)

Tree slug   Size: 70–80mm

A species typical of woods and rocky places, always requiring hard surfaces on which it can graze on algae and lichens. By day it shelters under loose bark or in deep crevices in rocks and walls, emerging in damp weather or at night to feed and climb. In the more oceanic parts of the British Isles it lives in exposed situations on mountains and sea cliffs, but in central and eastern England it is more narrowly restricted to sheltered places such as undisturbed woodland and disused stone quarries, avoiding intensively farmed, lowland areas where suitable habitats are few.

Probably native. There is little evidence of serious national decline, though in central and eastern areas of England there may be a few 10km-square extinctions caused by habitat destruction, and possibly also by atmospheric pollution damaging epiphytes on tree trunks and walls.

European: nearly throughout, except in northern Scandinavia.

(Ellis: 260; K. & C.: 138)

LIMACIDAE: *LEHMANNIA*

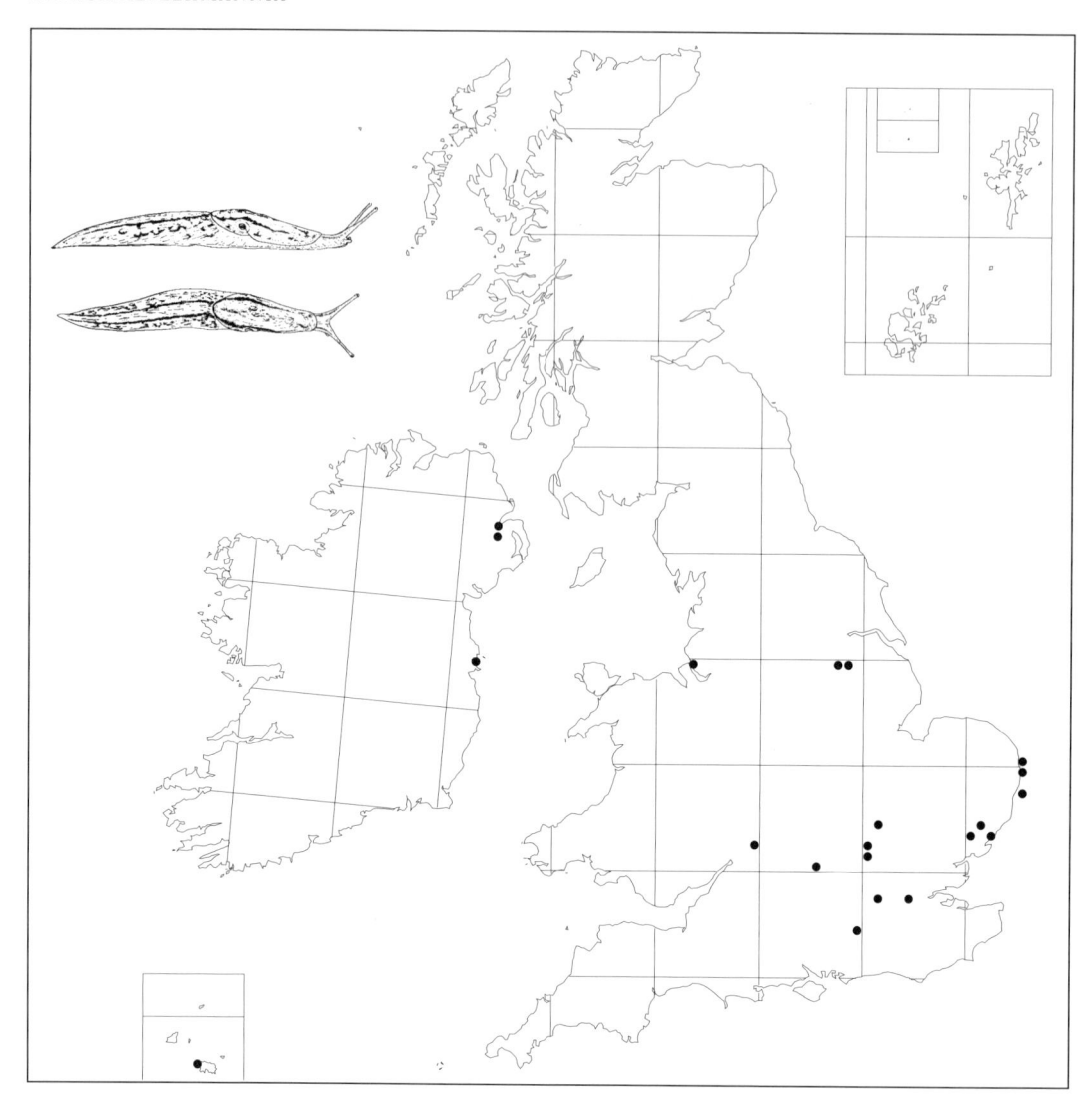

**Lehmannia valentiana** (Férussac, 1821) (*Limax valentianus* Férussac, *L. poirieri* Mabille)

Greenhouse slug   Size: 55–70mm

Unlike *L. marginata*, this is a ground-dwelling not a climbing slug. It lives in humanly disturbed places, sheltering under stones, pieces of wood and damp ground litter.

Introduced. *L. valentiana* has long been known in the British Isles from glasshouses in nurseries and botanic gardens, at Kew, Swansea, Belfast and elsewhere. In 1981 it was discovered in a garden and in adjacent waste ground by the sea at Portmarnock, Co. Dublin. Since then it has been found in open habitats in numerous places, north to the East Midlands, Merseyside and Belfast. It has probably been overlooked and is likely to be commoner than the few records suggest.

Originally Iberian peninsula; now widely spread by man. Found in greenhouses in most parts of Europe, and reported living in the open in France, Belgium, the Netherlands, Denmark and S. Sweden.

(K. & C.: 211)

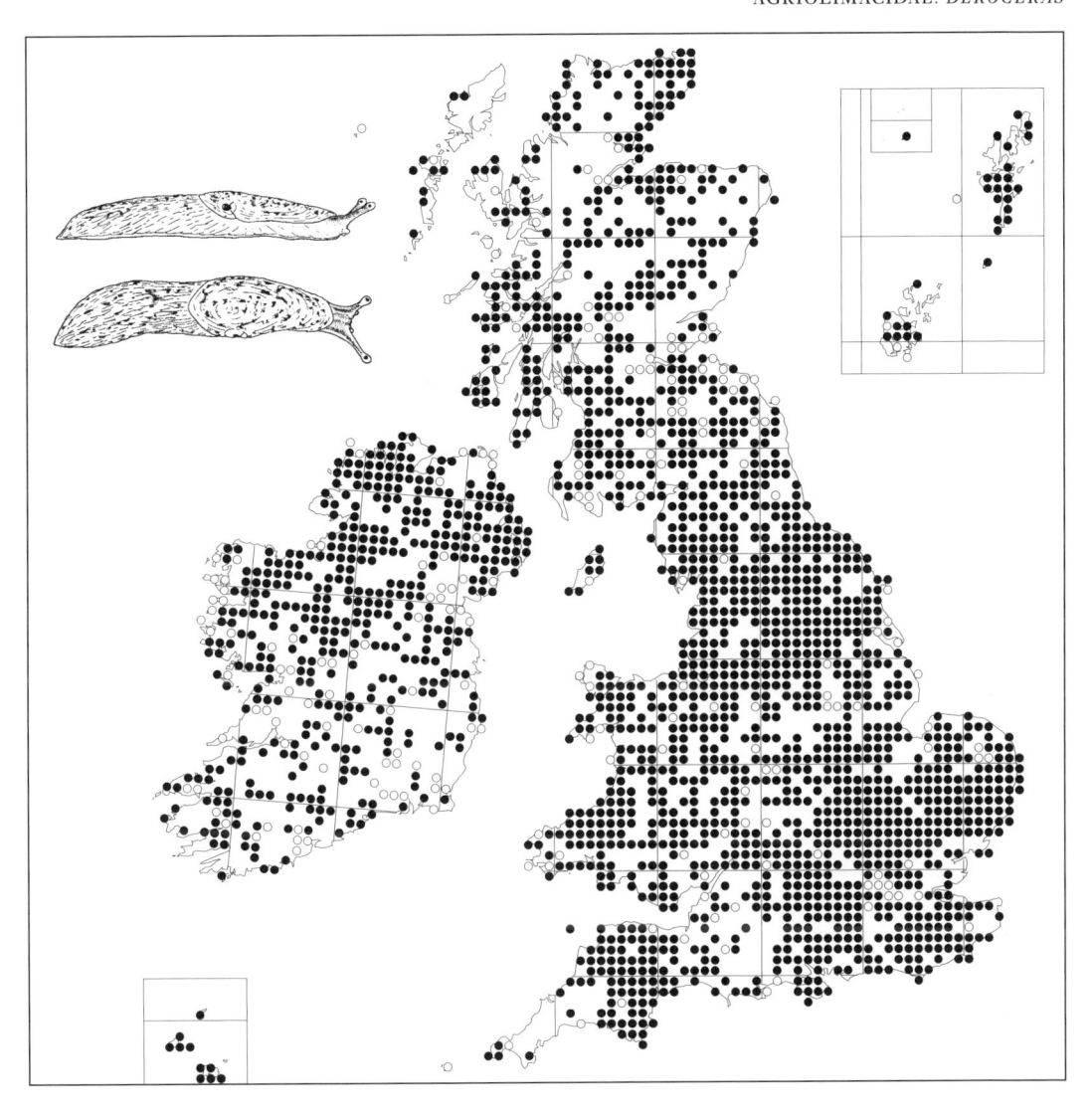

**Deroceras laeve** (Müller, 1774) (*Agriolimax laevis* (Müller))

Marsh slug    Size: 15–25mm

A typical wetland species, found on ground litter or low down on emergent vegetation in marshy places: fens, water meadows, river banks, lake shores, dune slacks, mountain flushes. It is virtually amphibious and can survive prolonged flooding. Occasionally it occurs in woods on heavy, badly drained soils well away from permanent water (Oldham, 1929a).

Probably native. Though a wetland species, *D. laeve* shows little evidence of significant national decline. Newly created habitats are often colonized after a few years.

Holarctic; throughout most of Europe.

(Ellis: 255; K. & C.: 142)

## AGRIOLIMACIDAE: *DEROCERAS*

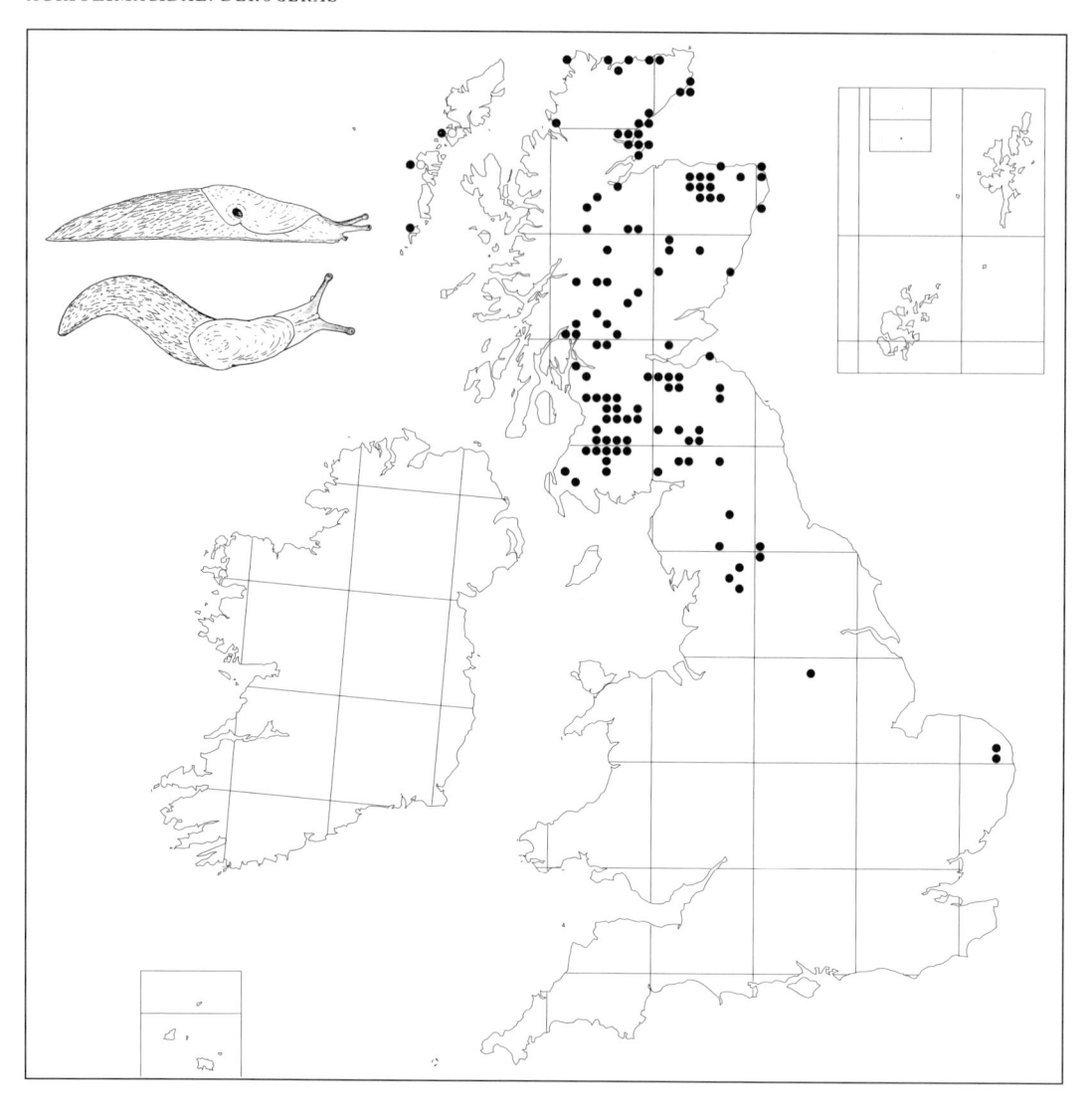

**Deroceras agreste** (L., 1758) (*Agriolimax agrestis* (L.) of Quick, 1960)

Size: 35–45mm

This is a slug of moist, grassy places, mainly away from human settlements and cultivation. In Scotland it particularly favours tall herbage with umbellifers along roadsides or in marshy ground by streams, but it also occurs in upland pastures and under stones on hillsides (Ellis, 1967; Kellock, 1970). Exceptionally, it inhabits birch woods. The isolated sites in the Norfolk Broads are in very wet places, in reed swamp and fen carr.

Probably native. A species likely to have been more widespread in Lateglacial and early Postglacial times. Though its northerly distribution is not in doubt, it remains underrecorded because of its external resemblance to the ubiquitous *D. reticulatum*. In the north of Ireland, search in suitable habitats has failed to detect it. There is no evidence of recent decline.

W. Palaearctic; in Europe mainly northern (to beyond the Arctic Circle) and montane.

Note that records of *Agriolimax agrestis* made prior to the 1940s refer to *Deroceras reticulatum*, not to the present species.

(K. & C.: 143)

AGRIOLIMACIDAE: *DEROCERAS*

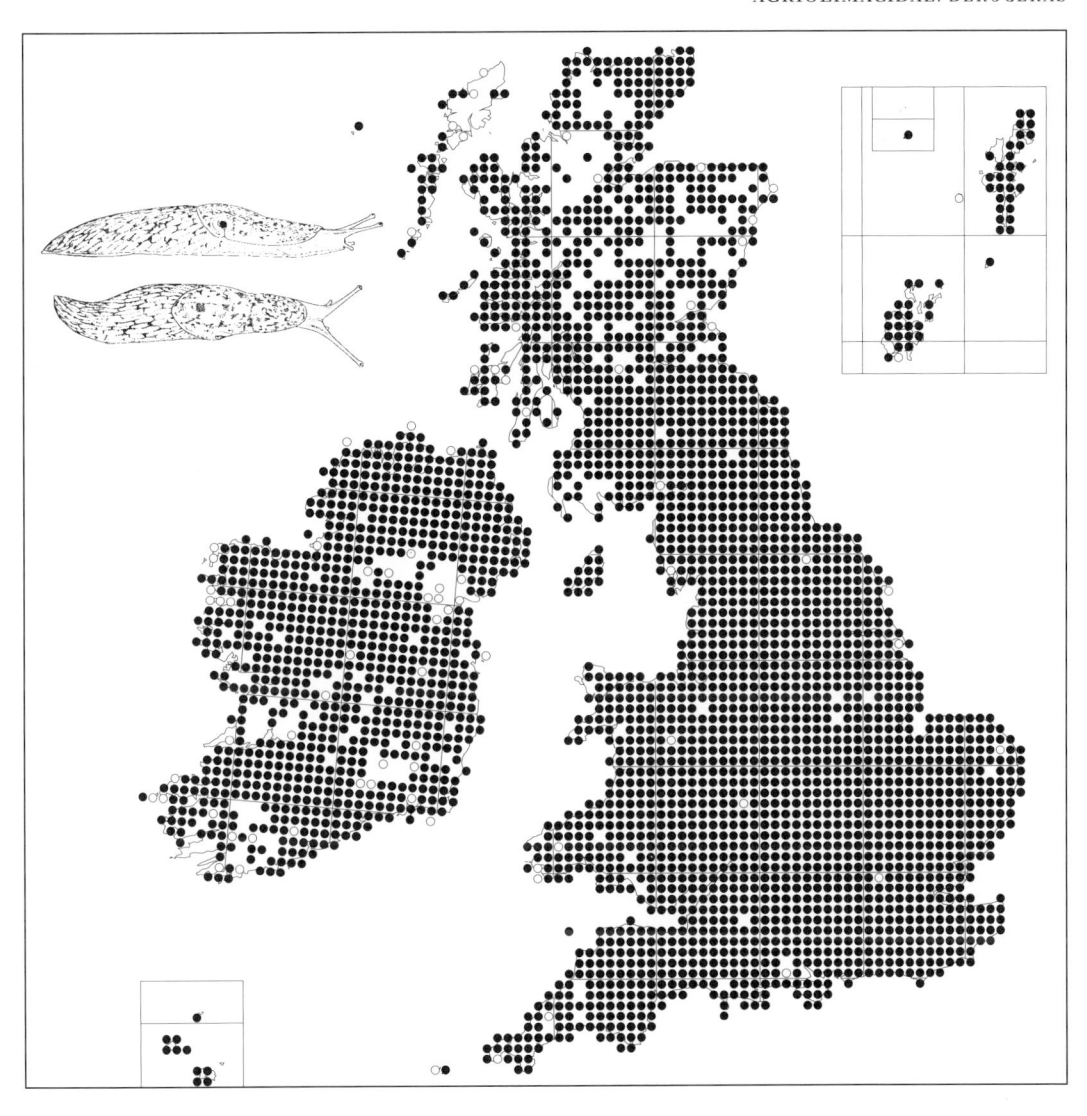

**Deroceras reticulatum** (Müller, 1774) (*Agriolimax reticulatus* (Müller), *A. agrestis* of older British authors, *non* Linnaeus)

Field (or milky) slug    Size: 35–50mm

This almost ubiquitous slug lives in open places, in pastures and arable fields, roadside verges, waste ground and urban gardens. Though not a burrower, it shelters under stones and ground litter (including human rubbish) and can be a serious agricultural and horticultural pest. It is rare in woods and absent from some harsh non-calcareous uplands tolerated by *Arion ater* and *A. intermedius*.

Probably native. No evidence of recent distributional change. The spread of the species in the highland zone may be due to man.

European: nearly throughout, but scarce in N. Scandinavia, where it is mainly synanthropic.

(Ellis: 255; K. & C.: 143)

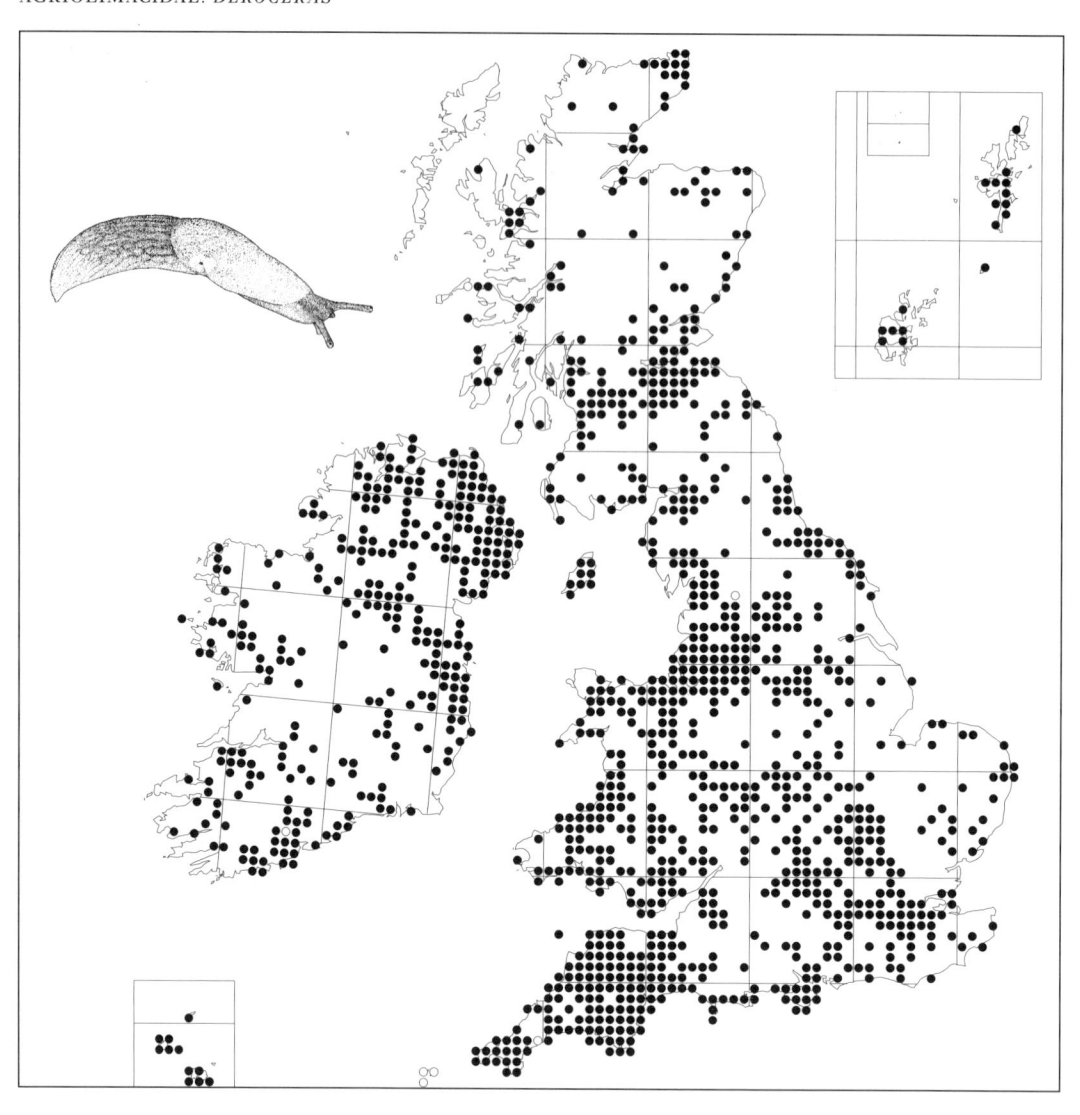

**Deroceras panormitanum** (Lessona & Pollonera, 1882) (*Agriolimax caruanae* (Pollonera), *Deroceras caruanae* (Pollonera))

Caruana's slug, Sicilian slug   Size: 25–35mm

A species typical of disturbed habitats, and associated with roadside rubbish, farmyards and gardens. It often shelters under stones, pieces of wood, cardboard and other litter in bare or sparsely vegetated waste ground. It can be a pest. In climatically mild areas (notably in S.W. England and in parts of Ireland), *D. panormitanum* has also successfully established itself in wild places, such as woods and sea cliffs.

Probably introduced. The first certain records of this species in Britain date from 1931 (Sennen Cove (Cornwall); Swansea (Glamorgan); Darlington (Co. Durham)) and in Ireland from 1958 (near Cork; Makings, 1959). In the last twenty-five years it has spread rapidly and is now a common slug in many areas.

Originally Mediterranean; now widely spread by man in N.W. Europe.

(K. & C.: 143)

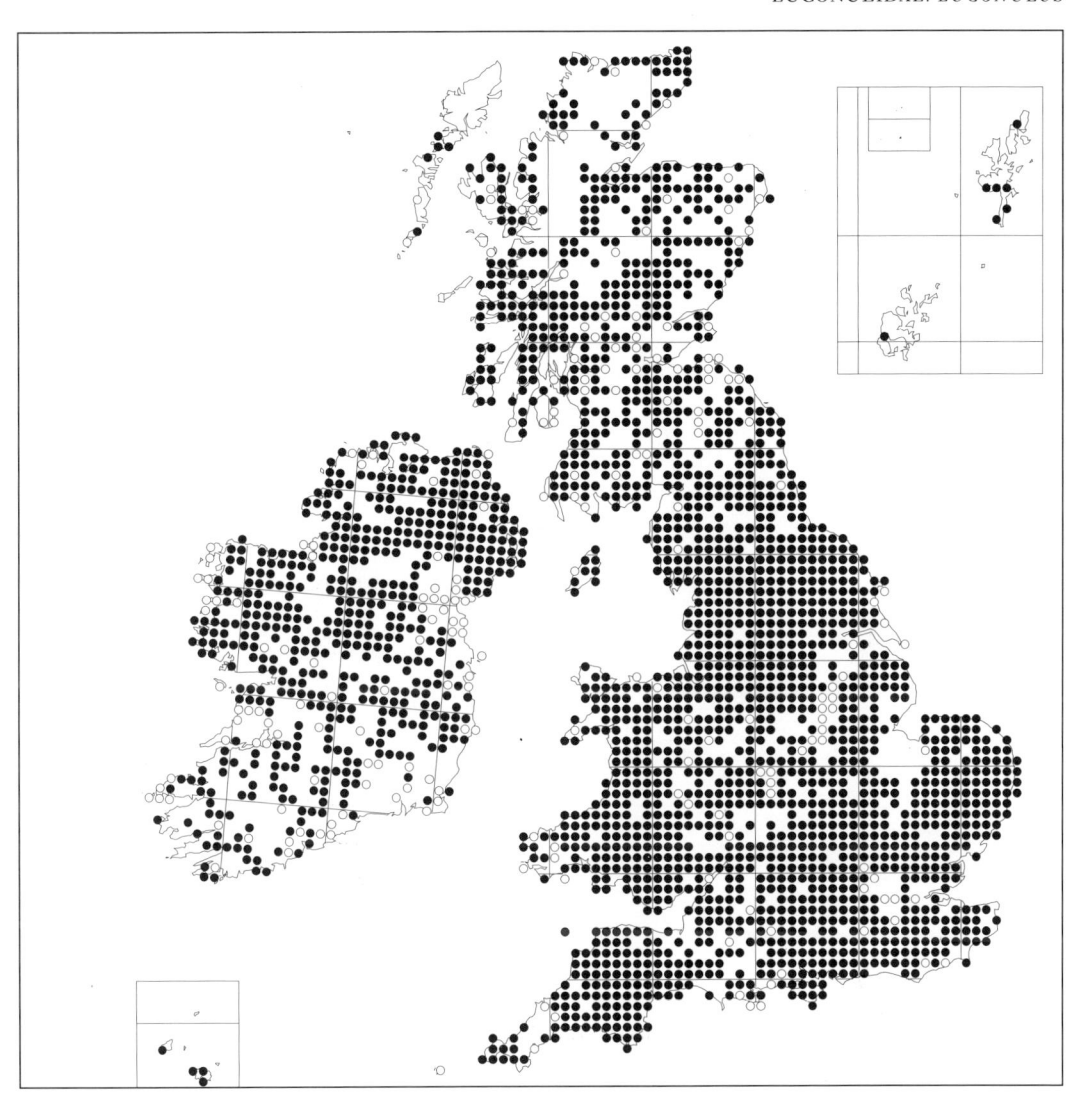

**Euconulus fulvus** aggregate (*Hyalinia fulva* of British authors)

Tawny glass snail

This map shows all records of *Euconulus* spp. The majority probably refer to *E. fulvus*, sensu stricto, rather than to *E. alderi*.

It should be noted that the morphological distinction between these forms is not always clear-cut, and some doubt remains about their validity as distinct species.

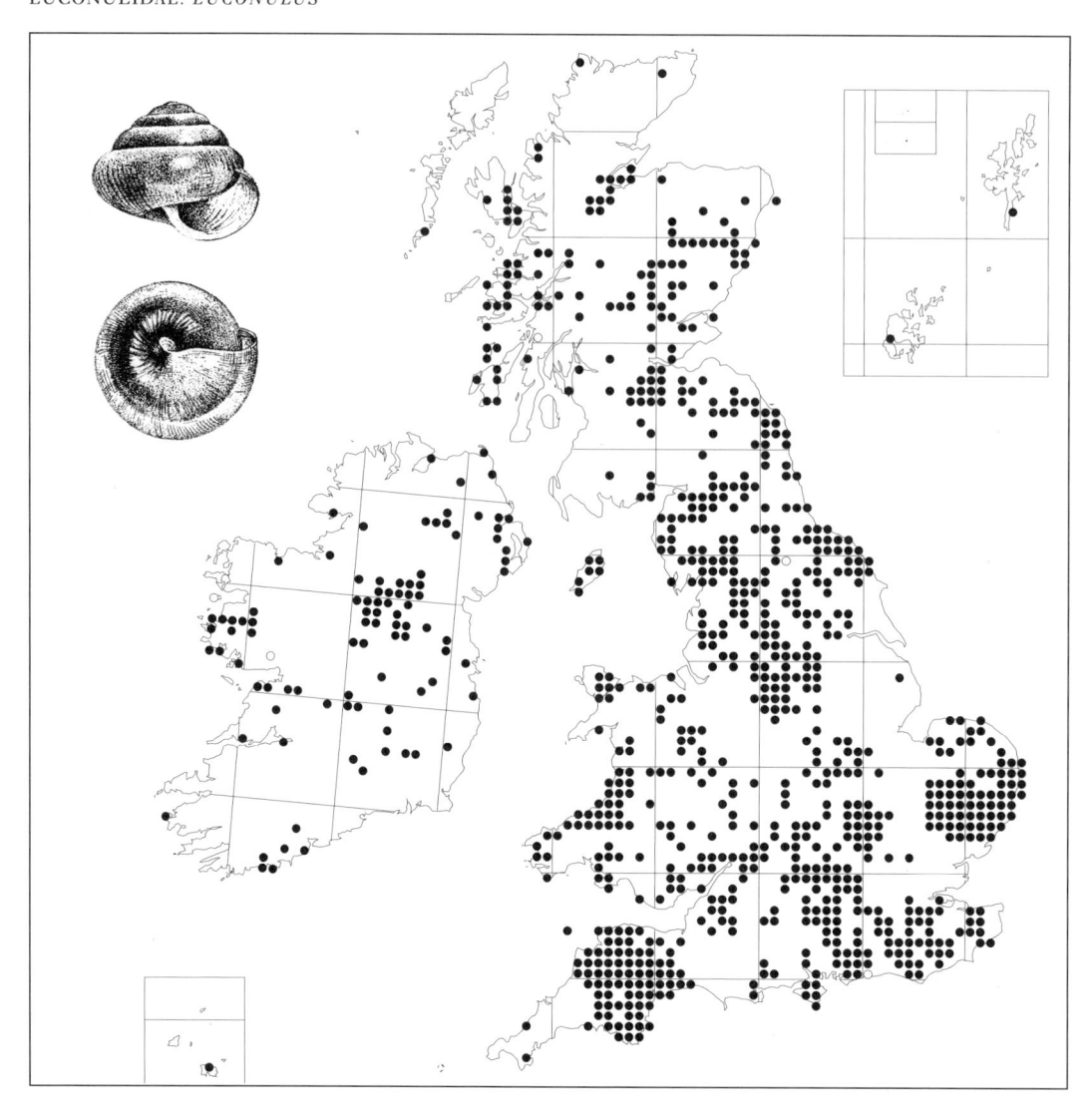

**Euconulus fulvus** (Müller, 1774).

Size: 2.8–3.5mm

Occurs in a wide variety of moist, sheltered and undisturbed places, such as ground litter in coniferous and deciduous woods, mossy hedge banks, dune slacks and marshes. It tolerates acid soils. In wet places it may be associated with *E. alderi*, or replaced by it.

Native (Lgl). Probably nearly ubiquitous and showing no evidence of geographical change, though underrecorded following the recognition of *E. alderi* in Britain in the 1970s.

Holarctic; throughout Europe to beyond the Arctic Circle.

(K. & C.: 148)

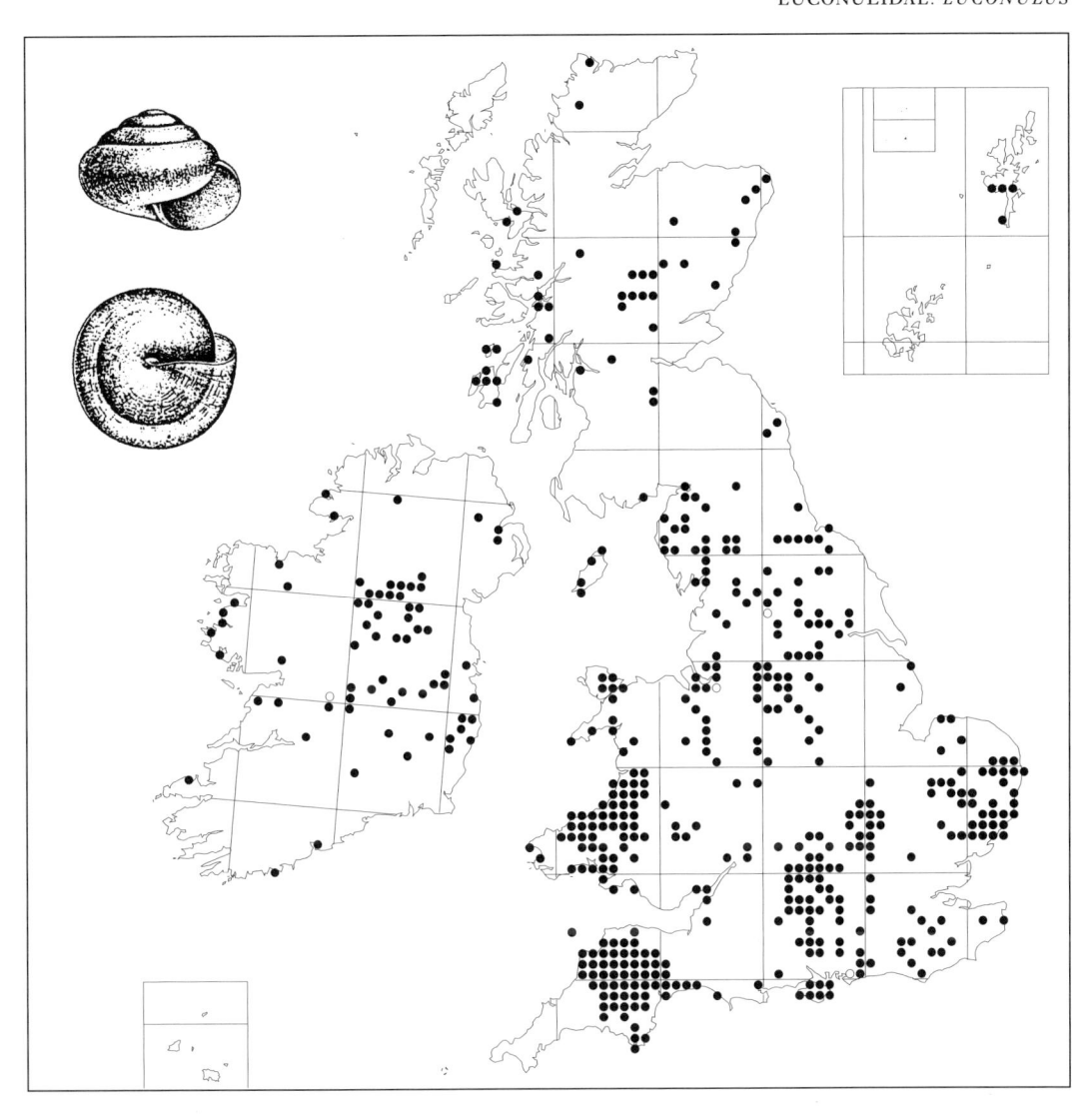

**Euconulus alderi** (Gray, 1840) (*E. fulvus* var. *alderi* Gray)

Size: 2.3–2.8mm

This snail lives in places similar to those inhabited by *E. fulvus* but generally wetter. It is typical of marshes or boggy woodland where it may occur to the exclusion of *E. fulvus*.

Native (Lgl). No clear evidence of national decline; probably suffering as a wetland species. Undoubtedly more local than *E. fulvus* in most parts of the British Isles though remaining seriously underrecorded.

Imperfectly known, but probably Holarctic.

(K. & C.: 149)

FERUSSACIIDAE: *CECILIOIDES*

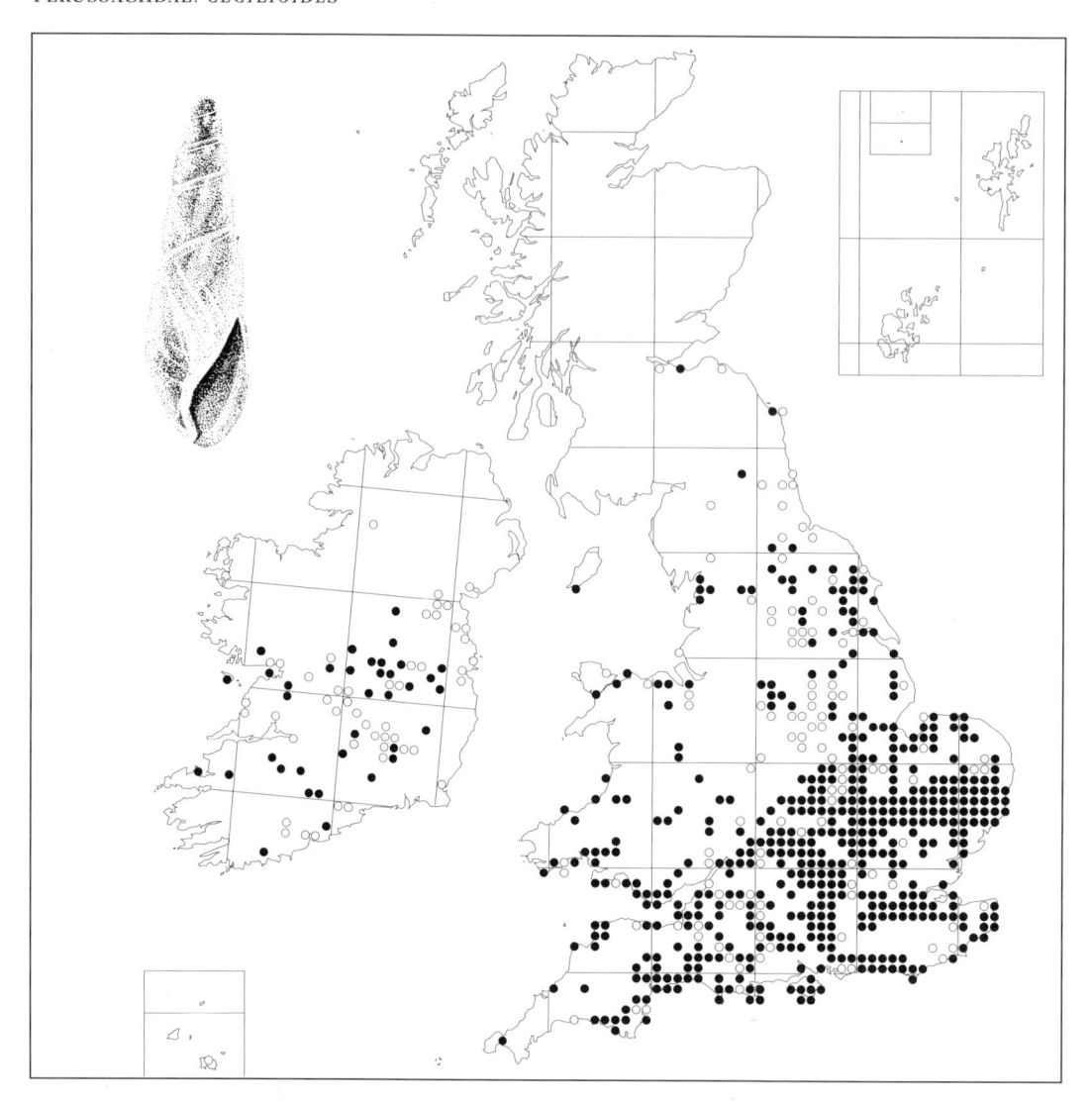

**Cecilioides acicula** (Müller, 1774)

Blind (or agate) snail   Size: 4.5–5.5mm

A blind, subterranean species, occurring at depths of two metres or more in unwooded calcareous places. It lives in the crevices of rocks and in cracks and rootlet holes in well-drained soils; the cavities of ancient, buried bones are also a favourite haunt. It has several times been reported from garden rockeries. It is a difficult species to find alive, though bleached, dead shells are often visible on the surface in the loose, friable soil of anthills, at the foot of walls, or in the screes of cliffs and quarries.

Probably introduced. Though many prehistoric fossil occurrences of *Cecilioides* have been claimed, probably all may be discounted as burrowings into geological or archaeological deposits. The species may still be extending its range through human agency.

Originally Mediterranean, but widely spread by man in C. and W. Europe, as far as southernmost Scandinavia.

(Ellis: 165; K. & C.: 149)

CLAUSILIIDAE: *COCHLODINA*

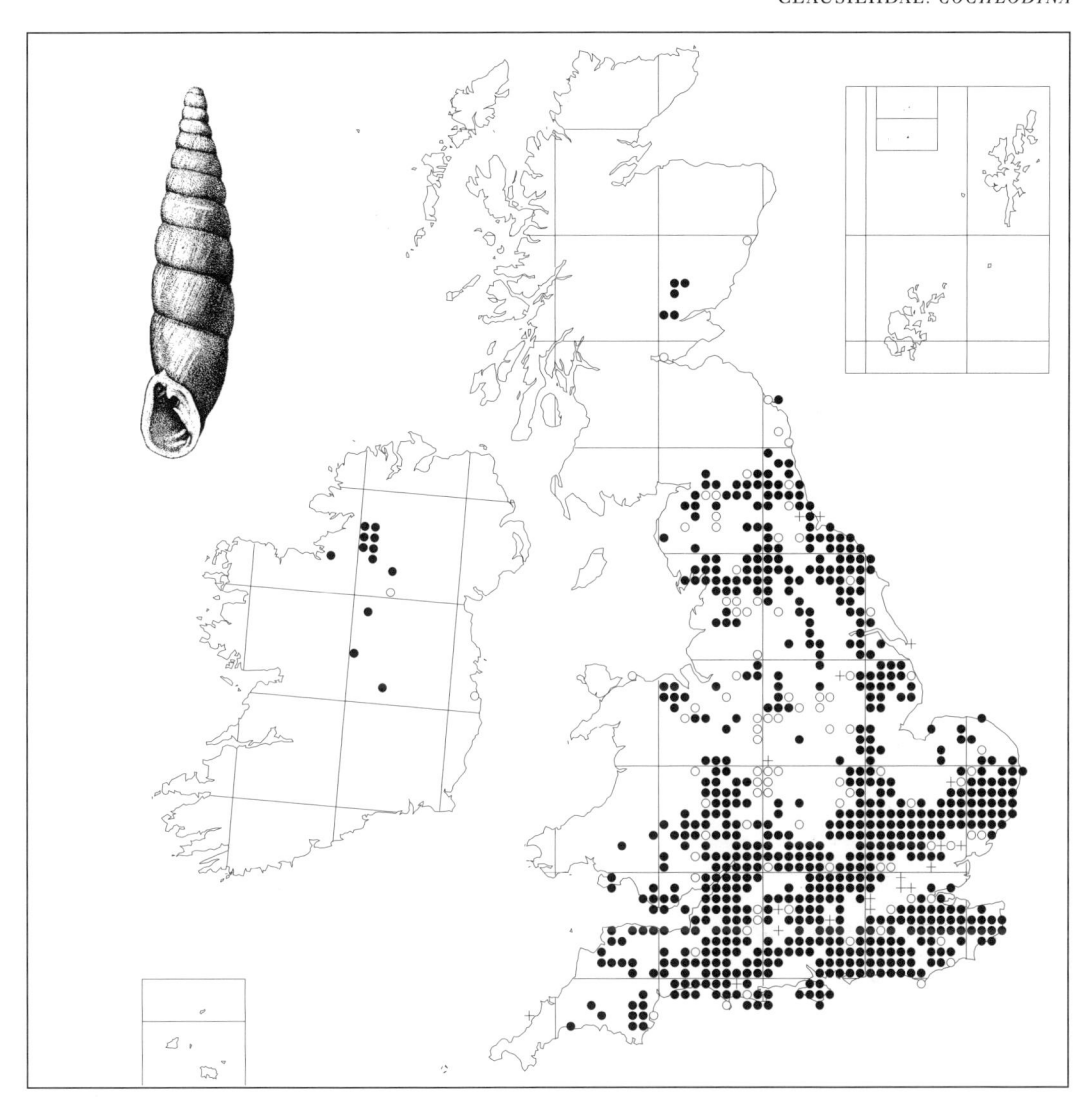

**Cochlodina laminata** (Montagu, 1803) (*Clausilia laminata* (Montagu), *Marpessa laminata* (Montagu))

Plaited door snail    Size: 15–17mm

This is a snail typical of deciduous woodland, living among moss, leaf litter and fallen timber, nearly always on base-rich soils. Old, overgrown quarries or wooded, limestone screes are favourite habitats. In moist weather it may climb weakly up vertical surfaces. It tolerates some degree of disturbance, and can be found in relatively modern, broad-leaved woods such as beech plantations and hazel coppices.

Native (E.Pgl). *C. laminata* shows little evidence of serious decline in the British Isles, though vulnerable in a few places through habitat loss. Certain isolated populations in Ireland are particularly at risk (Cawley, 1996a). RDB category (Ireland only): Rare.

European: widespread, north to C. Scandinavia.

(Ellis: 190; K. & C.: 155)

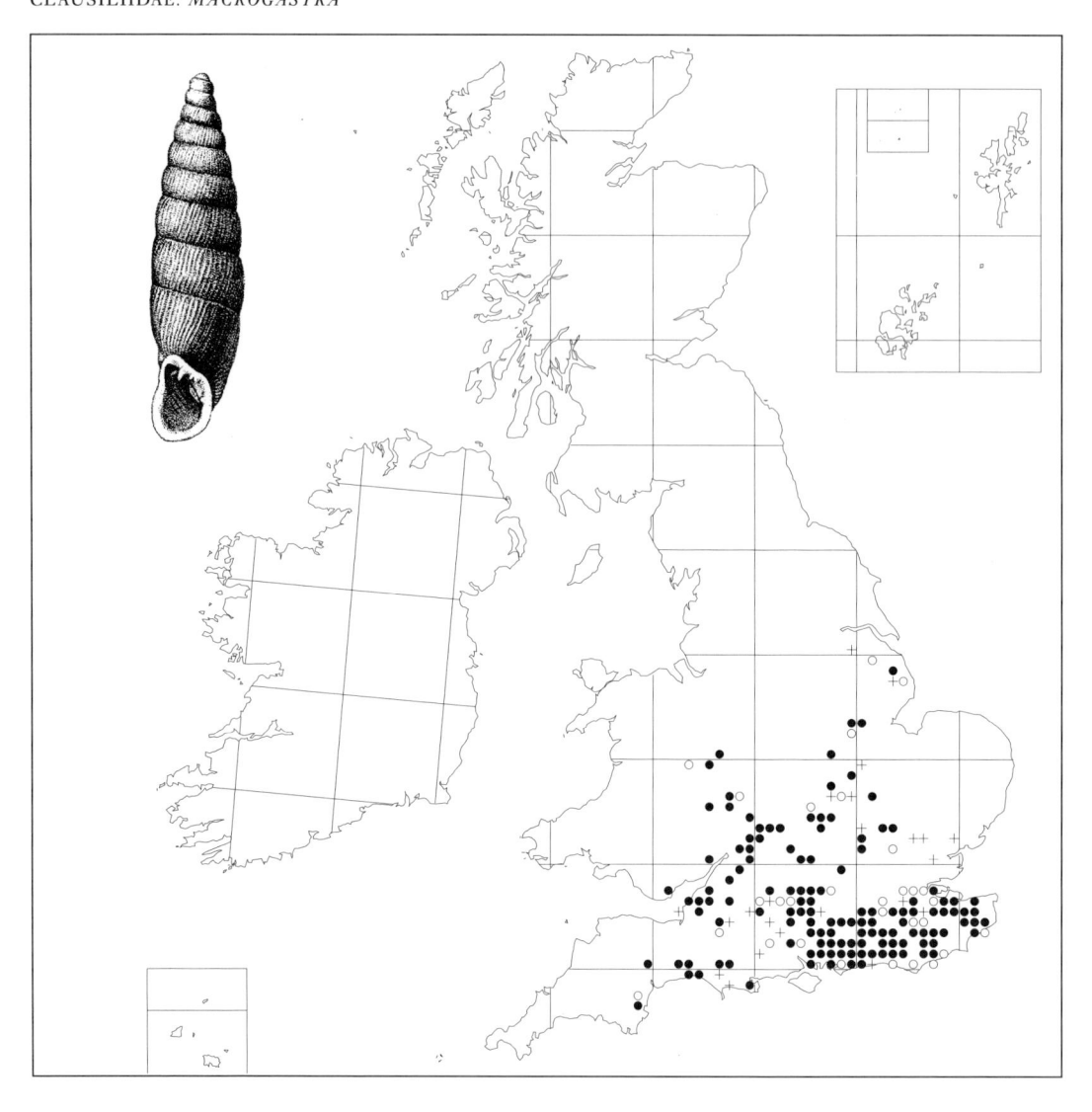

**Macrogastra rolphii** (Turton, 1826) (*Clausilia rolphii* Turton)

Rolph's door snail   Size: 11–14mm

Occurs locally in deciduous woods and copses, in mossy ground litter, among rocks and under fallen timber. It also lives in more open places in hedgerows and among grass and nettles on moist roadside banks. Unlike most clausiliids, it is not a climbing species. Though found mostly on chalk or limestone, it is tolerant of acid soils and in some woods in the Weald of Kent and Sussex occurs together with *Zonitoides excavatus*. It avoids conifer plantations, but is not uncommon in hazel and chestnut coppices and in some relatively modern plantations.

Native (E.Pgl). Fossils show that *M. rolphii* once possessed a slightly wider distribution in England, notably in Essex and Lincolnshire. There is little evidence of continued significant decline, though there may be a few extinctions at 10km-square level and some isolated colonies must be considered vulnerable.

W. European: mainly France, England, the Low Countries and N.W. Germany.

(Ellis: 189; K. & C.: 165)

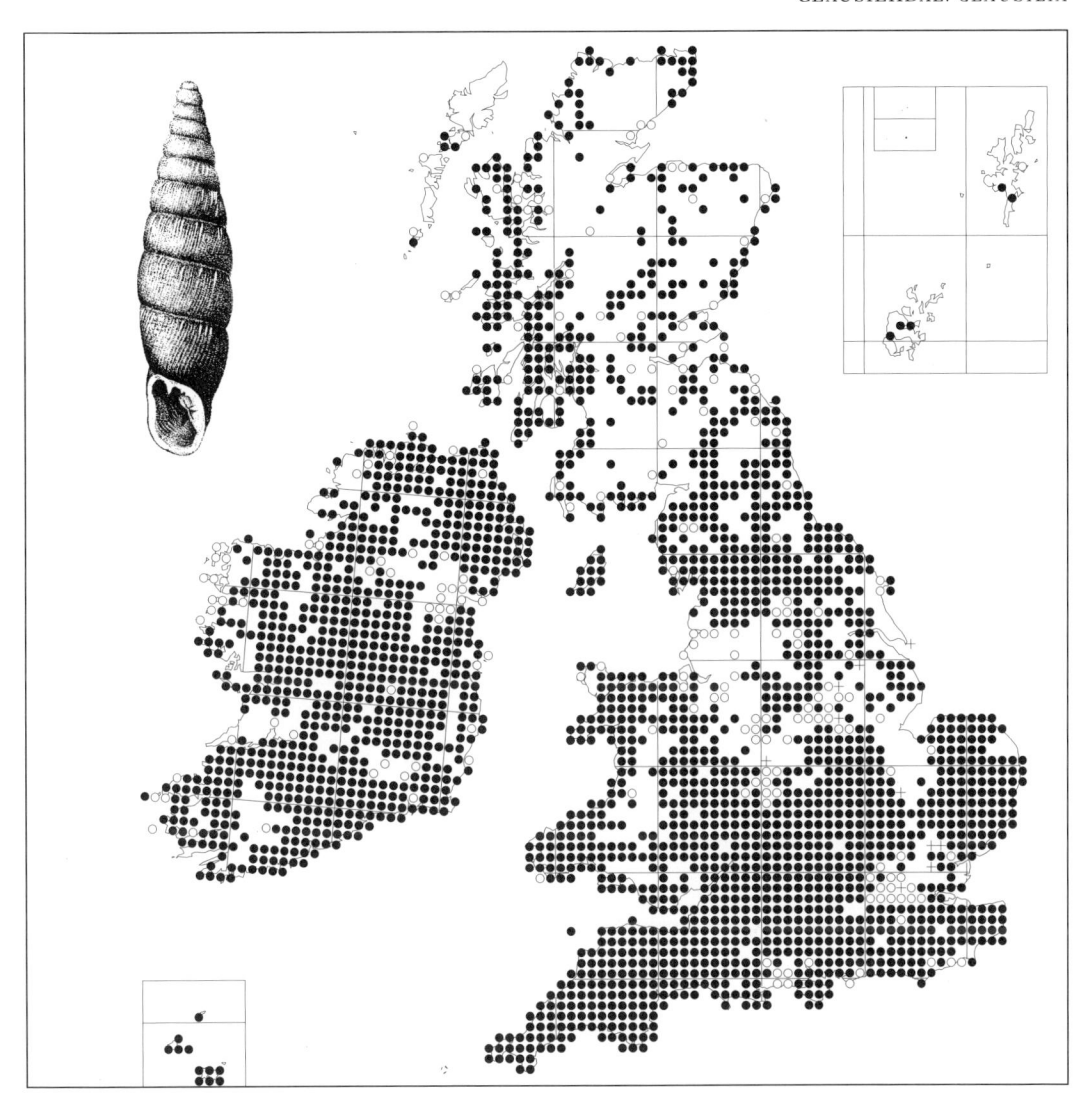

**Clausilia bidentata** (Ström, 1765) (*C. rugosa* of older British authors, *non* Draparnaud)

Common (or two-toothed) door snail   Size: 9–12mm

This is a common species in sheltered places, among ground litter and herbage, in woods, hedges, under walls and among rocks. In damp weather it climbs strongly up vertical surfaces. It prefers base-rich soils, especially in the northern parts of its range; in harsh acid upland areas, mortared stone walls may support isolated populations. Its absence from certain lowland areas of central and eastern England (mainly in S. Lancs., Cheshire, Staffs., Lincs. and S. Essex) appears to be due to a combination of natural lack of lime and of atmospheric pollution.

Native (E.Pgl). There is no evidence of major national change. However, when modern and old (mainly 1880–1914) records are compared, some clusters of 10km-square extinctions become apparent, concentrated in the Midlands and around London. It is probable that these reflect high $SO_2$ pollution, inhibiting the growth of lichens and other epiphytes on which *C. bidentata* feeds (Holyoak, 1978a).

N. European: widespread from the Alps north to C. Scandinavia (70°N in Norway).

(Ellis: 186; K. & C.: 166)

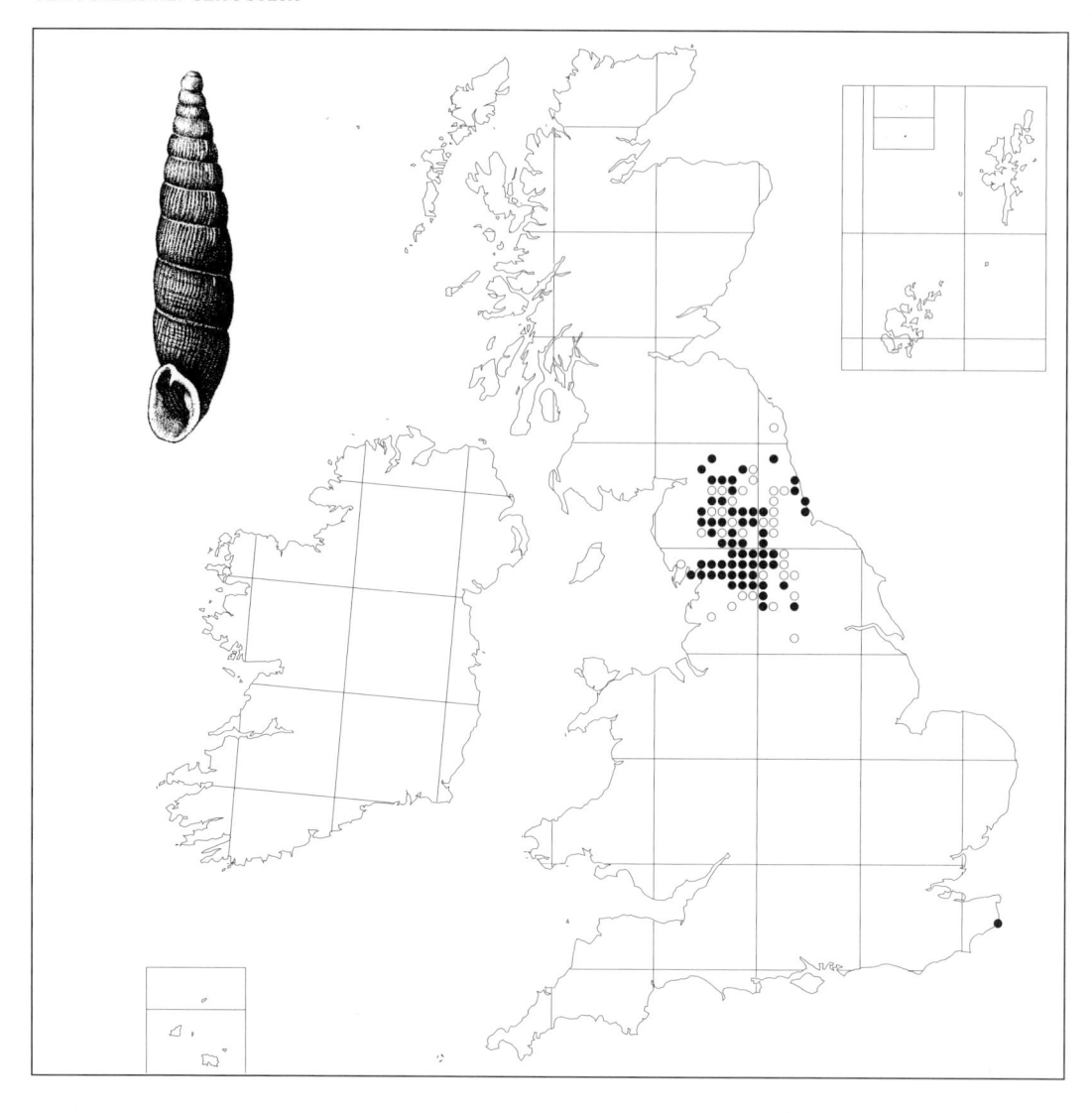

**Clausilia dubia** Draparnaud, 1805 (*C. cravenensis* Taylor)

Craven door snail    Size: 11–14 (occ.–16)mm

A species restricted to limestone rocks and walls, usually in exposed upland country. Most sites lie between 250m and 600m (Blackburn, 1941). In moist weather it emerges from crevices or from beneath ground litter to climb vertical surfaces, on which it feeds. In sheltered places it may be associated with *C. bidentata* but, unlike that species, is not normally found in dense woods.

Probably native (Pgl fossils of uncertain age only). The isolated colony at Dover Castle, Kent, known since 1873, is probably a modern introduction (Philp, 1984). Although it still flourishes in its strongholds in the northern Pennines, there are signs that some marginal populations have been lost over the past century. In particular, most of the 19th-century records from the Magnesian Limestone outcrop running from near Pontefract, Yorks., northwards through Durham to the R. Tyne have not been confirmed in recent years. This is an area of high atmospheric ($SO_2$) pollution, which has perhaps caused local extinction by damaging lichens and algae on which the species feeds.

Mainly C. European, with isolated outposts in N. England, Sweden and Finland.

A number of old records from southern England (Essex, Herts., Northants, Oxon.) and from Scotland (Midlothian, Fife, Harris, N. Uist) are inadequately documented and have been omitted from the map. Most are likely to refer to large *C. bidentata*.

(Ellis: 187; K. & C.: 166)

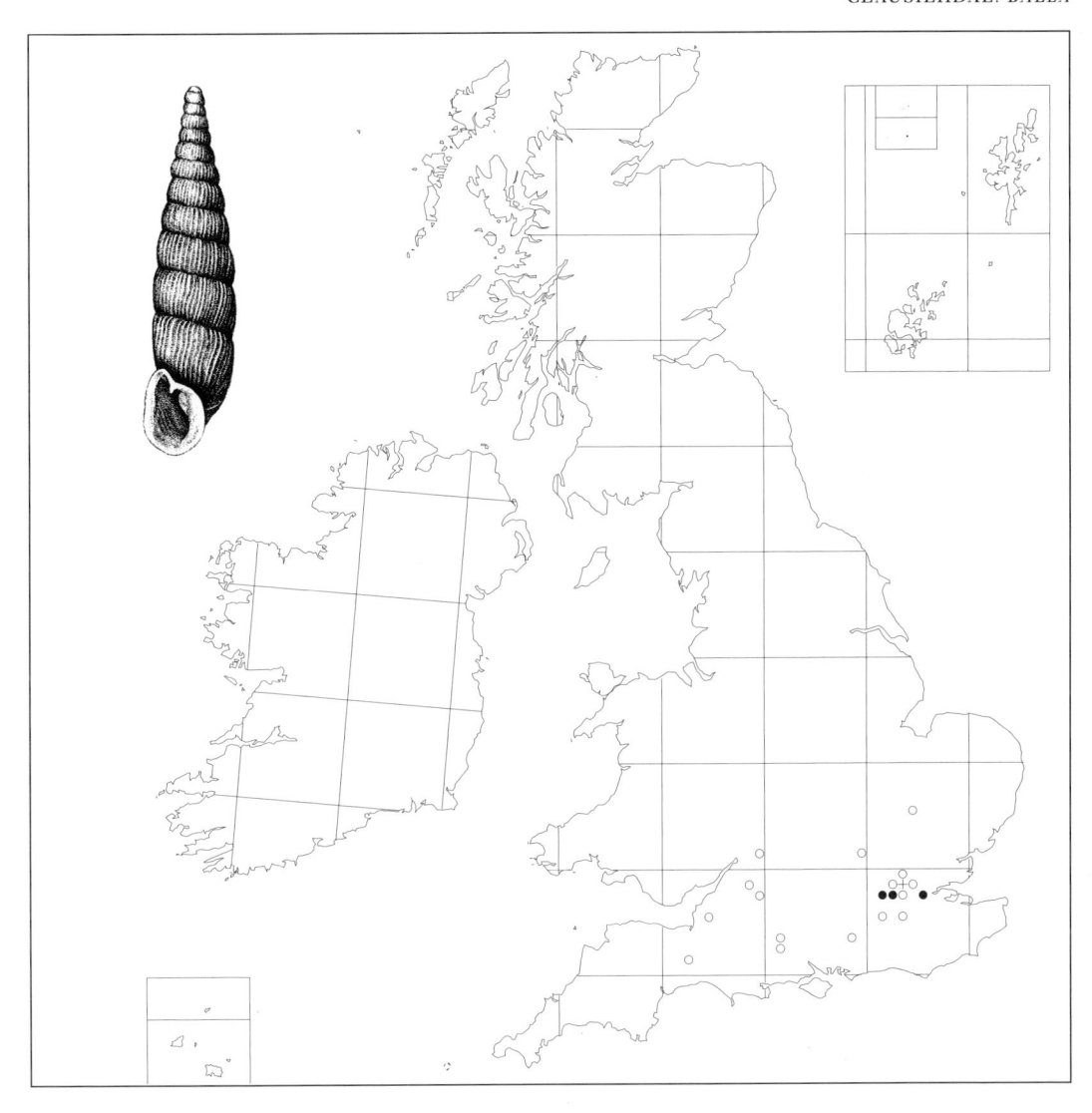

**Balea biplicata** (Montagu, 1803) (*Clausilia biplicata* (Montagu), *Laciniaria biplicata* (Montagu))

Thames door snail   Size: 16–18mm

All the known surviving colonies of this snail are on the banks of the R. Thames in or near London (Kew, Richmond, Isleworth, Chiswick, Purfleet) where it occurs in ground litter among nettles and under willows, though not in marshy places (Boycott, 1929). At Purfleet it lives on a dry overgrown chalk bank. It is commonly associated with human rubbish and with synanthropic molluscs like *Tandonia* and *Oxychilus draparnaudi*. It has several times been successfully introduced into gardens.

Introduced? (L.Pgl (Roman) fossils only). The history of this species in Britain is unclear. Its synanthropic habit suggests that it may be an old, accidental introduction. It appears now to be declining. In London most 19th-century localities have been obliterated by riverside development (Castell, 1962). Elsewhere in England attempts to refind it at several former localities have been unsuccessful, notably at Easton Grey, Wilts. (the type locality, 1803), Leckhampton, Glos. (1848) and Tring, Herts. (1870). The most recent records from outside London are from Bickenhall, Som. (1945; Turner, 1950) and Coe Fen, Cambs. (1949). RDB category: Rare.

Mainly C. European, but with outposts in England, N.E. France, the Low Countries and S. Scandinavia, possibly spread by man.

(Ellis: 184; K. & C.: 170)

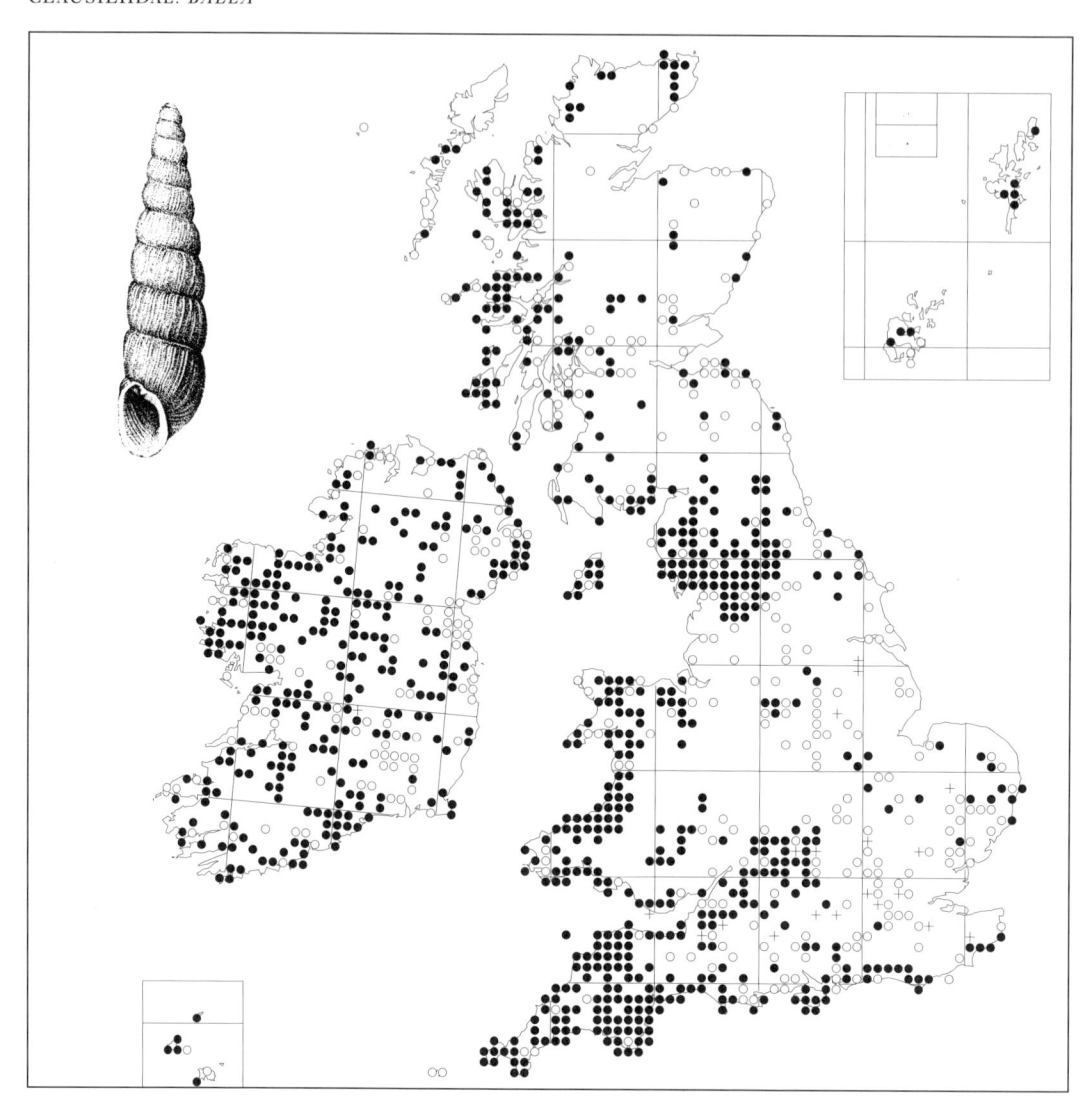

CLAUSILIIDAE: *BALEA*

**Balea perversa** (L., 1758)

Tree snail   Size: 8–10mm

A species typical of rocks, stone walls and trees, often on surfaces encrusted with lichens and other epiphytes. It is almost never found on the ground (Boycott, 1921). It favours dry, relatively exposed places and is rare in heavily shaded woods. On trees, it prefers those with a rough bark offering crannies in which to hide, or alternatively it shelters under loose, dead bark. In western areas it tolerates non-calcareous soils.

Native (E.Pgl). *B. perversa* has declined in many areas. It remains frequent in Ireland (where it is underrecorded) and in the more rocky, oceanic parts of western Britain. Though probably never as widespread further east where suitable habitats on rocks and walls are fewer, in lowland England many colonies have certainly disappeared within the past century. Around London it has not been seen for about eighty years. There is good evidence that the species is sensitive to the effects of atmospheric pollution (especially when on trees) and the area of its most marked decline in Britain corresponds well with that of the impoverishment of lichen floras ascribed to the same cause (Holyoak, 1978a).

European: widespread, but becoming local to the east, and coastal only beyond about 62°N in Scandinavia.

(Ellis: 184; K. & C.: 172)

**Testacella maugei** Férussac, 1819

Maugé's slug    Size: 60–100mm

This is a carnivorous, subterranean slug, living in burrows a few inches below the surface and emerging at night to feed on earthworms. It occurs mainly in rich garden soils (especially in old, well-manured nursery and vegetable gardens) but in Devon and Cornwall it has also been found in relatively wild places, among rocks and sea cliffs.

Probably introduced. *T. maugei* was first noted in a nursery at Clifton, Bristol, in 1812. The paucity of recent records partly reflects the difficulty of finding the species. It is also possible that the disappearance of traditionally managed kitchen gardens and the introduction of chemical fertilizers and pesticides may have caused some real decline, especially away from the extreme South West. In Ireland it has not been seen since the 1930s (Lissenhall, Co. Dublin, 1931; Fermoy, Co. Cork, 1933). In the London area it was last recorded in 1950 (Cheam, Surrey).

W. European and Atlantic, mainly near coasts: British Isles southwards to Gibraltar and Morocco, and Atlantic islands.

(Ellis: 132; K. & C.: 173)

TESTACELLIDAE: *TESTACELLA*

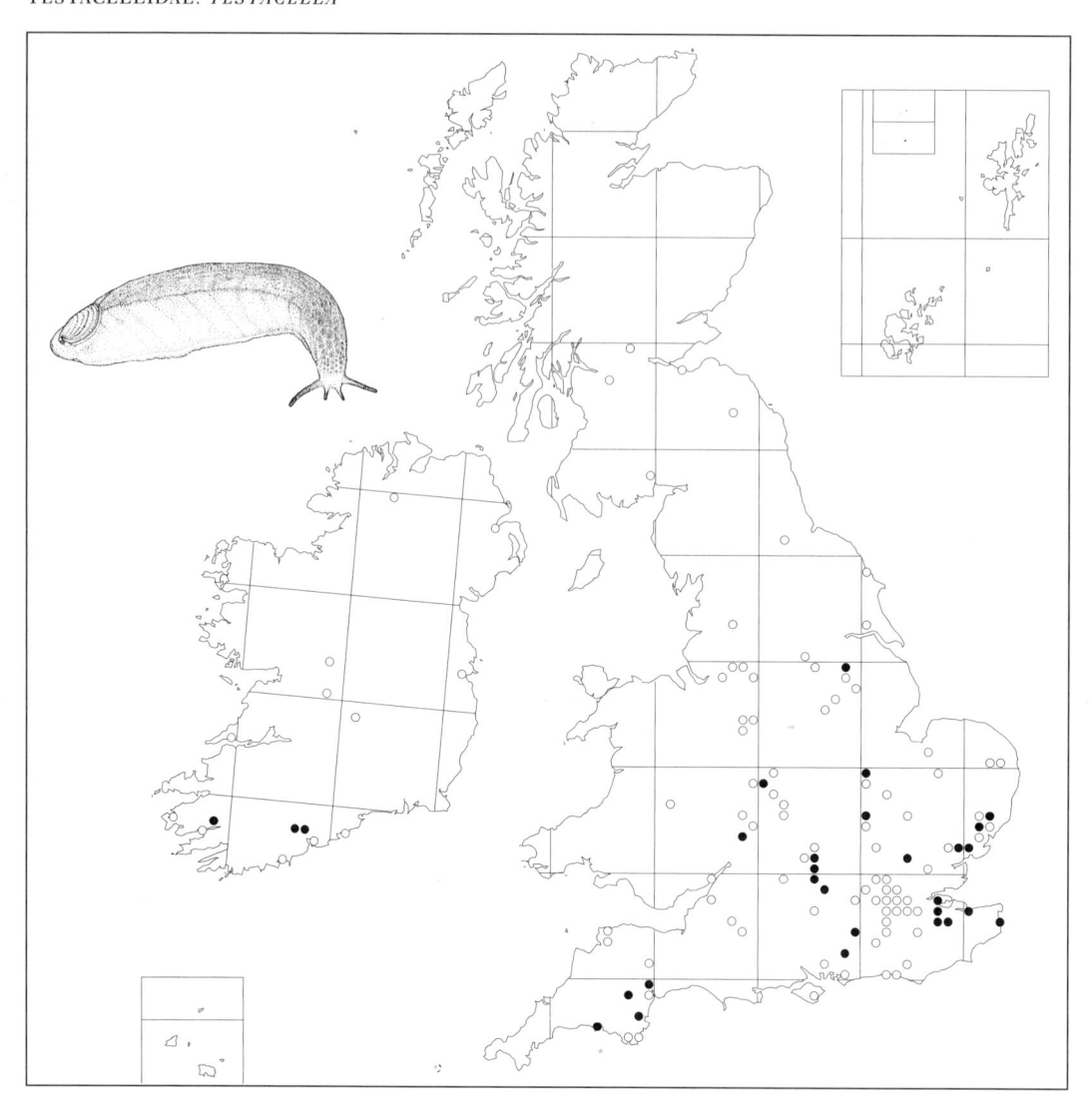

**Testacella haliotidea** Draparnaud, 1801

Shelled slug    Size: 70–100mm

A carnivorous, subterranean species, restricted almost entirely to gardens, including ornamental and vegetable gardens, nurseries and allotments; many 19th-century records were from the kitchen gardens of country houses. On south-western coasts (Devon, Co. Cork) it has occasionally been reported from relatively wild places near the sea.

Probably introduced. Like other *Testacella* species, *T. haliotidea* is seriously underrecorded. In recent years it has been noted more rarely than *T. scutulum*, but this may reflect only a problem of identification (see below).

W. European and Mediterranean: British Isles south to Spain, Italy, N. Africa and Atlantic islands; introduced into Germany.

The distinction between this and the following species (*T. scutulum*) is not satisfactorily established; they may prove to be only varieties of a single species.

(Ellis: 133; K. & C.: 173)

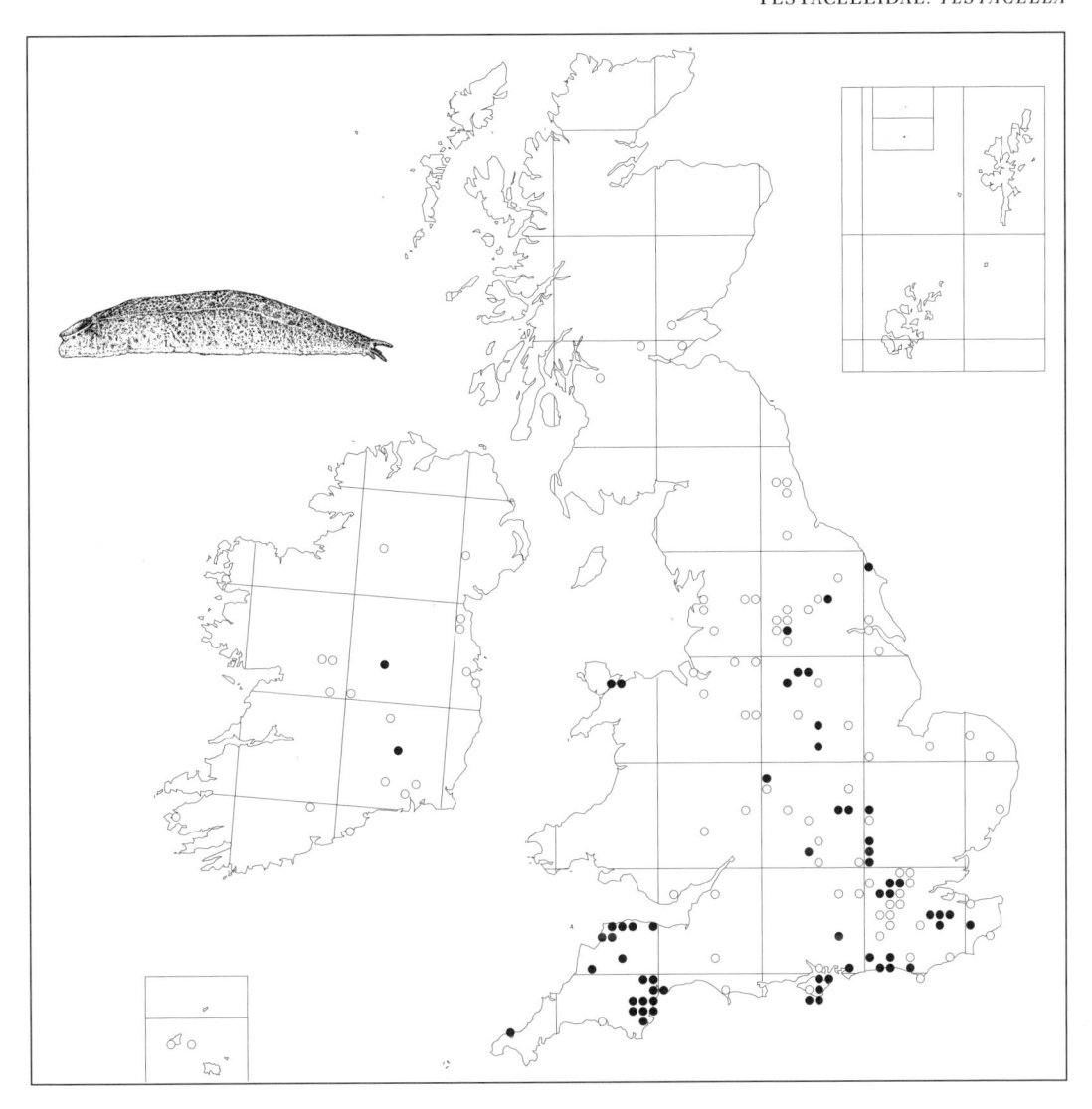

**Testacella scutulum** Sowerby, 1821 (*T. haliotidea* var. *scutulum* Sowerby)

Shield slug    Size: 70–100mm

Like other *Testacella* species, this is a carnivorous, subterranean slug, found mainly in richly organic soils in gardens and nurseries. It lives in burrows a few inches below the surface, emerging at night to feed on earthworms; by day it can sometimes be found under flat stones bedded in the earth or sheltering under rubbish at the foot of walls. Occasionally it lives in roadside verges, and along the south coast it is not uncommon in wild places on cliffs and in woods.

Probably introduced. Although the number of recent records is rather small, this species is underrecorded and probably fairly frequent in lowland Britain and Ireland. Nevertheless the widespread use of pesticides and chemical fertilizers may have caused some extinctions.

W. European and Mediterranean: British Isles south to Spain, Italy and N. Africa.

The taxonomy of this species requires further investigation. In the 19th century, '*scutulum*' was generally believed to be only a yellow variety of *T. haliotidea*, and this view is again gaining support.

(Ellis: 134; K. & C.: 174)

BRADYBAENIDAE: *BRADYBAENA*

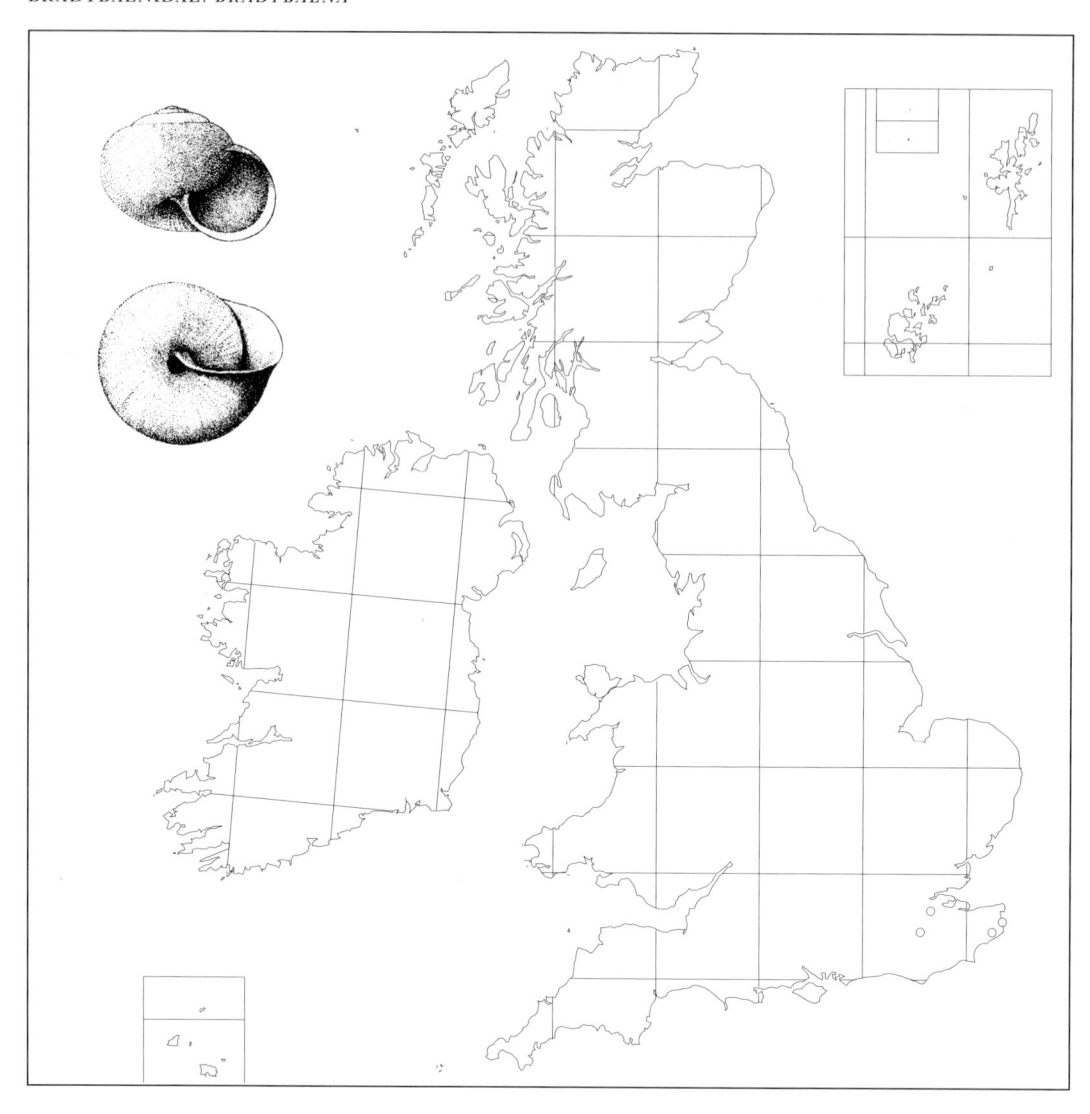

**Bradybaena fruticum** (Müller, 1774) (*Helix fruticum* (Müller), *Fruticicola fruticum* (Müller))

Bush snail   Size: 17–22mm

On the Continent, this is a common species of roadsides, river floodplains and scrubby waste ground. It favours tall, lush vegetation in rather moist, sheltered places, but avoids woods. The records from Kent appear to have all been from roadside banks.

Probably introduced. The Kent records are from Deal (about 1903), Penshurst (1911, 1917), Lydden (1908–1918) and Luddesdown (about 1914). *B. fruticum* cannot now be found at any of these places. The species has no Postglacial fossil history in Britain (though it lived here during the last interglacial period) and is likely to have been introduced accidentally in recent times.

Mainly Asian and E. European, reaching westwards to E. France and northwards to S. Scandinavia.

(Ellis: 192; K. & C.: 174)

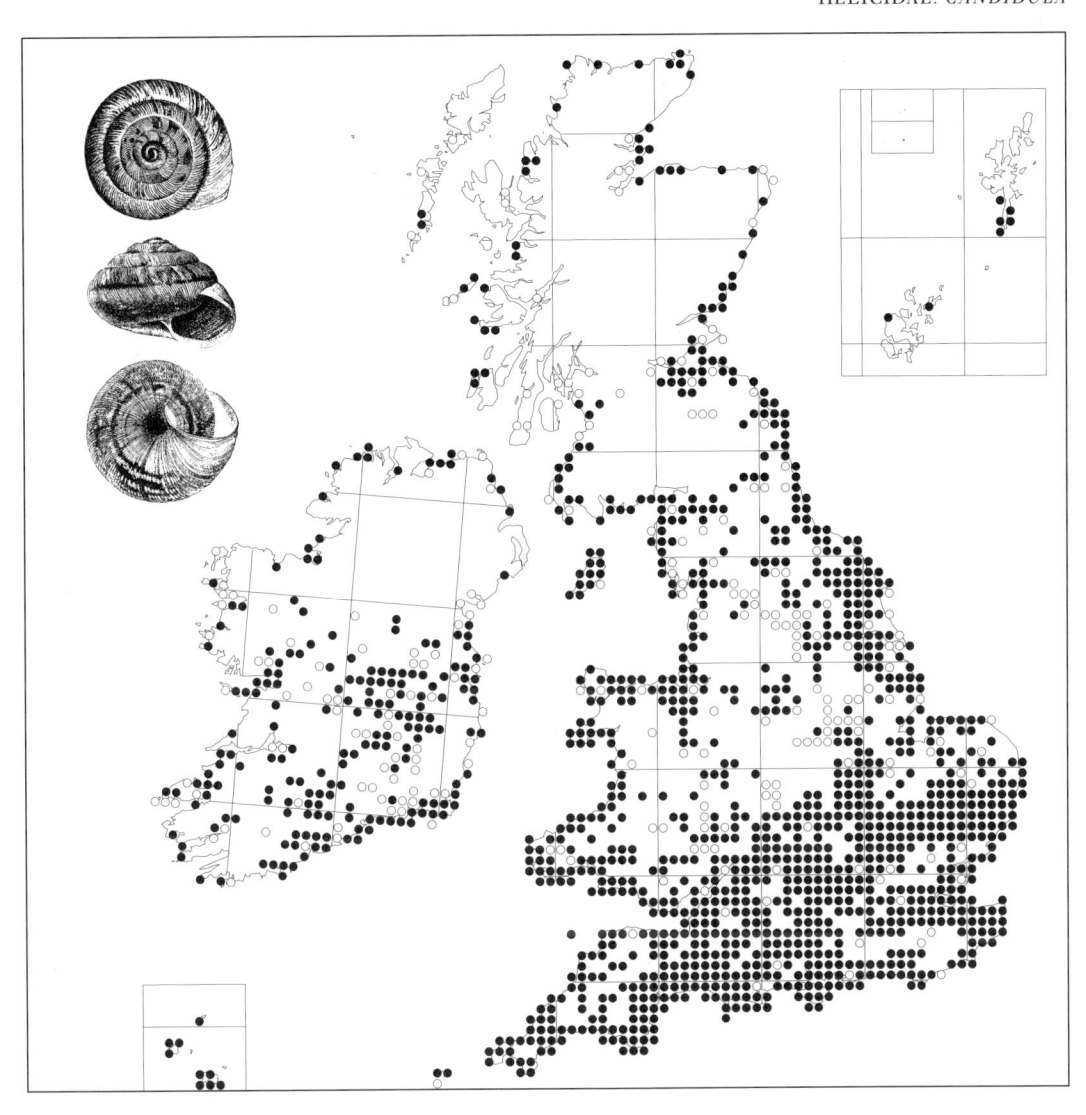

**Candidula intersecta** (Poiret, 1801) (*Helix caperata* Montagu, *Helicella caperata* (Montagu))

Wrinkled snail   Size: 7–12mm

This is a common snail in dry, base-rich grassland, usually in exposed sandy or stony situations. It is typical of chalk and limestone hillsides, stabilized dunes, sea cliffs, and waste ground of all kinds as in railway cuttings, roadside banks and quarries. In hot weather it may attach itself to plant stalks or to bare rocks and walls, and occasionally it climbs trees at the edges of woods. It is tolerant of cultivation and is often found in corn stubble.

Probably introduced. There are no certain pre-mediaeval fossil records of this species which is probably a relatively modern introduction from southern Europe. It behaves as a 'weed' in Britain, readily colonizing suitable man-made habitats on well-drained calcareous soils.

W. European: mainly France, British Isles, the Low Countries and coastal regions of Denmark and southernmost Sweden.

(Ellis: 201; K. & C.: 177)

HELICIDAE: *CANDIDULA*

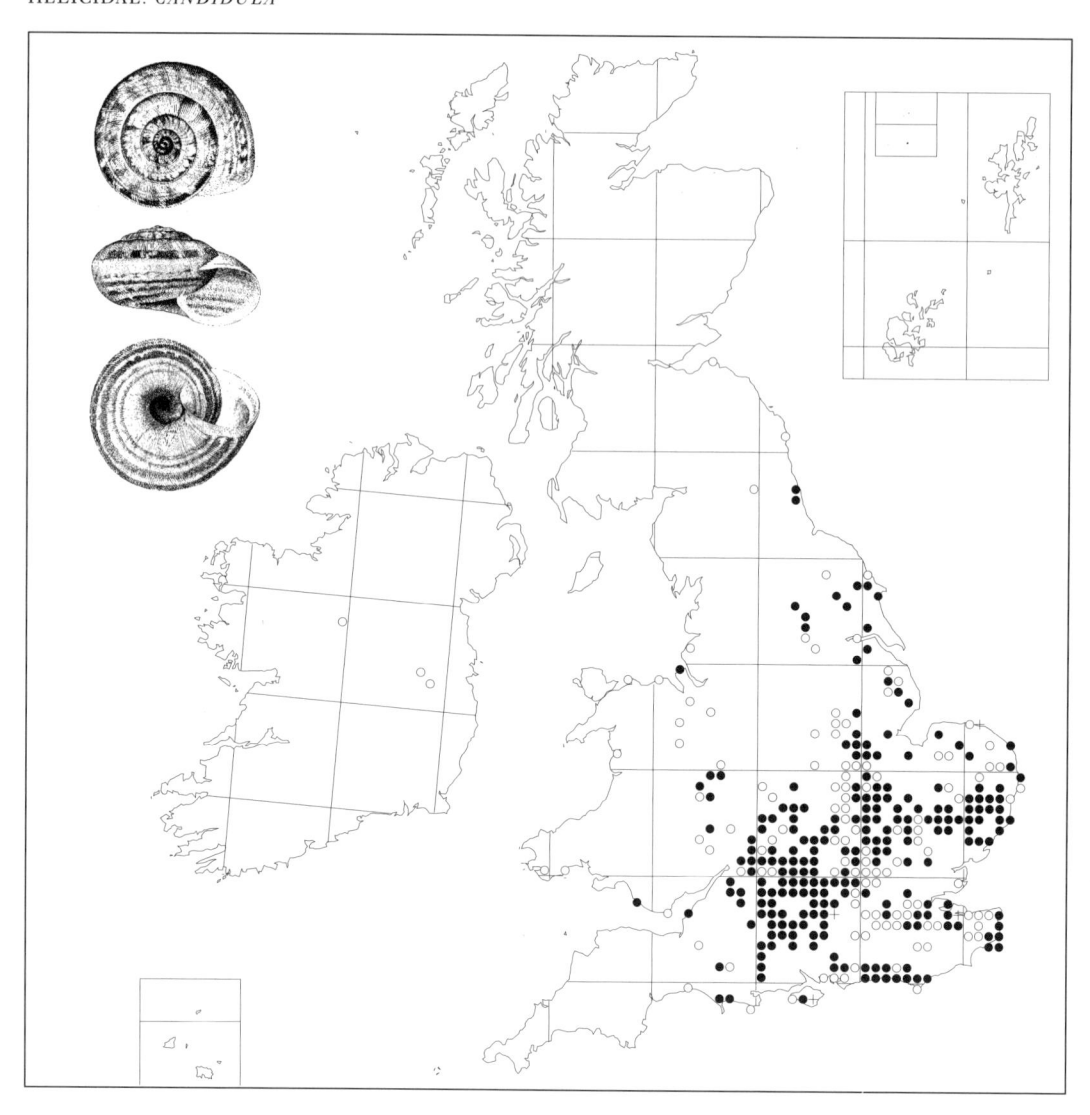

**Candidula gigaxii** (L. Pfeiffer, 1850) (*Helicella gigaxii* (Pfeiffer), *H. heripensis* (Mabille))

Size: 7–12mm

This species lives in dry, sunny places, mostly on chalk or limestone. Like the commoner *C. intersecta*, it is found in short grass on downs, in old quarries, railway cuttings and similar places, but it also inhabits taller, lusher herbage normally avoided by that species along roadsides and the edges of fields, ascending umbellifers and other weeds. It has several times been reported from corn stubble. Unlike *C. intersecta* it has no special association with the sea and is hardly ever found on sandhills.

Introduced? (L.Pgl fossils only). *C. gigaxii* was probably introduced into England as a 'weed' species, perhaps in the Romano-British period. It seems to be less common now than early this century. At a 10km-square level there have probably been some extinctions. In Pembrokeshire it was last seen in about 1912, in Denbighshire in 1921 and in Northumberland in 1934. The single Scottish record (Canty Bay, North Berwick) dates from 1930. The four Irish sites, in Kildare and Roscommon, were noted between 1918 and 1945 (Stelfox, 1958).

W. European: mainly Spain, France, British Isles and the Low Countries; accidentally introduced in a few places in Germany.

(Ellis: 199; K. & C.: 177)

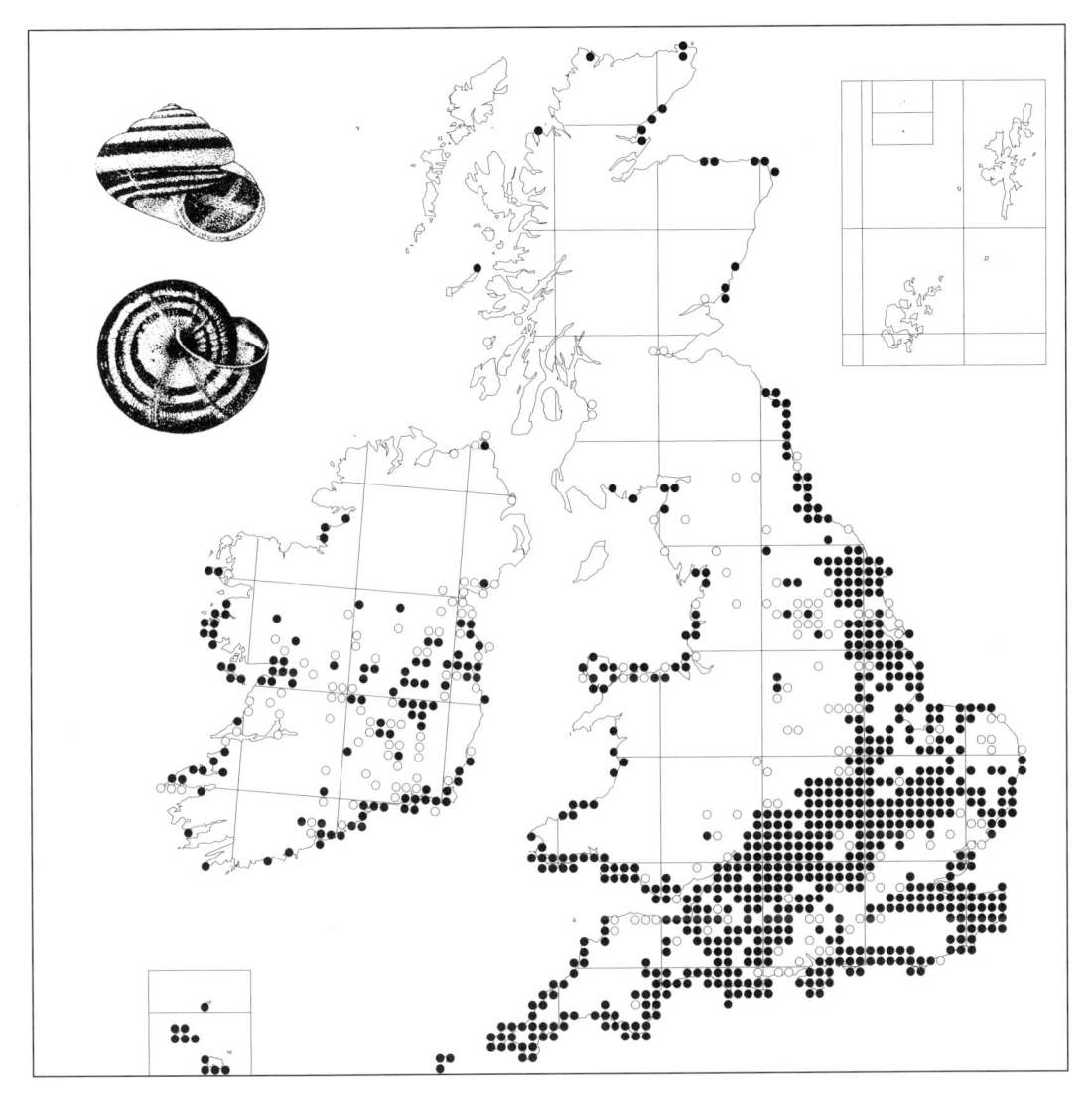

**Cernuella virgata** (da Costa, 1778) (*Helix virgata* (da Costa), *Helicella virgata* (da Costa))

Striped (or zoned) snail   Size: 9–18mm

A common snail of dry, sunny, calcareous places, in chalk downland, sea cliffs, sandhills, abandoned quarries and waste ground by fields, roads and railways. It likes short grass in exposed sandy or stony situations, but will also inhabit taller herbage, in hot weather attaching itself high up on the plants. It may also climb walls and trees. It is tolerant of cultivation and may abound in corn stubble in harvested fields (Boycott, 1921).

Probably introduced. This species is a relatively late, perhaps Romano-British, introduction into the British Isles (though it lived here during the last interglacial period). It behaves as a 'weed', colonizing suitable man-made habitats on well-drained, calcareous soils. Its continuing spread along coasts in northern Scotland is very recent (Kincardine, first record 1967; Sutherland, 1972; Caithness, 1974; Banff, 1984).

Mainly circum-Mediterranean, but spread by man northwards through western France to the British Isles and along the coasts of Belgium and Holland.

A closely related Continental species, *C. aginnica* (Locard), may possibly occur in England. Shells from Torquay, Devon, have been identified as this species, though anatomical confirmation is lacking (Kerney, 1976c; Clerx & Gittenberger, 1977). *C. neglecta* (Draparnaud), another Continental *Cernuella*, was briefly naturalized at Luddesdown, Kent, early this century.

(Ellis: 196; K. & C.: 177)

HELICIDAE: *HELICELLA*

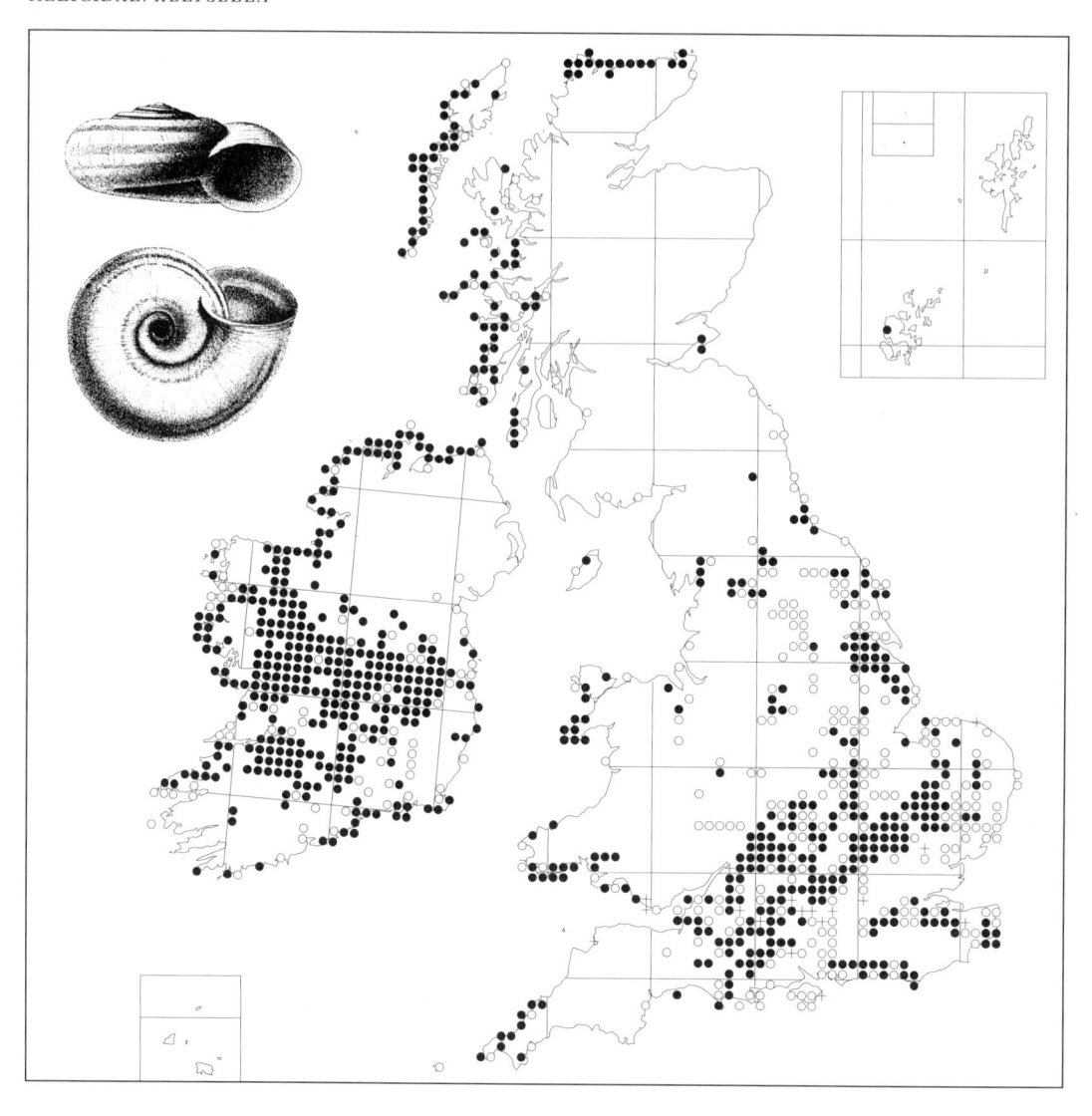

**Helicella itala** (L., 1758) (*Helix itala* L., *H. ericetorum* Müller)

Heath snail    Size: 12–20mm

A snail of dry, sunny, calcareous places. It is less tolerant of cultivation than the other common helicellines and in England is typical of old short-turfed chalk and limestone grassland. It also inhabits disused quarries, sea cliffs and sandhills. In hot, dry weather it may attach itself for long periods to the stalks of plants. In Ireland it extends into somewhat moister habitats than is usual in Britain and in the Carboniferous limestone country of the west may abound in lush herbage along roadsides.

Native (Lgl). Fossils show that *H. itala* was already present on the Chalk of southern England in the open landscape at the end of the glacial period. It was widely suppressed by Postglacial forest growth, expanding again from coastal and other refugia in the wake of human clearances from the Neolithic period onwards. It remains common by the sea in western and northern Britain, and in Ireland. In southern and eastern England however there is clear evidence that populations have declined strongly this century; in many areas where it was formerly common only weathered shells can now be found (Killeen, 1992). The ploughing up or reversion to scrub of much old chalk and limestone grassland may be partly responsible.

W. European: from the Mediterranean northwards to the British Isles and Denmark.

(Ellis: 194; K. & C.: 179)

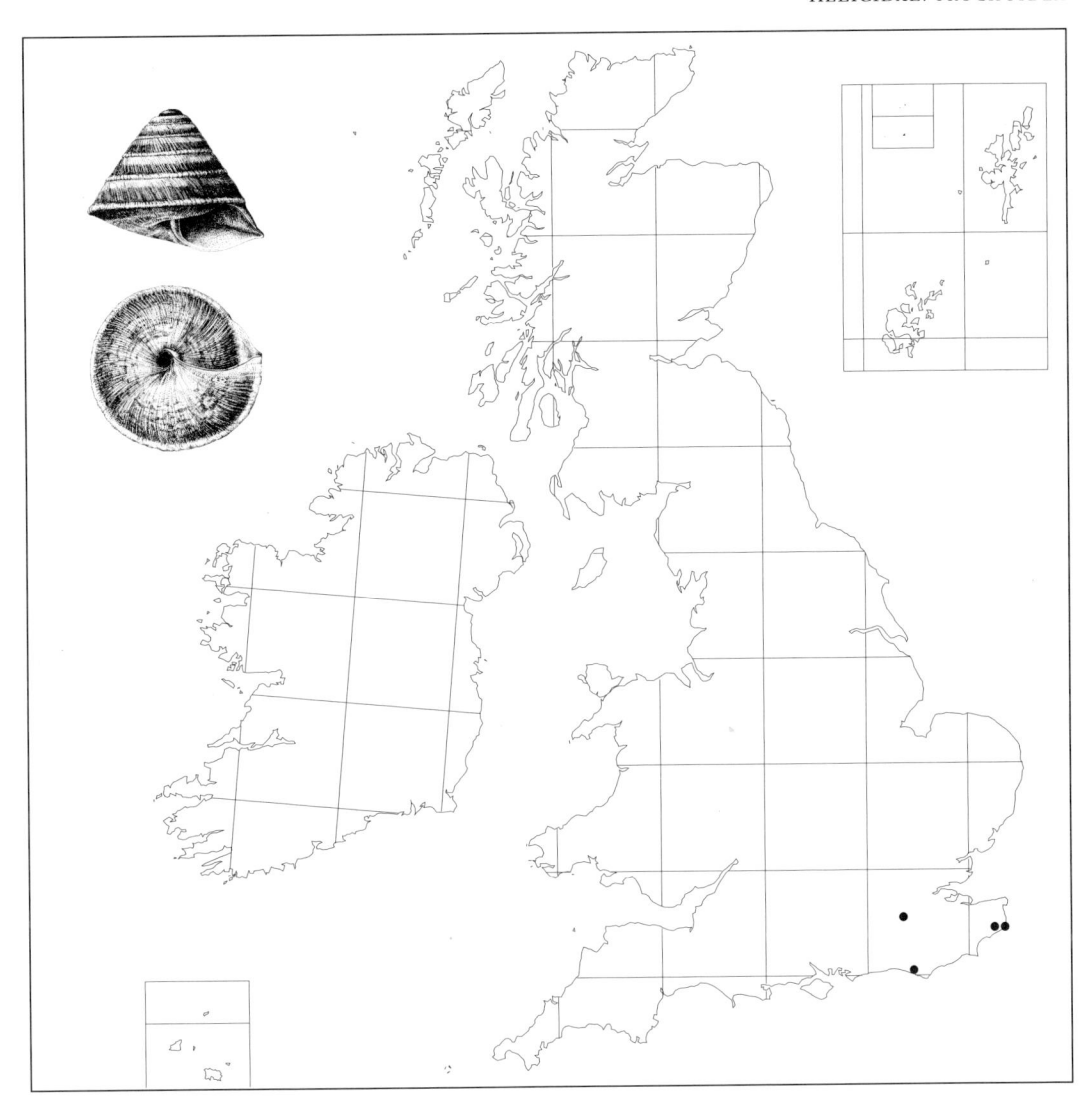

**Trochoidea elegans** (Gmelin, 1791) (*Helicella elegans* (Gmelin), *Helix terrestris* of Adams, 1896, and Step, 1901)

Top snail   Size: 8–10mm

This is a common species around the Mediterranean in dry, sunny, calcareous places, including stabilized dunes. The English sites are all in chalk grassland. It favours steep slopes on downs or roadsides, especially where the turf is broken and chalky rubble shows through. At Walmer it colonized nearly bare chalk in waste ground.

Introduced. This is a modern, accidental introduction from southern Europe, now naturalized in a few places. It was first found at Lydden, Kent, in 1890, and then at Chaldon, Surrey (1931), Walmer, Kent (1952 (Davis, 1954)) and Denton, Sussex (1967 (Jones, 1968)). Populations have been confirmed at all these places in recent years. The colonies are sharply defined and show little sign of spreading. Two other colonies were noted in the Dover area (Kent) early this century, at Kearsney and Shepherdswell (Kerney, 1978c).

Mediterranean; also introduced by man at a few sites in England, northern France and Belgium.

(Ellis: 204; K. & C.: 182)

HELICIDAE: *TROCHOIDEA*

**Trochoidea geyeri** (Soós, 1926) (*Helicella geyeri* (Soós))

Size: 6–8mm

On the Continent, *T. geyeri* lives in dry, open, calcareous places with a sparse grassy vegetation, especially among rocks and natural screes relatively undisturbed by man.

Extinct (fossil only). This species is known as a fossil in deposits of Lateglacial age on the Chalk in Wilts., Dorset, the Isle of Wight, Kent, Surrey, Sussex, Berks. and Cambs., always associated with grassland faunas (Sparks, 1953; Kerney, 1963). In general it died out very early in the Postglacial period as suitable open habitats were covered by forests. At Gwithian (Cornwall) it occurs in Bronze Age levels within coastal dune sands – a niche today occupied by other species of helicellines.

A local European species, with scattered populations known from Spain, France, Switzerland, Belgium, Germany, Austria and S. Sweden (Isle of Gotland).

(K. & C.: 182)

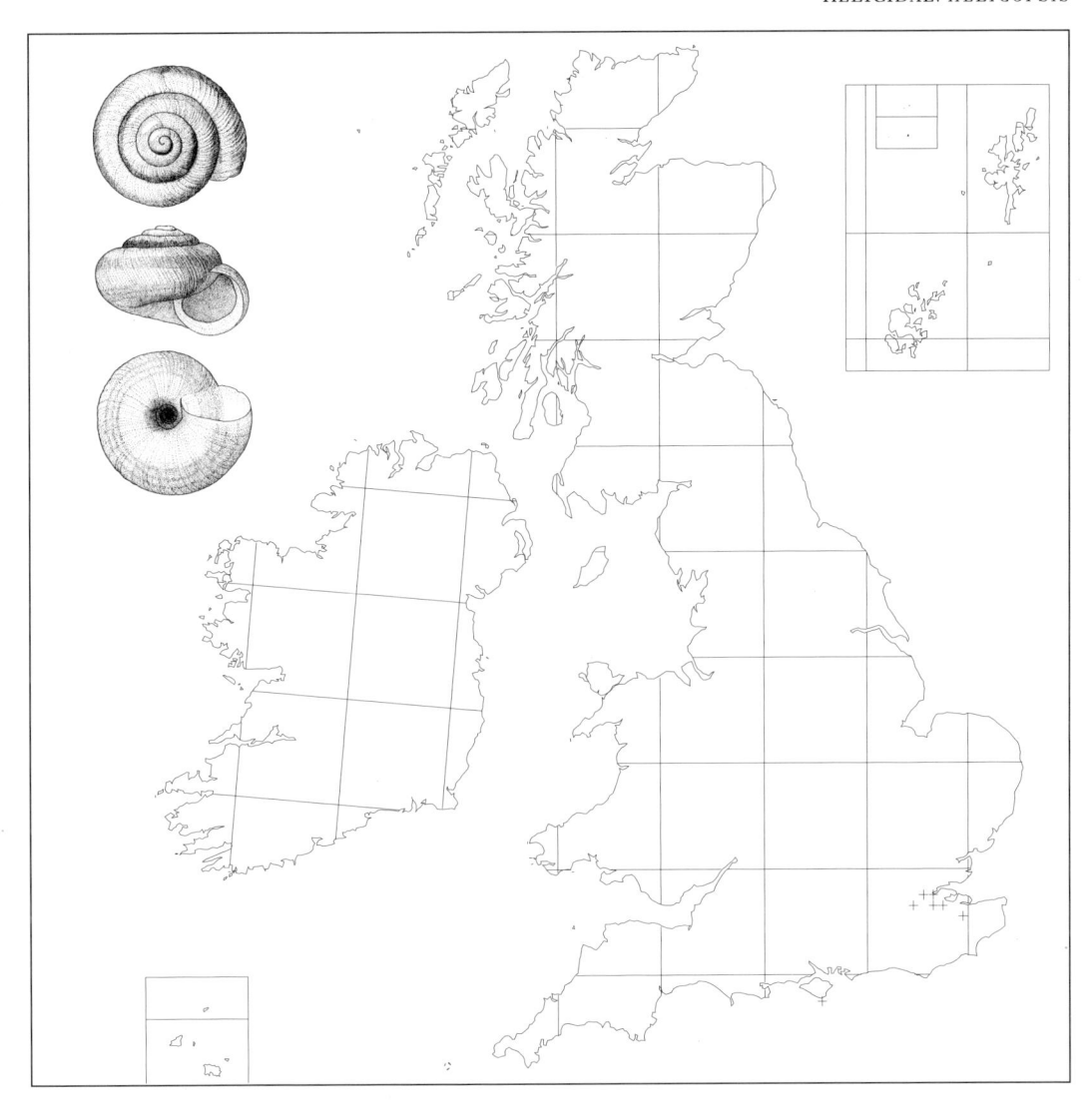

**Helicopsis striata** (Müller, 1774) (*Helicella striata* (Müller))

Size: 6–9mm

On the Continent, this species lives in sunny, open, calcareous places of a generally undisturbed character, though it occasionally colonizes roadsides or the edges of cultivated fields. It prefers well-established, stable grassland to rocky or stony ground (cf. *Trochoidea geyeri*) and in central Europe is often found on soils developed over Pleistocene loess.

Extinct (fossil only). *H. striata* has been found in a few deposits in Kent and the Isle of Wight, all poorly dated but probably of both Lateglacial and earliest Postglacial ages (Sparks, 1953; Preece, 1977). Like *T. geyeri*, *H. striata* no doubt disappeared because of habitat changes brought about by the spread of forests.

C. and E. European: a declining species, local in E. France, central Germany, Austria, the Czech Republic, Poland and Hungary; also in S. Sweden (Isle of Öland).

(K. & C.: 183)

HELICIDAE: *COCHLICELLA*

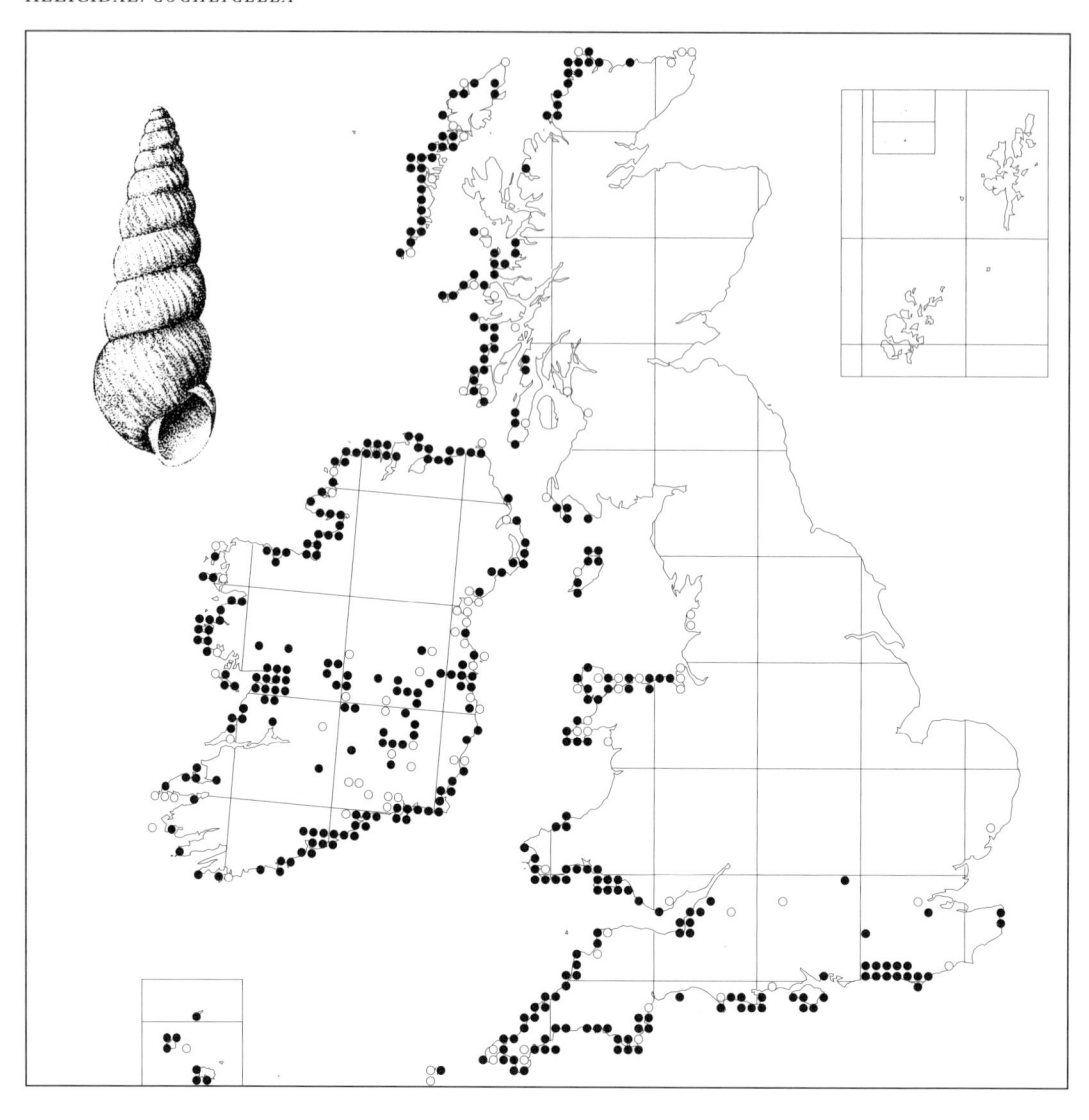

**Cochlicella acuta** (Müller, 1774) (*C. barbara* of some older British authors, *non* Linnaeus)

Pointed snail   Size: 10–18 (occ.–25)mm

A species found in abundance in dry, grassy, calcareous places. In Britain it is almost exclusively maritime, inhabiting cliffs, sandhills and waste ground and rarely straying more than a mile or two inland (Aubertin *et al.*, 1931). In the south of Ireland, unlike in Britain, it is found sporadically up to fifty miles from the sea, on dry roadside banks, in old quarries and similar places across the Carboniferous limestone plain.

Introduced? (L.Pgl fossils only). *C. acuta* is probably a relatively modern (late prehistoric) introduction which has established itself in suitable frost-free situations, mainly along coasts but also inland where conditions allow. It has probably now more or less reached its thermal limits, with only slight changes caused by short-term climatic fluctuations. In Caithness it appears to be extinct (last seen 1904); however, there is evidence that it has spread eastwards along the Sussex and Kent coasts during the present century, and this may perhaps be linked with a rise in sea temperatures noted in the English Channel (Leersnyder & Hoestlandt, 1958). Chance inland populations in England (Bath (Som.), before 1912; Aldbourne (Wilts.), 1929; Newbourn (Suffolk), 1933; Guildford (Surrey), 1981; Southfleet (Kent), 1987; West Wycombe (Bucks.), 1988) have rarely survived more than a few seasons.

Mediterranean region, Atlantic islands and coasts of Europe as far north as the British Isles and Belgium.

(Ellis: 204; K. & C.: 183)

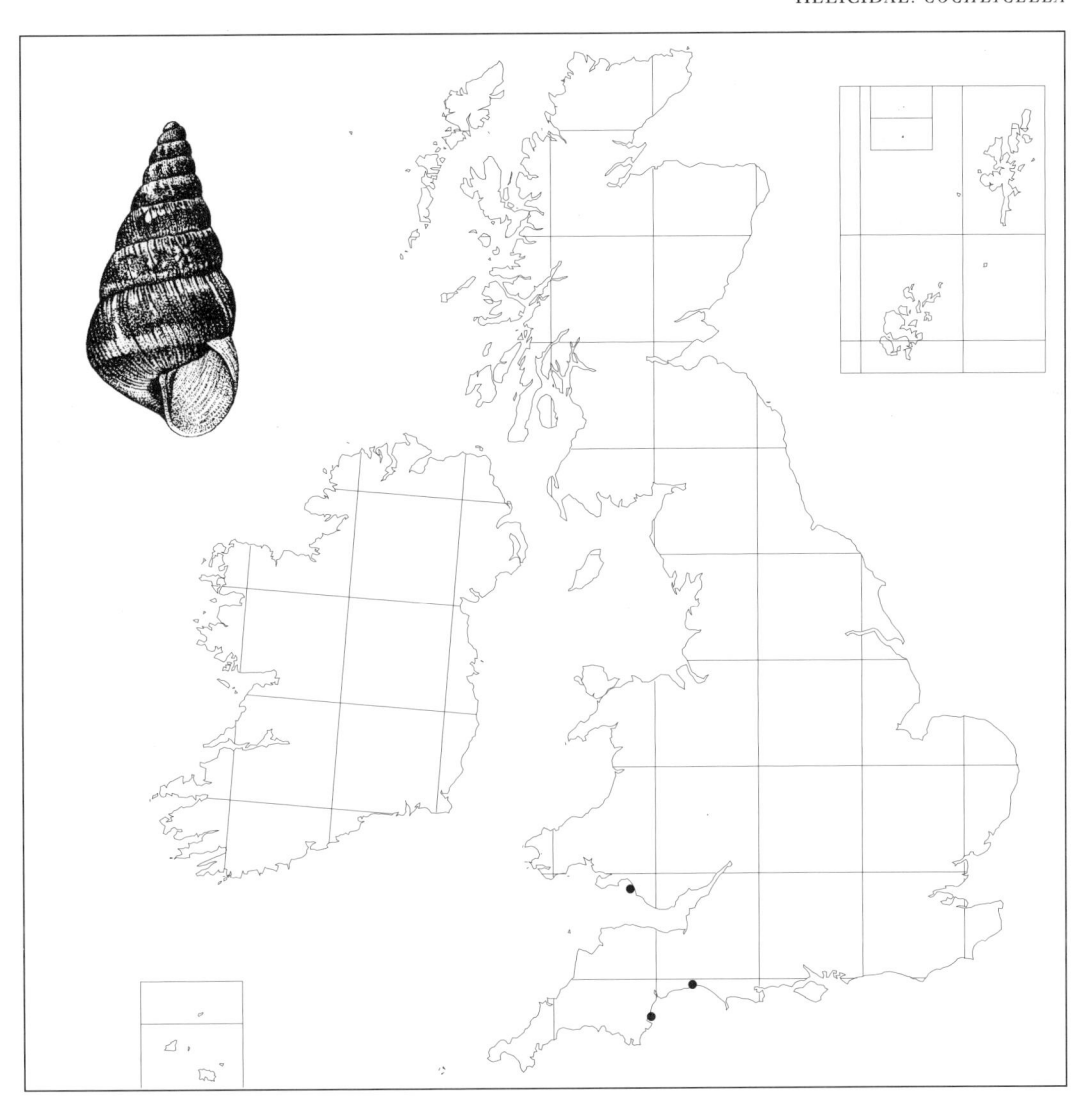

**Cochlicella barbara** (L., 1758) (*C. ventricosa* (Draparnaud))

Size: 10–12mm

This common Mediterranean species lives in dry, open, calcareous places, mostly by the sea. Like *C. acuta*, it is probably frost sensitive. The British populations are on the coast: in Devon, on a grassy roadside bank, and S. Wales in sandhills.

Introduced. *C. barbara* was discovered at Torquay, Devon, in 1975. Although the colony is small, the presence of old weathered shells showed that it had been established there for some years. Another, much larger colony was found at Kenfig, Glamorgan, in 1996 (Boyd, 1997). In addition *C. barbara* was deliberately introduced into Dorset (with Torquay specimens) in 1977; by the mid-1980s this population was well established but had not spread.

Mediterranean: also accidentally introduced in a few places in northern France, Belgium and England.

(K. & C.: 184)

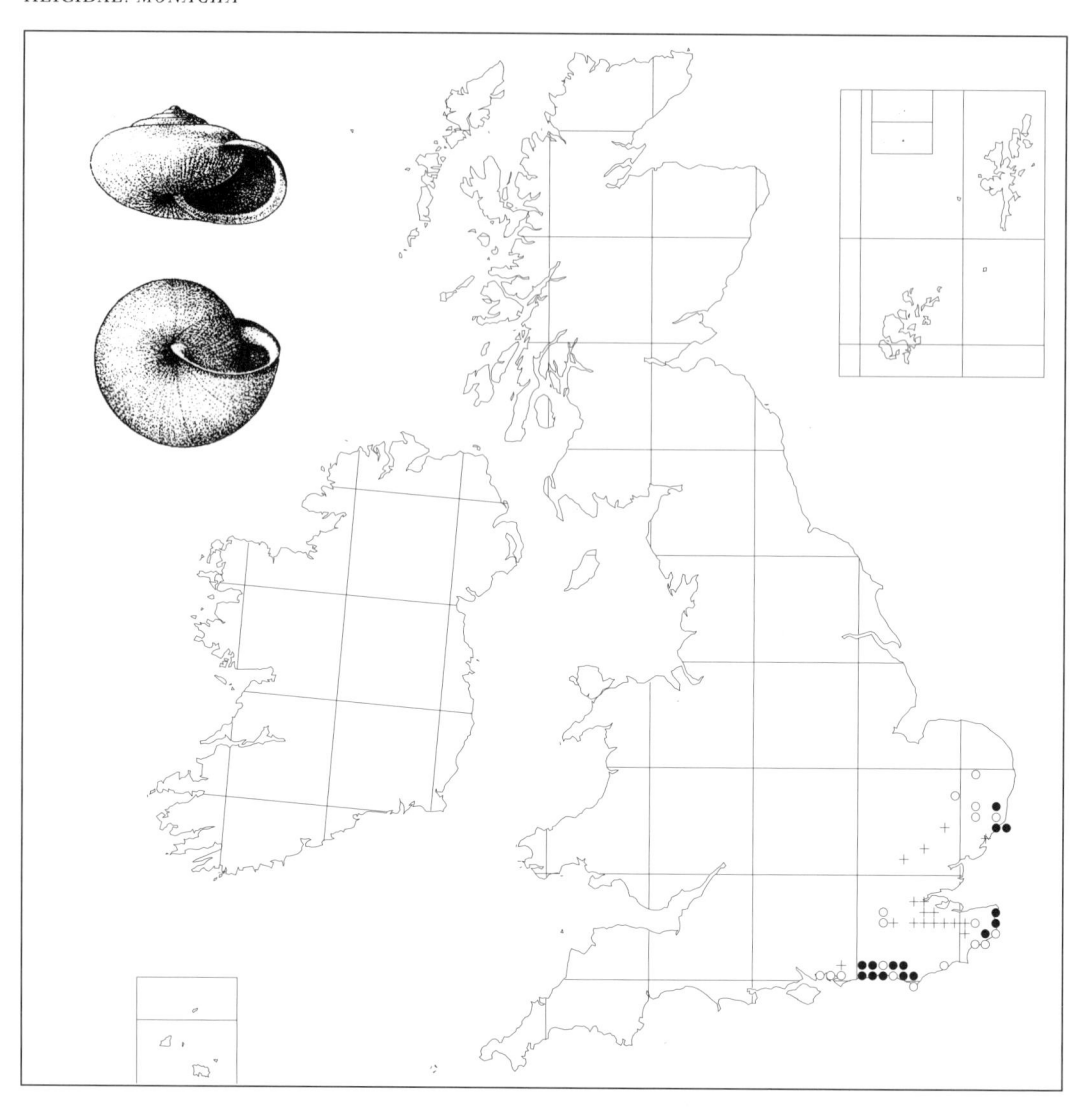

**Monacha cartusiana** (Müller, 1774) (*Helix cartusiana* Müller, *Theba cartusiana* (Müller))

Carthusian (or Chartreuse) snail   Size: 10–15mm

Restricted to open, unshaded habitats on calcareous soils. Most colonies are in dry, short-turfed chalk grassland (often grazed by sheep or cattle), but it is also found in moister places among nettles and other weeds of waste ground on roadside banks, by ploughed fields, and in alluvial pastures near the sea. In E. Kent (Deal) it is locally abundant in dry, grassy vegetation on sandhills.

Introduced? (L.Pgl (Neolithic) fossils only). This species was probably introduced to Britain from southern Europe as a 'weed' of cultivation by prehistoric farmers. Fossils show that it once possessed a wider distribution in S.E. England, being frequent on the chalk of the North Downs and on calcareous soils in East Anglia. Its retreat may perhaps be linked with a decline in summer temperatures since the Postglacial optimum (Kerney, 1970a). In this century it has disappeared from further areas, probably because of changes in farming practices. It was last seen in Norfolk in 1890, in W. Suffolk in 1908, in Surrey in 1910, and in Hants in 1959; in E. Suffolk it survives precariously (Bishop & Bishop, 1973; Killeen, 1992). Adventive colonies have been reported elsewhere from time to time. Most populations must be considered at risk. RDB category: Rare.

Principally Mediterranean and S.E. European; also France, Germany (west of Rhine valley), the Low Countries and England.

(Ellis: 206; K. & C.: 184)

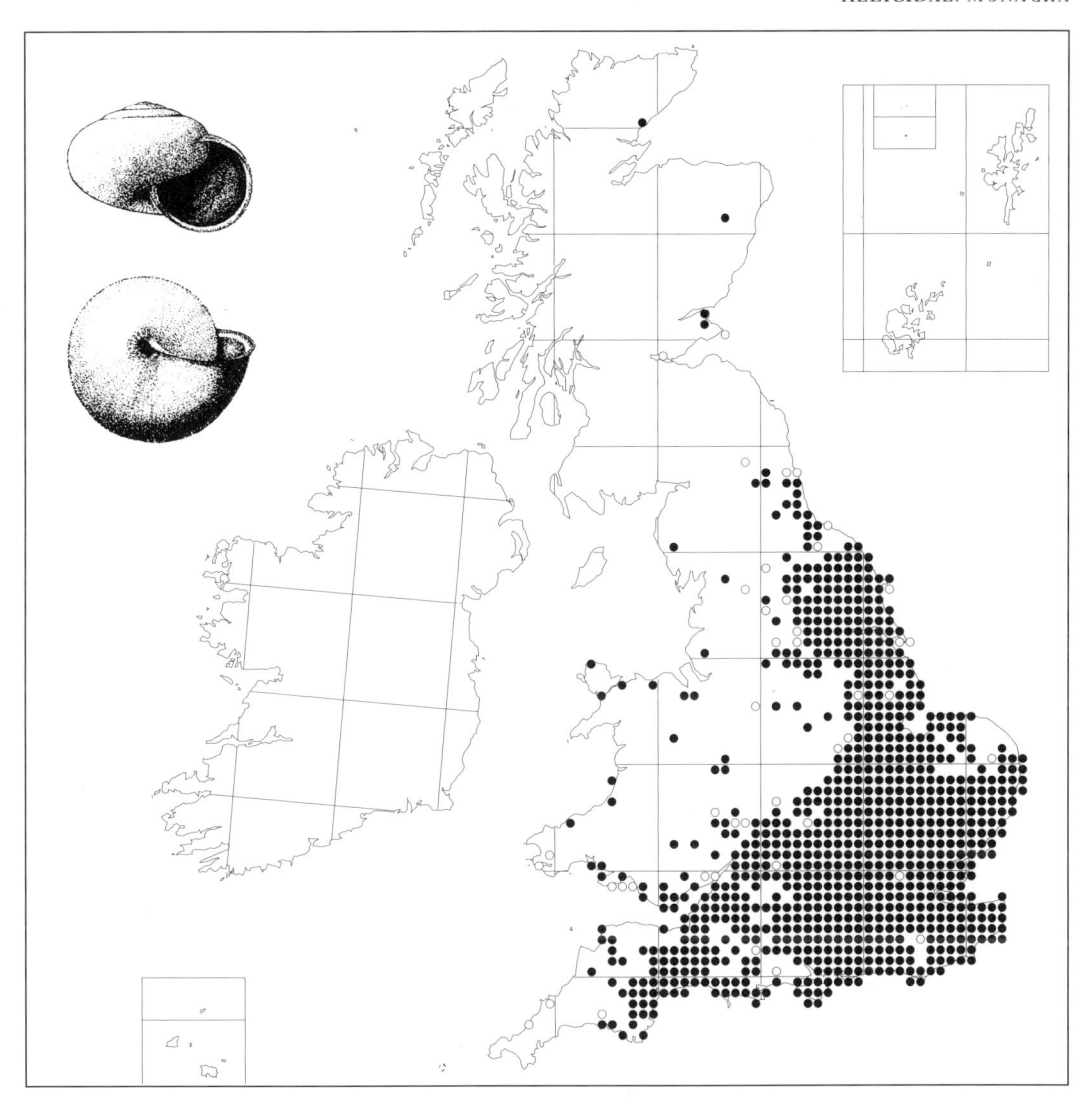

HELICIDAE: *MONACHA*

**Monacha cantiana** (Montagu, 1803) (*Helix cantiana* Montagu, *Theba cantiana* (Montagu))

Kentish snail    Size: 17–20mm

This is a characteristic snail of waste ground, favouring well-drained, base-rich soils. It is typical of roadsides, the edges of fields, old quarries and railway embankments, living among tall grasses, umbellifers, nettles and similar weeds. Occasionally it is found on sandhills. Immature snails, in particular, often attach themselves high up on the vegetation for long periods.

Introduced. This 'weed' species came to Britain possibly in late Roman times, though it remained rare until the mediaeval period (Kerney, 1970a). During the present century it has continued to spread, populations establishing themselves at isolated sites well to the north and west of its main compact area of distribution, for example in Shropshire (first noted 1909), Fife (1920), Staffs. (1943), Cornwall (1947), Pembs. (1957), Derbys. (1969), Anglesey (1974), Sutherland (1975), Aberdeenshire (1980), Cumberland (1980), Cardiganshire (1983), Carmarthenshire (1985) and Lancs. (1994). Attempts to introduce the species to Ireland have been unsuccessful.

Principally Mediterranean, but spread by man in N. France, the Low Countries, N.W. Germany and Britain.

(Ellis: 206; K. & C.: 184)

HELICIDAE: *ASHFORDIA*

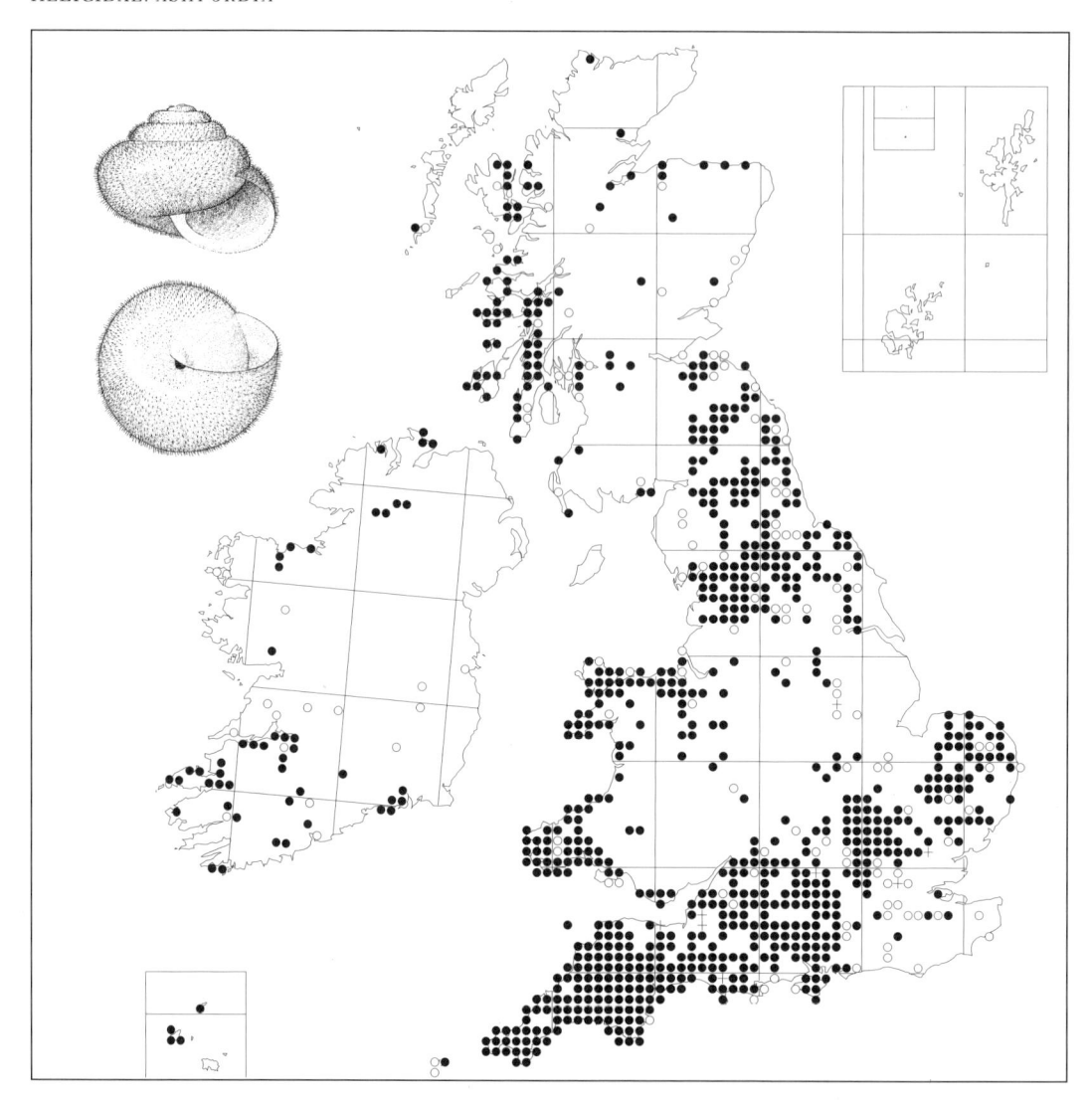

**Ashfordia granulata** (Alder, 1830) (*Helix granulata* Alder, *Monacha granulata* (Alder))

Silky snail   Size: 7–9mm

This is a snail of damp herbage. It is rare in woods, preferring open, unshaded places. In central and eastern England it is virtually restricted to marshy ground in river valleys, where it is associated with succineids and other semi-aquatic snails, but in northern and western parts of the British Isles it is found much more freely away from water on grassy roadside banks, in old quarries and waste ground, and on sandhills and sea cliffs.

Native (E.Pgl). In spite of its discontinuous distribution, there is no fossil evidence that *A. granulata* once possessed a wider range. Nor is there evidence of significant recent national decline, though the species has doubtless become scarcer in some intensively farmed areas of lowland England and may have disappeared from a few 10km squares. In Ireland it appears always to have been very local.

Mainly British Isles, but known also from a few places on the coast of N.W. France and in N. Spain.

(Ellis: 213; K. & C.: 185)

190

HELICIDAE: *PERFORATELLA*

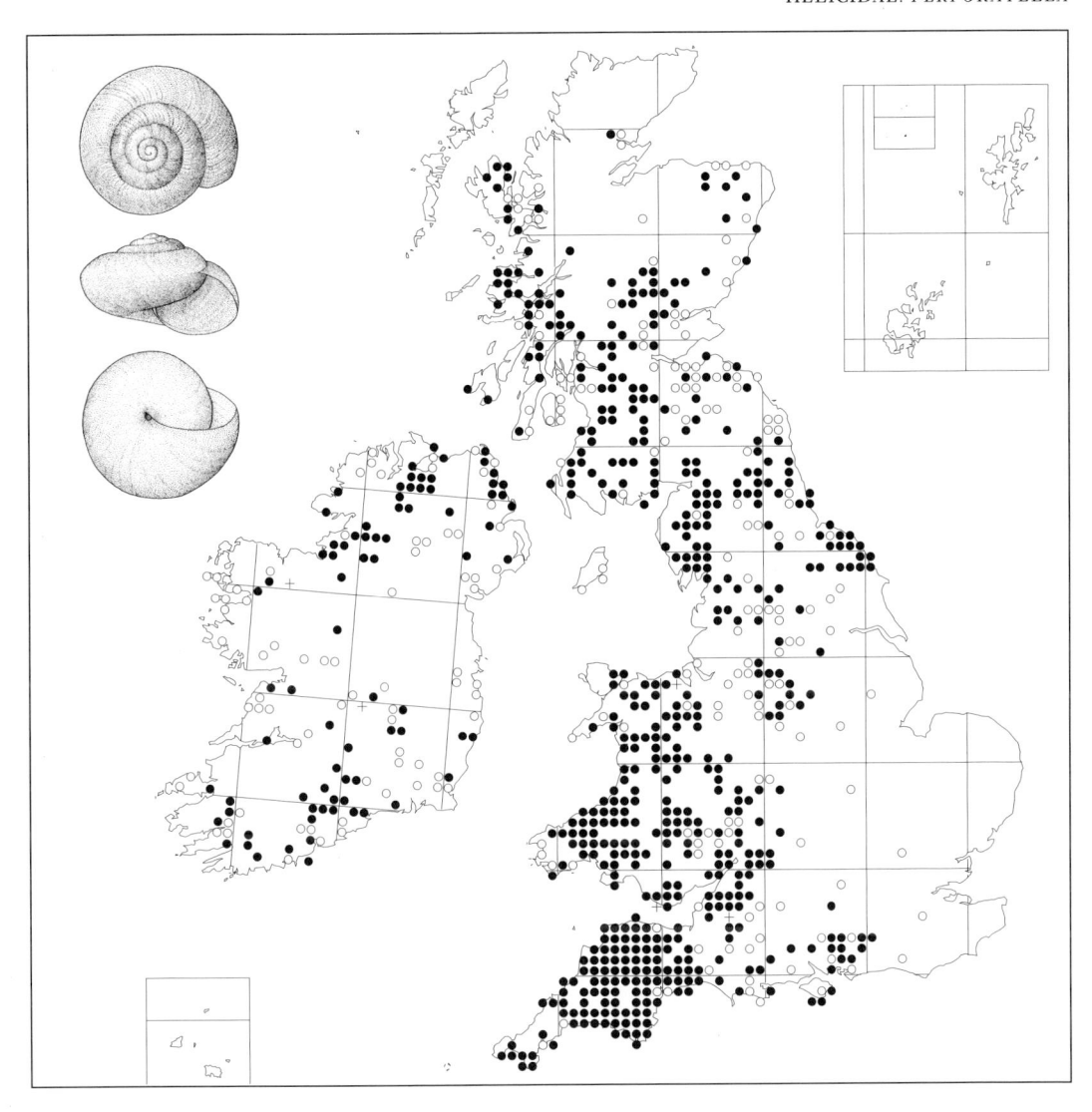

**Perforatella subrufescens** (Miller, 1822) (*Hygromia fusca* (Montagu), *H. subrufescens* (Miller), *Zenobiella subrufescens* (Miller))

Brown snail    Size: 8–10mm

A snail of old, broad-leaved woodland (for example, of oak, hazel, ash, elm or beech), especially in moist, sheltered situations on valley sides. It tolerates poor, acid soils and is often associated with greater woodrush (*Luzula sylvatica*). Though found mostly in ground litter it has several times been noted high up on the trunks, branches and leaves of trees (Alkins, 1925). It is also found in more open (but always relatively wild) places, in ancient hedgerows, among rocks and sea cliffs, on lush grassy roadside banks, and occasionally in marshes.

Native (E.Pgl). This is a snail typical of undisturbed habitats. Though still plentiful in many places in the highland zone, it is undoubtedly receding from the English lowlands, partly perhaps through long-term climatic change but principally because of habitat destruction. It was last recorded in Bucks. in 1834, E. Sussex in 1852, Kent in 1888, Northants. in 1896, Lincs. in 1912 and Oxon. in 1925. Though it is underrecorded in Ireland, there also it appears to be retreating from the central limestone plain.

W. European Atlantic: mainly British Isles and coastal regions of western France.

(Ellis: 216; K. & C.: 185)

HELICIDAE: *PERFORATELLA*

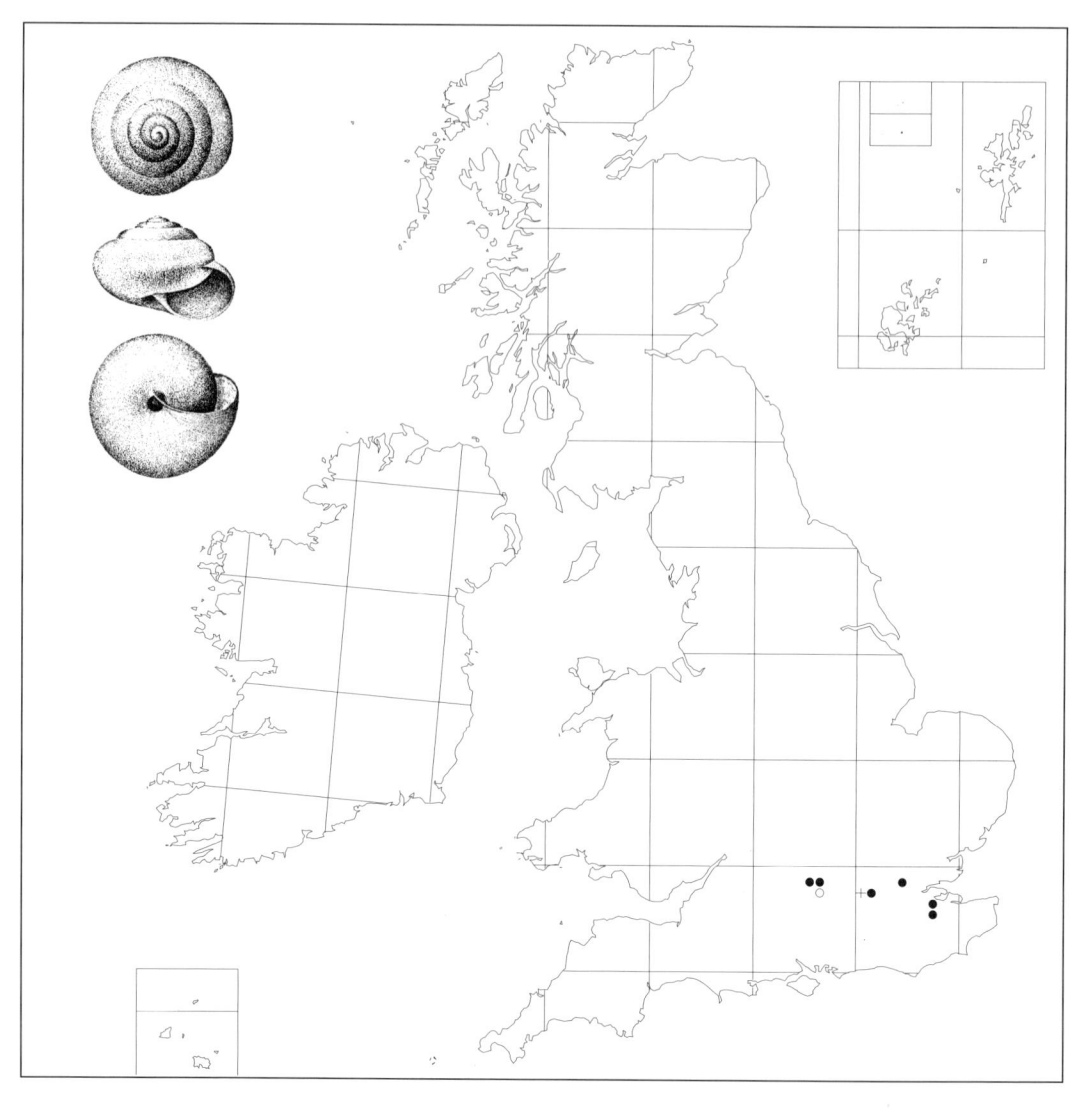

**Perforatella rubiginosa** (Rossmässler, 1838)
(*Pseudotrichia rubiginosa* (Rossmässler))

Size: 6–8mm

This rare species lives in marshes on river floodplains. Typically it is found among flood debris in muddy, poorly vegetated ground shaded by willows, associated with *Lymnaea truncatula*, *Oxyloma pfeifferi*, *Zonitoides nitidus* and *Deroceras laeve*. It is restricted to places subject to periodic inundation.

Probably native (L.Pgl (Neolithic) fossils only). *P. rubiginosa* was recognized in Britain only in 1981, by the R. Thames at Syon Marsh, Brentford, Middx. (Verdcourt, 1982; Naggs, 1983a,b). It has since been found living at a few further sites along the Thames and Medway. Though undoubtedly underrecorded, searches elsewhere suggest that the species is not common. It must be considered vulnerable because of the specialized nature of the habitat, easily destroyed by flood-control schemes or by riverside developments. A population at Pangbourne, Berks., from which specimens were collected in 1955, has now disappeared. RDB category: Vulnerable.

Largely E. European and Siberian; in Europe mainly along the Danube and Rhine valleys, and across the N. German plain to S. Sweden.

*P. rubiginosa* has a shell similar to that of *Trichia plebeia*, with which it may be confused in the absence of anatomical confirmation; a few mapped records of *T. plebeia* may therefore refer to the present species.

(K. &. C.: 190)

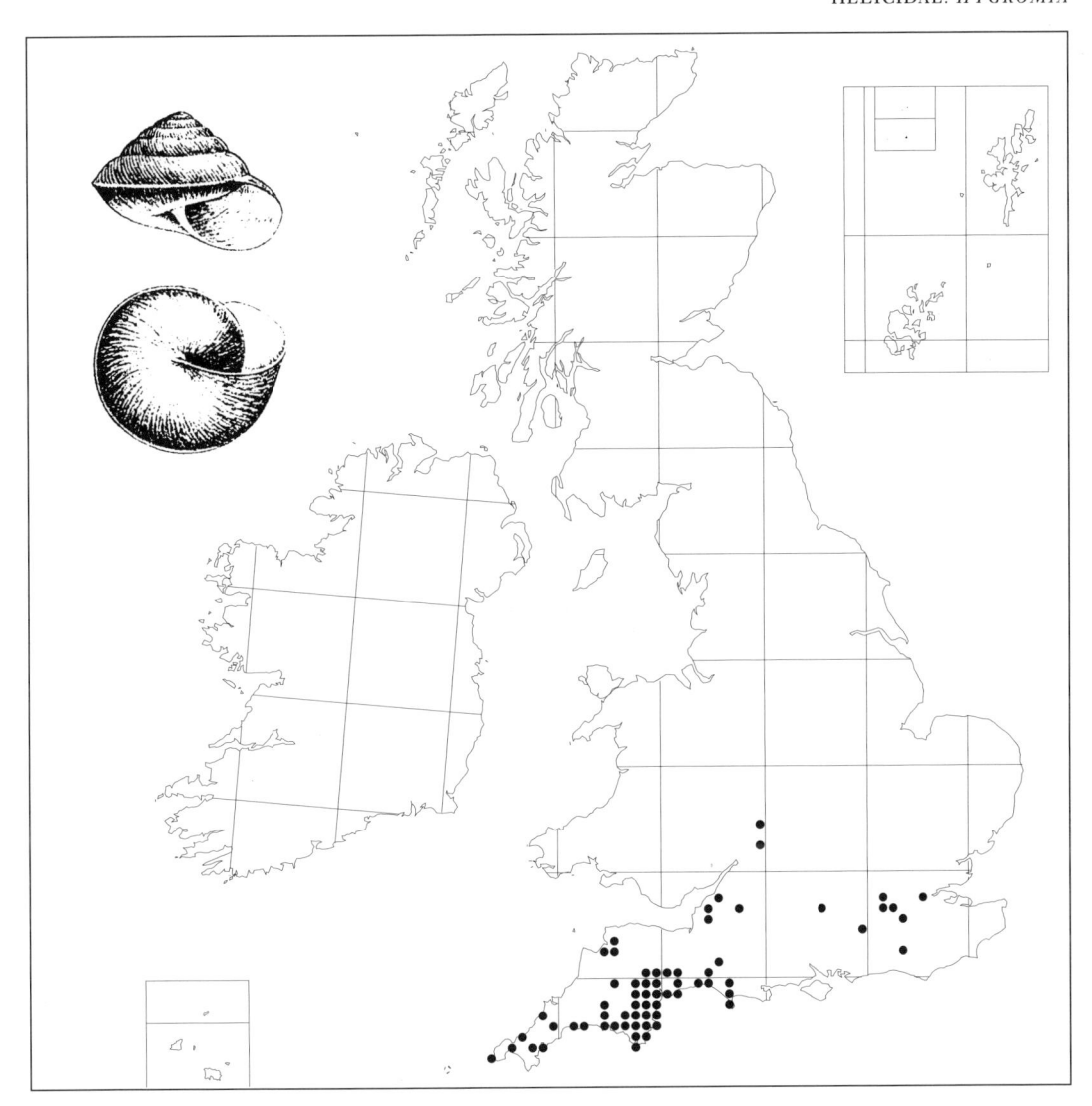

**Hygromia cinctella** (Draparnaud, 1801)

Girdled snail   Size: 10–12mm

A snail of waste ground on well-drained, base-rich soils. In England it lives among grass, nettles, ivy, umbellifers and similar plants on roadsides, under stone walls, and in gardens and rockeries (Biggs, 1957).

Introduced. This alien species was first noted in 1950 at Paignton, Devon, where it had possibly been introduced with nursery or garden plants (Comfort, 1950; 1951). It is now established over a considerable area in Devon and Cornwall and is spreading rapidly. Since 1982 populations, mostly in gardens or waste ground, have been detected in Som., Dorset, Glos., Worcs., Berks., Surrey, Kent and Sussex.

A common Mediterranean species, accidentally introduced in a few places in central and northern Europe (England, France, Switzerland, Austria, Hungary).

(K. & C.: 190)

HELICIDAE: *HYGROMIA*

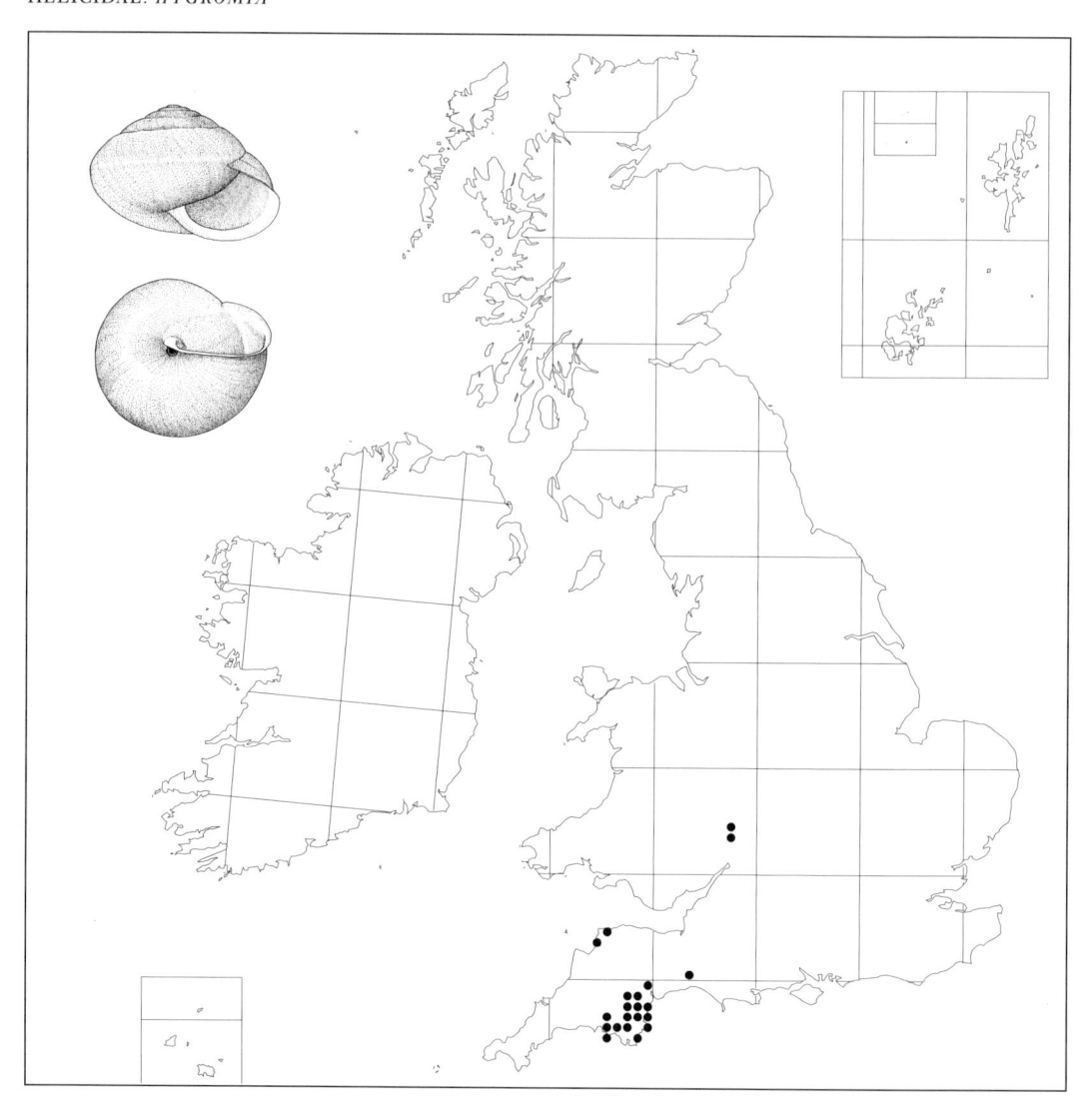

**Hygromia limbata** (Draparnaud, 1805) (*Helix limbata* Draparnaud)

Hedge snail    Size: 12–15mm

Inhabits moist, grassy places, among herbage and ground litter on sheltered roadside banks, in overgrown ditches, old stone quarries and gardens. It often climbs the stems of umbellifers and similar plants.

Introduced. A naturalized colony was discovered at Combeinteignhead, S. Devon, in 1917 (Huggins, 1922; Stratton, 1954). It is now well established in this area and has spread considerably. In 1965 it was found near Little Malvern, Worcestershire, in gardens into which it had probably been introduced accidentally with plants. In 1991 it was detected also in N. Devon and is likely to spread further. There is an early report of this species from the London area (1837), doubtless a short-lived casual.

S.W. European: Spain, southern and western France, England (spread by man).

(Ellis: 215; K. & C.: 191)

HELICIDAE: *TRICHIA*

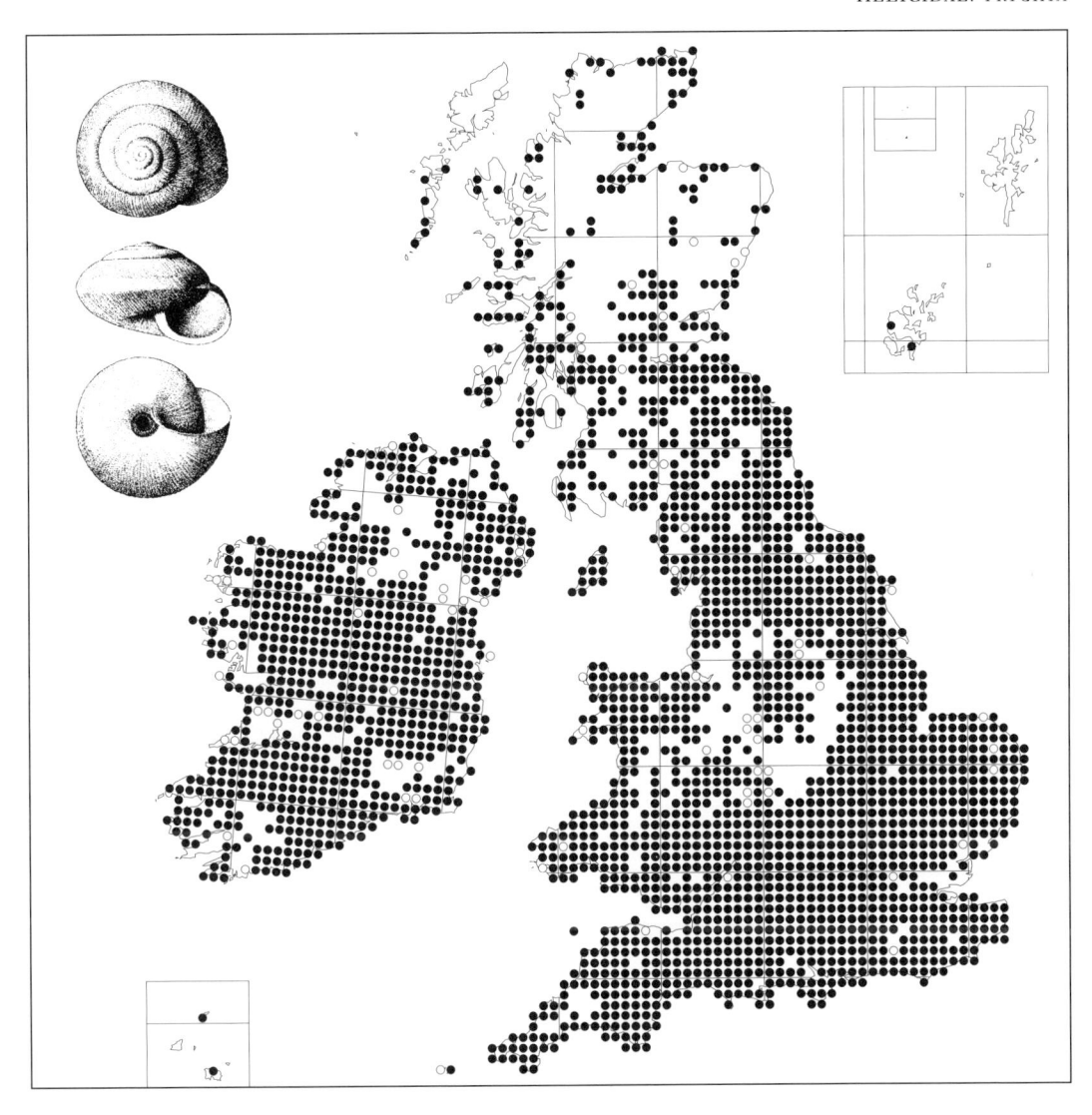

**Trichia striolata** (C. Pfeiffer, 1828) (*Helix striolata* Pfeiffer, *Hygromia striolata* (Pfeiffer), *Helix* (or *Hygromia*) *rufescens* of older British authors, *non* Pennant)

Strawberry snail   Size: 11–15mm

This is a common snail of waste ground. It lives among tall grasses, nettles and similar weeds (often among human rubbish) on roadsides, under walls, along the edges of woods and fields, in quarries and in gardens, where it is occasionally a pest. In the south of England it may occur in old woodland and other semi-natural habitats, but in the highland zone it is more or less restricted to the neighbourhood of settlements where lime and shelter have been provided by man. It is strikingly absent from a large area of apparently suitable country in central England, mainly on Triassic and Upper Carboniferous rocks.

Native (E.Pgl). Though already present here in the Boreal period, *T. striolata* has become much commoner through human activity. In Scotland, where it was rare in the 1930s, it is still extending its range. North of the Great Glen it was first noted at the following dates: W. Inverness-shire, 1954; Outer Hebrides, 1962; Caithness, Orkney and Easter Ross, 1966; E. Inverness-shire, 1967; Skye, 1968; W. Sutherland, 1970; Wester Ross, 1973; E. Sutherland, 1974.

British Isles, N. France, and lowlands of C. and S. Germany; in the Alps range uncertain owing to confusion with other species.

(Ellis: 210; K. & C.: 194)

HELICIDAE: *TRICHIA*

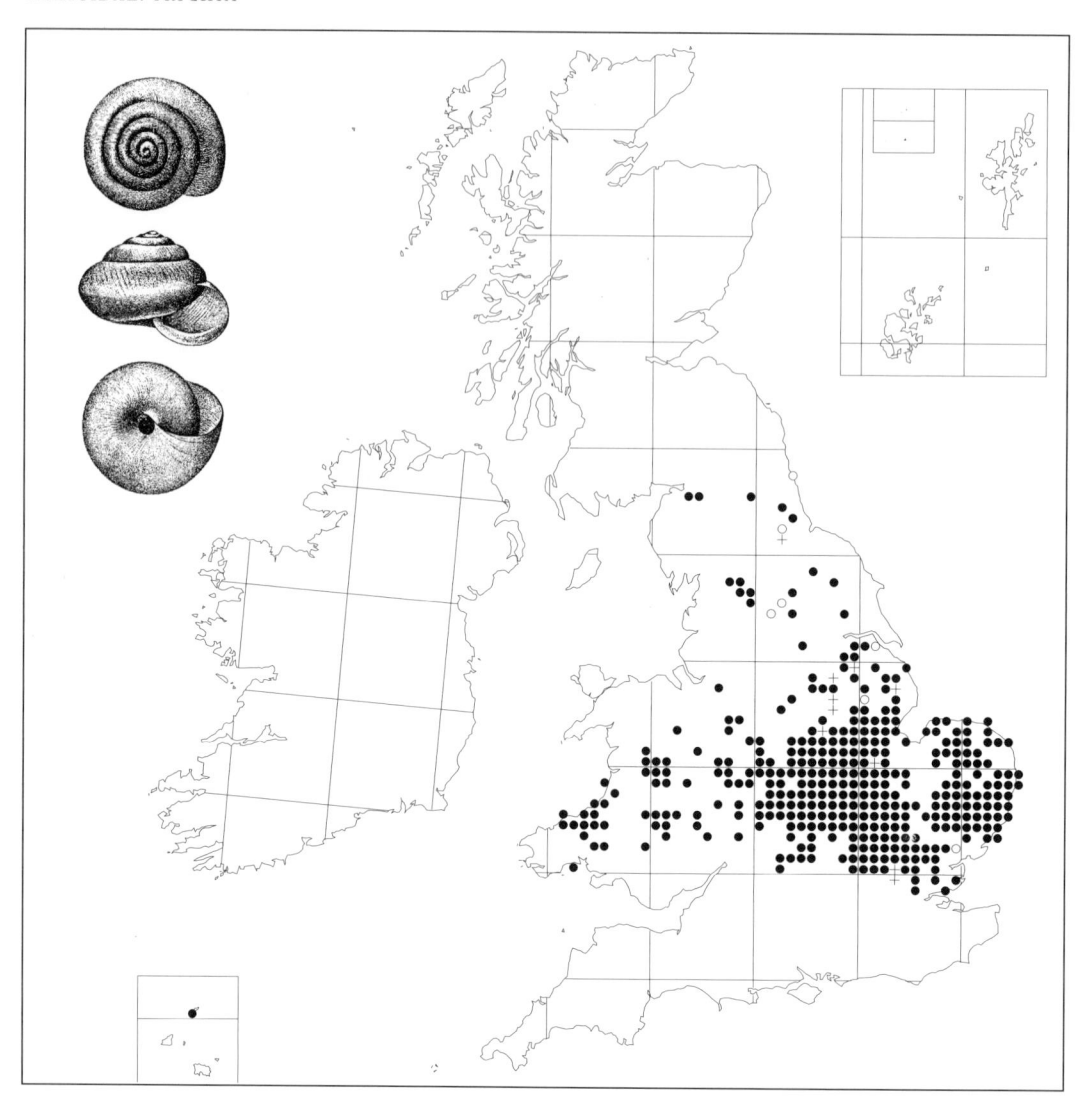

**Trichia plebeia** (Draparnaud, 1805) (*T. hispida* var. *liberta*, and *Hygromia liberta* of British authors, *non* Westerlund)

Size: 7–9mm

A snail of moist herbage and ground litter in sheltered places: woods, fields, roadside verges, under walls and in waste ground. In general it occupies the same kinds of places as *T. hispida*, but is rarely found with it.

Native (E.Pgl). No clear evidence of recent distributional change. Postglacial fossils are common in the Midland counties.

Mainly C. European: England, France, E. Spain, S. Germany, Switzerland, Austria, Czech Republic.

There is uncertainty about the precise taxonomic relationship between *T. plebeia* and *T. hispida*, which are not normally associated and which replace each other geographically in parts of central England (especially in Warwicks., Leics. and Northants.). Clear-cut anatomical differences exist, but some populations are difficult to name on shell characters alone (Naggs, 1985). A few records may be misidentifications of *Perforatella rubiginosa*.

(Ellis: 208; K. & C.: 194)

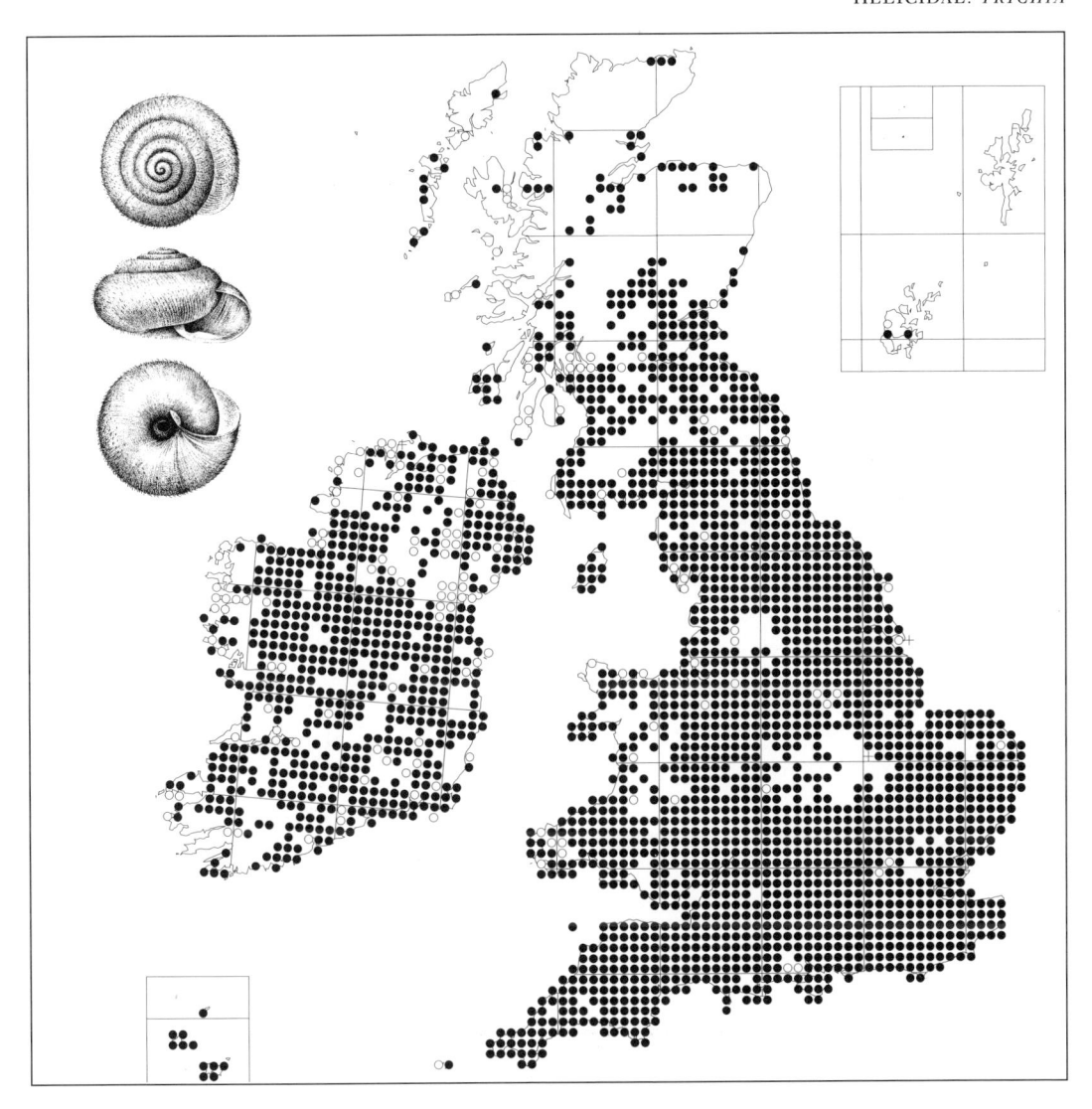

**Trichia hispida** (L., 1758) (*Helix hispida* L., *Hygromia hispida* (L.))

Hairy snail    Size: 6–10mm

A species of ground litter and herbage in moist, generally well-vegetated places: roadside verges, fields, marshes, the base of walls and waste ground. Unlike *T. striolata*, it is rare in gardens. It also tends to avoid deeply shaded woods. Some varieties can live in fairly dry, exposed places, tucked in at the roots of plants in stabilized dunes or short-turfed limestone grassland. *T. hispida* is always commonest on base-rich soils, especially in the northern parts of its range.

Native (Lgl). No clear evidence of recent distributional change, though in northern Scotland, where the species is largely confined to disturbed habitats, it may still be spreading through human activity. Its rarity round Manchester may be due to pollution (cf. *Cepaea nemoralis*).

European, reaching its northern limits in S. Scandinavia, where it has been spread by man.

*T. hispida* is very variable in shell form. It is possible that two or more species are included under this name in the British Isles. Records of *plebeia*, which has the best claim to be considered as a distinct species (Naggs, 1985), are shown separately; some records mapped as *hispida*, especially from the English Midlands, should probably be more correctly assigned to *plebeia*.

(Ellis: 208; K. & C.: 191)

HELICIDAE: *PONENTINA*

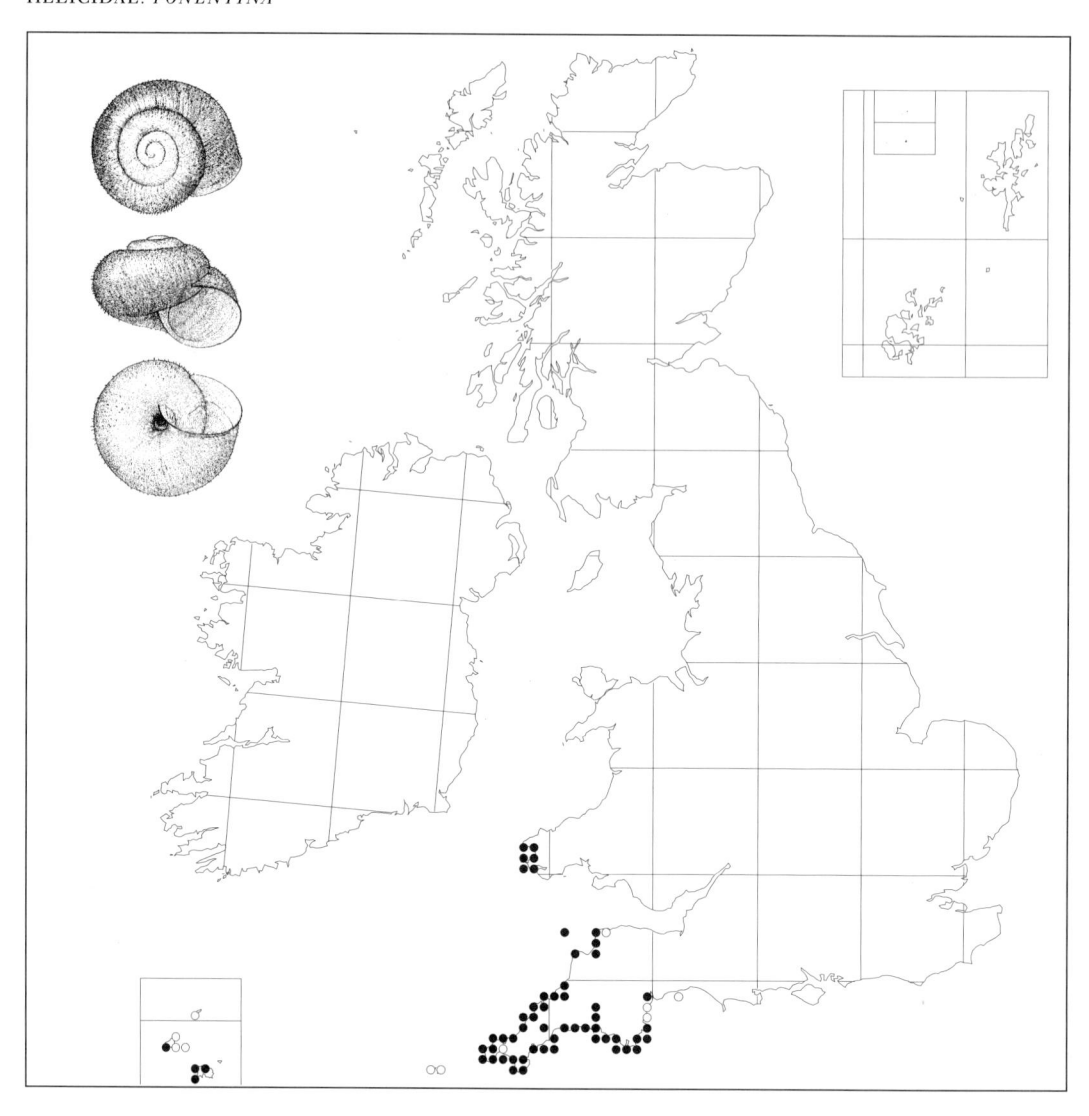

**Ponentina subvirescens** (Bellamy, 1839) (*Hygromia subvirescens* (Bellamy), *Trichia subvirescens* (Bellamy), *Helix* (or *Hygromia*) *revelata* of British authors, *non* Férussac)

Green snail   Size: 5–7mm

Lives in short grass and heather on cliffs by the sea (occasionally inland along estuaries), nearly always on well-drained, non-calcareous soils developed over granite, sandstones or slates. It likes stony places with patchy vegetation. The following plants are typical of its habitats: cock's-foot (*Dactylis glomerata*), lesser hawkbit (*Leontodon taraxacoides*), sea campion (*Silene maritima*), sheep's sorrel (*Rumex acetosella*) and thrift (*Armeria maritima*). Few other shelled molluscs are normally present (Oldham, 1933).

Probably native (L.Pgl fossils only). Though unconfirmed at a few old locations, the species remains frequent within its range and shows no evidence of decline.

S.W. European: Atlantic coasts of Morocco, Portugal, Spain, France, S.W. Britain.

(Ellis: 211; K. & C.: 197)

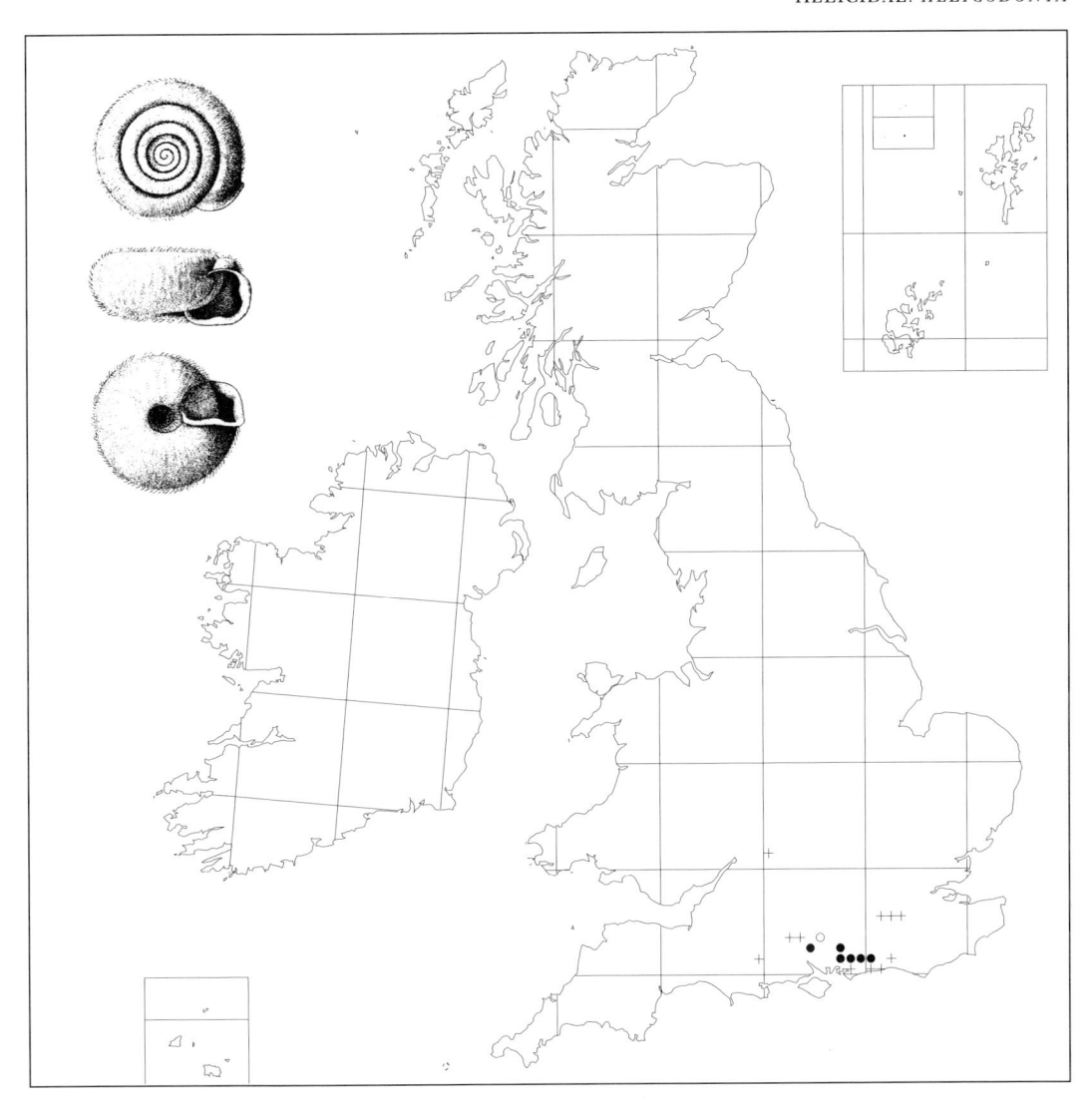

### Helicodonta obvoluta (Müller, 1774) (*Helix obvoluta* Müller)

Cheese snail   Size: 11–14mm

A very local species of chalk beech woods, found in leaf litter and under logs and fallen branches; occasionally it climbs weakly up tree trunks. It is also recorded from mature ash woods, and from old ash/hazel scrub. Nearly all the surviving sites are on the steep escarpment slope of the South Downs (Beeston, 1919; Cameron, 1972).

Probably native (L.Pgl fossils only). Fossils dating from the Neolithic period onwards (mainly from archaeological contexts) show that *H. obvoluta* formerly possessed a wider distribution in southern England, having perhaps been favoured by the higher summer temperatures of the later part of the Postglacial thermal optimum (cf. *Ena montana*). Its range extended north to the Cotswolds (Chedworth, Glos.; Roman period) and west to Dorset. On the North Downs in Surrey its disappearance may be fairly recent, judging from the presence of dead shells in beech woods at Mickleham (Castell, 1962). In central Hampshire it survives at one site near Winchester, known since 1883. Within its main, narrow range along the face of the South Downs, from Petersfield in the west to near Amberley in the east, it remains common at a number of places. Clear felling and replanting present a threat to some populations. RDB category: Rare.

Principally C. European, with outposts in N. France, England, the Netherlands and N. Germany.

(Ellis: 218; K. & C.: 199)

HELICIDAE: *ARIANTA*

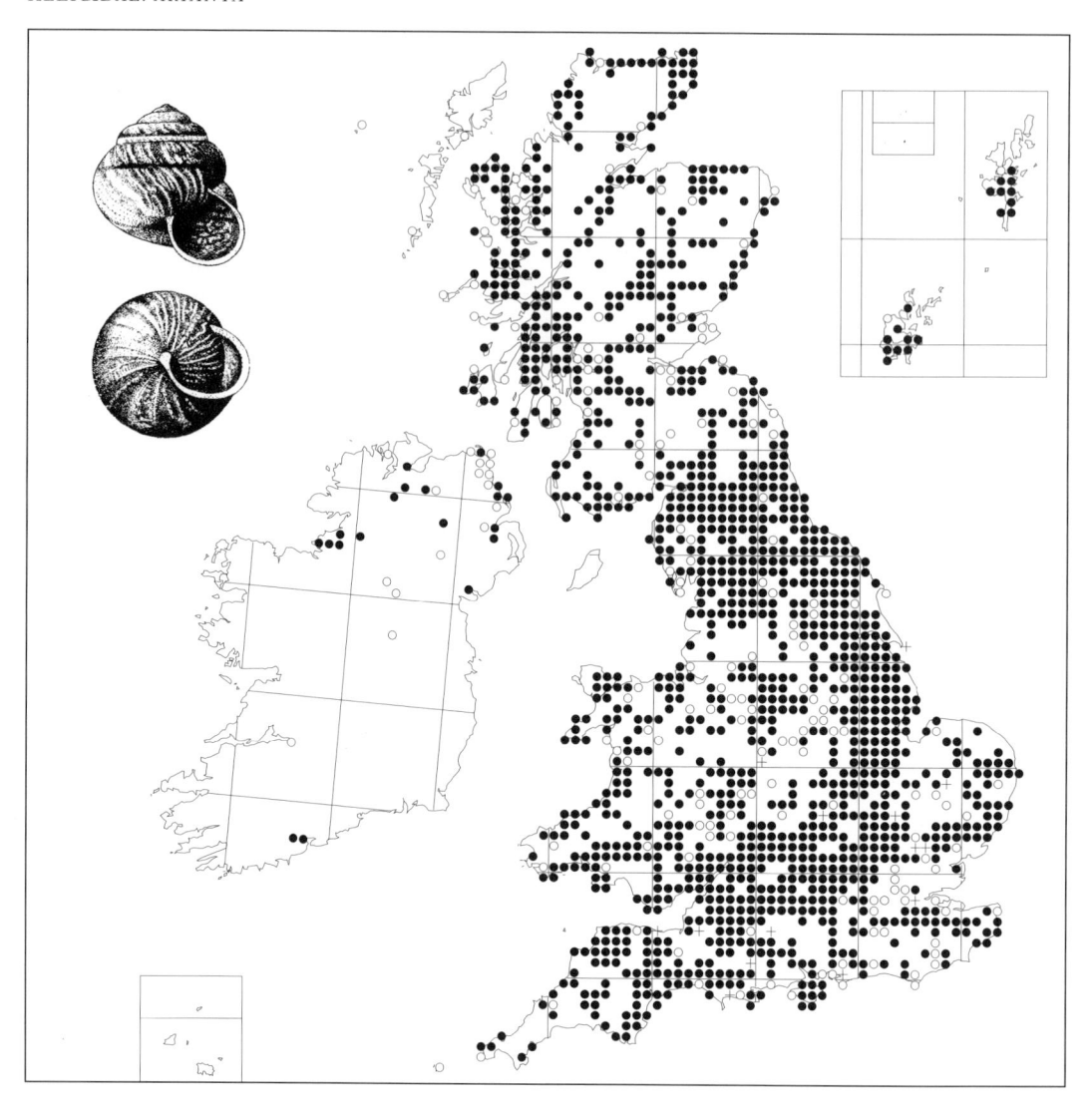

**Arianta arbustorum** (L., 1758) (*Helix arbustorum* L.)

Copse snail    Size: 17–23mm

A species of herbage and ground litter in moist, sheltered places. It inhabits woods, sea cliffs, lush roadsides and river floodplains, often among nettles and similar tall, rank vegetation. In S.E. England it is commonest on base-rich soils but in the highland zone it may be found also in poor, acidic places, and may ascend to great altitudes (1,160m on Ben Lawers). In the extreme north of Scotland it untypically inhabits sandhills (Oldham, 1929b). The few Irish records are mostly from old native woodland, where it is associated with such species as *Perforatella subrufescens*. (Cawley, 1996a).

Native (Lgl). There is little evidence of significant distributional change, though the species is certainly rarer than formerly in some intensively farmed areas of lowland England where suitable uncultivated refuges are now sparse. RDB category (Ireland only): Rare.

European: throughout to beyond the Arctic Circle, but rare south of the Alps and in the extreme west.

(Ellis: 220; K. & C.: 199)

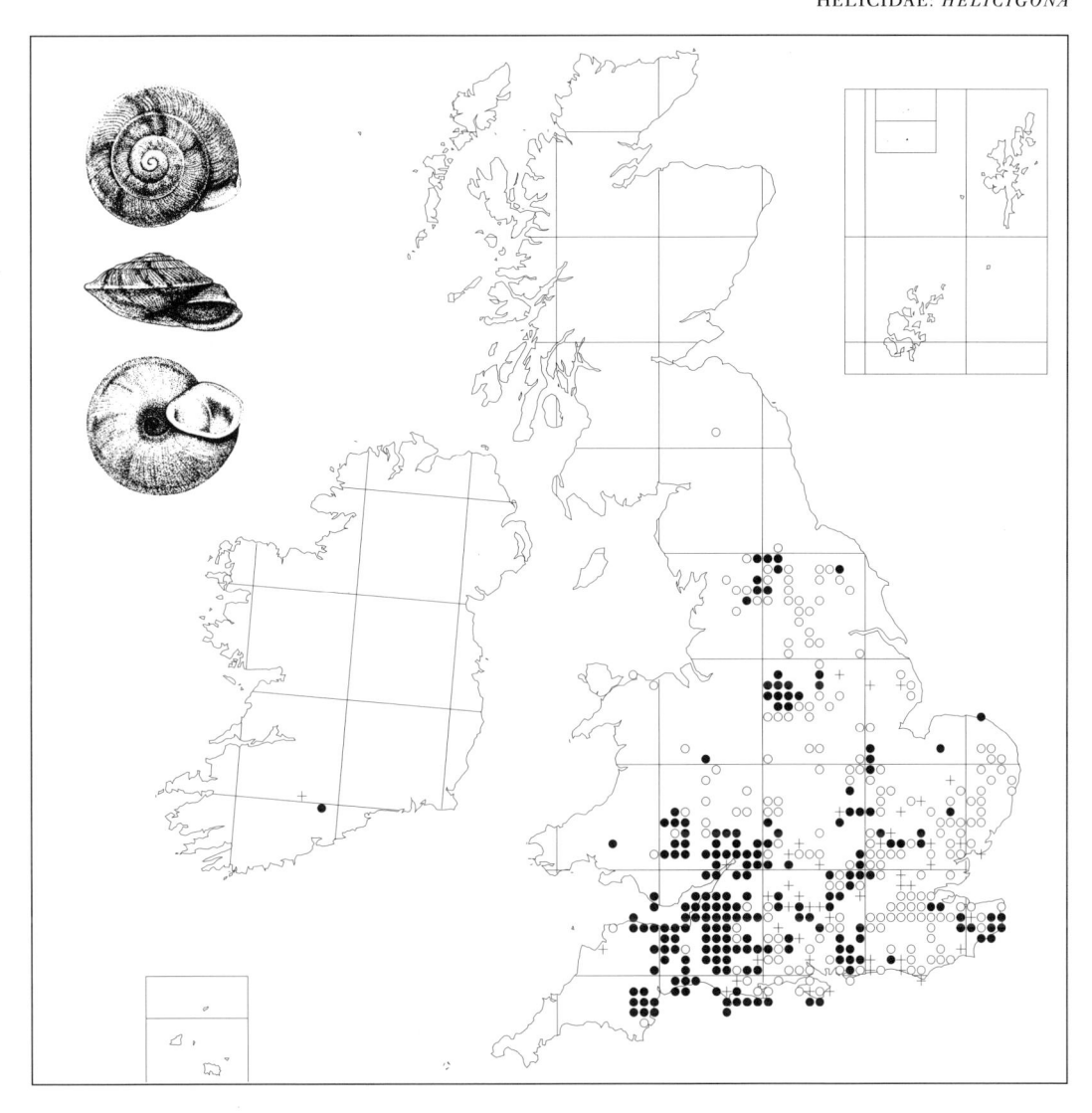

**Helicigona lapicida** (L., 1758) (*Helix lapicida* L.)

Lapidary snail   Size: 15–20mm

This is a snail characteristic of limestone rocks, quarries and stone walls. It needs deep crevices in which to hide, emerging at night or in wet weather to climb and graze. It is found also in deciduous woods (especially of beech) and in old hedges, always on well-drained, calcareous soils.

Native (E.Pgl). In suitable rocky habitats in western counties and in the Peak District, *H. lapicida* remains common. Elsewhere it is receding. It appears to be extinct or nearly so in E. Yorks., Lincs., Notts., Leics., Warwicks., Cambs., Norfolk, Suffolk (Killeen, 1992) and Essex. The destruction of old hedgerows (its principal habitat in East Anglia) has contributed to its decline. It is possible also that atmospheric pollution affecting epiphytes on grazing surfaces may have played some part, judging from its disappearance from woodland round London and from walls and woods on the Magnesian Limestone in N.W. England where only dead shells can now be found. The unique Scottish population (Hawick, Roxburgh; 1872–1890) was probably an introduction. The unique Irish colony on limestone crags at Fermoy, Co. Cork (Phillips, 1914) has been known since the 1870s and may be native; the nearby fossil occurrence (Castletownroche) is unfortunately not securely dated. RDB category (Ireland only): Endangered.

W. and C. European: widespread north to the British Isles and S. Scandinavia (rare in the N. European plain).

(Ellis: 219; K. & C.: 199)

HELICIDAE: *THEBA*

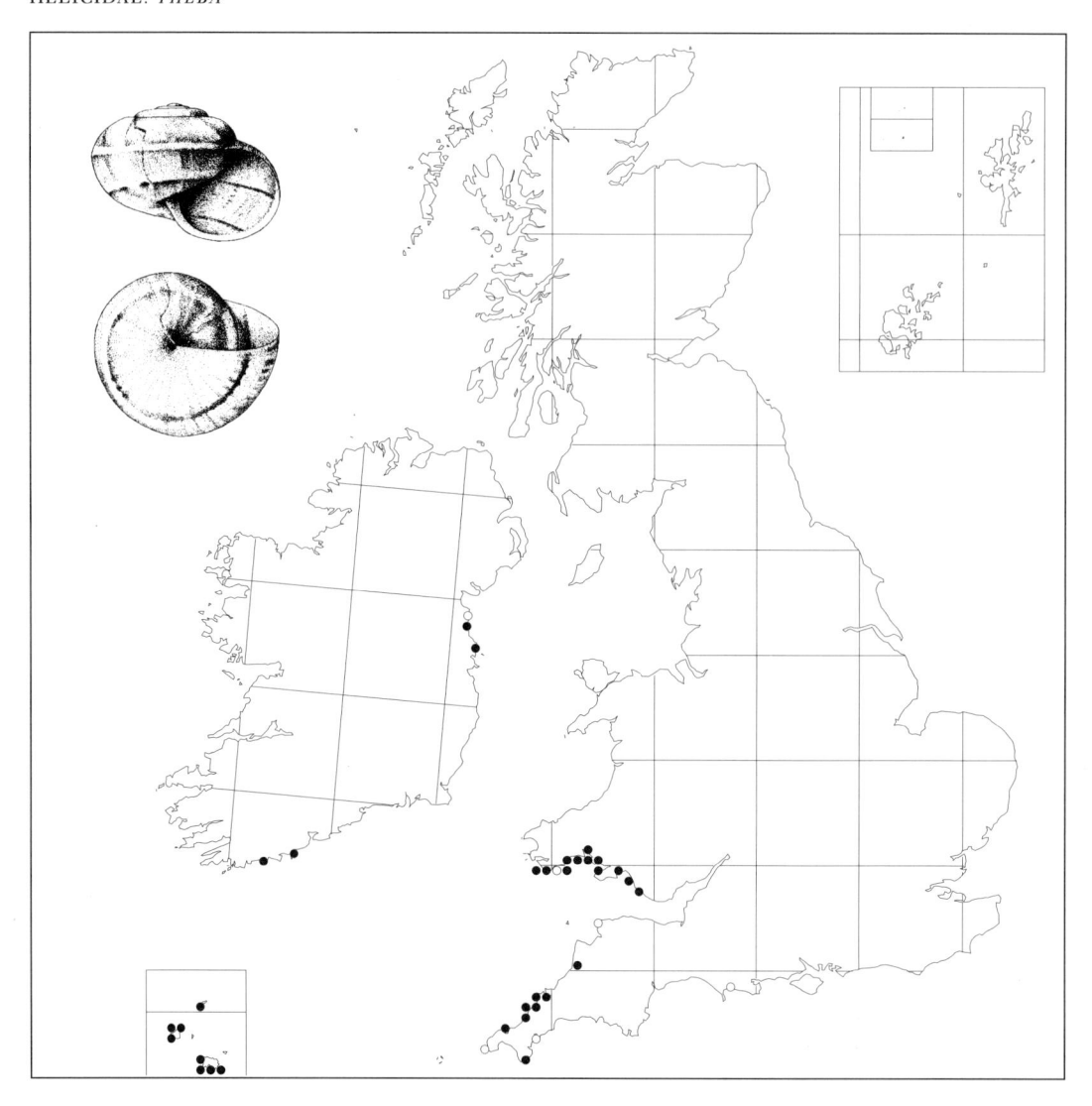

**Theba pisana** (Müller, 1774) (*Helix pisana* Müller, *Euparypha pisana* (Müller))

White (or sandhill) snail   Size: 16–22mm

Locally common in windswept, calcareous places by the sea, either on sandhills or in sparsely vegetated, sandy waste ground. In dunes it can live in nearly bare sand poorly fixed by grasses. In dry weather it attaches itself in clusters to the stems of the taller plants, such as wild carrot (*Daucus carota*), sea beet (*Beta vulgaris* ssp. *maritima*) or sea radish (*Raphanus maritimus*). Its commonest molluscan associates are *Cernuella virgata* and *Cochlicella acuta* (Barrett, 1972; Humphreys, 1976).

Introduced. This is a common Mediterranean pest species, accidentally introduced by man to Britain and Ireland and now well established in a number of frost-free, coastal localities. Its presence in S.W. England, S. Wales and near Dublin was noted early in the 19th century. Many new colonies have been reported in Cornwall, S. Wales and Ireland (Co. Cork) in recent years and the species is undoubtedly spreading (Humphreys *et al.*, 1982; Meredith, 1994). There are also old records from near Weymouth, Dorset (1799) and from Woolacombe Sands, N. Devon (1844). Chance populations are occasionally reported inland in southern England, especially in hot summers, but these always succumb to winter frosts.

Originally circum-Mediterranean, but spread by man in many parts of the world, including the Atlantic coasts of Europe north to the British Isles and the Low Countries.

(Ellis: 222; K. & C.: 202)

HELICIDAE: *CEPAEA*

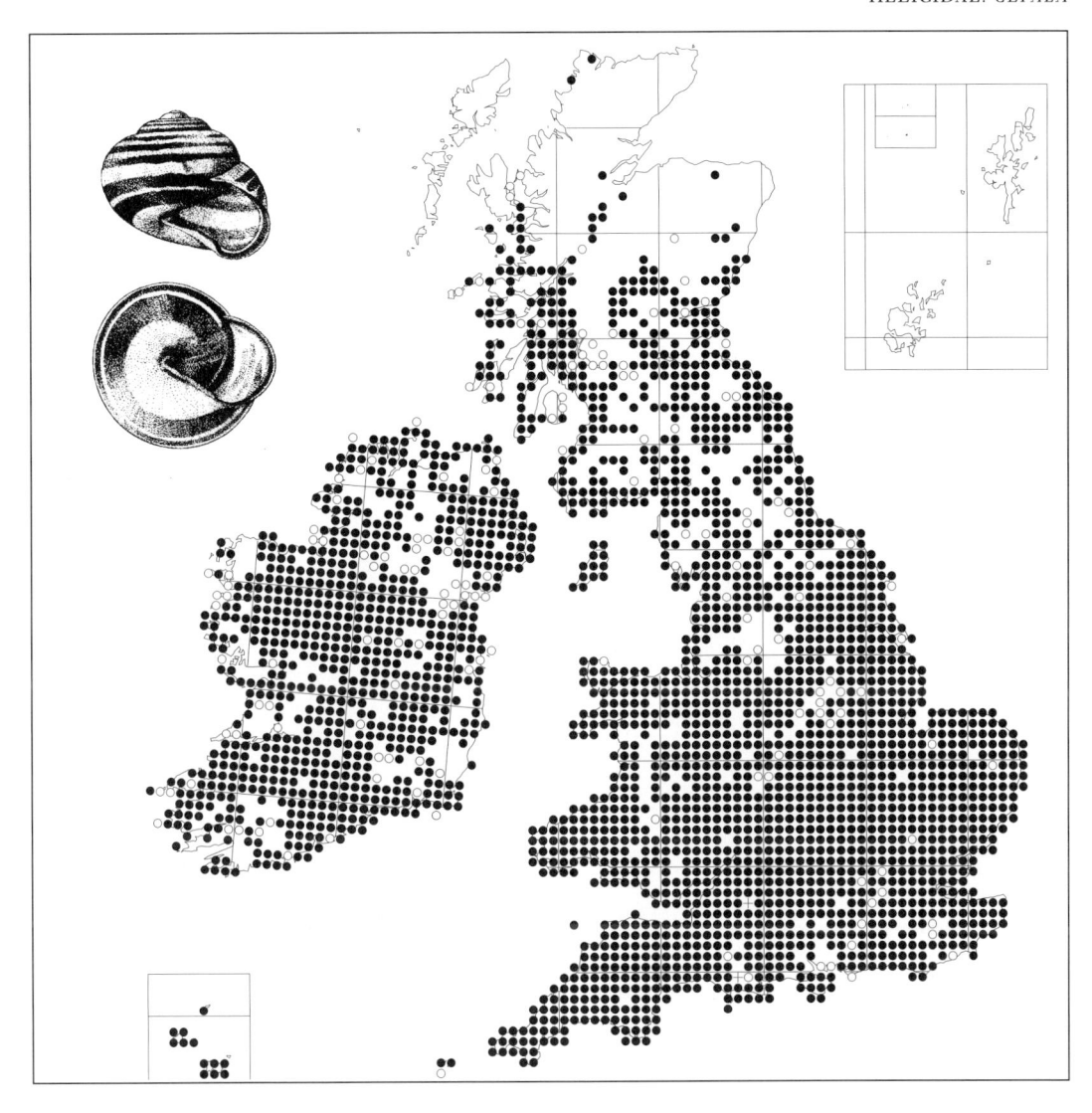

**Cepaea nemoralis** (L., 1758) (*Helix nemoralis* L.)

Grove (or brown-lipped) snail    Size: 18–25mm

This well-known snail lives in all kinds of moderately moist places, mainly on base-rich soils: grassy fields, roadsides, old quarries, woods, waste ground, sea cliffs and sandhills. It shelters among herbage and under ground litter but, in moist weather, will climb the taller plants and sometimes trees and walls. It attains 600m in Scotland (Ben Lawers) and 900m in N. Wales (Snowdon). At its northern limits it shows signs of having been spread by man. The isolated colonies on the W. Sutherland coast may be recent introductions.

Native (E.Pgl). There is little evidence of significant change at a 10km-square level. However tetrad mapping in the Isle of Wight (Preece, 1980) and Suffolk (Killeen, 1992) suggests that *C. nemoralis* has declined locally in comparison with *C. hortensis*. Its absence from the industrial country of S. Lancs., already commented on a century ago, may be due to atmospheric pollution and consequent soil acidification.

W. European: widely distributed, north to coastal areas of S. Scandinavia.

The field ecology and population genetics of this polymorphic species have been much studied. For a comprehensive review with references to previous literature see Jones *et al.* (1977).

(Ellis: 228; K. & C.: 203)

HELICIDAE: *CEPAEA*

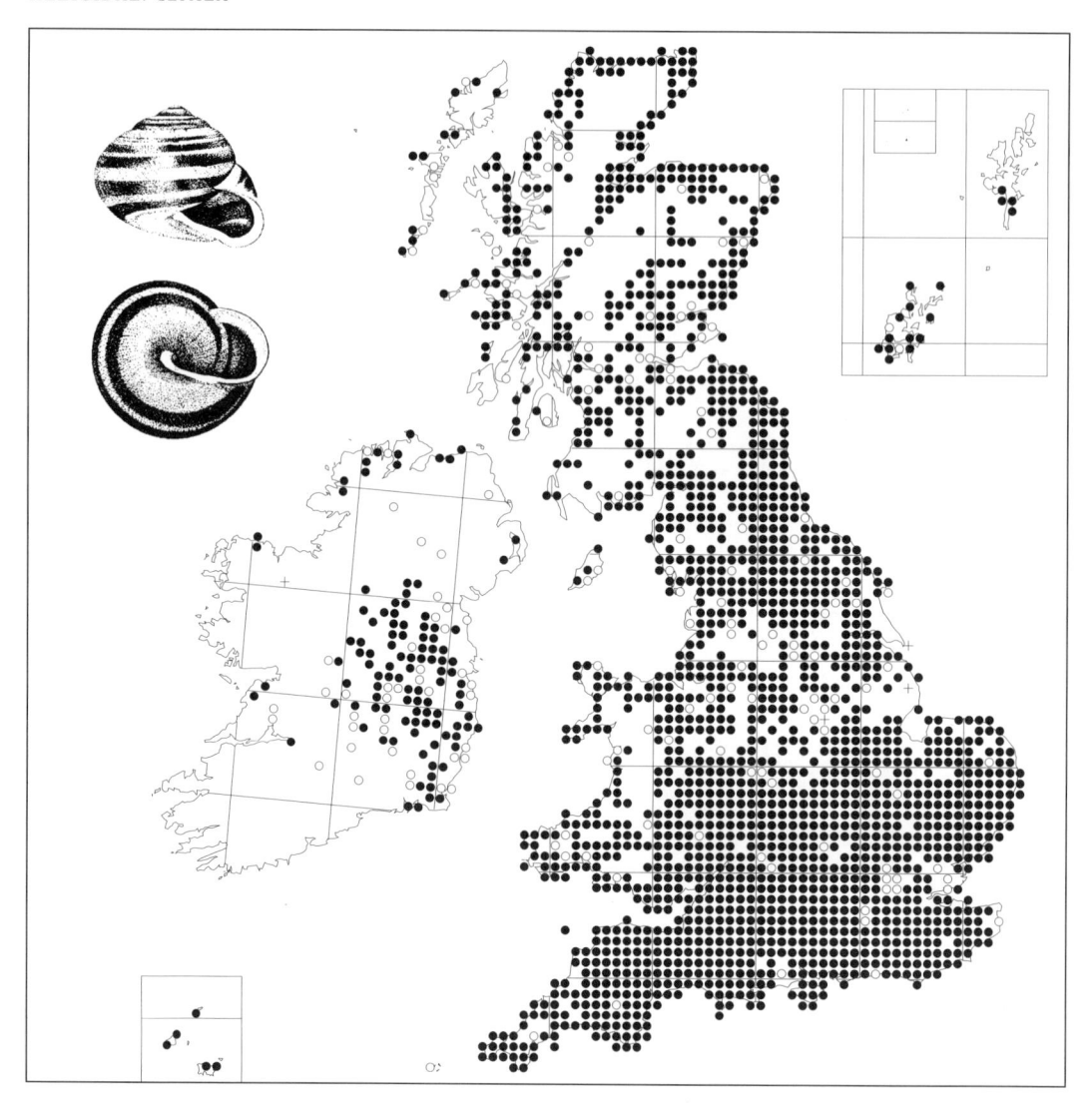

**Cepaea hortensis** (Müller, 1774) (*Helix hortensis* Müller)

White-lipped snail    Size: 16–20mm

Like *C. nemoralis*, this species lives in a variety of moderately humid, sheltered places, under ground litter and among grass and herbage by roads, in woods, among rocks and cliffs, and sometimes in large gardens. In the north of Scotland, beyond the range of *C. nemoralis*, it lives also in maritime turf in stable dunes (Oldham, 1929b). In the highland zone it attains nearly 1,000m. *C. hortensis* tends to favour moister and more shaded places than *C. nemoralis*; it is especially fond of lush roadsides among nettles, umbellifers and similar plants. The two species are nevertheless quite often found together.

Native (E.Pgl). No evidence of significant recent change in distribution.

W. and C. European, north to S. Scandinavia (to the Arctic Circle on the coast of Norway).

Like *C. nemoralis*, this species has been the subject of much research: for a review of the literature, see Jones *et al.* (1977).

(Ellis: 225; K. & C.: 204)

**Helix aspersa** Müller, 1774 (*Cryptomphalus aspersus* (Müller))

Garden (or common) snail   Size: 30–40mm

This familar snail inhabits a wide variety of sheltered places, generally on base-rich soils: hedge banks, sea cliffs, sandhills, old quarries, graveyards and neglected waste ground of all kinds, especially among rubbish, long grass, nettles and other weeds. In the south it is also not uncommon in deciduous woods. It requires crevices for hibernation and may occur abundantly beneath ivy on walls, or under bricks and stones in quite small suburban gardens (where it can be a troublesome pest). In the north it is mainly coastal.

Introduced. *H. aspersa* was introduced into England probably early in the Romano-British period, by accident in the course of trade with the European mainland. It has since spread north to the Orkneys. Lack of lime and low winter temperatures are the main factors preventing further penetration in central and northern Britain. Its distribution appears not to have changed significantly since the 19th century, though marginal populations may appear and disappear and there are signs of minor recession around the conurbations of Birmingham, Manchester and Glasgow. Around Greater Manchester, where it was regarded as extinct by about 1880, it is now re-establishing itself, principally in gardens.

Mainly Mediterranean, but reaching northwards to the British Isles, the Rhine valley and the Low Countries. Introduced in many parts of the temperate world.

(Ellis: 235; K. & C.: 205)

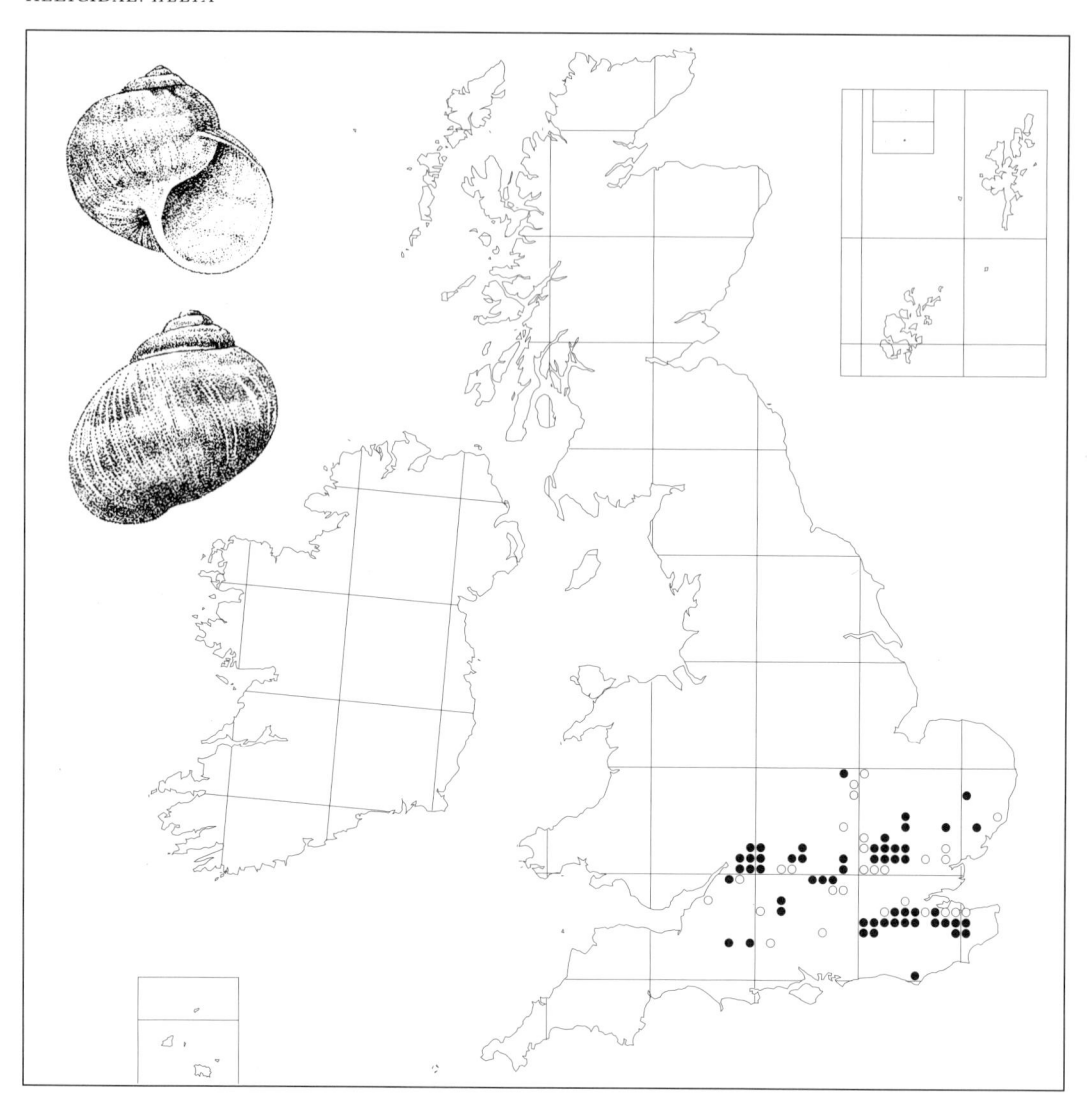

**Helix pomatia** L., 1758

Roman snail, vine snail, edible snail    Size: 35–45mm

The Roman snail lives in undisturbed, grassy or bushy places on well-drained, calcareous soils: hedge banks, uncultivated downland, the margins of fields, dry open woods, old chalk quarries. It requires a loose soil in which it can burrow in order to hibernate and lay its eggs. Unlike *H. aspersa*, it has no special association with human settlements and is not normally found in gardens.

Introduced? (Poorly dated L.Pgl fossils only). *H. pomatia* is likely to have been introduced into England, perhaps deliberately, there being no evidence to disprove the popular idea that the Romans brought it here for food. Its distribution probably owes much to man, and some of the more isolated colonies no doubt originated as releases. There are additionally many documented records over the past 150 years (not shown here) of deliberate introductions in various parts of England, Scotland and Ireland; these rarely survived for long. Within its main area of distribution there are some slight signs of 20th century decline. Intensive farming has destroyed some populations; others are being depleted by restaurateurs and amateur cooks. *H. pomatia* has poor powers of natural dispersal, and its large size and relatively slow rate of reproduction make individual colonies vulnerable (Pollard, 1974; Welch & Pollard, 1975; Alexander, 1994).

Mainly C. and S.E. European, but extending westwards to France and England, and north to the coasts of the southern Baltic.

(Ellis: 232; K. & C.: 205)

## MARGARITIFERIDAE: *MARGARITIFERA*

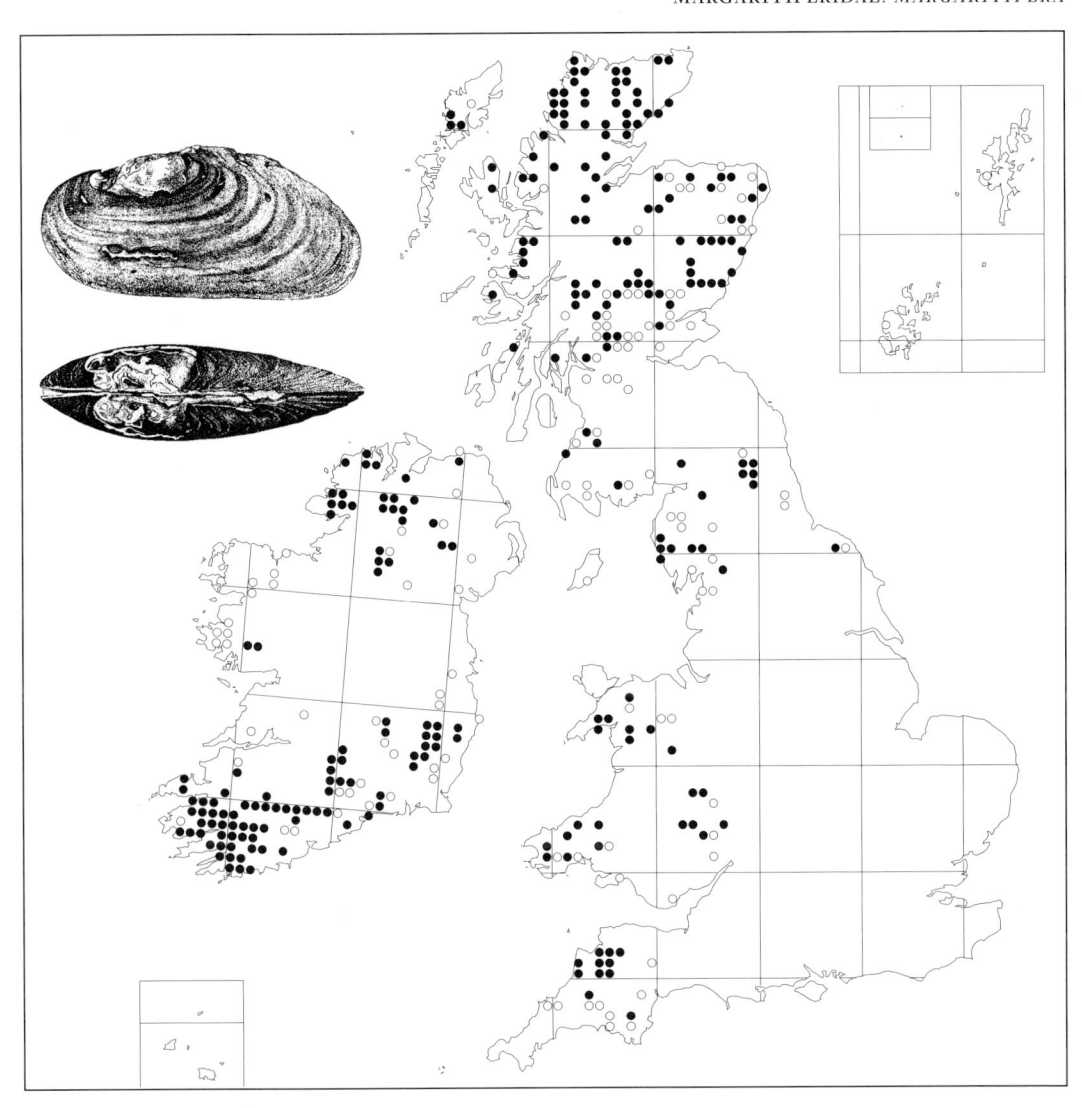

**Margaritifera margaritifera** (L., 1758) (*Unio margaritifer* (L.), *Margaritana margaritifera* (L.), *Margaritifera durrovensis* Phillips)

Freshwater (or Scottish) pearl mussel
Size: 90–140mm

Restricted to quickly flowing, permanent rivers and streams, mainly in the highland zone. It requires clean, cool, well-oxygenated water, free from mud and suspended matter. It likes sandy substrates, often burrowing into the interstices between pebbles or boulders. Nearly all its habitats are in soft water, with the notable exception of some rivers in southern Ireland (R. Nore, R. Barrow; form *durrovensis*) (Chesney *et al.*, 1993). It is rarely associated with other molluscs, apart from *Ancylus fluviatilis*.

Probably native (poorly dated or L.Pgl fossils only). Though still fairly common in its strongholds in Scotland and southern Ireland (Lucey, 1993), the pearl mussel is declining and has become extinct in some 10km squares. The causes are various. Destruction by pearl fishers must be blamed in a few cases. Physical changes to the habitat which alter the bottom and increase turbidity downstream, for example by dredging, can also be damaging. Lastly, populations may be reduced or destroyed by pollution; a suspected cause in some rivers is enrichment by phosphates and nitrates derived from farmland. The species is particularly vulnerable because of its unusual longevity (one hundred years or more) and slow reproduction (Cranbrook, 1976; Young & Williams, 1983).

Holarctic circumpolar; in Europe now largely restricted to the northern highland zone, to 71°N in Norway.

(Ellis (Bivalves): 20)

MARGARITIFERIDAE: *MARGARITIFERA*

**Margaritifera auricularia** (Spengler, 1793) (*Unio sinuatus* Lamarck, *Pseudunio auricularia* (Spengler))

Size: 140–170mm

In southern Europe, *M. auricularia* inhabits the deeper, quieter stretches of large lowland rivers, usually in hard water (cf. *M. margaritifera*). Few living populations have been reported during the past century.

Extinct (fossils only). This large and handsome bivalve formerly lived in the River Thames. Many shells have been found in dredgings or in foreshore deposits between Mortlake and Bermondsey in London (Jackson & Kennard, 1909). Direct radiocarbon dating has shown these to be between 4,000 and 5,000 years old (Preece *et al.*, 1983). The cause of the disappearance of the species from Britain is unknown but climatic changes are likely to have played some part. *M. auricularia* also occurred in the Thames during the last interglacial period.

An endangered species, recorded only from France, N. Italy, Spain, Portugal and Morocco. As a Postglacial fossil it is known additionally from England, Germany, Czech Republic, Switzerland and C. Italy.

(Ellis (Bivalves): 18)

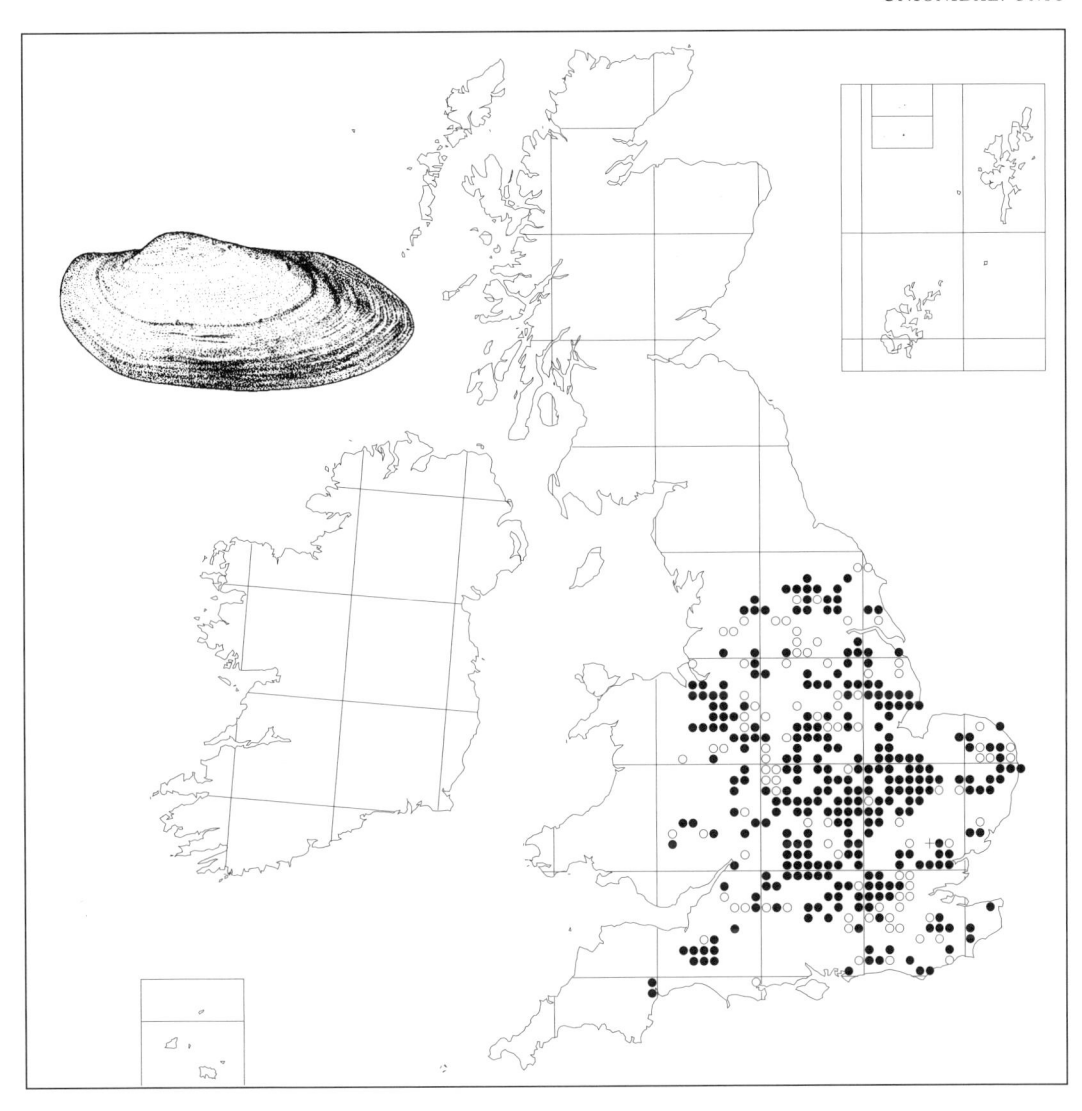

**Unio pictorum** (L., 1758)

Painter's mussel   Size: 75–100 (occ.–140)mm

Essentially a lowland species, found in hard water in slow-flowing rivers, canals, drainage dykes, lakes and reservoirs. It is rare in small closed ponds. It requires a firm, silty bottom in which to bury itself, avoiding both soft mud and stony or gravelly substrates.

Probably native (poorly dated or L.Pgl fossils only). *U. pictorum* shows no signs of serious recent decline. Its distribution owes much to the construction of canals and canalized rivers, and a few sites have consequently been lost through the abandonment and silting-up of parts of the canal system. The species has also sometimes been introduced into reservoirs, ornamental lakes and flooded gravel pits, either deliberately, or accidentally with fish (on which the parasitic larvae of *Unio* depend).

European, but rare in the Mediterranean region, and not beyond 60°N in Scandinavia.

(Ellis (Bivalves): 24)

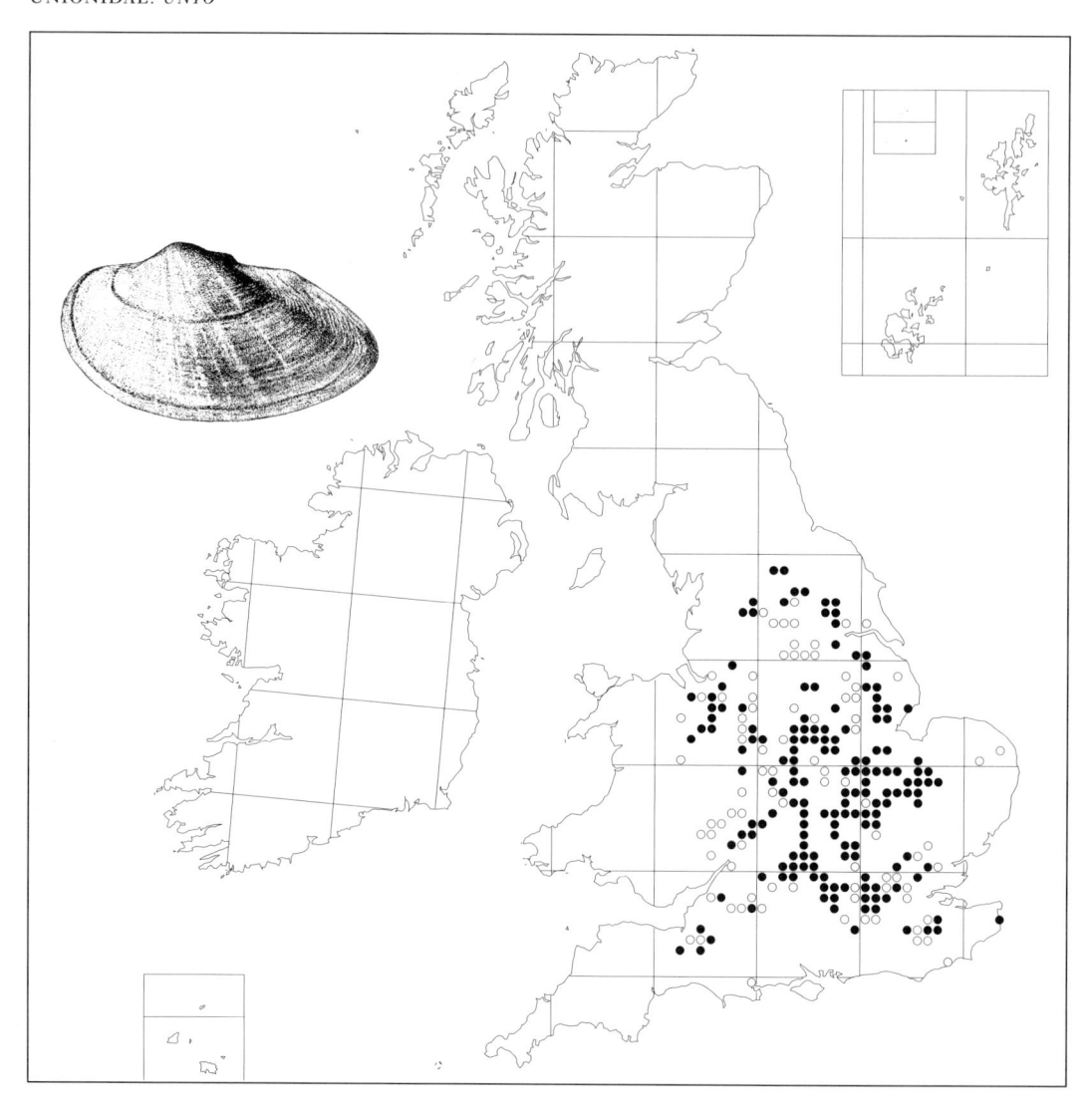

UNIONIDAE: *UNIO*

**Unio tumidus** Philipsson, 1788

Swollen river mussel   Size: 70–90 (occ.–120)mm

A hard-water species found in quiet lowland rivers and canals, mostly within the interconnected 'canal basin' of England and Wales. It appears to require cleaner, better-oxygenated water than other species of *Unio* or *Anodonta*. Though occasionally present in lakes and large ponds, it is much rarer than *U. pictorum* in such places. It needs a firm bottom, avoiding stony substrates and deep, soft mud.

Probably native (poorly dated or L.Pgl fossils only). The distribution clearly owes much to man. *U. tumidus* is perhaps a little rarer than formerly and some outlying stations have not been confirmed in recent years. Losses can mostly be ascribed to the draining or change in character of canal habitats.

European, mainly in the lowlands between the Alps and the Baltic sea (to 61°N in Sweden).

Ellis (Bivalves): 26

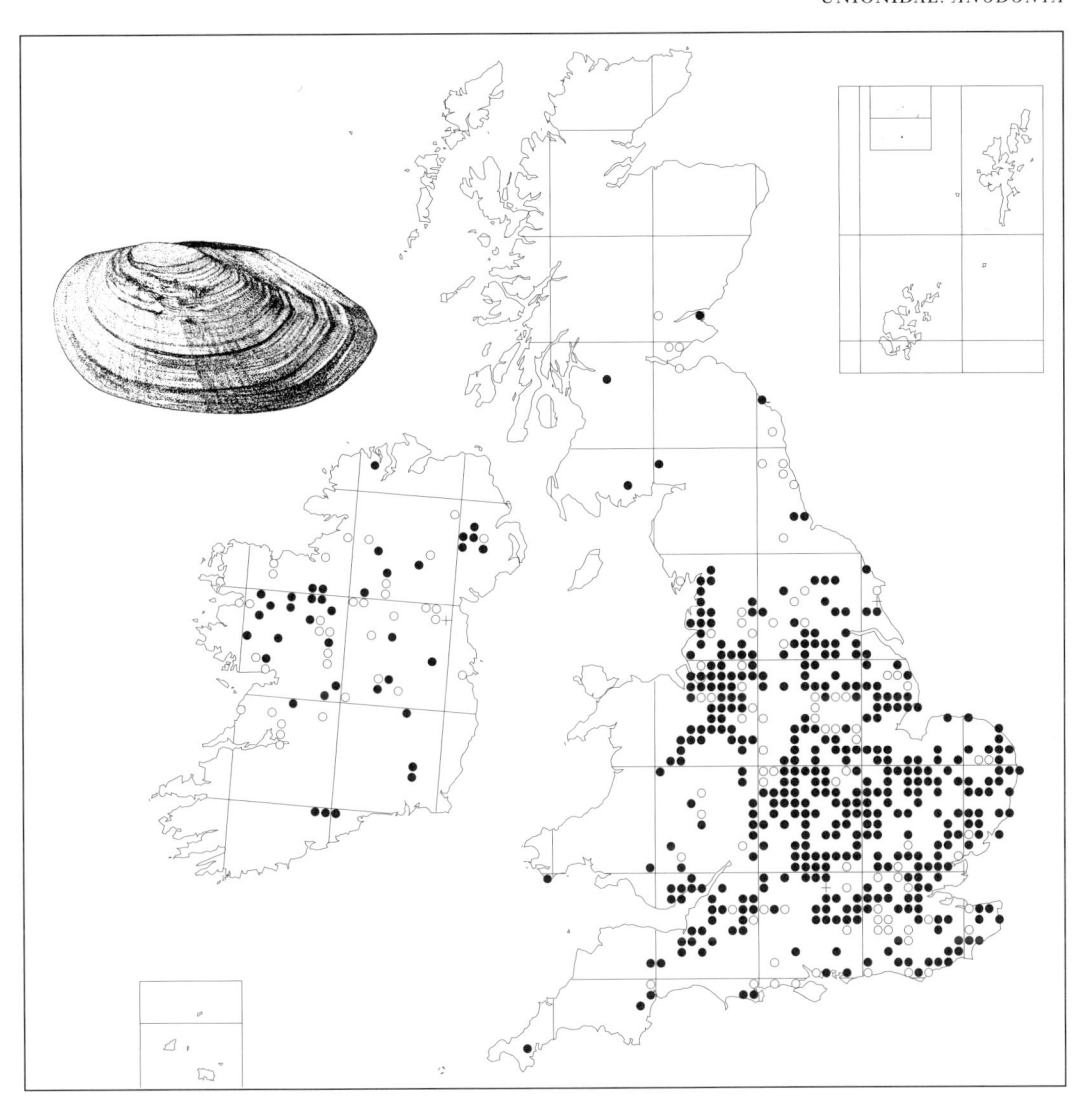

## Anodonta cygnea (L., 1758)

Swan mussel    Size: 100–150 (occ.–200)mm

A typical lowland species, found in large bodies of quiet or slowly moving water in lakes, drainage dykes, canals and rivers. It often occurs in reservoirs, flooded gravel pits, or ornamental lakes such as those in the parks of country houses. It avoids gravelly or rocky substrates, requiring a firm, muddy bottom, often bare of vegetation, in which to burrow.

Native (E.Pgl). Though slightly underrecorded, *A. cygnea* remains frequent throughout its range. It has been introduced by man into many artificial water bodies, either deliberately, or accidentally by stocking with fish parasitized by the larvae.

European: widespread in lowlands, though absent from most of Scandinavia and from Finland.

There has been much confusion in the past between the two common species of *Anodonta*. In particular, some of the old Irish records of '*cygnea*' shown here may belong to *A. anatina*, though the majority are correct.

(Ellis (Bivalves): 32)

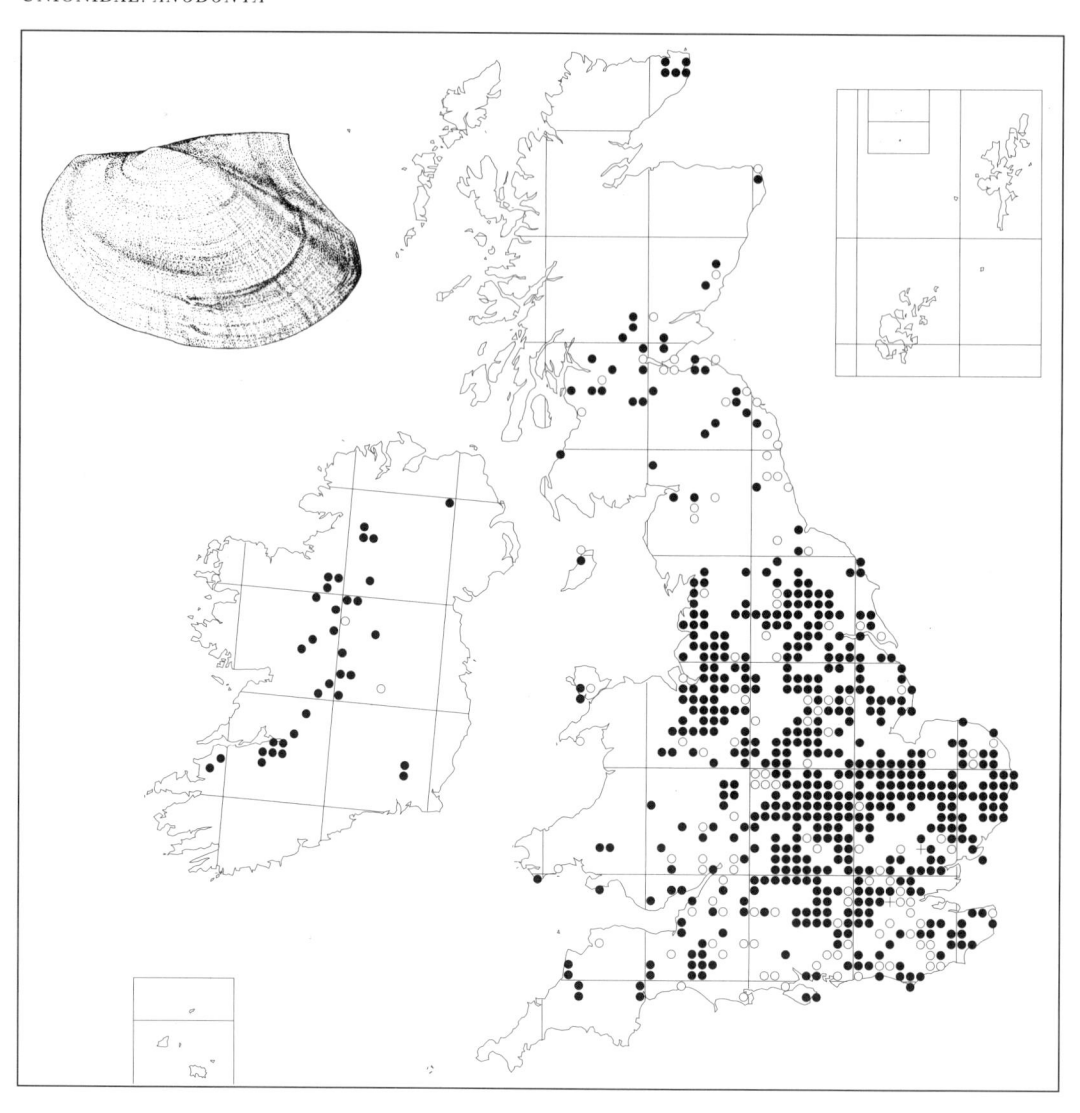

UNIONIDAE: *ANODONTA*

**Anodonta anatina** (L., 1758) (*A. piscinalis* Nilsson)

Duck mussel   Size: 70–130mm

Lives in sizeable bodies of hard water mainly in the lowland zone: rivers, canals, drainage ditches and lakes. It is found in similar places to those for *A. cygnea*, although it is commoner in running water and rarer in closed lakes and ponds. The two species are nevertheless often found together.

Probably native (poorly dated or L.Pgl fossils only). *A. anatina* gives no evidence of national decline. The distribution has been altered by man, like that of *A. cygnea*, by stocking habitats with fish and the construction of canals. The isolated stations in the highland zone, especially in northern Scotland (River of Wick, Caithness (Meiklejohn, 1972); Loch of Strathbeg, Aberdeenshire), show every sign of having originated naturally.

European: widespread in lowlands, northwards to S. Scandinavia.

(Ellis (Bivalves): 34)

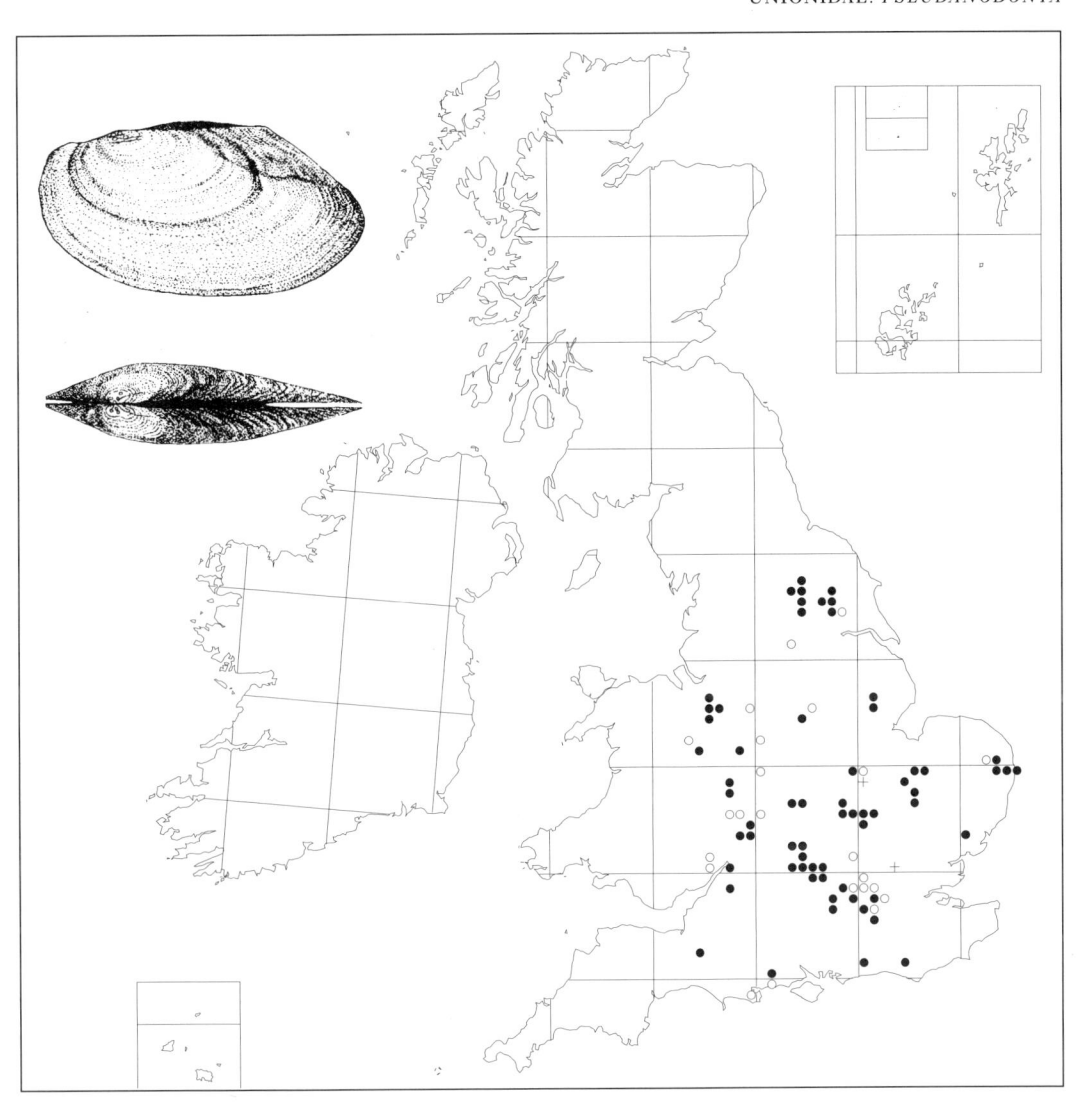

UNIONIDAE: *PSEUDANODONTA*

**Pseudanodonta complanata** (Rossmässler, 1835)
(*Anodonta minima* of British authors, *non* Millet, *A. elongata* Holandre, *Pseudanodonta rothomagensis* Locard)

Compressed river mussel   Size: 55–80mm

This local species is restricted to hard water in slow lowland rivers and occasionally in canals. Unlike *Anodonta cygnea* and *A. anatina*, it is unknown in closed lakes and ponds. In rivers it is usually associated with other large species of mussel though always present in much smaller numbers.

Probably native (poorly dated or L.Pgl fossils only). It was first recognized in Britain as a distinct species in 1910 (Foxall & Overton, 1912). It is often overlooked and therefore somewhat underrecorded, but there is no clear evidence that it has declined significantly.

W. European: widespread but local in lowlands between the Alps and S. Scandinavia.

(Ellis (Bivalves): 36)

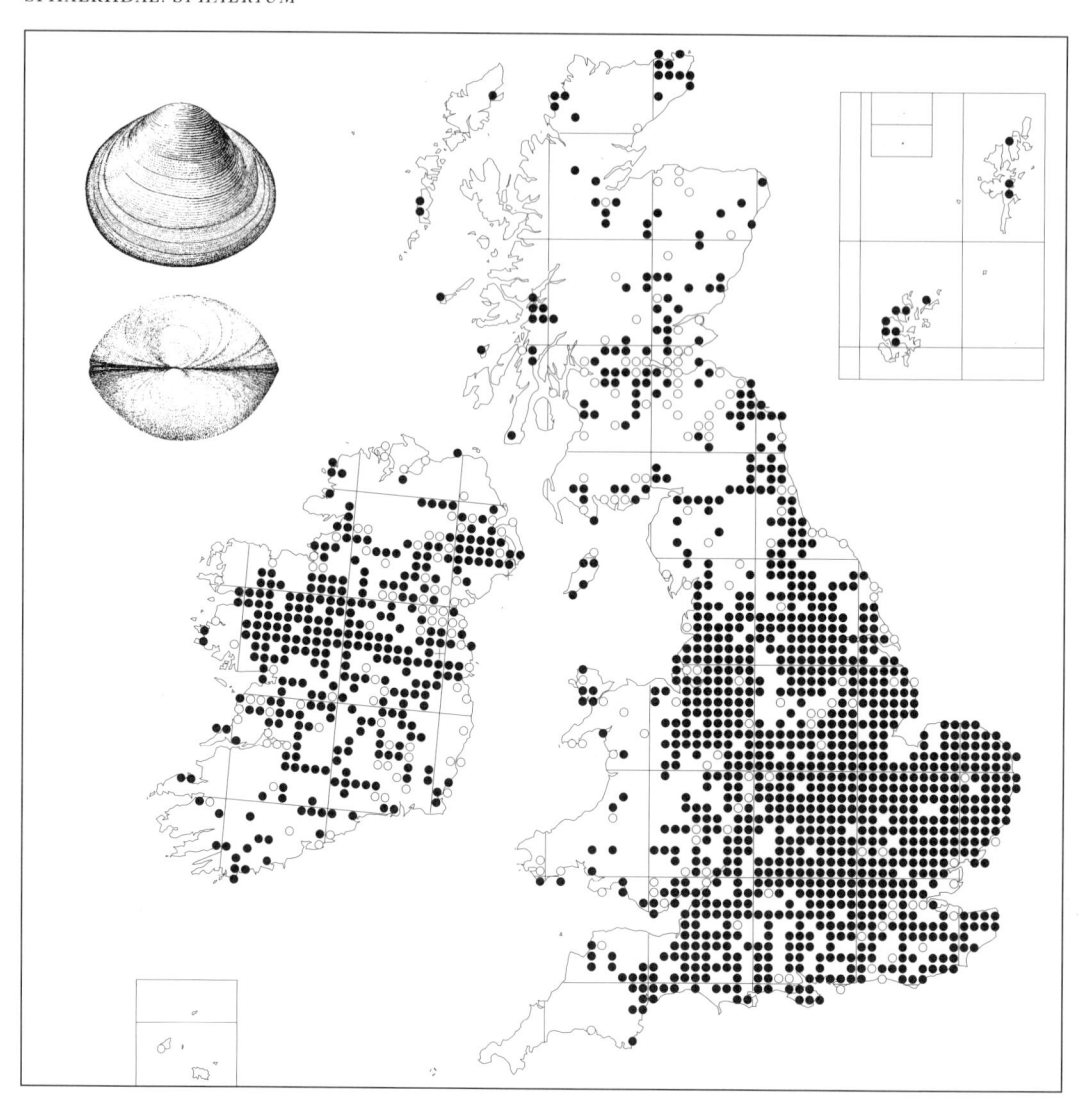

SPHAERIIDAE: *SPHAERIUM*

**Sphaerium corneum** (L., 1758)

Horny orb mussel    Size: 10–14mm

A common species in a variety of aquatic habitats, including rivers, quiet streams, lakes, ponds and ditches. It lives in running or stagnant water, hard or soft, avoiding only places subject to desiccation. Though mainly a lowland species, it occurs also in some highland lochs. *Sphaerium* is unusual among freshwater bivalves in that it leaves the substrate to crawl on aquatic vegetation, allowing it to inhabit places where the bottom sediment is foul because of rotting plant debris or pollution; it is therefore not infrequent in places where *Pisidium* or unionids are absent.

Native (Lgl). No evidence of distributional change.

Palaearctic (probably Holarctic); extending over the whole of Europe, though rare and sporadic in N. Scandinavia.

(Ellis (Bivalves): 44)

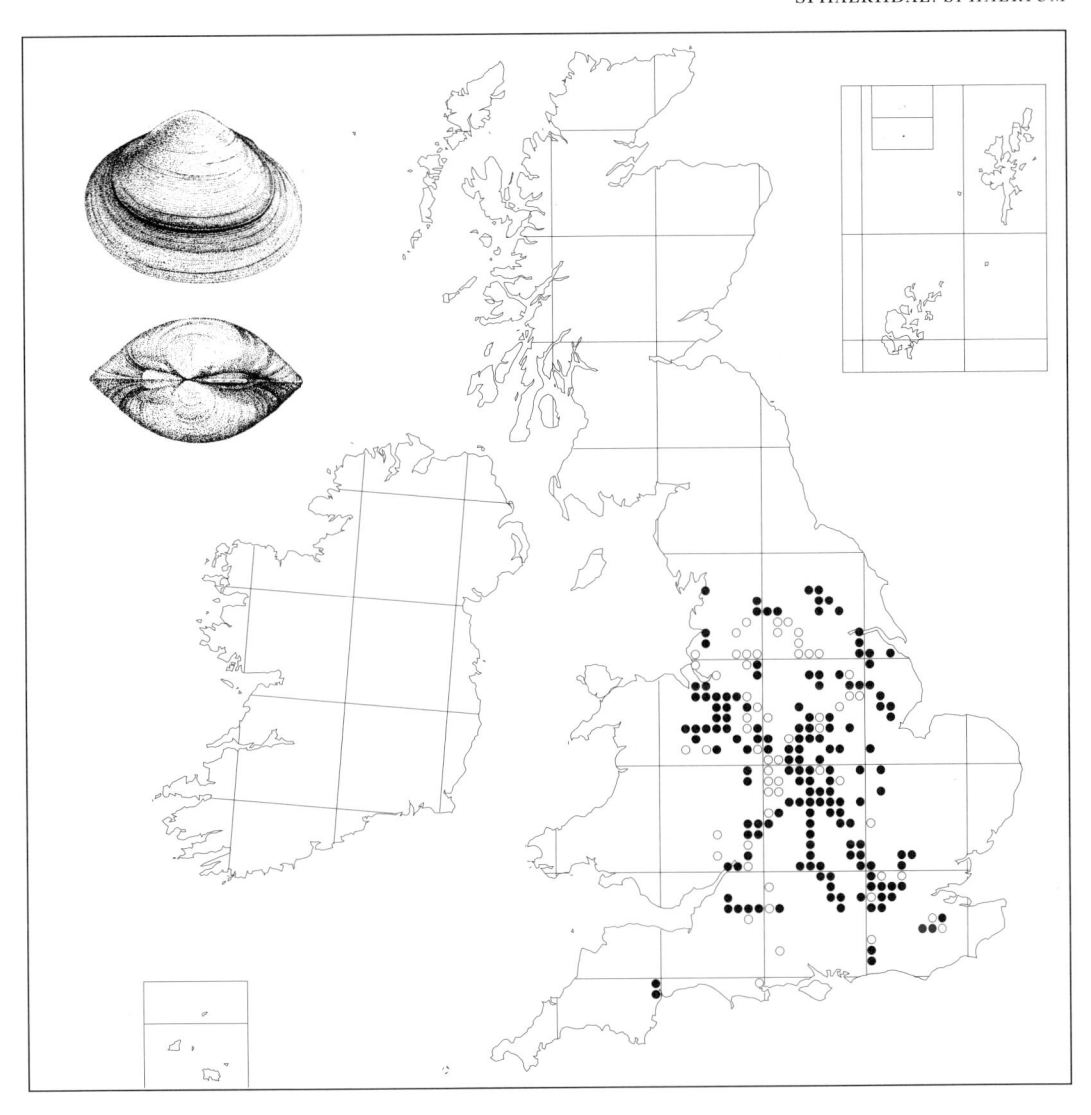

## Sphaerium rivicola (Lamarck, 1818)

Nut (or river) orb mussel   Size: 17–22mm

A bivalve restricted to canals and large rivers. It is a hard-water species, typical of large and well-oxygenated water bodies with a rich diversity of molluscs, nearly always including *Viviparus viviparus*. It inhabits deeper water than *S. corneum* and therefore often escapes notice unless dredging is undertaken.

Probably native (L.Pgl fossils only). Its 'canal basin' distribution in the English lowlands owes much to man. There is no evidence of serious decline, though a few populations (e.g. in canals near Birmingham and in the industrial North-West) may have been lost through pollution or habitat change.

Mainly C. and E. European: common in the N. European lowlands but absent from the Alpine region and from Scandinavia.

(Ellis (Bivalves): 42)

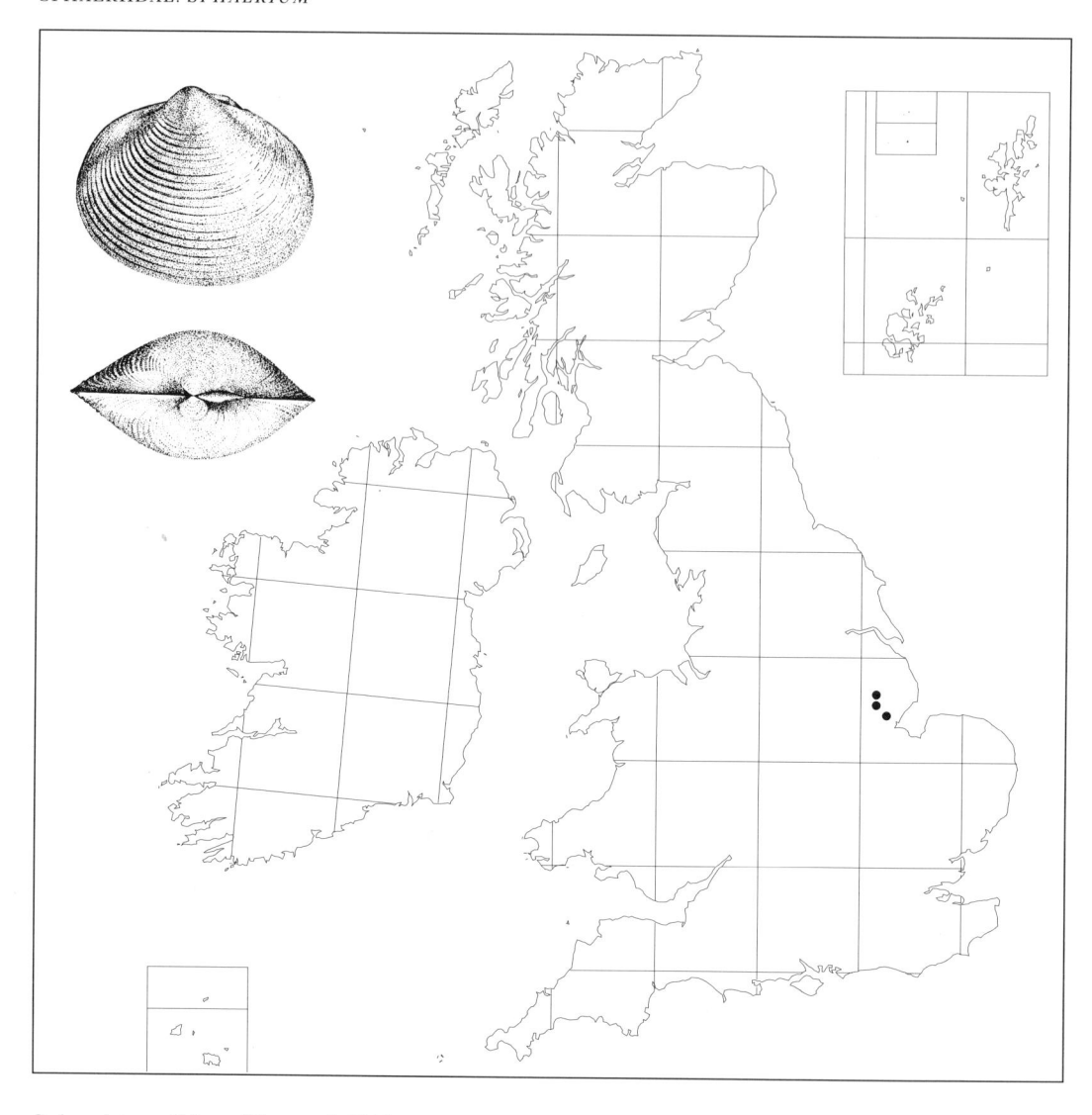

SPHAERIIDAE: *SPHAERIUM*

**Sphaerium solidum** (Normand, 1844)

Size: 8–11mm

This rarity occurs along about 15km of the River Witham in Lincolnshire, both in the main channel and close by in deep drains opening into it via sluices (Redshaw & Norris, 1974). The Witham is a canalized river, artificially deepened and straightened. In spite of its turbidity and the relative sparseness of the aquatic flora it has a molluscan fauna of the richest kind, including *Theodoxus fluviatilis*, *Viviparus viviparus*, *Unio tumidus* and *Sphaerium rivicola*. Like *S. rivicola*, *S. solidum* prefers relatively deep water and most of the specimens so far recovered have been of dead (though fresh) shells.

Probably native. *S. solidum* was discovered in Britain only in 1968. Though no Postglacial fossils are known it is unlikely to be a recent introduction. (It lived here also earlier in the Pleistocene). Its restriction to one stretch of a single river puts it at risk from pollution. RDB category: Endangered.

Mainly C. and S.E. European: local across the N. German plain west to the Low Countries, France and England (not in Scandinavia).

(Ellis (Bivalves): 41)

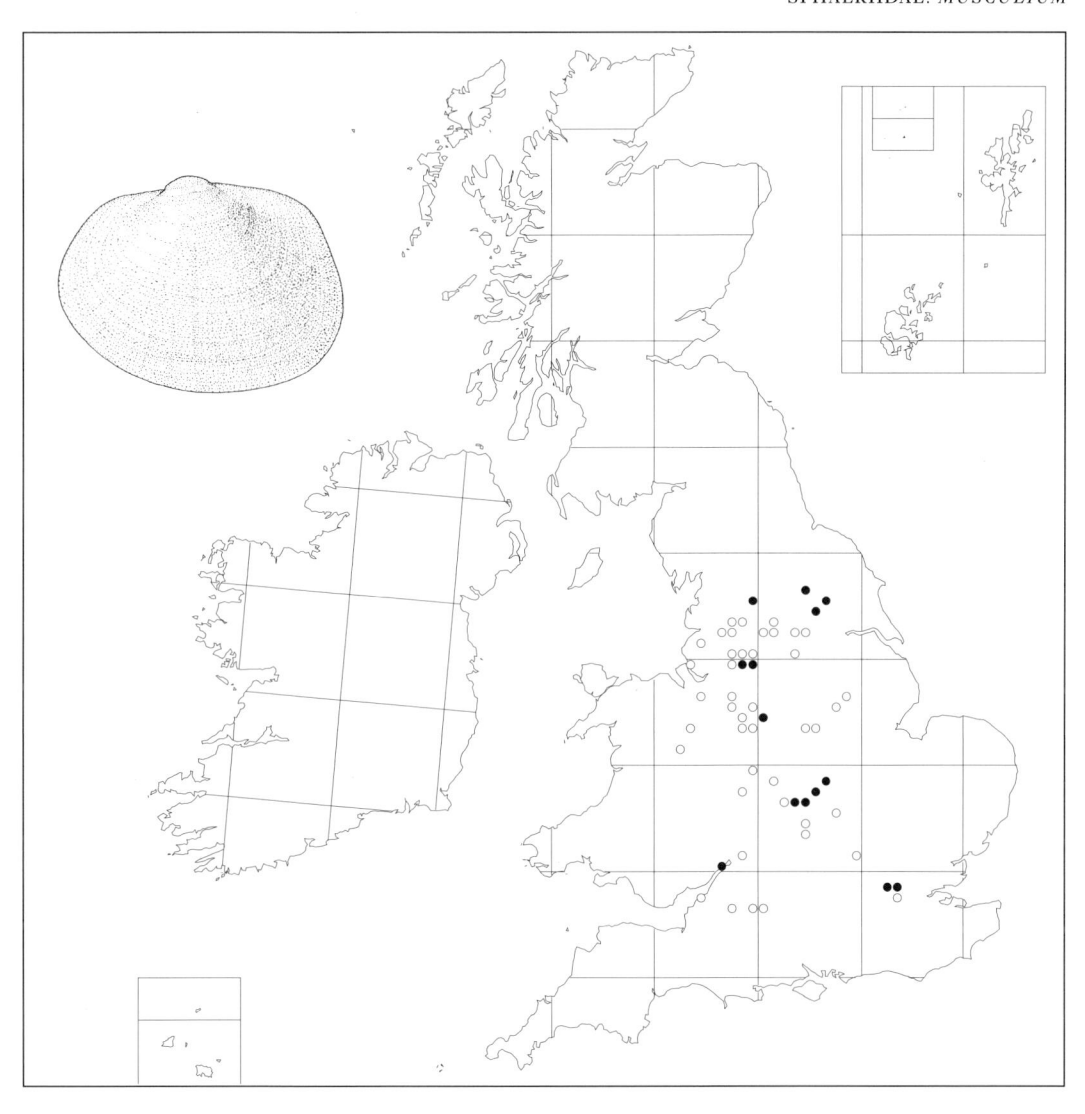

SPHAERIIDAE: *MUSCULIUM*

**Musculium transversum** (Say, 1829) (*Sphaerium transversum* (Say), *S. pallidum* Gray, *S. ovale* of British authors, *non* Férussac)

Oblong orb mussel    Size: 11–15mm

In Britain this is a species found exclusively in canals and canalized rivers, where it is usually associated with the rich molluscan faunas typical of such places. It likes a muddy substrate and, like other *Sphaerium* and *Musculium*, can tolerate anaerobic bottom conditions (e.g. in urban canals).

Introduced. *M. transversum* has no geological history in Britain and is doubtless an accidental introduction from N. America. It was first reported with certainty (as *Sphaerium pallidum*) in 1856 from the Grand Union Canal at Kensal Green, London. By 1900 it had been found in many places within the 'canal basin' of England and Wales, especially in industrial areas of the North West. For unknown reasons it has since declined to become a rarity. It still survives in London, in the Grand Union and Regent's Canals (Kerney, 1971).

N. American; in Europe known with certainty only from the British Isles and (formerly) the Netherlands.

(Ellis (Bivalves): 48)

217

SPHAERIIDAE: *MUSCULIUM*

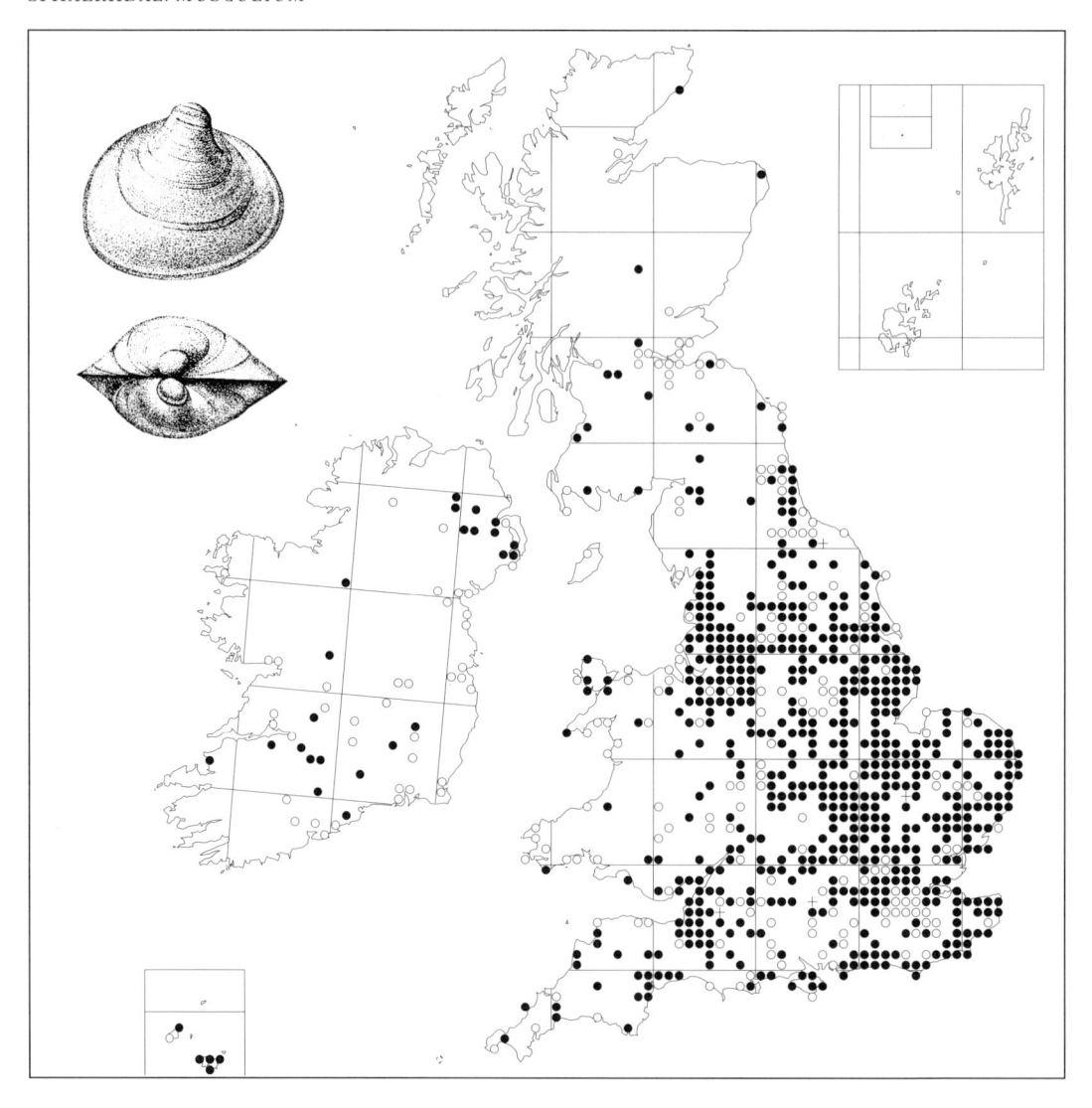

**Musculium lacustre** (Müller, 1774) (*Sphaerium lacustre* (Müller))

Lake (or capped) orb mussel   Size: 8–13mm

A species found in aquatic habitats of all kinds, in hard or soft water, running or stagnant. However, it is especially frequent in small closed ponds and swampy ditches; in such places it is often associated with *Aplexa hypnorum*, *Lymnaea palustris*, *Anisus leucostoma* or *Pisidium personatum*. Unlike *Sphaerium corneum*, it will tolerate occasional desiccation. It is sometimes common in weed-choked marsh drains over anaerobic substrates devoid of other bivalves, probably because its habit of crawling on plants makes it indifferent to poor bottom conditions.

Native (Lgl). In Britain *M. lacustre* shows little significant decline, though it is probably a little scarcer than formerly because of agricultural drainage and the removal of field ponds. In Ireland (where it has always been uncommon) there is perhaps a hint of more serious decline; around Dublin, an area for which there are numerous old records, it has not been seen since the early years of this century.

Palaearctic (probably Holarctic); throughout most of Europe, north to S. Scandinavia.

(Ellis (Bivalves): 47)

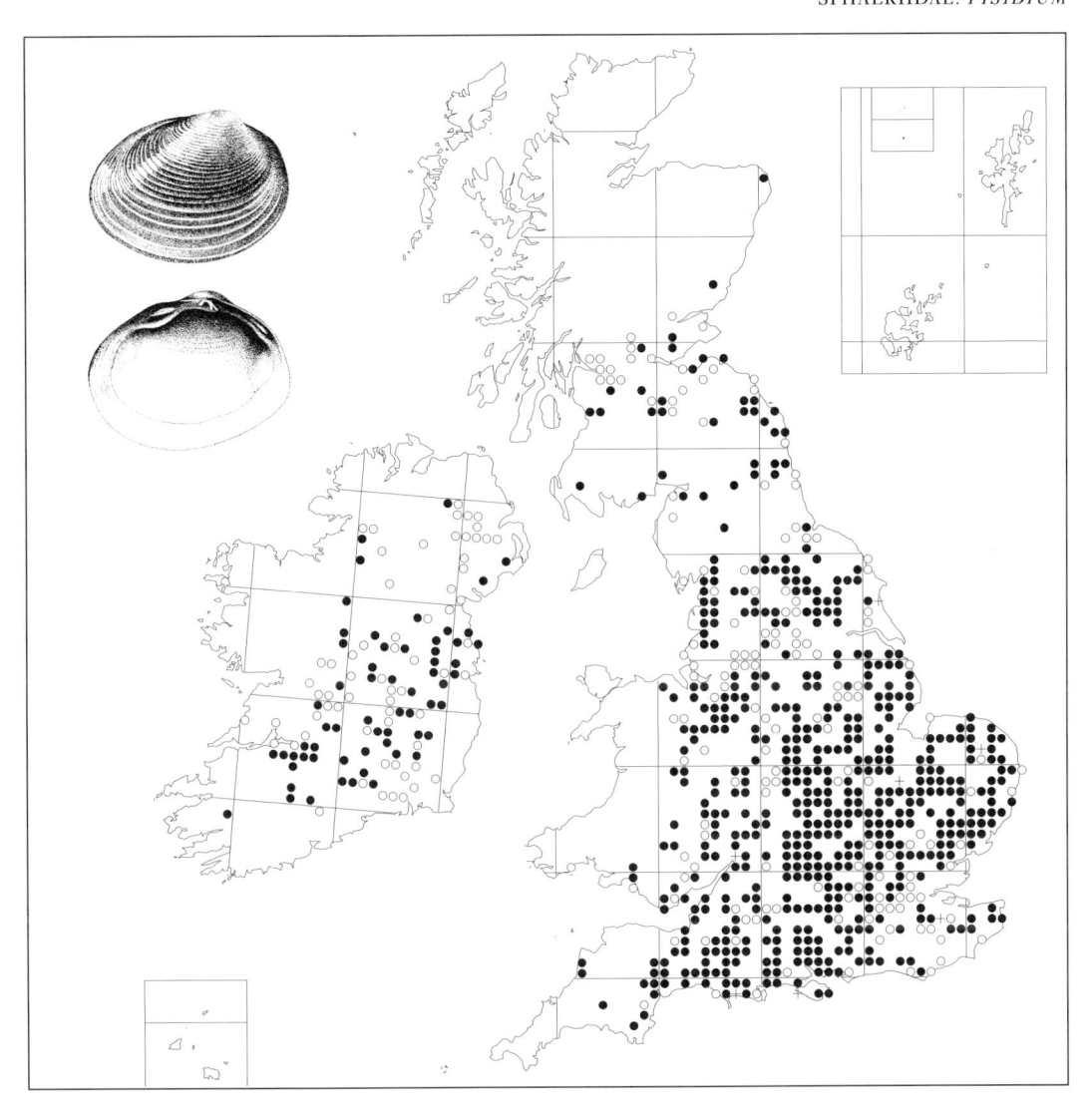

SPHAERIIDAE: *PISIDIUM*

**Pisidium amnicum** (Müller, 1774)

River (or giant) pea shell   Size: 7–11mm

A lowland species, found in clean, moderately hard, running water in rivers, canals, sizeable streams and drainage ditches; occasionally it inhabits large lakes. It is almost invariably associated with several other species of *Pisidium*, usually including *henslowanum*, *subtruncatum* and *nitidum*.

Native (Lgl). No evidence of significant regional change.

Palaearctic and eastern N. America; in Europe to beyond the Arctic Circle (rare in most of Scandinavia).

It should be noted that the high proportion of old records in Ireland and Scotland has no general significance; in Ireland it reflects the work of the pioneers of the genus *Pisidium* (A. W. Stelfox, R. A. Phillips and C. Oldham) early this century, and in southern Scotland intensive recording mostly by D. K. Kevan in the years 1930–60. The same bias is evident also on most of the other *Pisidium* maps.

(Ellis (Bivalves): 53)

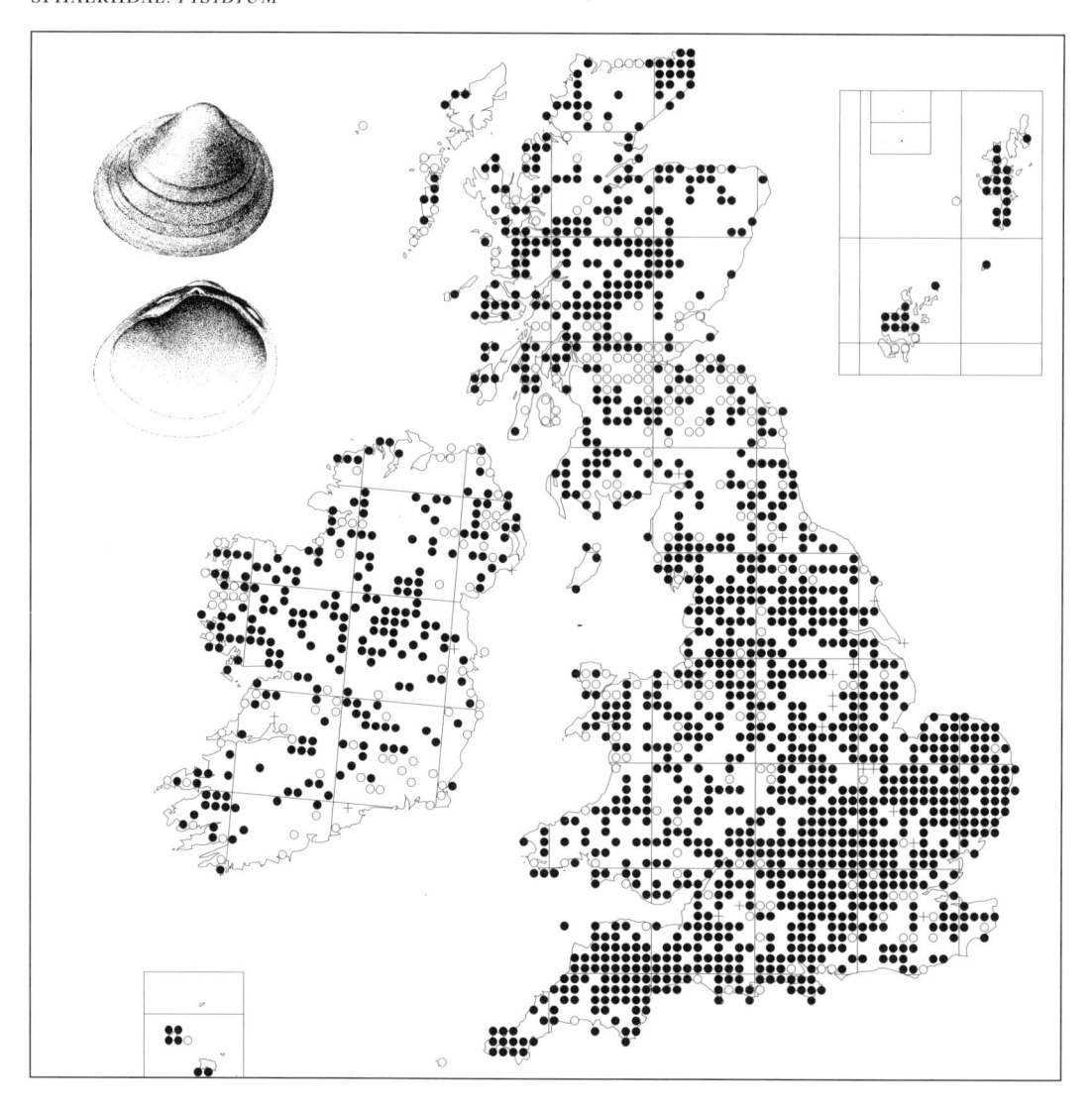

SPHAERIIDAE: *PISIDIUM*

**Pisidium casertanum** (Poli, 1791) (*P. cinereum* Alder)

Size: 3.0–4.5 (occ.–6.0)mm

This is an exceptionally catholic (and variable) species, found on muddy or silty substrates in all kinds of aquatic habitats, in hard and soft water alike, from the richest rivers and canals to the meanest pools and ditches. In places subject to desiccation it occurs with no other *Pisidium* except *personatum*. In the Scottish highlands it ascends to over 1,000m. The adaptability of *P. casertanum* is clearly related to its ubiquity.

Native (Lgl). No evidence of distributional change.

Holarctic; probably cosmopolitan, and perhaps the world's most widely distributed non-marine mollusc.

(Ellis (Bivalves): 56)

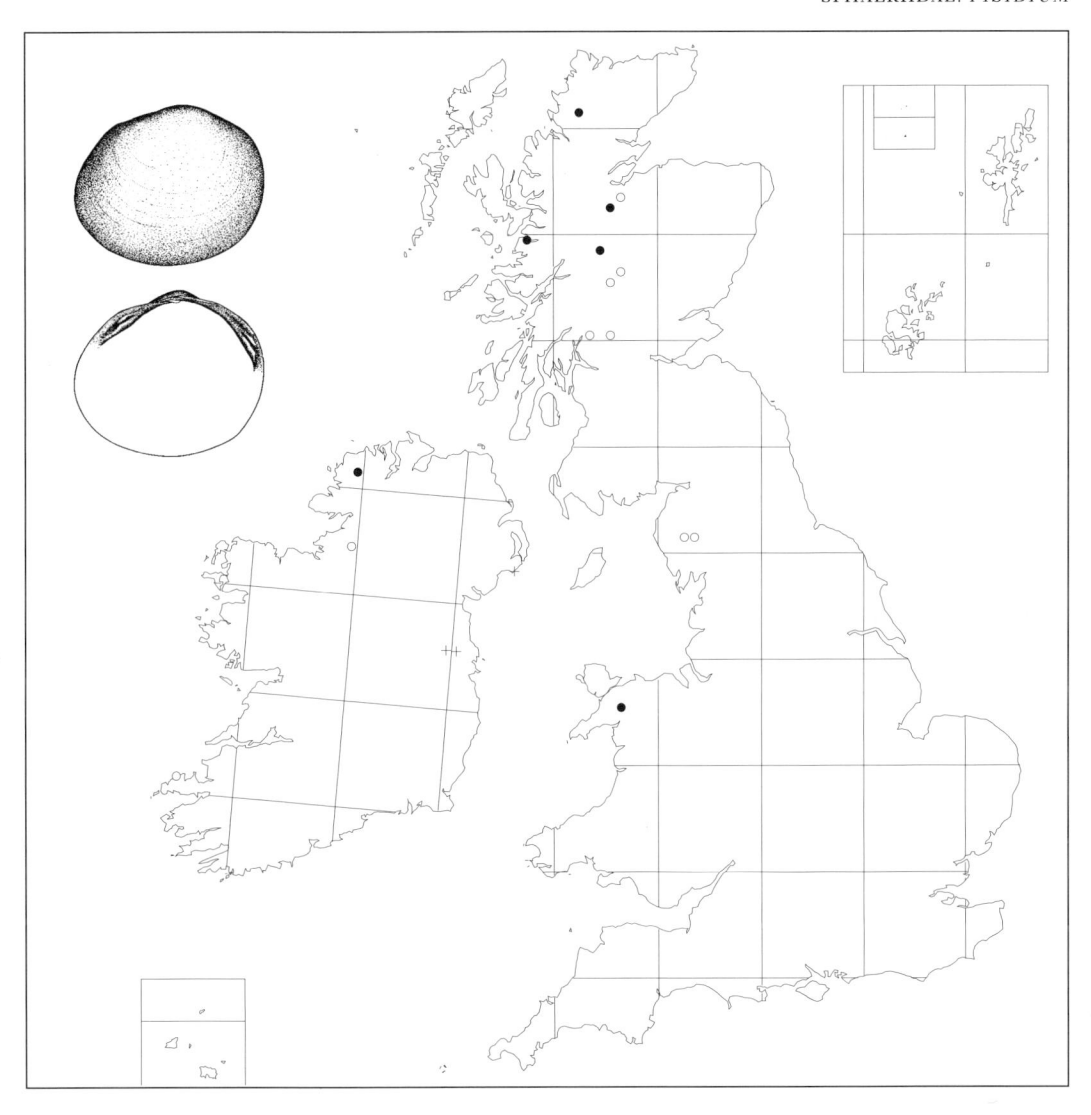

SPHAERIIDAE: *PISIDIUM*

**Pisidium conventus** Clessin, 1877

Size: 2.2–2.8mm

A truly Arctic species, found where temperatures are always low. It inhabits cold mountain tarns, often north-facing and deeply shaded. It reaches the following elevations: 754m in Wales (Snowdon, Caernarvonshire), 691m in England (Helvellyn, Westmorland), 701m in Ireland (Brandon Mountain, Co. Kerry) and about 335m in Scotland (Grampians, Perthshire). At lower elevations it lives in the colder, deeper parts of large lakes (e.g. Loch Lomond; Hunter & Slack, 1958), sometimes at remarkable depths (Loch Ness, −125m; Loch Morar, −300m). The bottom of Loch Morar is the deepest possible habitat for a freshwater mollusc in the British Isles.

Native (Lgl). A relict species, no doubt widely distributed in the British Isles at the close of the glacial period. Whether its decline continues is uncertain. Probably it is secure in most of its recorded refuges, though several have not been visited in recent years (e.g. in the Lake District; Oldham, 1938). It is, in any case, underrecorded and may be more common in the highland zone than the sparse records suggest.

Holarctic circumpolar; in Europe mainly in the Alps and in the northern highland zone, to beyond the Arctic Circle.

(Ellis (Bivalves): 59)

221

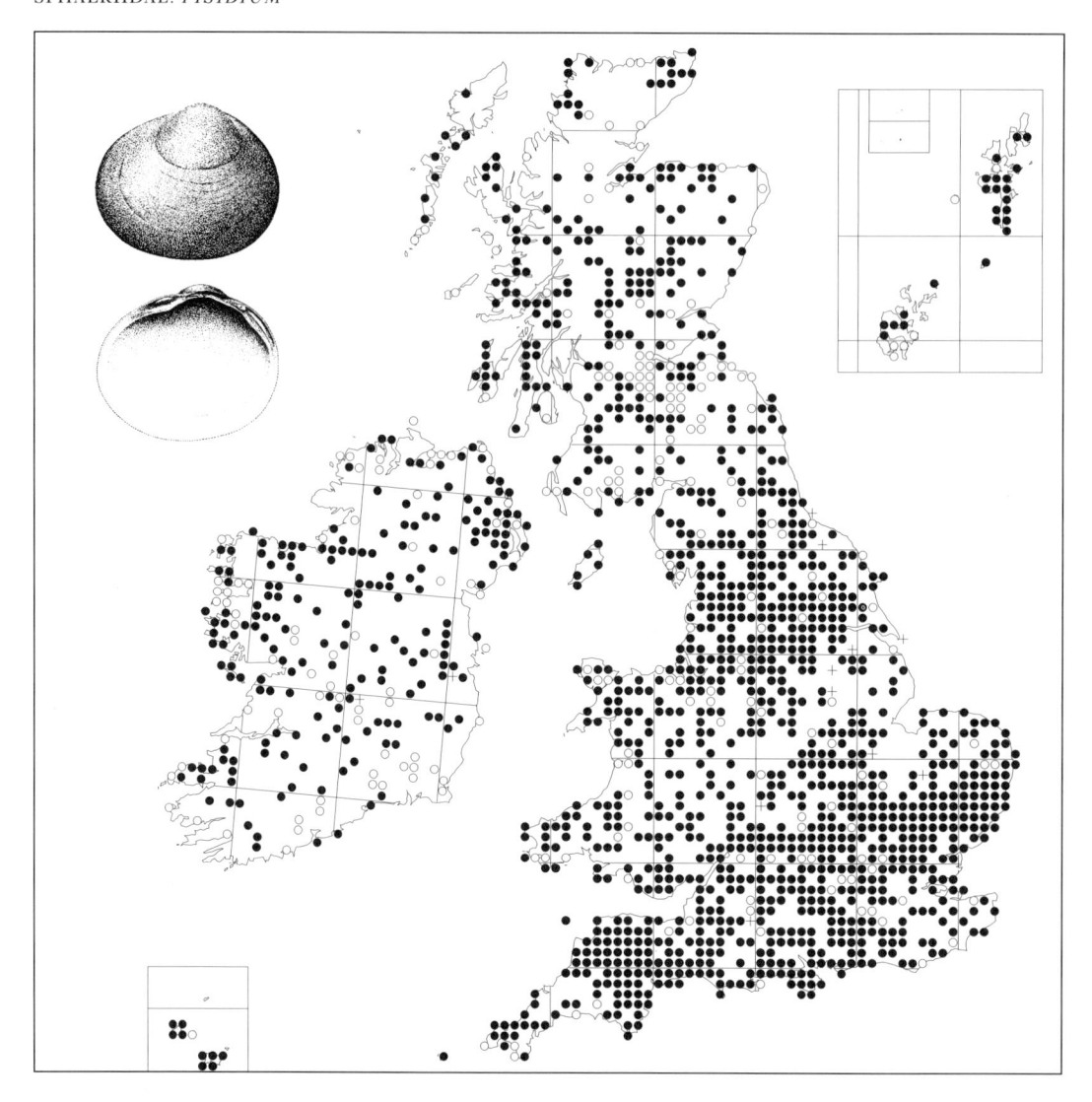

SPHAERIIDAE: *PISIDIUM*

**Pisidium personatum** Malm, 1855

Size: 2.5–3.5mm

A species restricted to the poorest aquatic habitats. It inhabits grassy ponds, hillside flushes and stagnant, roadside ditches, especially in places liable to dry up in the summer; occasionally it is found away from standing water, for example under ground litter in marshy woodland. Usually it is the only bivalve present, though sometimes it is associated with *P. casertanum*. Frequent gastropod associates are *Lymnaea truncatula*, *L. palustris* and *Anisus leucostoma*. It occurs in hard and soft water alike, from sea-level to over 900m. It also lives in deep lakes, but only in the profundal zone, not in the richer littoral zone where other species of *Pisidium* replace it; in Windermere it is recorded from between −30m and −60m, and in Loch Ness (with *P. conventus*) from −125m.

Native (E.Pgl). No evidence of distributional change. Man-made habitats are rapidly colonized.

Mainly European (closely related forms occur in N. America); in Europe nearly throughout, but absent from N. Scandinavia.

(Ellis (Bivalves): 61)

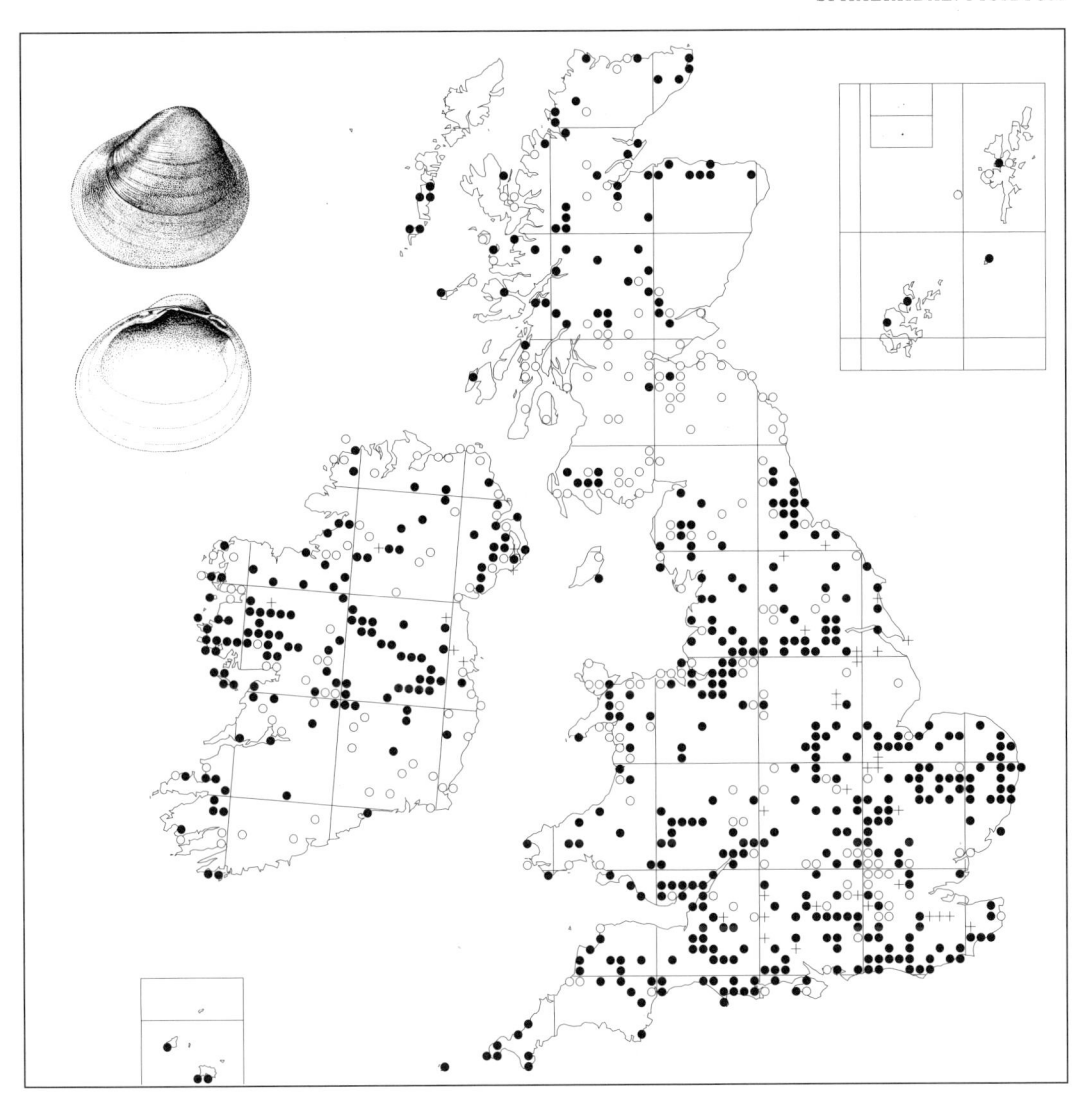

SPHAERIIDAE: *PISIDIUM*

**Pisidium obtusale** (Lamarck, 1818)

Size: 2.5–3.5mm

A species characteristic of small bodies of stagnant water, though not usually those subject to desiccation (cf. *P. personatum*). Ditches and small ponds choked with floating grasses or emergent marsh plants are typical habitats, in which it is often the only species of *Pisidium*. In canals, large lakes or rivers, and in flowing water generally, it is rare. It tolerates very soft water and can be found in *Sphagnum*-filled pools in peat bogs. In Scotland it is recorded at over 700m (Perthshire).

Native (Lgl). Little evidence of serious decline, though certainly scarcer than formerly in intensively farmed lowland areas because of habitat destruction.

European (probably Holarctic); throughout Europe, to beyond the Arctic Circle.

(Ellis (Bivalves): 63)

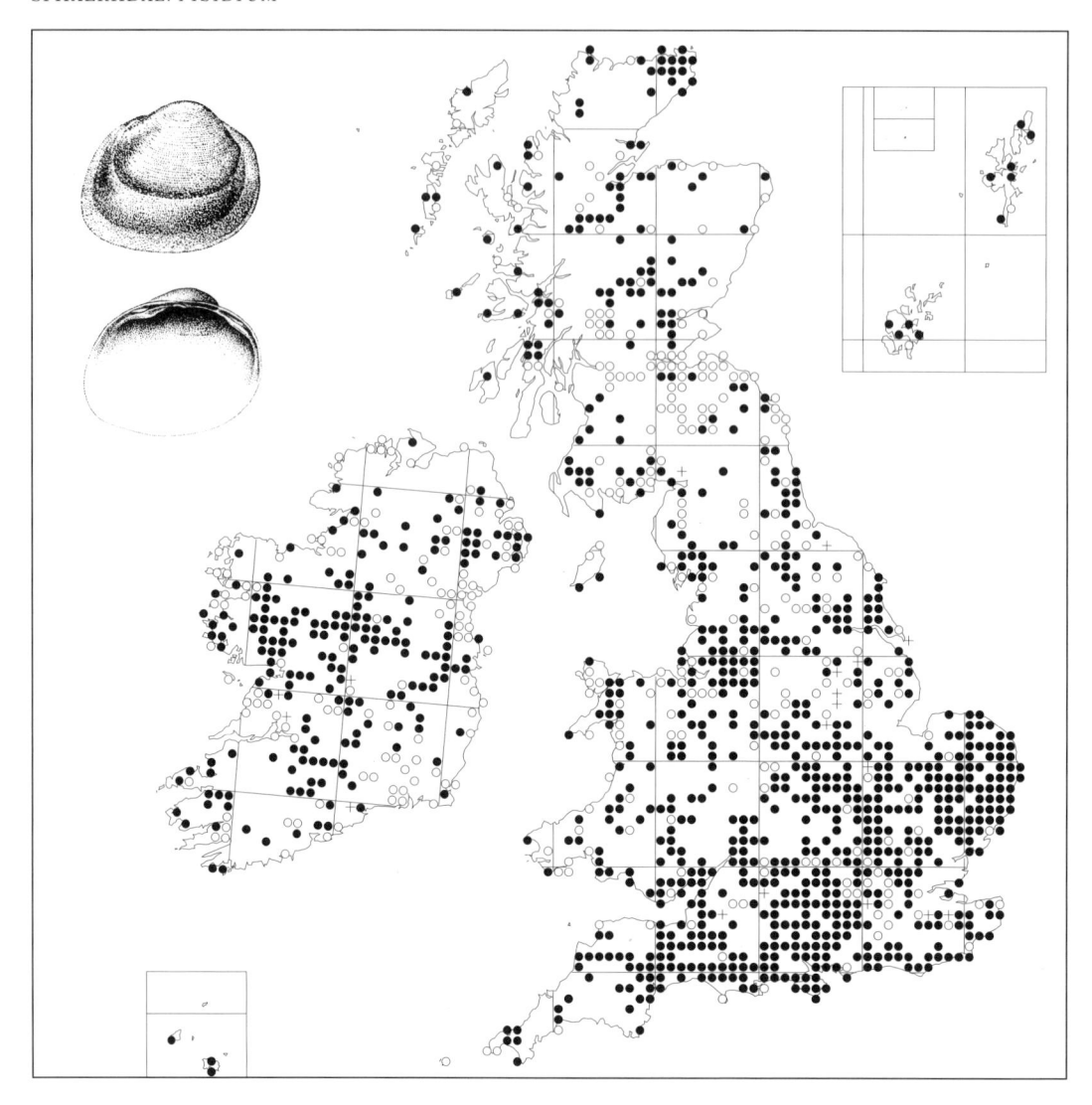

SPHAERIIDAE: *PISIDIUM*

**Pisidium milium** Held, 1836

Rosy pea shell   Size: 2.5–3.5mm

Occurs in a wide variety of aquatic habitats, avoiding only swampy places and those liable to dry up. It is frequent in rivers, canals, lakes and marsh drains. It is perhaps most characteristic of clean, quiet water with a good aquatic flora, especially in ponds, alongside such species as *Gyraulus crista*, *Hippeutis complanatus* and *Acroloxus lacustris*. It is tolerant of soft water.

Native (Lgl). No evidence of significant distributional change.

Holarctic; in Europe nearly throughout.

(Ellis (Bivalves): 65)

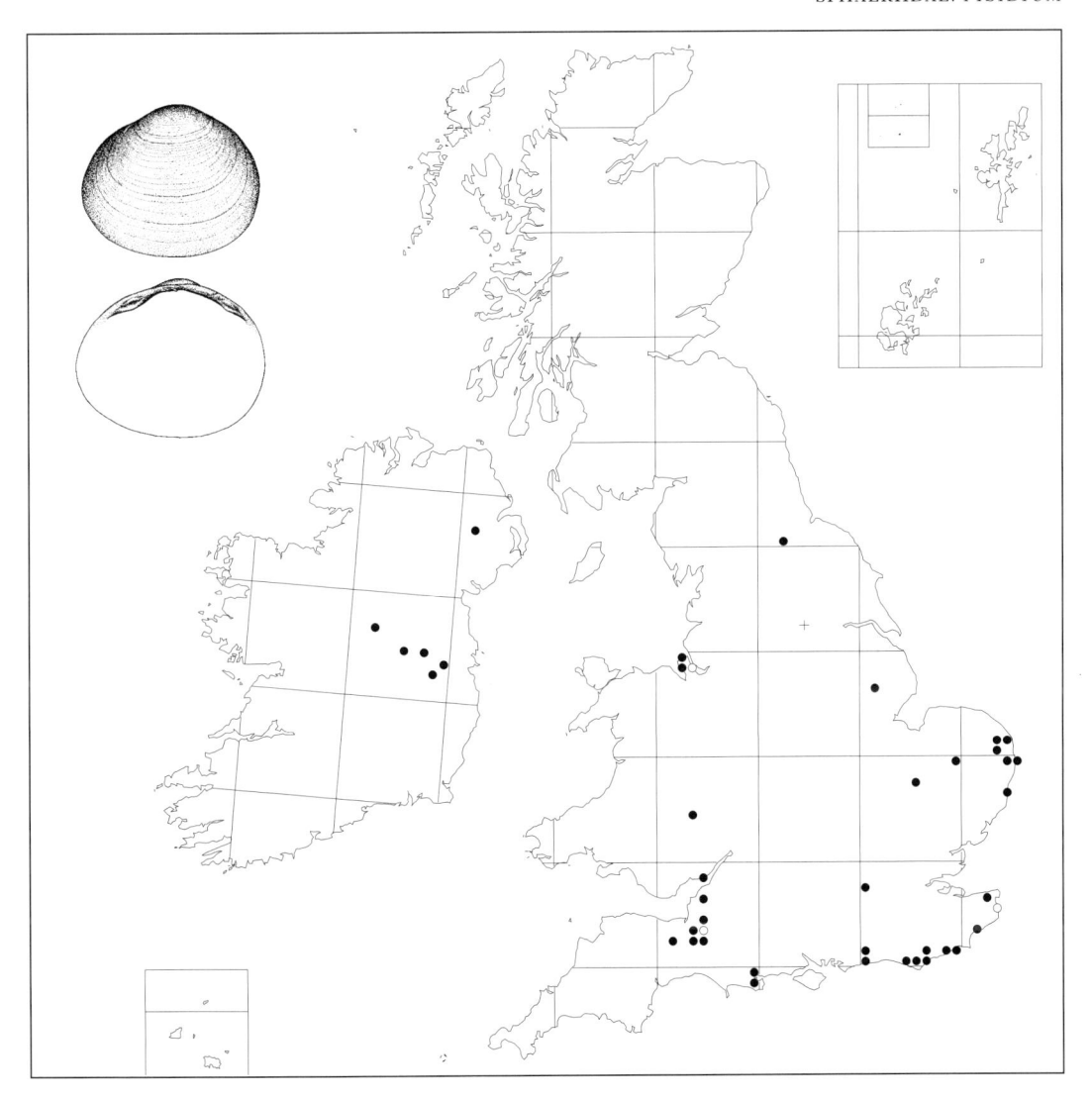

**Pisidium pseudosphaerium** Schlesch, 1947

Size: 2.5–3.2mm

A very local species found in lowland marsh drains and ponds, and in Ireland in abandoned canals (Royal Canal, Lagan Canal; Kerney, 1969). It requires clear, clean water in stagnant places choked with aquatic plants, often over a richly organic bottom. It occurs either alone or associated with *P. obtusale* and *P. milium*, but rarely with other species of *Pisidium* (Kuiper, 1949; Dance, 1956; 1957). In southern England it has been found in places containing rare, relict species (like *Valvata macrostoma*, *Anisus vorticulus* and *Segmentina nitida*), notably in the East Anglian broadland and the Pevensey Levels and Lewes Levels, Sussex.

Native (E.Pgl). This is a species of specialized habitat, easily destroyed by drainage, dredging, or eutrophication caused by nitrate or phosphate enrichment. Most of its habitats are of artificial origin; nevertheless it shows signs of having a relict distribution, suggesting that passive dispersal does not occur easily outside limited areas. For these reasons the species must be considered at risk, even though probably not so rare as the records suggest. RDB category: Rare.

European: local, mainly in lowlands between the Alps and S. Scandinavia.

(Ellis (Bivalves): 67)

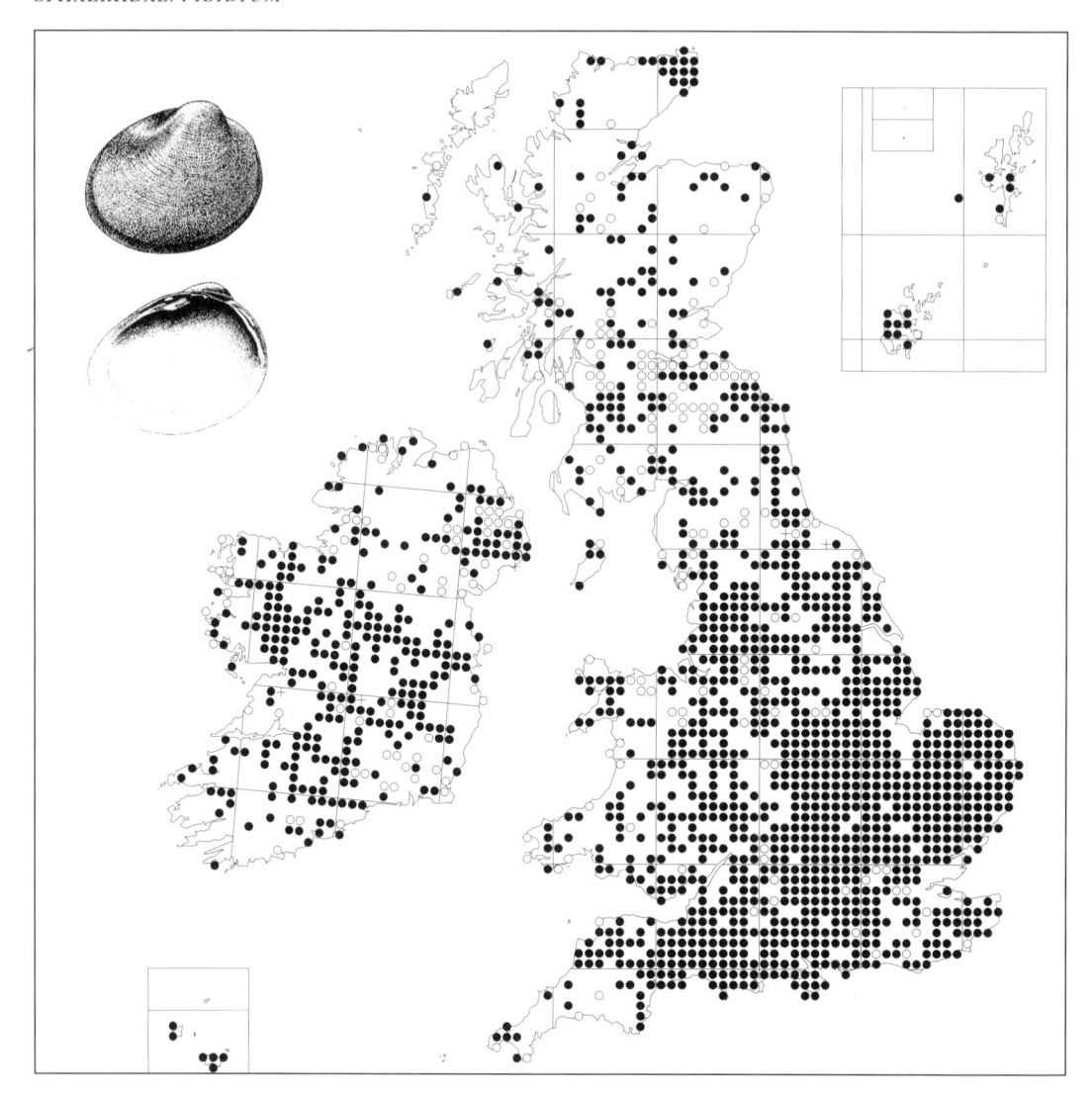

SPHAERIIDAE: *PISIDIUM*

**Pisidium subtruncatum** Malm, 1855

Size: 2.7–4.0mm

A common *Pisidium* in a variety of aquatic habitats, including rivers, streams, canals, lakes, ponds and drainage ditches. It has perhaps a slight preference for running water, avoiding small, swampy pools and places subject to desiccation. It tolerates soft water. It is mainly a lowland species, rare above 300m.

Native (Lgl). No evidence of distributional change. The species is tolerant of mild pollution.

Holarctic; in Europe occurring nearly throughout, though scarce in the Mediterranean region.

(Ellis (Bivalves): 69)

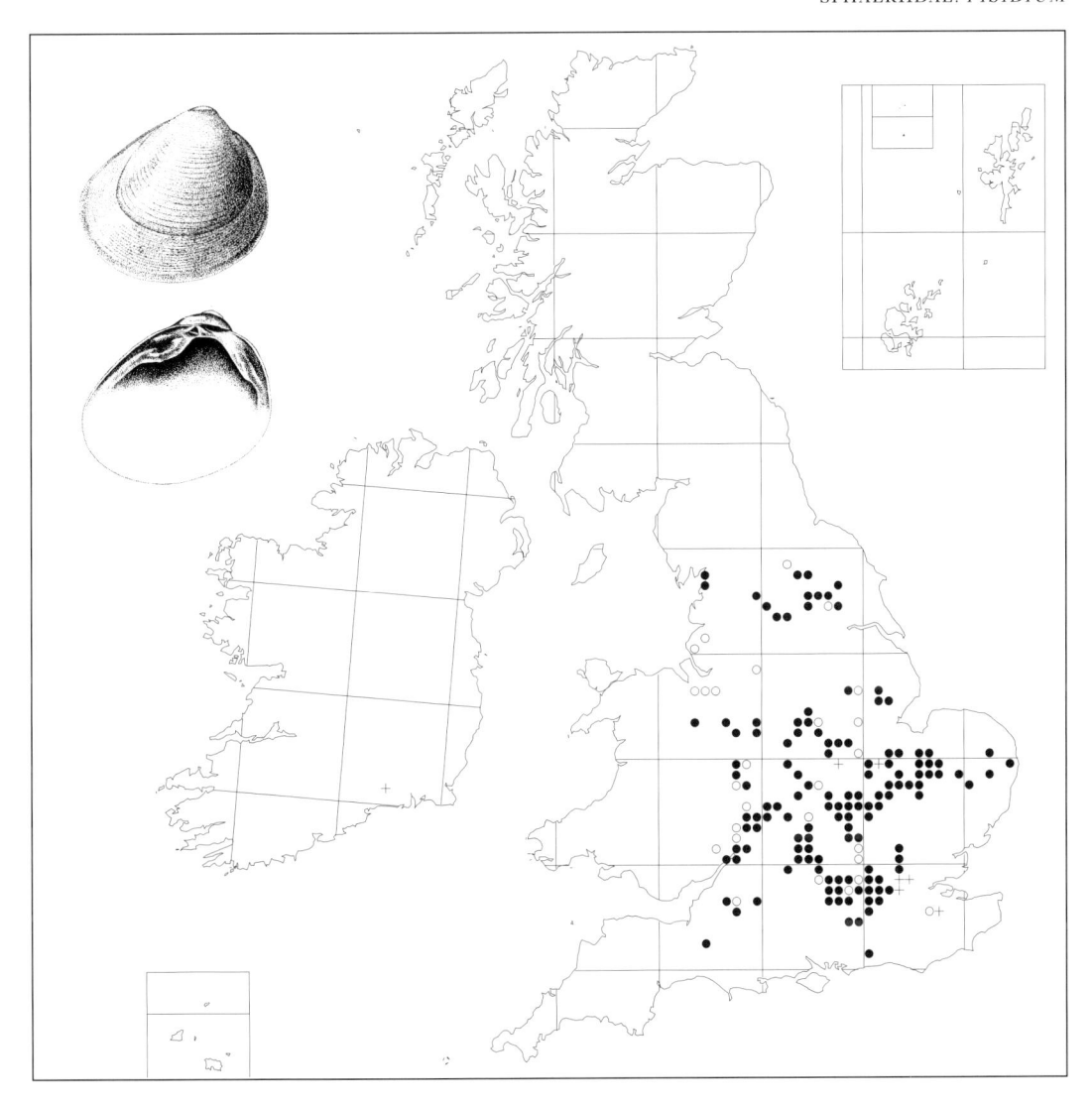

SPHAERIIDAE: *PISIDIUM*

**Pisidium supinum** A. Schmidt, 1851

Size: 3.3–4.5mm

A species restricted to large lowland rivers and canals. It is to be found in clean, slowly moving, well-oxygenated hard water, usually with a varied molluscan fauna including several species of *Pisidium*. Normally it inhabits clean mud or silt, but in large rivers it is sometimes found also on coarse sandy substrates.

Native (E.Pgl). Like *Unio tumidus* and *Sphaerium rivicola* this is a 'canal basin' species whose distribution owes much to man. In Britain there is little evidence of recent distributional change, though a few sites may have been lost through bottom pollution, especially in the lower Thames and in canals in the North West. In Ireland shells of uncertain age dredged from the R. Suir (Fiddown, Co. Waterford (Phillips, 1916)) show that *P. supinum* formerly occurred in that country.

Palaearctic and N.E. North America; in Europe mainly in the lowlands between the Alps and S. Scandinavia (to 60°N in Sweden).

There is some doubt as to whether *P. supinum* is a distinct species, or only a hypertrophied form of *P. henslowanum* developed under optimum conditions, analogous to *P. casertanum* var. *ponderosa* Stelfox and *P. nitidum* var. *crassa* Stelfox. However, the fact that mixed populations of *henslowanum* and *supinum* sometimes occur tells against this view.

(Ellis (Bivalves); 71)

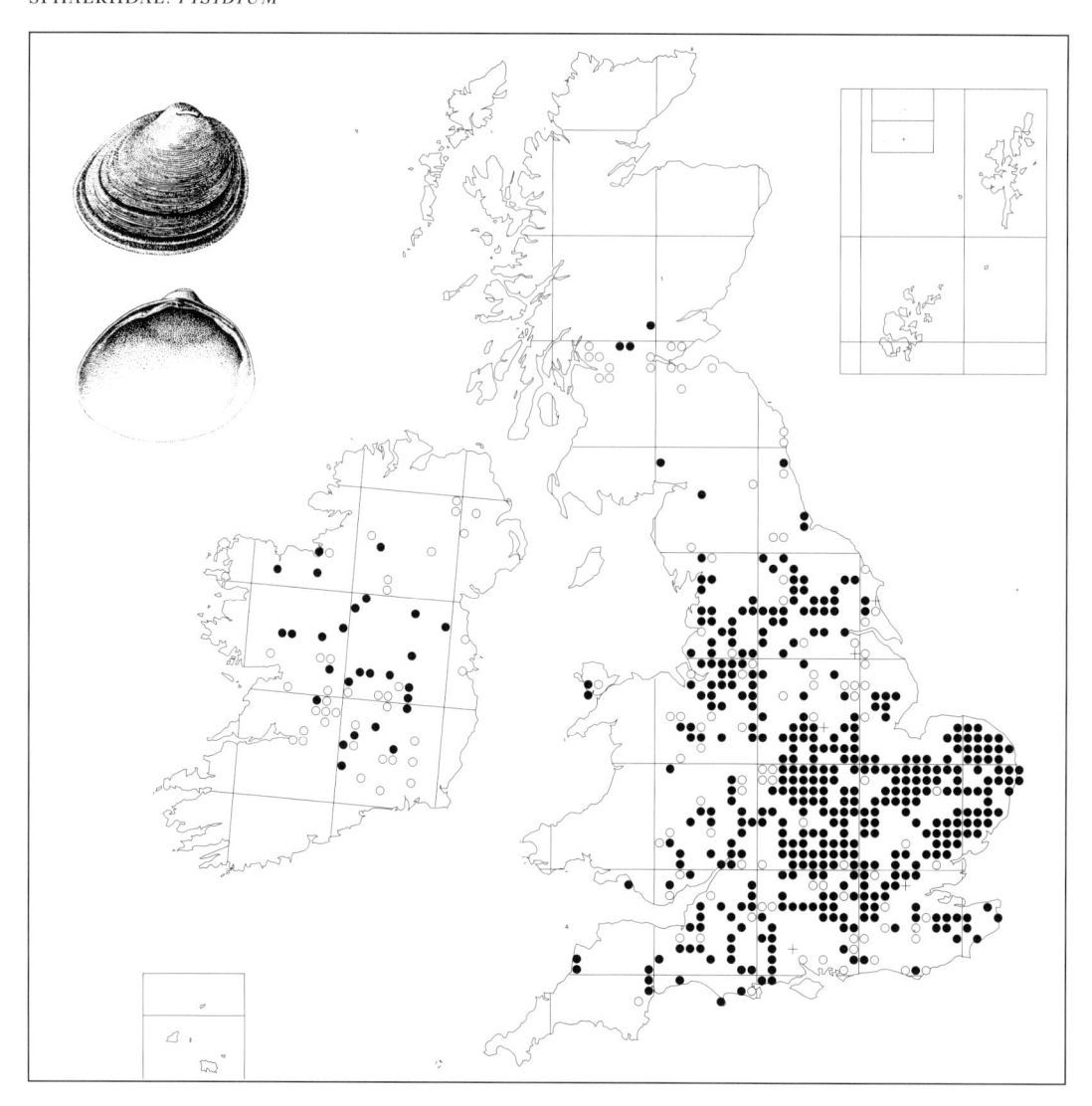

SPHAERIIDAE: *PISIDIUM*

**Pisidium henslowanum** (Sheppard, 1823)

Size: 3.5–5.0mm

A species found in moderately hard water in lowland rivers, canals, marshland drains and sizable lakes. It has a distinct preference for running water and is almost never found in small closed ponds. It is usually associated with a diverse aquatic fauna, including several other species of *Pisidium*.

Native (Lgl). No evidence of significant distributional change.

Palaearctic and N.E. North America (probably introduced); widespread in Europe, though rare in the northern highland zone and not reaching the Arctic Circle.

(Ellis (Bivalves): 72)

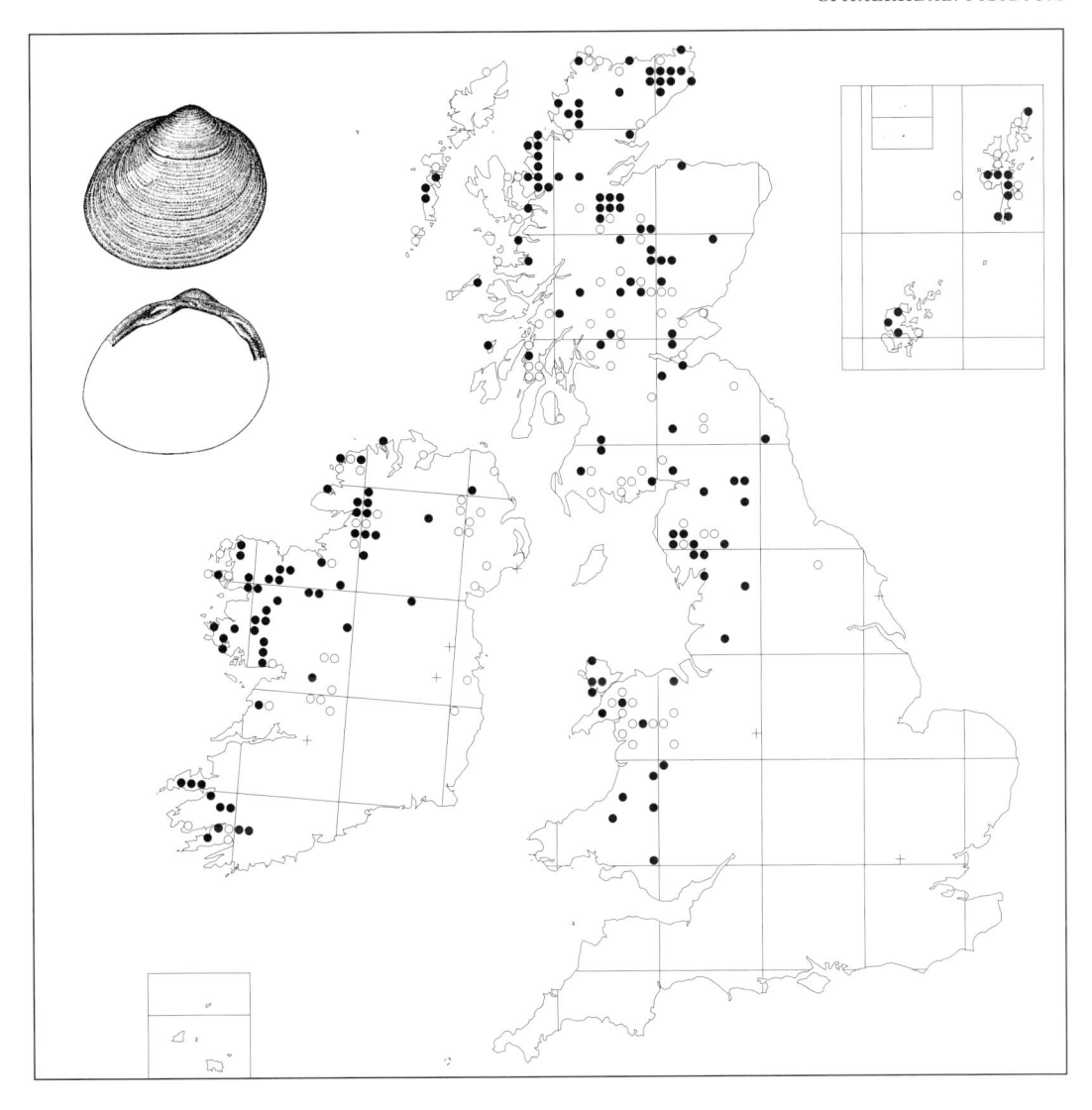

## Pisidium lilljeborgii Clessin, 1886

Size: 3.0–4.5mm

Virtually restricted to lakes in the highland zone, ranging from small lochans to major bodies of water (Loch Lomond, Lough Derg, Lough Neagh). In mountain corrie lakes it occurs at elevations of over 700m and in Windermere it extends from the littoral zone down to −40m. Very exceptionally, it occurs in rivers (R. Teifi, Carmarthen; Dance, 1970b) and in canals (Llangollen, Denbighshire; Toome, Co. Antrim). It lives in both hard and soft water (though mostly in the latter) and tolerates a wide range of bottom conditions, from organic mud to well-washed, gritty sand disliked by most other species of *Pisidium*.

Its most frequent associates are *P. hibernicum* and *P. nitidum*.

Native (Lgl). This boreal, relict species was more widely distributed at the close of the last glacial period, when it occurred in the English lowlands. Within its present restricted range it appears secure and shows no evidence of more recent decline. A few sites may have been lost by the conversion of mountain lakes into reservoirs.

Holarctic circumpolar; in Europe mainly in the Alpine chains and in the northern highland zone, to 71°N.

(Ellis (Bivalves): 74)

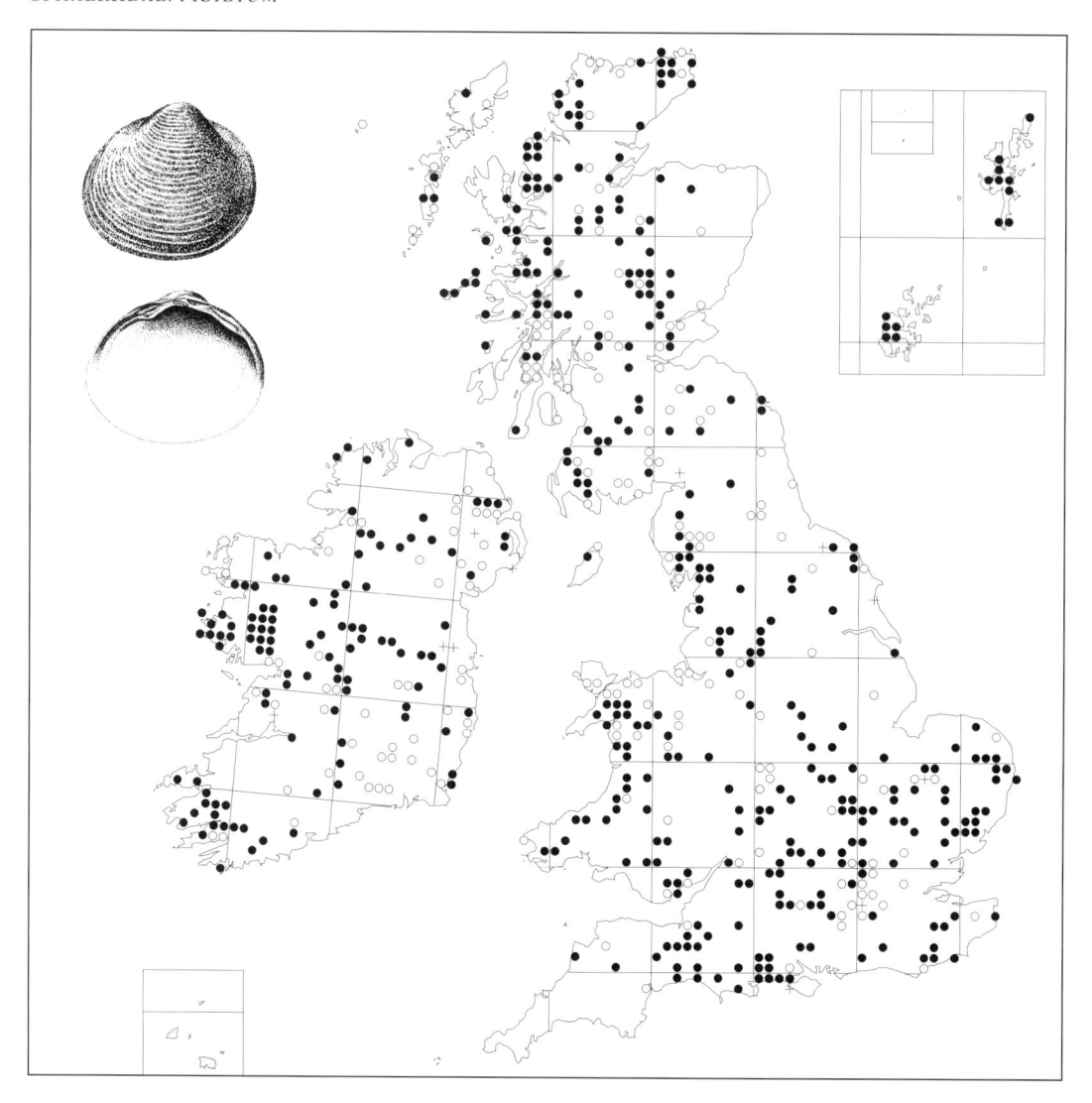

SPHAERIIDAE: *PISIDIUM*

**Pisidium hibernicum** Westerlund, 1894

Size: 2.5–3.5mm

A species typical of clean, clear water, hard or soft, in lakes, canals, rivers and marsh drains. It is rare in small closed ponds, and avoids places choked with weeds or subject to desiccation. Although it occurs in a variety of rich lowland habitats, generally in small numbers compared with associated species of *Pisidium*, it is more truly characteristic of lakes in the highland zone, where it often flourishes alongside *P. lilljeborgii*. It occurs at heights of up to 700m.

Native (Lgl). No evidence of significant recent decline.

Palaearctic (closely related forms in N. America); in Europe mainly north of the Alps, reaching northernmost Scandinavia.

(Ellis (Bivalves): 76)

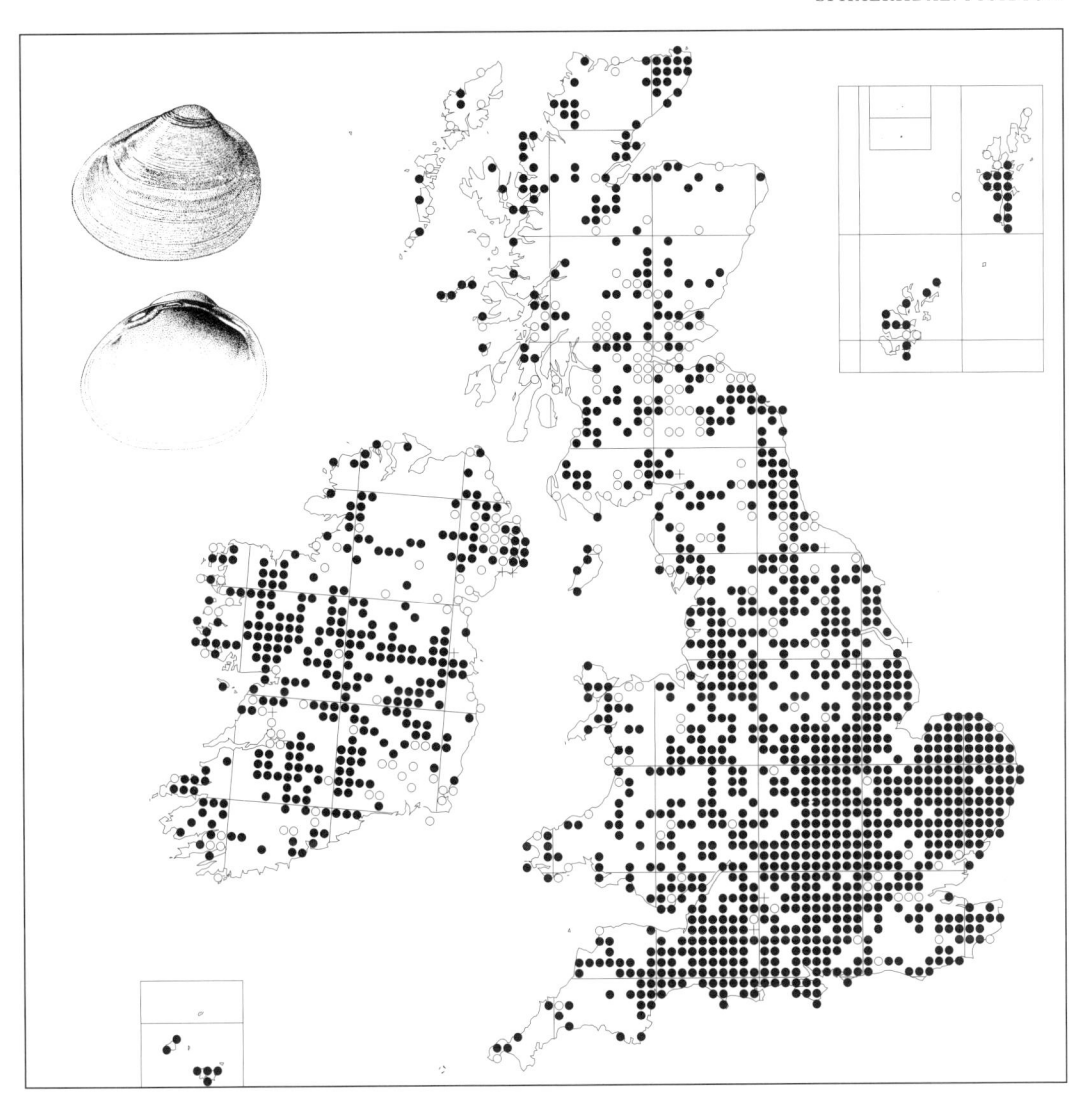

**Pisidium nitidum** Jenyns, 1832

Size: 3–4mm

A common bivalve in hard or soft water, running or stagnant, in rivers, streams, canals, lakes, ponds and ditches. It likes relatively clean, unpolluted bottom sediments with good oxygenation, avoiding swampy, anaerobic places and those liable to dry up. It is mainly a lowland species, but is found in mountain lakes at heights of up to *c*.490m.

Native (Lgl). No evidence of geographical change. Holarctic; in Europe common throughout.

(Ellis (Bivalves): 78)

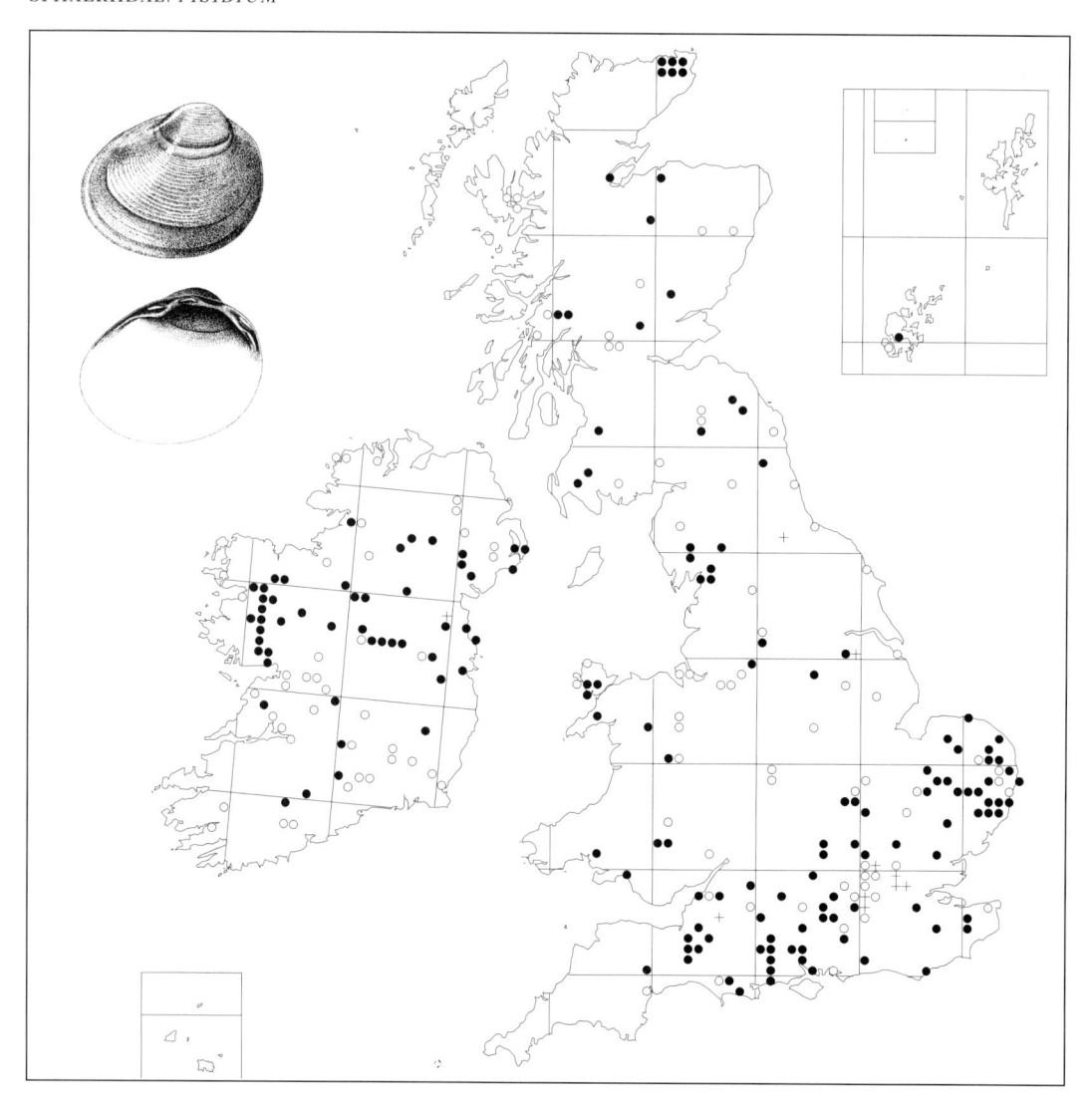

SPHAERIIDAE: *PISIDIUM*

**Pisidium pulchellum** Jenyns, 1832

Size: 3.5–4.0mm

A local species, occurring in slow streams, marsh drains, lakes and ponds; it is rare in major rivers or canals, in which it is usually present in much smaller numbers than other *Pisidium* species. It likes clean, still water, especially over muddy (sometimes organic) substrates. It prefers hard water and is mainly a lowland species, but is found also in upland lakes at moderate elevations.

Native (E.Pgl). No clear sign of overall change.

Palaearctic; in Europe mainly in the lowlands between the Alps and the Baltic, and rare in Scandinavia.

(Ellis (Bivalves): 80)

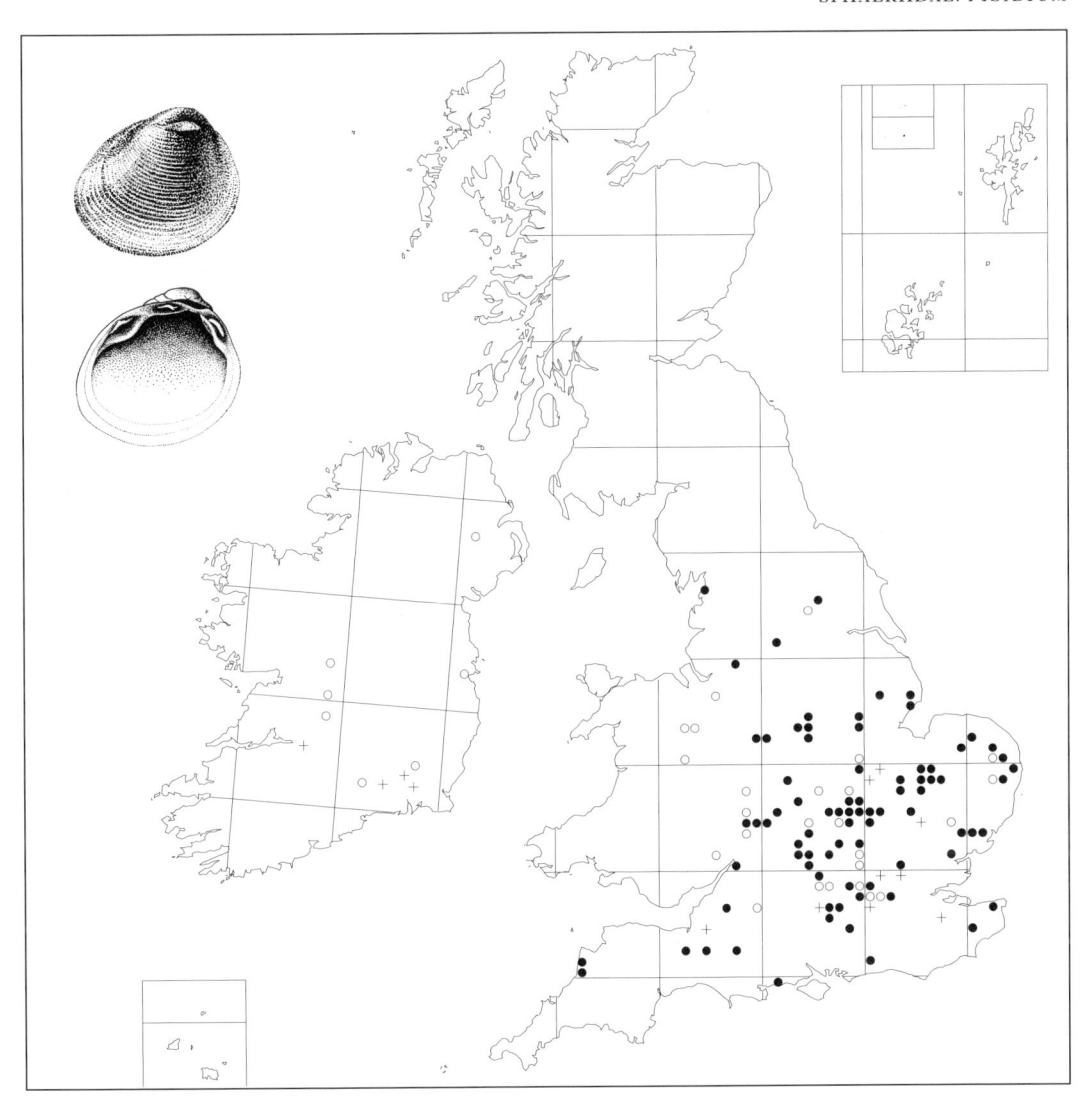

SPHAERIIDAE: *PISIDIUM*

**Pisidium moitessierianum** Paladilhe, 1866 (*P. torquatum* Stelfox, *P. parvulum* of British authors, *non* Classin)

Size: 1.5–2.0mm

A species restricted to canals and sizable lowland rivers. It requires slowly moving, well-oxygenated hard water over clean, unpolluted substrates varying from fine mud to gritty sand. It is generally associated with rich molluscan faunas, usually including *P. supinum*.

Native (E.Pgl). Owing to its minute size, *P. moitessierianum* is somewhat underrecorded. Nevertheless, there are signs of recent recession and the pattern of Postglacial fossils suggests that this is part of a longer-term decline. In bottom sediments from rivers and canals it is often represented only by bleached dead shells, contrasting with the fresh shells of the associated species of *Pisidium*. It was last seen living in Wales over sixty years ago. The isolated population in Northern Ireland (Lagan Canal, Moira, Co. Antrim) has certainly disappeared through habitat change. In Ireland it was last noted in 1924. RDB category (Ireland only): Extinct.

European (possibly Palaearctic); lowlands of Europe from the Mediterranean to S. Scandinavia.

(Ellis (Bivalves): 82)

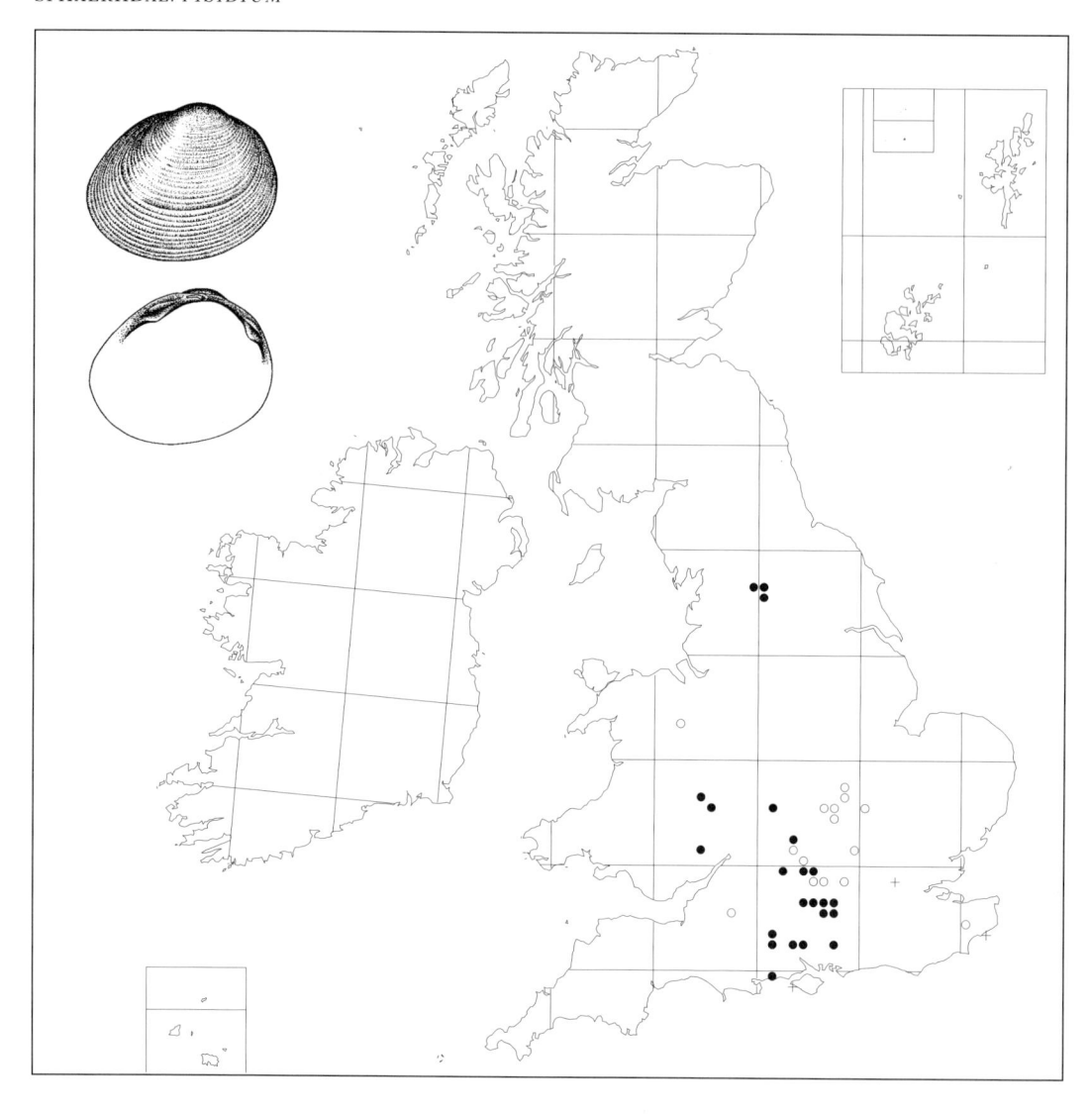

SPHAERIIDAE: *PISIDIUM*

**Pisidium tenuilineatum** Stelfox, 1918

Size: 1.5–1.8mm

An extremely local species found in canals and lowland rivers north to Yorks. (R. Wharfe). Very occasionally it occurs in large ponds (South Harting, Sussex (Kerney, 1970b); Aymestrey, Herefs.). It favours fine silty or muddy substrates in clean, hard, unpolluted water. On the Continent it is recorded also from limestone spring pools, a type of habitat in which it has not been found living in Britain; however, in Kent and the Isle of Wight, it occurs as a fossil in early Postglacial deposits which formed in such places.

Native (E.Pgl). Though doubtless underrecorded owing to its minute size, this species shows signs of decline and must be considered potentially at risk. North of London, in Bucks., Beds. and Northants, only dead shells have been recovered in recent years in spite of careful search; this area includes the type-locality of the species (Grand Union Canal at Marsworth, Bucks.; Stelfox, 1918). RDB category: Rare.

W. Palaearctic; in Europe local in lowland areas from the Mediterranean to southernmost Sweden.

(Ellis (Bivalves): 84)

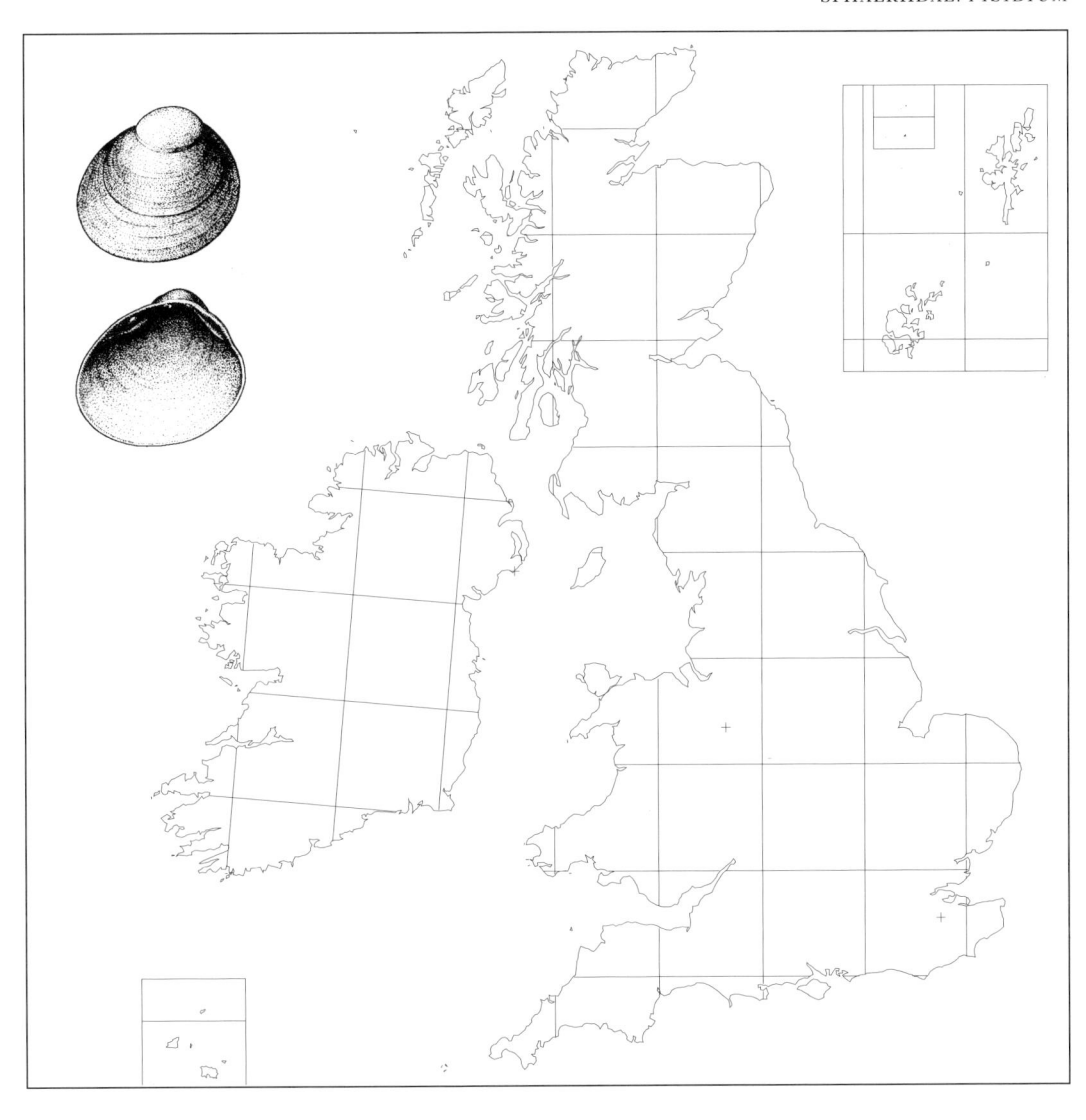

**Pisidium vincentianum** B. B. Woodward, 1913 (*P. stewarti* Preston)

Size: 2–3mm

This is a cold-water species, living, at the present time, only in lakes, never in running water. In the Himalayas in Tibet it has been found at 4,420m, the highest known altitude for any freshwater mollusc.

Extinct (fossil only). *P. vincentianum* has been found in lake deposits of Lateglacial or earliest Postglacial date in Kent, Shropshire and Co. Down (Dance, 1961; Stelfox *et al.*, 1972). It is also known from several earlier deposits belonging to the last glacial period.

Western Siberia and Himalayas (*P. stewarti*). Extinct in Europe: known as a fossil in Pleistocene and early Postglacial deposits in England, France, the Low Countries, Germany, Denmark, Czech Republic and Switzerland.

(Ellis (Bivalves): 86)

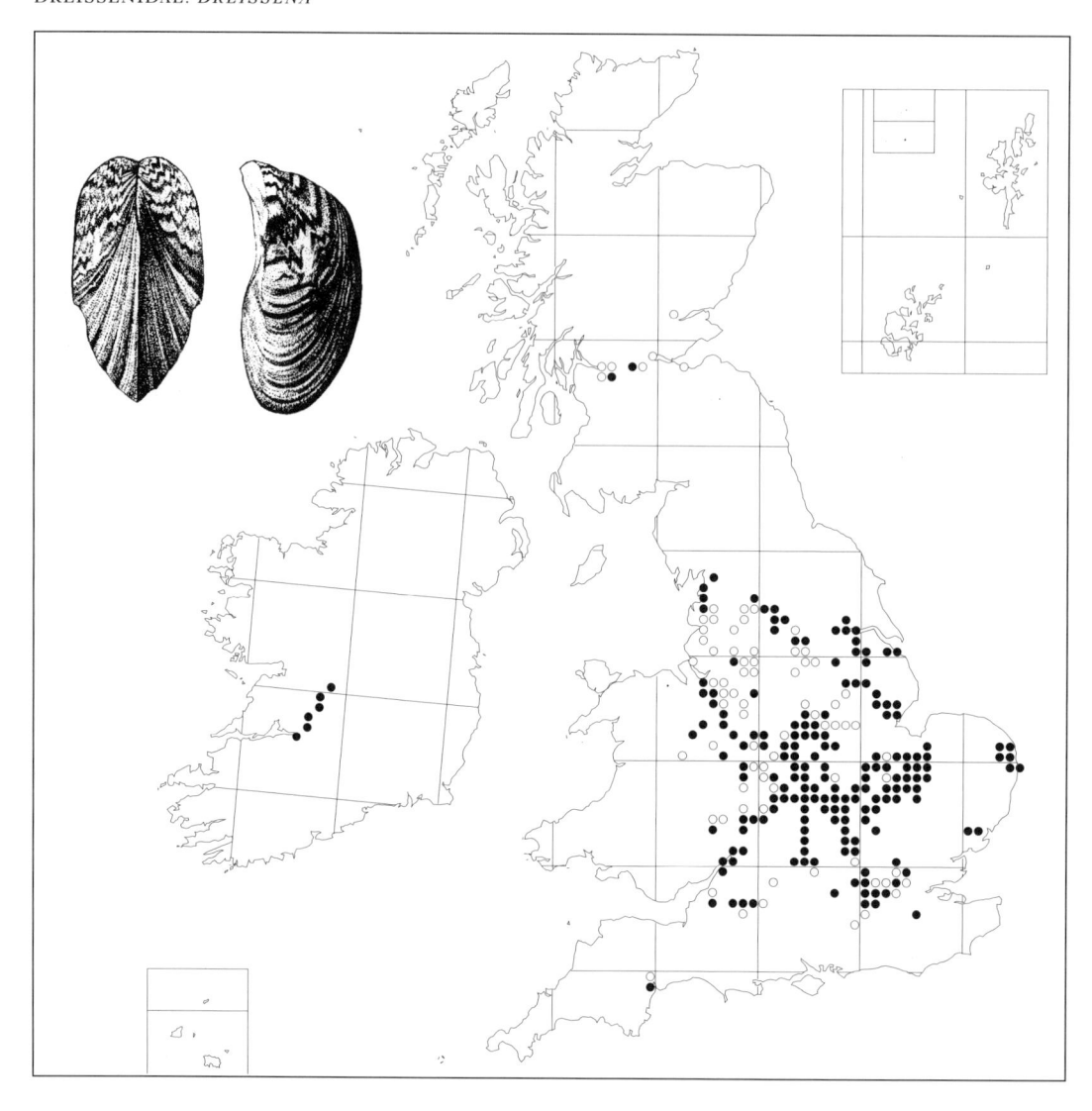

DREISSENIDAE: *DREISSENA*

**Dreissena polymorpha** (Pallas, 1771)

Zebra mussel   Size: 25–40mm

Restricted to clean, well-oxygenated water in lowland rivers, canals and reservoirs. It is strongly gregarious and, like marine mussels (Mytilidae), attaches itself to stones or other hard surfaces, including the masonry of locks and bridges; sometimes it encrusts the shells of living *Unio* or *Anodonta*. It can live inside water mains and was formerly a serious pest in London waterworks. It can tolerate slightly brackish water.

Introduced. This alien species was first noted in Britain in the early 1820s in the London Docks and at Wisbech, Cambs. It was almost certainly introduced with timber from the Baltic. Its subsequent rapid spread through the interconnected 'canal basin' of England and Wales is well documented and by the 1850s the distribution approximated to that of today (Kerney & Moreton, 1970). Occurrences in physically isolated parts of the canal system, notably near Exeter and in Scotland (first noted near Edinburgh in 1833), probably represent separate introductions from abroad. In 1997 it was noted rapidly spreading in the R. Shannon (including Lough Derg), which it appears to have invaded via Limerick Docks (McCarthy & Fitzgerald, 1997). Though *Dreissena* remains common in many places in Britain, there is evidence of local decline this century, especially around Manchester where it is nearly extinct. Other disappearances are due to the abandonment of waterways (e.g. the Grantham Canal).

Originally confined to the region of the Caspian and Black Seas; now spread through much of lowland Europe. Introduced also in N. America.

(Ellis (Bivalves): 88)

# MAPS OF ENVIRONMENTAL FACTORS

The British Isles are highly diverse, both climatically and topographically. The molluscan fauna reflects this diversity. Six of the more important environmental factors are mapped here, of which three are climatic, one topographic and one geological; the remaining environmental factor is a measure of atmospheric pollution.

Climatic data were digitized from transparent map overlays already published by the Institute of Terrestrial Ecology (1978). Maps were then processed on a SysScan mapping workstation and plotted with a high-resolution pen plotter.

The other environmental maps have been produced by the same procedure as the maps of species distribution, using data values on a 10km grid. Data for two of the maps (maximum altitude and calcareous rocks) were taken from an ITE database of land characteristics (Ball, Radford & Williams, 1983) for Great Britain. For Ireland, maximum altitude in each 10km square was read from Irish Ordnance Survey maps, and the occurrence of calcareous rocks was transferred from the overlays in Perring & Walters (1962), converted to the Irish grid.

The final map, of atmospheric sulphur dioxide, was prepared from data kindly supplied in machine-readable medium by Warren Spring Laboratory.

ATLAS OF MOLLUSCS

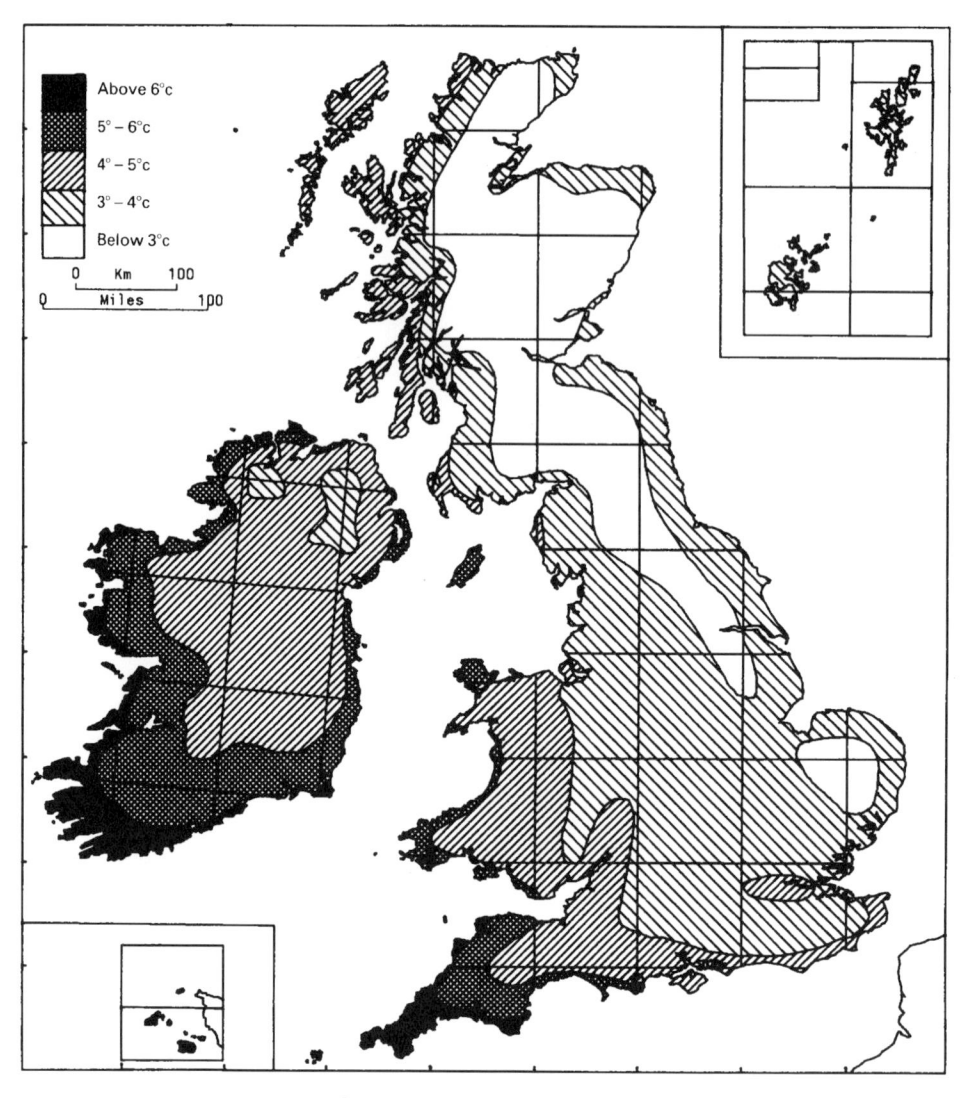

**January mean temperature**
Values are corrected to sea level. Isopleths are spaced at intervals of 1°C, except for the extreme south-west of Ireland and England, where small areas have a January mean in excess of 7°C.

MAPS OF ENVIRONMENTAL FACTORS

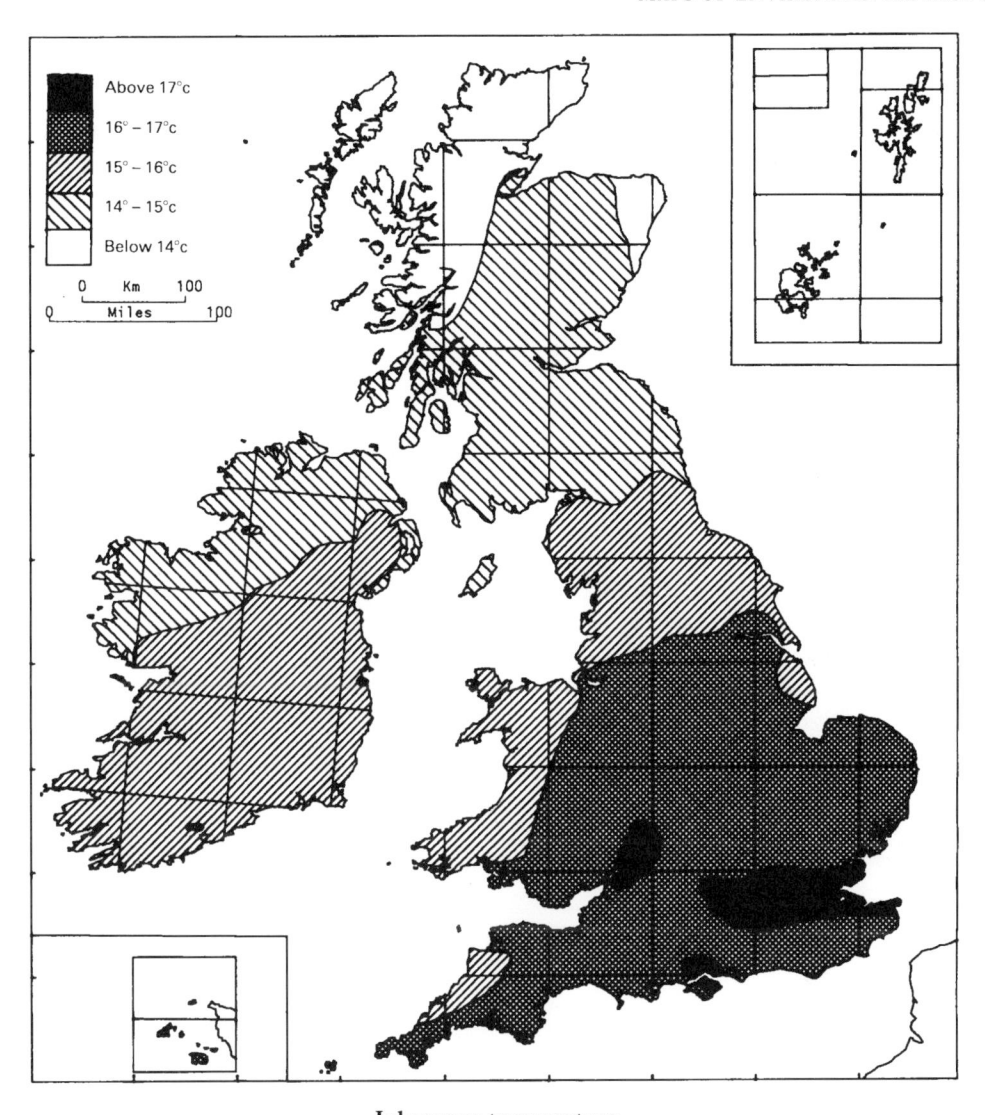

**July mean temperature**
Values are corrected to sea level. Isopleths are spaced at intervals of 1°C.

ATLAS OF MOLLUSCS

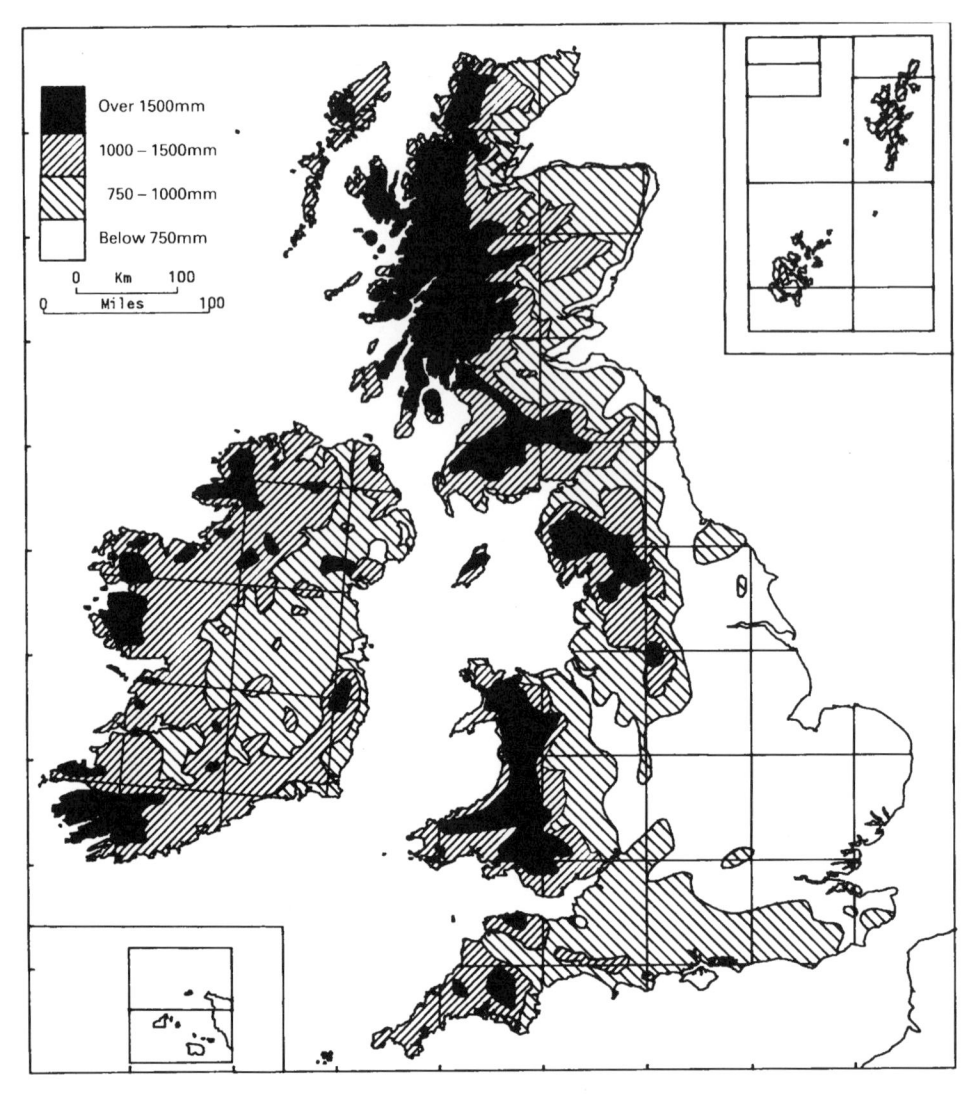

**Annual rainfall**
The map shows total precipitation, of which a small amount falls as snow, especially in the higher Scottish mountains.

MAPS OF ENVIRONMENTAL FACTORS

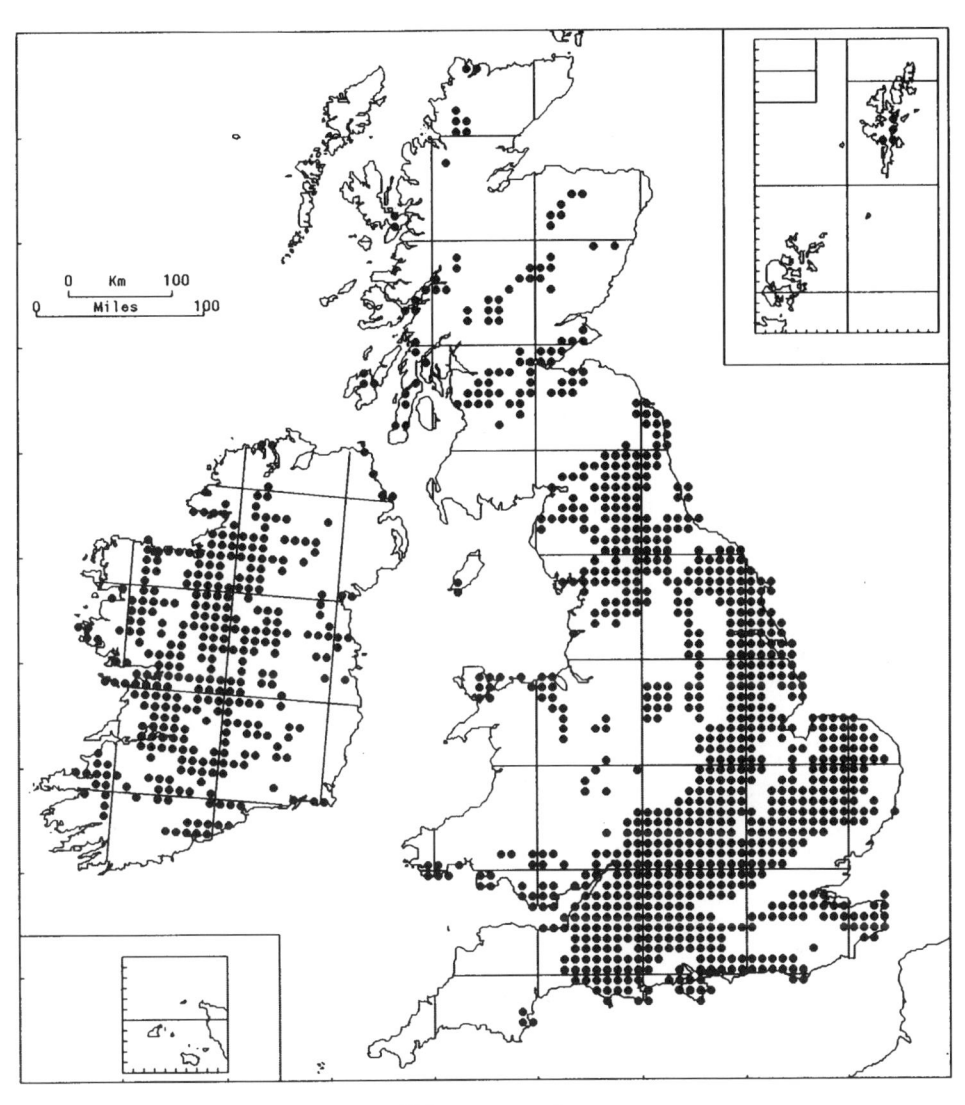

**Calcareous rocks**
The map indicates grid squares with chalk, limestone or metamorphic calcareous rock underlying at least 5 per cent of the land area of the square.

ATLAS OF MOLLUSCS

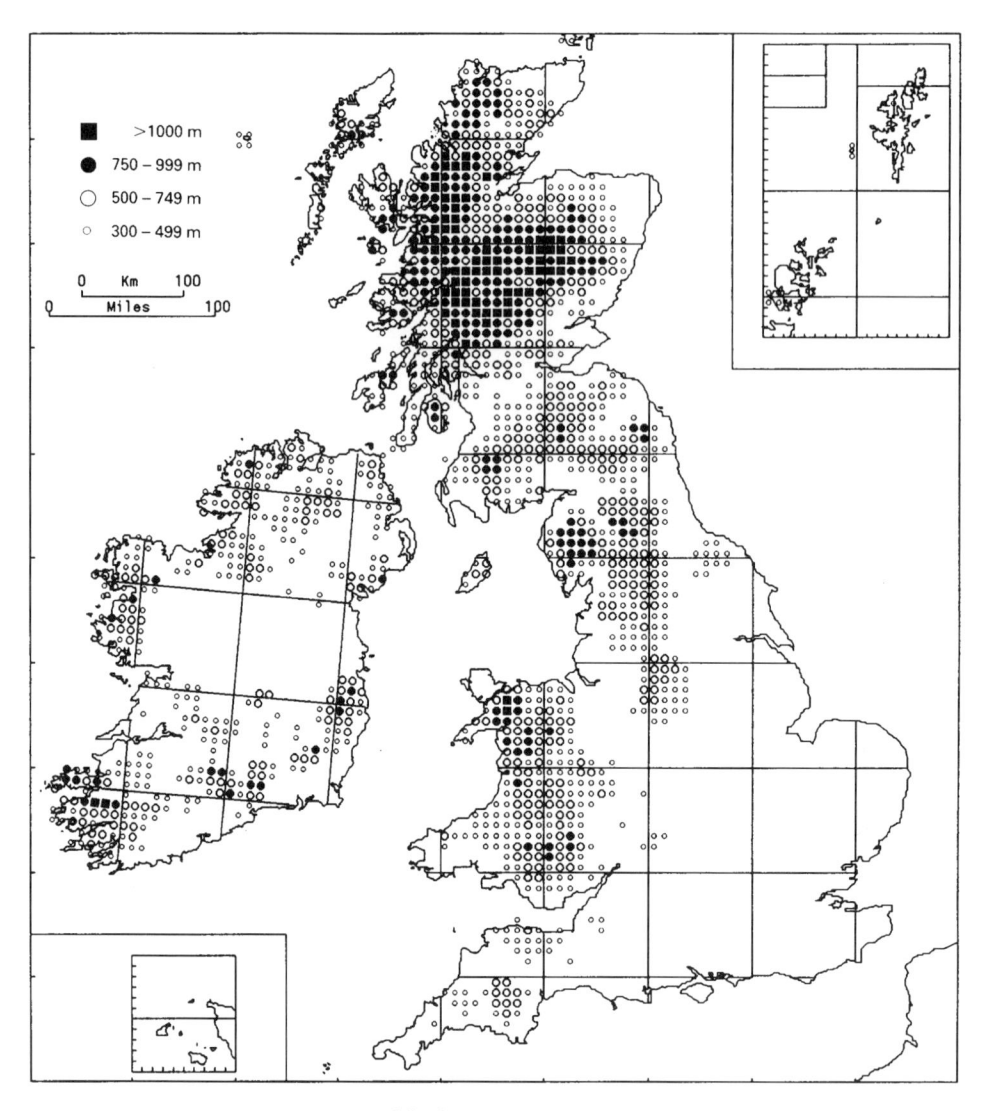

**Maximum altitude**
10km squares of the National Grid are classified according to the maximum altitude of land occurring in that square.

MAPS OF ENVIRONMENTAL FACTORS

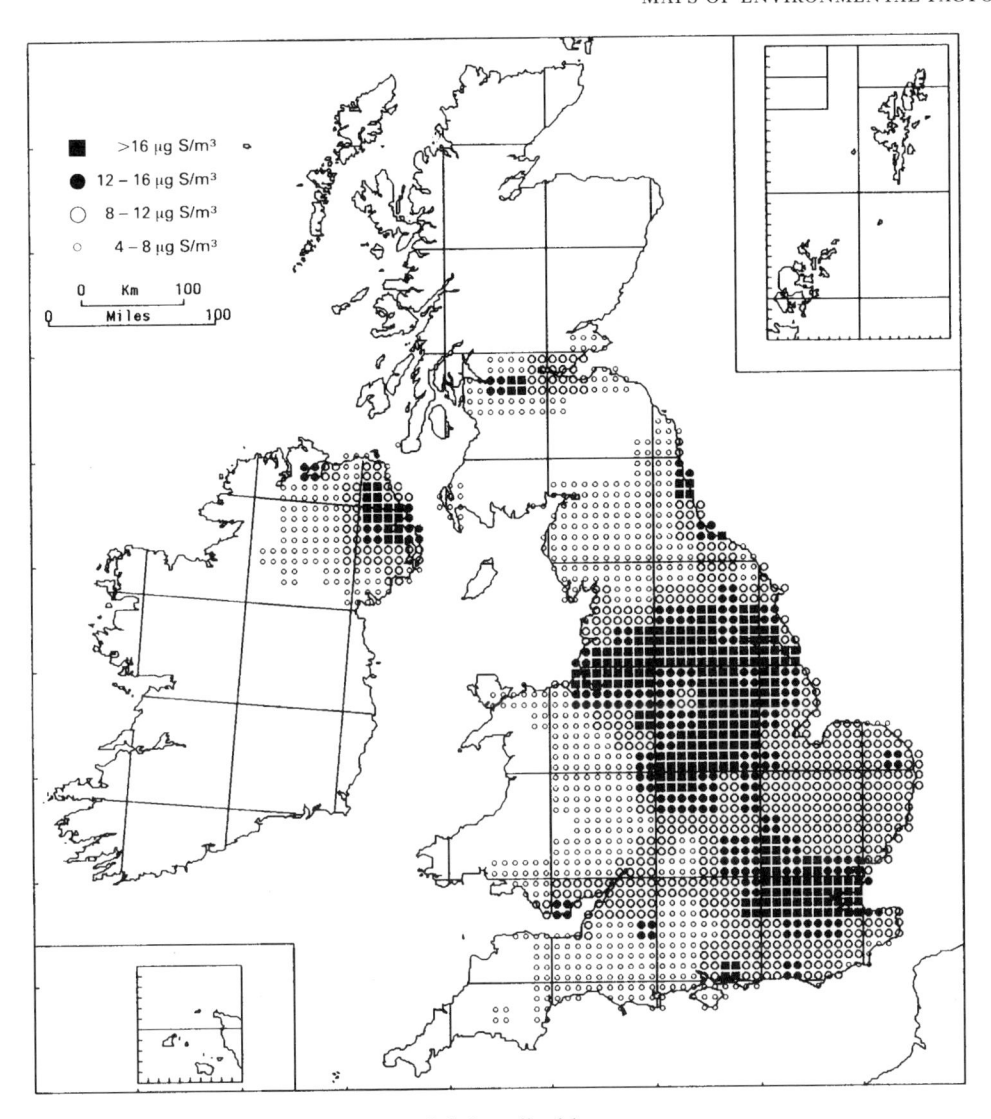

**Sulphur dioxide**

The map shows the estimated mean annual concentration of sulphur dioxide in the atmosphere in 1987, expressed as the mass of sulphur per cubic metre. Only the United Kingdom has been mapped; comparable data were not available for the Irish Republic. Estimates were made by Warren Spring Laboratory on the basis of a mathematical model, which relies on a rather limited number of measuring stations (mainly urban) together with information on emissions from factories and other sources. Emissions of sulphur dioxide have fallen sharply during the period of the Conchological Society's mapping scheme; in 1960, atmospheric concentrations would have been much higher.

# LIST OF LOCALITIES CITED IN THE TEXT

Localities mentioned in the text are given below with their co-ordinates. Where possible, the 10km square or squares are given but for larger areas (e.g. Grampians, Pennines) the 100km squares are given. The map on page 247 gives the numerical equivalents of the 100km-square alphabetical codes, used below.

Aghada, W86
Aldbourne, SU27
Amberley, TQ01
Aran Islands, L71, 80, 81, 90
Ardingly, TQ32
Attleborough, SP39
Aymestrey, SO46

Bala Lake, SH83, 93
Balcombe, TQ33
Barking Creek, TQ48
Bath, ST76
Beckton, TQ48
Belfast, J37
Ben Lawers, NN64
Bermondsey, TQ37
Bexhill, TQ70
Bickenhall, ST21
Blarney, W67
Bossington, SU33
Brandon Mountain, Q41
Braunton Burrows, SS43
Buntingford, TL33
Burgh Castle, TG40
Burnham Beeches, SU98
Burren, The, M, R
Burwell, TF37

Caerphilly, ST18
Caerwys, SJ17
Canty Bay, NT58
Castlethorpe, SE90
Castletownroche, R60
Chaldon, TQ35
Channel Islands, WV
Cheam, TQ26
Chedworth, SP01
Chislet Marshes, TR16
Chiswick, TQ27
Christchurch, SZ19
Clifton, ST57
Cloonascragh Bog, M82

Coe Fen, TL45
Coire Garbhlach, NN89
Coleraine, C83
Combeinteignhead, SX97
Cork City, W67
Cotswolds, SO, SP, ST
Crosby Gill, NY61

Darlington, NZ21
Dawlish Warren, SX97
Deal, TR35
Denton, TQ40
Dooaghtry, L76
Dover Castle, TR34
Droylsden, SJ99
Dudley, SO99
Dukinfield, SJ99
Durham, NZ24
Durlston Head, SZ07
Durness, NC46

Easby Abbey, NZ10
Easton Grey, ST88
Edgeworth, SO90
Edinburgh, NT27
Epping Forest, TQ49
Exeter, SX99
Exminster, SX98

Farlington Marshes, SU60
Fenland, The, TF, TL
Fermoy, W89
Fiagh Bog, R99
Fiddown, S41
Fleet, SU85
Fleet, The, SY58, 67, 68
Folkestone, TR23
Forth & Clyde Canal, NS
Fretherne, SO70

Gait Barrows, SD47
Geal Charn, NN47

244

## LIST OF LOCALITIES CITED IN THE TEXT

Glasgow Necropolis, NS66
Glen Tilt, NN96, 97
Gloucester & Berkeley Canal, SO70, 71, 81
Grampians, NN
Grand Canal, N, O
Grangemouth, NS98
Grantham Canal, SK63, 73, 83
Grassington, SD96
Grays, TQ67
Great Harrowden, SP86
Greenwich, TQ37
Groomsport, J58
Guildford, TQ40
Gwithian, SW54

Haresfield Beacon, SO80
Harris, Island of, NB, NG
Hawick, NT51
Helvellyn, NY31
Huddersfield Canal, SE01, 11
Humewood Castle, S98

Inverkeithing, NT18
Isleworth, TQ17

Kearsney, TR24
Kenfig, SS78
Kennington, SP50
Kensal Green, TQ28
Kew, TQ17

Lagan Canal, J06, 16, 26
Lake District, NY, SD
Leckhampton, SO19
Lewes Levels, TQ40
Limerick, R55
Lismore, Island of, NM83, 84
Lissenhall, O14
Little Malvern, SO74
Loch of Harray, HY21, 31
Loch Lomond, NN30, NS38, 39
Loch Morar, NM79, 89
Loch Ness, NH41, 52, 53
Loch Skene, NT11
Loch of Stenness, HY21
Loch of Strathbeg, NK05
London Docks, TQ37
Lough Derg, M80, R78, 79, 89
Lough Neagh, H, J
Llangollen, SJ14
Llyn Mair, SH64
Llyn Trawsfynedd, SH63, 73
Luddesdown, TQ66
Luton, TL01
Lydden, TR24

Manchester, SJ89
Marsworth, SP91
Mendips, ST
Mickleham, TQ15
Moira, J16
Monkton Marshes, TR26

Mortlake, TQ27
Mount Stewart, J56

Newbourn, TM24
New Forest, SU, SZ
Norfolk Broads, TG
North Uist, NF

Oglethorpe, SE44
One Tree Hill, TQ55
Orton Fell, NY60
Oulton Broad, TG40, 50
Oxford, SP50

Paignton, SX86
Pangbourne, SU67
Peak District, SK
Pendleton, SJ89
Pennines, SD, SE, NY, NZ
Penshurst, TQ54
Peterborough, TL19
Petersfield, SU72
Pevensey Levels, TQ60
Pitlochry, NN95
Plumstead, TQ47
Pollboy Bog, M82
Pontefract, SE42
Portland, Isle of, SY67
Portmarnock, O24
Potton, TL24
Purfleet, TQ57

Regent's Canal, TQ28, 38
Richmond, TQ17
River Barrow, S
River Deben, TM25, 35
River Great Ouse, TF, TL
River Lea, TL30, TQ38, 39
River Nore, S
River Shannon, R
River Suir, S
River Teifi, SN
River Tyne, NZ06, 16, 26, 36
River Waveney, TM
River Wharfe, SD96, SE05, 06
River of Wick, ND25, 35
River Witham, TF
River Yare, TG
Royal Canal, N, O

Saltfleet, TF49
Sandwich, TR35
Scales Tarn, SD39
Scawby, SE90
Schull, V93
Sennen Cove, SW32
Shannon Estuary, R
Shelmaliere Commons, S91
Shepherdswell, TR24
Skye, Island of, NG
Snowdon, SH65
Southfleet, TQ66

South Harting, SU72
Staines, TQ07
Sunbiggin Tarn, NY60
Swansea, SS69
Syon Marsh, TQ17

Temple Demesne, N89
Thetford, TL88
Thompson Common, TL99
Thornham, TF74
Toome, H99
Torquay, SX96
Tring, SP91

Walmer, TR34
Waltham Abbey, TL30

Walton-on-Thames, TQ06
Weald, The, TQ
Went Vale, SE51
Westbere Marshes, TR26
West Wycombe, SU89
Whitchurch, ST17
Whitecliff Bay, SZ68
Wicken Fen, TL56, 57
Widdybank Fell, NY83
Winchester, SU42
Windermere, NY30, SD39
Wisbech, TF40
Wise Fen Tarn, SD39
Woody Bay, SS64
Woolacombe Sands, SS44

NATIONAL GRIDS AND VICE-COUNTIES

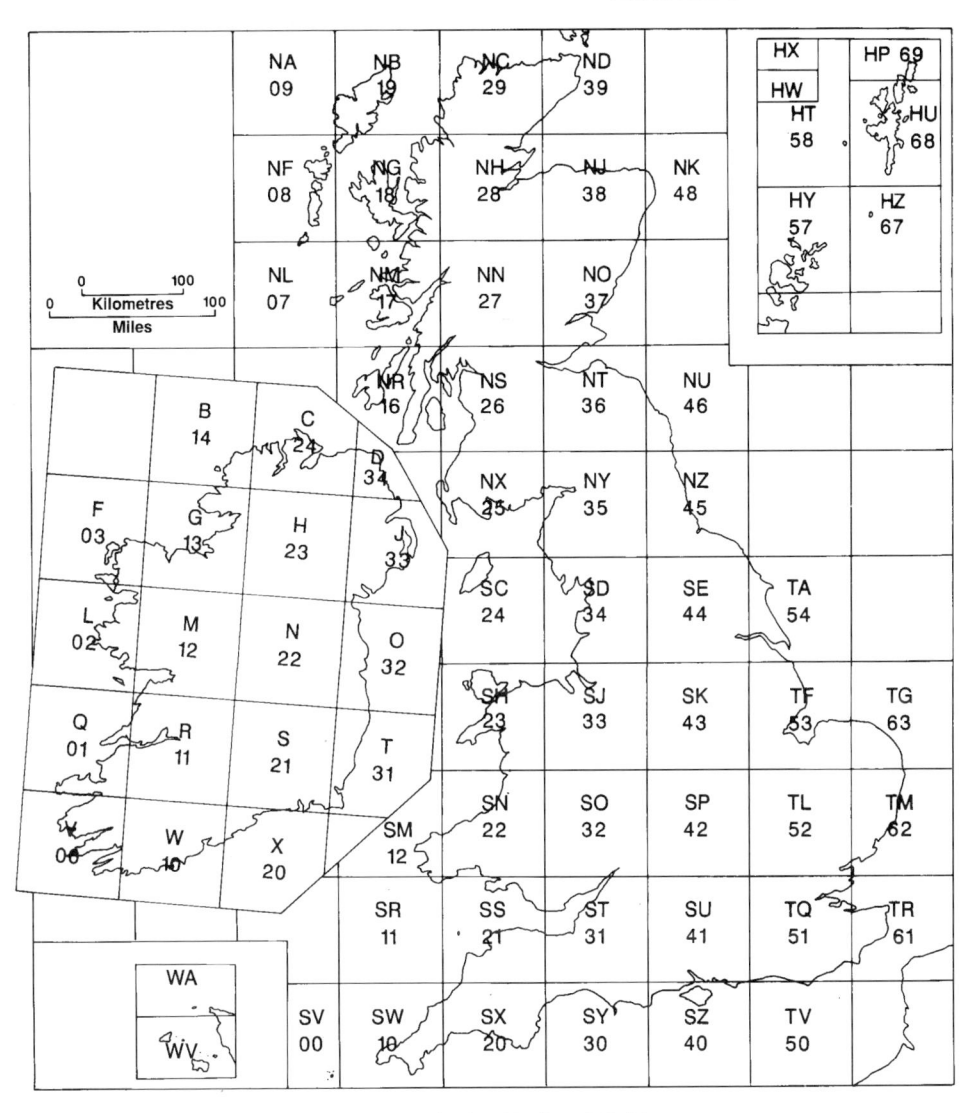

**British and Irish National Grids**

The map shows the numerical equivalents of the 100km-square alphabetical codes used in the list of localities cited in the text.

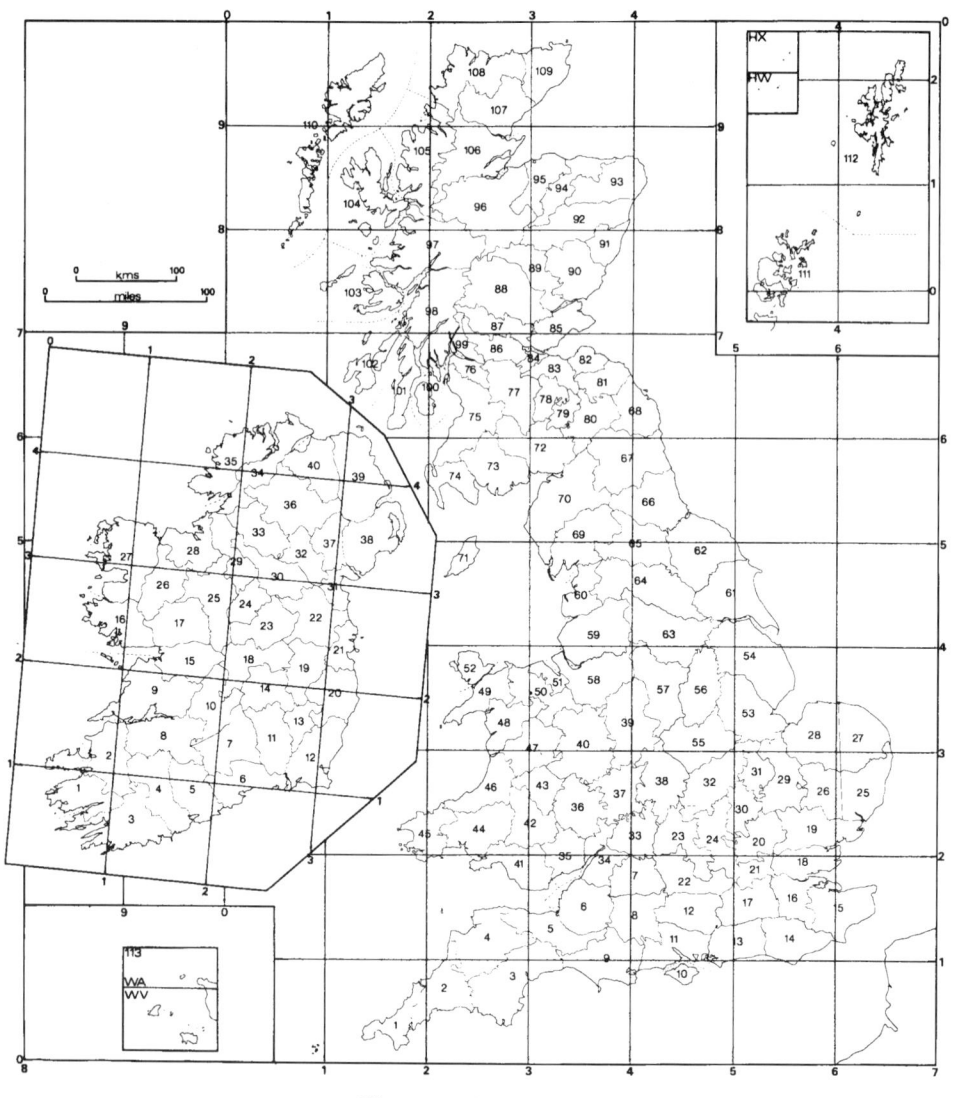

ATLAS OF MOLLUSCS

## Vice-counties

This map shows the geographical recording units which have been employed for over a century and are still in use today. Reference to them is made in the text. For Great Britain their boundaries (which often diverge significantly from those of modern administrative counties) are defined authoritatively by Dandy (1969). The Irish vice-counties established by Babington and Praeger (Praeger, 1896) are essentially similar to the existing Irish counties, though subdivided in the case of Kerry, Cork, Tipperary, Galway, Mayo and Donegal. Brief definitions of the British and Irish vice-counties can also be found in Ellis (1951).

**England and Wales**

1 West Cornwall (with Scilly)
2 East Cornwall
3 South Devon
4 North Devon
5 South Somerset
6 North Somerset
7 North Wiltshire
8 South Wiltshire
9 Dorset
10 Isle of Wight
11 South Hampshire
12 North Hampshire
13 West Sussex
14 East Sussex
15 East Kent
16 West Kent
17 Surrey
18 South Essex
19 North Essex
20 Hertfordshire
21 Middlesex
22 Berkshire
23 Oxfordshire
24 Buckinghamshire

25 East Suffolk
26 West Suffolk
27 East Norfolk
28 West Norfolk
29 Cambridgeshire
30 Bedfordshire
31 Huntingdonshire
32 Northamptonshire
33 East Gloucestershire
34 West Gloucestershire
35 Monmouthshire
36 Herefordshire
37 Worcestershire
38 Warwickshire
39 Staffordshire
40 Shropshire (Salop)
41 Glamorgan
42 Breconshire
43 Radnorshire
44 Carmarthenshire
45 Pembrokeshire
46 Cardiganshire
47 Montgomeryshire
48 Merionethshire
49 Caernarvonshire
50 Denbighshire
51 Flintshire
52 Anglesey
53 South Lincolnshire
54 North Lincolnshire
55 Leicestershire (with Rutland)
56 Nottinghamshire
57 Derbyshire
58 Cheshire
59 South Lancashire
60 West Lancashire
61 South-east Yorkshire
62 North-east Yorkshire
63 South-west Yorkshire
64 Mid-west Yorkshire
65 North-west Yorkshire
66 Durham
67 South Northumberland
68 North Northumberland (Cheviot)
69 Westmorland with North Lancashire

70 Cumberland
71 Isle of Man

**Scotland**

72 Dumfriesshire
73 Kirkcudbrightshire
74 Wigtownshire
75 Ayrshire
76 Renfrewshire
77 Lanarkshire
78 Peeblesshire
79 Selkirkshire
80 Roxburghshire
81 Berwickshire
82 East Lothian (Haddington)
83 Midlothian (Edinburgh)
84 West Lothian (Linlithgow)
85 Fifeshire (with Kinross)
86 Stirlingshire
87 West Perthshire (with Clackmannan)
88 Mid Perthshire
89 East Perthshire
90 Angus (Forfar)
91 Kincardineshire
92 South Aberdeenshire
93 North Aberdeenshire
94 Banffshire
95 Moray (Elgin)
96 East Inverness-shire (with Nairn)
97 West Inverness-shire
98 Argyll Main
99 Dunbartonshire
100 Clyde Isles
101 Kintyre
102 South Ebudes
103 Mid Ebudes
104 North Ebudes
105 West Ross
106 East Ross
107 East Sutherland
108 West Sutherland
109 Caithness
110 Outer Hebrides
111 Orkney Islands
112 Shetland Islands (Zetland)

113 Channel Islands

**Ireland**

H.1 South Kerry
H.2 North Kerry
H.3 West Cork
H.4 Mid Cork
H.5 East Cork
H.6 Waterford
H.7 South Tipperary
H.8 Limerick
H.9 Clare
H.10 North Tipperary
H.11 Kilkenny
H.12 Wexford
H.13 Carlow
H.14 Leix (Queen's County)
H.15 South-east Galway
H.16 West Galway
H.17 North-east Galway
H.18 Offaly (King's County)
H.19 Kildare
H.20 Wicklow
H.21 Dublin
H.22 Meath
H.23 Westmeath
H.24 Longford
H.25 Roscommon
H.26 East Mayo
H.27 West Mayo
H.28 Sligo
H.29 Leitrim
H.30 Cavan
H.31 Louth
H.32 Monaghan
H.33 Fermanagh
H.34 East Donegal
H.35 West Donegal
H.36 Tyrone
H.37 Armagh
H.38 Down
H.39 Antrim
H.40 Londonderry

# LIST OF RECORDERS

G. B. T. Abbott, M. E. Adam, F. A. Adams, D. Aldridge, K. N. A. Alexander, W. E. Alkins, M. J. Allen, K. G. Allenby, A. Allison, P. S. Anderson, R. Anderson, Elizabeth Andrews, S. Angus, S. G. Appleyard, J. Armitage, P. Armitage, G. A. Arnold, M. A. Arnold, M. J. Ashdown, R. Atkins, N. K. Atkinson, Angela Avens, S. E. R. Bailey, R. E. Baker, M. Baldwin, G. A. S. Barnacle, R. S. K. Barnes, B. W. Barrett, J. F. M. de Bartolomé, J. A. B. Bass, C. J. Bayne, D. Beaumont, K. E. Beckett, M. Bell, D. P. Bentley, F. G. Berry, Kathryn Berry, H. E. J. Biggs, D. Bilton, E. O. Bishop, M. J. Bishop, Dorothy Blezard, E. Blezard, M. R. Block, D. E. Bolton, J. P. Bowdrey, H. J. M. Bowen, D. Boxley, R. Boyce, Janet Boyd, J. H. Bratton, J. Briggs, D. S. Brown, Ellen Brown, P. U. Buckle, P. F. Burns, A. Buse, A. J. Cain, P. Cambridge, R. A. D. Cameron, J. Campbell, J. A. Carman, R. Carr, Pamela Carter, I. M. Cassells, C. P. Castell, M. Cawley, A. Chalkley, J. Charles, J. L. Charlish, Elizabeth Charter, A. O. Chater, June Chatfield, D. K. Clements, R. Clinging, P. R. Cobb, B. F. Coles, Louise Collins, M. Collins, G. M. Collis, B. Colville, M. E. Conway, A. Cook, J. Cooper, J. Cooter, G. B. Corbet, Alison Corley, D. C. F. Cotton, R. H. Cowie, Lord Cranbrook, D. J. Cranmer, G. I. Crawford, T. J. Crawford-Sidebotham, C. T. Cribb, R. J. Croome, I. C. Cross, T. E. Crowley, S. P. Dance, M. B. Davidson, P. Davies, Stella Davies, Susan Davies, D. S. Davis, J. Daws, I. K. Dawson, J. J. Day, E. Dearing, B. P. Dennison, T. R. E. Devlin, R. A. Dines, C. R. P. Diver, R. Dixon, R. M. Dobson, D. Doogue, H. E. M. Dott, M. J. D'Oyly, E. P. Drage, C. D. Drake, C. M. Drake, R. J. Driscoll, N. Dudley, C. R. duFeu, J. H. Duffus, Jo Dunn, W. F. Edwards, Mary Elliott, A. E. Ellis, W. A. Ely, K. S. Erskine, I. M. Evans, J. G. Evans, N. J. Evans, B. C. Eversham, J. G. Farthing, S. T. A. Featherstone, C. Felton, I. D. Finney, M. Fishpool, Eleanor Fogan, Marjorie Fogan, A. P. Foster, A. P. Fowles, Jane Fox, P. M. Freeman, R. Fresco-Corbu, Larch Garrad, J. Gerlach, D. J. Gibbs, D. I. Gibson, T. S. H. Gibson, D. Gilbertson, N. Giles, C. L. Gillard, E. A. Glennie, M. Goodchild, Kathleen Goldie-Smith, C. B. Goodhart, D. A. Gowans, A. Graham, T. Gray, D. Green, J. J. D. Greenwood, A. T. Gregg, S. J. Gregory, L. W. Grensted, S. J. Grove, J. P. Guest, Marion Gunning, D. W. Guntrip, J. B. Hall, N. Hammond, G. H. Harper, J. F. Harper, G. J. Harris, J. I. Harris, D. R. B. Harrison, M. B. Hart, Barbara Hart-Jones, E. I. Harvey, D. M. Hawker, R. D. Hawkins, C. Hayton, M. P. Headen, D. Heppell, P. B. Heppleston, R. T. Herbert, Susan Hewitt, A. J. Hicklin, R. C. Higgins, K. Hill, Rosemary Hill, Pat Hill-Cottingham, Sandra Hogg, Rosalind Holt, D. T. Holyoak, P. M. Horsfield, Elizabeth Howe, D. Howlett, M. R. Hughes, J. Humphreys, J. Hunnisett, J. Hurley, Ann Hurst, J. M. C. Hutchinson, T. Huxley, E. G. Ing, A. G. Irwin, J. W. Jackson, M. Jackson, N. Jackson, T. J. James, Margaret Jaques, T. J. Jennings, A. J. Jeram, A. W. Jones, J. S. Jones, P. H. Jones, D. H. Keen, E. Kellock, M. Kennedy, D. K. Kevan, I. J. Killeen, P. A. Kingsbury, T. D. Knight, G. Kroon, M. Ladle, Olivia Lassière, J. C. Lavin, P. Lee, Lady Christina Letanka, Jan Light, Nicole Limondin, D. Lindley, P. List, J. E. Llewellyn-Jones, L. Lloyd-Evans, D. C. Long, G. E. H. Long, R. H. Lowe, J. Lucey, T. T. Macan, R. MacDonald, Ann McDonnell, C. R. McLeod, Nora McMillan, C. McNeill-Ritchie, Alison Maddock, P. S. Maitland, P. Makings, V. S. Mallet, M. Marklove, Dorothy Marriott, R. W. Marriott, F. Marshall, J. Mathias, F. May, Lady Sophie Meade, A. D. J. Meeuse, R. G. Meiklejohn, B. J. Meloy, Hazel Meredith, R. Meyers, Amanda Millar, D. J. P. Miller, Annie Milles, J. O'N. Millott, E. Milne-Redhead, D. Minchin, P. S. Mobsby, C. Moore, Evelyn Moorkens, P. B. Mordan, B. D. Moreton, I. K. Morgan, C. Moriarty, M. J. Morphy, R. W. Morrell, B. S. Morton, Sally Murrell, G. Musker, G. N. Myers, F. C. Naggs, B. S. Nau, P. F. Newell, Jennifer Newton, N. W. Nicholson, A. Norris, Mandy North, W. F. Norton, R. J. O'Connor, M. O'Grady, M. O'Meara, A. P. H. Oliver, A. G. H. Osborn, T. Pain, D. Painter, M. Palles-Clark, Margaret Palmer, J. K. Partridge, K. Paterson, Jean Paton, C. R. C. Paul, J. F. Peake, D. Pearman, W. A. Pearson, J. D. Peek, F. H. Perring, C. W. Pettitt, A. D. P. van Peursen, B. Philp, E. G. Philp, D. G. Pickrell, J. Piekarczyk, G. W. Pitchford, C. W. Plant, Elizabeth Platts, E. Pollard, D. E. Pomeroy, R. C. Preece, Mary Pugh, A. W. Punter, H. E. Quick, Beryl Rands, D. G. Rands, P. J. Raven, B. Rawlinson, Julie Rayner, P. Reavell, E. J. Redshaw, Jennifer Rees, J. A. Rennie, Jane Reynolds, S. M. Rhind, D. T. Richardson, N. J. Riddiford, G. Riley, T. H. Riley, E. Robinson, Helena Ross, M. W. Rowe, Jill Royston, A. J. Rundle, N. W. Runham, Mary Saul, D. R. Saunders, R. Saville, Hilary Scott, C. Scotter, D. R. Seaward, Mary Seddon, Frances Seeley, T. Serjeant, M. E. A. Shardlow, Anne Shaw, H. H. Shephard, F. W. Shotton, T. B. Silcocks, Rosemary Simpson, M. Sinclair, J. F. Skinner, G. Slator, A. G. Smith, Betty Smith, C. J. Smith, I. F. Smith, Shelagh Smith, A. South, G. G. Spencer, Melanie Spirit, M. Spray, A. I. Spriggs, A. W. Stelfox, D. I. Steward, A. McG. Stirling, L. W. Stratton, M. Street, A. T. Sumner, Caroline Sutton, M. D. Sutton, C. M. Swaine, R. M. Tallack, P. Tattersfield, D. A. J. Taylor, M. A. Taylor, M. Telfer, P. Temple, B. Thistleton, I. J. L. Tillotson, R. Tindal, P. Topley, J. U. Tuck, Stella Turk, C. Turner, P. A. Ventris, B. Verdcourt, G. J. M. Visser, P. M. Wade, H. W. Waldén, D. M. Walker, Margaret Walker, R. B. Walker, I. D. Wallace, T. J. Wallace, M. R. Wallis, A. Walton, M. V. Walton, J. W. Ward, A. A. Wardhaugh, P. D. Warris, T. Warwick, A. R. Waterston, R. Watkin, Helen Weideli, M. D. Weideli, Susan Westwood, P. F. Whitehead, D. A. A. Whiteley, G. Whitfield, V. Wilkin, H. Williams, R. B. Williams, R. B. G. Williams, M. J. Willing, P. J. Wilson, P. T. Wimbleton, F. R. Woodward, T. M. Worsfold, D. R. Worth, J. Wright, J. F. Wright, J. O. Young, J. Z. Young, M. R. Young.

# BIBLIOGRAPHY

Adams, L. E., 1884. *The collector's manual of British land and freshwater shells.* London.

———, 1896. *The collector's manual of British land and freshwater shells.* 2nd edition. Leeds.

Alder, J., 1837. Notes on the land and fresh-water Mollusca of Great Britain, with a revised list of species. *Mag. Zool. Bot.* **2**: 101–119.

Alexander, K. N. A., 1994. The Roman snail *Helix pomatia* L. in Gloucestershire and its conservation. *Gloucestershire Naturalist* no. 7: 9–14.

Alkins, W. E., 1925. Note on the habits of *Hygromia fusca* (Montagu). *J. Conchol., Lond.* **17**: 243–247.

Allen, D. E., 1976. *The naturalist in Britain.* London.

Anderson, R., 1974. *Vitrina (Semilimax) pyrenaica* (Férussac) in Co. Antrim. *Ir. nat. J.* **18**: 51.

———, 1991. Evidence of recent spread in *Semilimax pyrenaicus* (Férussac) and *Arion owenii* Davies. *Ir. nat. J.* **23**: 510.

———, 1992. A second Co. Down station for *Semilimax pyrenaicus* (Férussac). *Ir. nat. J.* **24**: 172.

———, 1996. *Physa gyrina* (Say), a North American freshwater gastropod new to Ireland. *Ir. nat. J.* **25**: 248–253.

——— & Norris, A., 1974. *Boettgerilla pallens* Simroth new to Ireland. *J. Conchol., Lond.* **28**: 207–208.

Aubertin, D., Ellis, A. E. & Robson, G. C., 1931. The natural history and variation of the pointed snail, *Cochlicella acuta* (Müll.). *Proc. zool. Soc.* no. 67 (1930): 1027–1055.

Ball, D. F., Radford, G. L. & Williams, W. M., 1983. *A land characteristics data bank for Great Britain* (Bangor Occasional Paper No. 13). Bangor.

Ballantine, W. J. & Bradley, D. J., 1963. A note on the Irish involute *Lymnaea.* *Proc. malac. Soc. Lond.* **35**: 86–88.

Bank, R. A. & Butot, L. J. M., 1984. Some more data on *Hydrobia ventrosa* (Montagu, 1803) and "*Hydrobia*" *stagnorum* (Gmelin, 1791) with remarks on the genus *Semisalsa* Radoman, 1974. *Malak. Abh. Mus. Tierk. Dresden* **10**: 5–15.

———, ——— & Gittenberger, E., 1979. On the identity of *Helix stagnorum* Gmelin, 1791, and *Turbo ventrosus* Montagu, 1803 (Prosobranchia, Hydrobiidae). *Basteria* **43**: 51–60.

Barrett, B. W., 1972. *Theba pisana* in Guernsey, 1971. *J. Conchol., Lond.* **27**: 391–396.

Beeston, H., 1919. Field notes on *Helicodonta obvoluta* Müller. *J. Conchol., Lond.* **16**: 31–36; 44–50.

Biggs, H. E. J., 1957. Notes on *Hygromia cinctella* (Draparnaud). *J. Conchol., Lond.* **24**: 177–178.

Bishop, M. J., 1976. *Hydrobia neglecta* Muus in the British Isles. *J. moll. Stud.* **42**: 319–326.

——— & Bishop, S., 1973. A new Suffolk locality for the snail *Monacha cartusiana* (Müller). *Suffolk Natural History* **16**: 47–48.

Blackburn, E. P., 1941. Distribution of *Clausilia cravenensis* Taylor (*suttoni* Westerlund) in Britain. *J. Conchol., Lond.* **21**: 289–300.

Boycott, A. E., 1921. Oecological notes. *Proc. malac. Soc. Lond.* **14**: 128–130; 167–172.

———, 1927. Further notes on *Vitrina major* in Britain. *Proc. malac. Soc. Lond.* **17**: 141–144.

———, 1929. The habitat of *Clausilia biplicata* Mont. *J. Conchol., Lond.* **18**: 340–343.

———, 1934. The habitats of land Mollusca in Britain. *J. Ecol.* **22**: 1–38.

———, 1936a. The habitats of freshwater Mollusca in Britain. *J. animal Ecol.* **5**: 116–186.

———, 1936b. *Neritina fluviatilis* in Orkney. *J. Conchol., Lond.* **20**: 199–200.

———, 1938. Experiments on the artificial breeeding of *Limnaea involuta*, *Limnaea burnetti* and other forms of *Limnaea peregra*. *Proc. malac. Soc. Lond.* **23**: 101–108.

———, 1939. Distribution and habitats of *Ena montana* in England. *J. Conchol., Lond.* **21**: 153–159.

——— & Oldham, C., 1930. The food of *Geomalacus maculosus*. *J. Conchol., Lond.* **19**: 36.

———, Oldham, C., & Waterston, A. R., 1932. Notes on the lake *Lymnaea* of south-west Ireland. *Proc. malac. Soc. Lond.* **20**: 105–127.

Boyd, J., 1997. New colony of *Cochlicella barbara* at Kenfig. *Conchologists' Newsletter* no. 140: 781–782.

Bratton, J. H. (ed.), 1991. *British Red Data Books*: 3. *Invertebrates other than insects.* Joint Nature Conservation Committee: Peterborough.

Brown, D. S., 1977. *Ferrissia* – a genus of freshwater

limpet new for Britain. *Conchologists' Newsletter* no. 62, 23–25.

Cameron, R. A. D., 1972. The distribution of *Helicodonta obvoluta* (Müll.) in Britain. *J. Conchol., Lond.* **27**: 363–369.

———, 1992. *Vertigo moulinsiana, V. lilljeborgi* and others in Shropshire. *J. Conchol., Lond.* **34**: 186.

——— & Letanka, C., 1976. *Abida secale* (Draparnaud) (Gastropoda: Chondrinidae) in north-west England. *J. Conchol., Lond.* **29**: 81–85.

Castell, C. P., 1962. Some notes on London's molluscs. *J. Conchol., Lond.* **25**: 97–117.

Castillejo, J., 1998. *Guia de las babosas Ibericas.* Santiago.

Cawley, M., 1996a. Notes on some non-marine Mollusca from Co. Sligo and Co. Leitrim, including a new site for *Vertigo geyeri* Lindholm. *Ir. nat. J.* **25**: 183–185.

———, 1996b. *Tandonia rustica* (Millet) (Mollusca: Gastropoda) new to Ireland. *Ir. nat. J.* **25**: 302.

Chater, A. O., 1985. *Vertigo lilljeborgi* living in Cardiganshire. *J. Conchol., Lond.* **32**: 147–148.

Chatfield, J. E., 1977. *Helicodiscus singleyanus* (Pilsbry) (Pulmonata: Endodontidae) found in the British Isles. *J. Conchol., Lond.* **29**: 137–140.

Cherrill, A. J. & James, R., 1985. The distribution and habitat preferences of four species of Hydrobiidae in East Anglia. *J. Conchol., Lond.* **32**: 123–133.

Chesney, H. C. G., Oliver, P. G. & Davis, G. M., 1993. *Margaritifera durrovensis* Phillips, 1928; taxonomic status, ecology and conservation. *J. Conchol., Lond.* **34**: 267–299.

Clerx, J. P. M. & Gittenberger, E., 1977. Einiges über *Cernuella* (Pulmonata, Helicidae). *Zool. Meded. Leiden* **52**: 27–56.

Coles, B. & Colville, B., 1979. *Catinella arenaria* (Bouchard-Chantereaux) and *Vertigo geyeri* Lindholm, from a base-rich fen in north-west England. *J. Conchol., Lond.* **30**: 99–100.

——— & ———, 1980. A glacial relict mollusc. *Nature, Lond.* **286**: 761.

Colville, B., 1992. *Assiminea grayana* Fleming new to the Irish fauna. *J. Conchol., Lond.* **34**: 256.

———, 1994a. *Vertigo angustior* Jeffreys, 1830 living in Scotland. *J. Conchol., Lond.* **35**: 89.

———, 1994b. A second site for *Vertigo geyeri* Lindholm living in mainland Britain. *Conchologists' Newsletter* no. 130: 376–377.

———, Lloyd-Evans, L. & Norris, A., 1974. *Boettgerilla pallens* Simroth, a new British species. *J. Conchol., Lond.* **28**: 203–207.

Comfort, A., 1950. *Hygromia cinctella* (Draparnaud) in England. *J. Conchol., Lond.* **23**: 99–100.

———, 1951. Distribution of *Hygromia cinctella* (Draparnaud) at Paignton. *J. Conchol., Lond.* **23**: 136.

Cooper, J. E., 1924. Note on *Planorbis stroemi*, Westerlund (=*acronicus*, Férussac), living in the Thames. *Proc. malac. Soc. Lond.* **16**: 15.

Cotton, D. C. F., 1996. First Irish record of a living population of the river snail *Viviparus viviparus* (L.) (Gastropoda: Prosobranchia) in the Shannon catchment, Co. Leitrim. *Ir. Nat. J.* **25**: 278–280.

Cranbrook, Earl of, 1976. The commercial exploitation of the freshwater pearl mussel, *Margaritifera margaritifera* L. (Bivalvia: Margaritiferidae) in Great Britain. *J. Conchol., Lond.* **29**: 87–91.

Dance, S. P., 1956. A new Sussex locality for *Pisidium pseudosphaerium* Favre. *J. Conchol., Lond.* **24**: 91–92.

———, 1957. Notes on the *Pisidium* fauna of the Pevensey Levels district, with special reference to *P. pseudosphaerium* Favre. *J. Conchol., Lond.* **24**: 195–199.

———, 1961. On the genus *Pisidium* at Upton Warren. *Phil. Trans. R. Soc.* B **244**: 418–421.

———, 1969. Re-discovery of the wall whorl snail in south Wales. *Nature in Wales* **11**: 161–163.

———, 1970a. Trumpet ram's-horn snail in north Wales. *Nature in Wales* **12**: 10–14.

———, 1970b. *Pisidium lilljeborgii* Clessin in the River Teifi, west Wales. *J. Conchol., Lond.* **27**: 177–181.

———, 1972. *Vertigo lilljeborgi* Westerlund in north Wales. *J. Conchol., Lond.* **27**: 387–389.

Dandy, J. E., 1969. *Watsonian vice-counties of Great Britain.* pp. 1–38; 2 maps. Ray Society publication no. 146. London.

Davies, S. M., 1977. The *Arion hortensis* complex, with notes on *A. intermedius* Normand (Pulmonata: Arionidae). *J. Conchol., Lond.* **29**: 173–187.

———, 1979. Segregates of the *Arion hortensis* complex (Pulmonata: Arionidae) with the description of a new species, *Arion owenii. J. Conchol., Lond.* **30**: 123–128.

———, 1987. *Arion flagellus* Collinge and *A. lusitanicus* Mabille in the British Isles: a morphological, biological and taxonomic investigation. *J. Conchol., Lond.* **32**: 339–354.

Davis, A. G., 1952. *Truncatellina cylindrica* (Férussac) in Norfolk. *J. Conchol., Lond.* **23**: 269–270.

———, 1954. *Helicella elegans* (Gmelin) at Walmer, Kent. *J. Conchol., Lond.* **24**: 19.

———, 1955. *Truncatellina cylindrica britannica* in Dorset and Isle of Wight. *J. Conchol., Lond.* **24**: 61–62.

Dean, J. D., 1920. Occurrence of *Physa gyrina* Say in Great Britain. *J. Conchol., Lond.* **16**: 127.

——— & Kendall, C. E. Y., 1909. *Vertigo alpestris* (Alder): its distribution in north Lancashire and Westmorland, and its association with *Vertigo pusilla* Müller. Supplementary note. *J. Conchol., Lond.* **12**: 309, pl. 4.

Diver, C. *et al.*, 1939. A discussion on the variation of *Lymnaea* in shell-form and anatomy with special reference to *L. peregra, L. involuta*, and allied forms. *Proc. malac. Soc. Lond.* **23**: 303–315.

Dixon, R. & Watson, J. W., 1858. *A descriptive manual of British land and fresh-water shells.* Darlington.

Drake, C. M. (ed.), 1997. *Vertigo moulinsiana* – surveys and studies commissioned in 1995–6. *English Nature Research Reports* no. 217.

Ellis, A. E., 1926. *British snails.* Oxford.

———, 1928. *Planorbis vorticulus*, Troschel, in West Sussex. *Proc. malac. Soc. Lond.* **18**: 127.

———, 1931. Notes on some Norfolk Mollusca. *J. Conchol., Lond.* **19**: 177–178.

———, 1932. Further localities for *Planorbis vorticulus* Troschel. *J. Conchol., Lond.* **19**: 258–259.

———, 1951. Census of the distribution of British non-marine Mollusca. 7th edition. *J. Conchol., Lond.* **23**: 171–244.

———, 1967. *Agriolimax agrestis* (L.): some observations. *J. Conchol., Lond.* **26**: 189–196.

———, 1978. *British freshwater bivalve Mollusca.* Linnean Society of London, synopses of the British fauna (N.S.) no. 11.

Evans, J. G., 1972. *Land snails in archaeology.* London & New York.

Evans, N. J., 1978. *Limax pseudoflavus* Evans: a critical description and comparison with related species. *Ir. nat. J.* **19**: 231–236.

Ferry, B. W., Baddeley, M. S. & Hawksworth, D. L., 1973. *Air pollution and lichens.* London.

Fogan, M., 1969. *Vitrina (Semilimax) pyrenaica* (Férussac) in Kerry North. *Ir. nat. J.* **16**: 175.

Foxall, W. H. & Overton, H., 1912. *Pseudanodonta rothomagensis* Loc. in Britain. *J. Conchol., Lond.* **13**: 274.

Fretter, V. & Graham, A., 1978. The prosobranch molluscs of Britain and Denmark. Part 3. *J. moll. Stud.*, supplement no. 5.

Gittenberger, E. & Bank, R. A., 1996. A new start in *Pyramidula* (Gastropoda: Pyramidulidae). *Basteria* **60**: 71–78.

Gray, J. E., 1840. *A manual of the land and fresh-water shells of the British islands...by William Turton, M.D.* London.

Guntrip, D. W., 1986. *Toltecia pusilla* (Lowe, 1831) living in Britain. *J. Conchol., Lond.* **32**: 200–201.

Harris, G. J., 1985. *Pseudamnicola confusa* rediscovered in the Thames estuary. *J. Conchol., Lond.* **32**: 147.

Harting, J. E., 1875. *Rambles in search of shells, land and freshwater.* London.

Hawksworth, D. L. & Rose, F., 1970. Qualitative scale for estimating sulphur dioxide air pollution in England and Wales using epiphytic lichens. *Nature, Lond.* **227**: 145–148.

Hingley, M. R., 1979. The colonization of newly-dredged drainage channels on the Pevensey levels (East Sussex), with special reference to gastropods. *J. Conchol., Lond.* **30**: 105–122.

Holyoak, D. T., 1978a. Effects of atmospheric pollution on the distribution of *Balea perversa* (Linnaeus) (Pulmonata: Clausiliidae) in southern Britain. *J. Conchol., Lond.* **29**: 319–323.

———, 1978b. *Ferrissia wautieri* (Mirolli) (Pulmonata: Ancylidae) naturalized in Sussex, England. *J. Conchol., Lond.* **29**: 349–350.

———, 1983. Field records – Mollusca. *Ir. nat. J.* **21**: 188–189.

Horsley, J. W., 1915. *Our British snails.* London.

Huggins, H. C., 1922. The south Devon race of *Hygromia limbata* (Drap.). *J. Conchol., Lond.* **16**: 297–301.

Humphreys, J., 1976. Field observations on *Theba pisana* (Müller) (Gastropoda: Helicidae) at St Ives, Cornwall. *J. Conchol., Lond.* **29**: 93–106.

———, Stephens, B. & Turk, S. M., 1982. A new British site for *Theba pisana* (Müller). *J. Conchol., Lond.* **31**: 73.

Hunter, W. R. & Slack, H. D., 1958. *Pisidium conventus* in Loch Lomond. *J. Conchol., Lond.* **24**: 245–246.

Hurley, J., 1981. A history of the occurrence of *Lymnaea glabra* (Gastropoda: Pulmonata) in Ireland. *Ir. nat. J.* **20**: 284–287.

Institute of Terrestrial Ecology, 1978. *Overlays of environmental and other factors for use with Biological Records Centre distribution maps.* Cambridge.

Jackson, J. W. & Kennard, A. S., 1909. On the former occurrence of *Unio (Margaritana) margaritifer* Linné in the River Thames. *J. Conchol., Lond.* **12**: 321–322.

——— & Taylor, F., 1904. Observations on the habits and reproduction of *Paludestrina taylori*. *J. Conchol., Lond.* **11**: 9–11.

Jeffreys, J. G., 1862. *British Conchology.* 1. *Land and freshwater shells.* London.

Jones, A. W., 1968. *Helicella elegans* (Gmelin) in Sussex. *Conchologists' Newsletter* no. 27: 68.

Jones, J. S., Leith, B. H. & Rawlings, P., 1977. Polymorphism in *Cepaea*: a problem with too many solutions? *Ann. Rev. ecol. Syst.* **8**: 109–143.

Kellock, E., 1970. *Agriolimax agrestis* (L.) in the Moray Firth area of Scotland. *J. Conchol., Lond.* **27**: 105–109.

Kennard, A. S. & Woodward, B. B., 1926. *Synonymy of the British non-marine Mollusca (recent and post-Tertiary).* British Museum (Natural History): London.

Kerney, M. P., 1963. Late-glacial deposits on the Chalk of south-east England. *Phil. Trans. R. Soc.* B **246**: 203–254.

———, 1967. Distribution mapping of land and freshwater Mollusca in the British Isles. *J. Conchol., Lond.* **26**: 152–160.

———, 1969. *Pisidium pseudosphaerium* Schlesch new to Ireland. *J. Conchol., Lond.* **27**: 25–26.

———, 1970a. The British distributions of *Monacha cantiana* (Montagu) and *Monacha cartusiana* (Müller). *J. Conchol., Lond.* **27**: 145–148.

———, 1970b. *Pisidium tenuilineatum* Stelfox in Sussex. *J. Conchol., Lond.* **27**: 115–116.

Kerney. M. P., 1971. Rediscovery of *Sphaerium transversum* (Say) in London. *Conchologists' Newsletter* no. 37: 209–210.

———, 1972. The British distribution of *Pomatias elegans* (Müller). *J. Conchol., Lond.* 27: 359–361.

———, 1976a. *Atlas of the non-marine Mollusca of the British Isles*. Institute of Terrestrial Ecology: Cambridge.

———, 1976b. Non-marine Mollusca from Faversham, Kent, figured by George Walker in 1784. *J. Conchol., Lond.* 29: 75–77.

———, 1976c. A helicid new to Britain? *Conchologists' Newsletter* no. 57: 506–507.

———, 1978a. *Semilimax pyrenaicus* (Férussac) in Co. Wicklow. *Ir. nat. J.* 19: 194.

———, 1978b. *Oxychilus helveticus* (Blum) (Pulmonata: Zonitidae) new to Ireland. *J. Conchol., Lond.* 29: 261–262.

———, 1978c. Rediscovery of *Trochoidea elegans* (Gmelin) near Dover. *Conchologists' Newsletter* no. 67: 114.

———, 1982a. Vice-comital census of the non-marine Mollusca of the British Isles (8th edition), *J. Conchol., Lond.* 31: 63–71.

———, 1982b. The mapping of non-marine Mollusca. *Malacologia* 22: 403–407.

———, 1983. *Lauria sempronii* (Charpentier) from a Neolithic flint mine in Sussex. *J. Conchol., Lond.* 31: 258.

———, 1986. A 19th-century record of *Limax maculatus* in the British Isles. *Conchologists' Newsletter* no. 97: 361.

———, 1987. John Morton's list of Northamptonshire Mollusca (1712). *J. Conchol., Lond.* 32: 289–295.

——— & Cameron, R. A. D., 1979. *A field guide to the land snails of Britain and north-west Europe*. London.

——— & Fogan, M., 1969. *Vitrea diaphana* (Studer) in Britain. *J. Conchol., Lond.* 27: 17–24.

——— & May, F., 1960. *Planorbis (Anisus) vorticulus* Troschel in the Thames valley. *J. Conchol., Lond.* 24: 403–404.

——— & Moreton, B. D., 1970. The distribution of *Dreissena polymorpha* (Pallas) in Britain. *J. Conchol., Lond.* 27: 97–100.

——— & Norris, A., 1972. *Lauria sempronii* (Charpentier) living in Britain. *J. Conchol., Lond.* 27: 517–518.

———, Cameron, R. A. D. & Jungbluth, J. H., 1983. *Die Landschnecken Nord- und Mitteleuropas*. Hamburg & Berlin.

Kevan, D. K., 1931. The occurrence of a rare snail (*Succinea oblonga*) in East Lothian and Stirlingshire. *Scottish Nat.* (1931), 185–186.

——— & Waterston, A. R., 1933. *Vertigo lilljeborgi* (West.) in Great Britain (with additional Irish localities). *J. Conchol., Lond.* 19: 296–313.

Killeen, I. J., 1983. *Vertigo angustior* Jeffreys living in Suffolk. *J. Conchol., Lond.* 31: 257.

———, 1992. *The land and freshwater molluscs of Suffolk*. Suffolk Naturalists' Society: Ipswich.

———, 1997. Survey of the terrestrial snail *Vertigo angustior* at three sites in England (Gait Barrows NNR, Florden Common and Martlesham Creek). *English Nature Research Reports* no. 228.

——— & Willing, M. J., 1997. Survey of ditches in East Anglia and South-east England for the freshwater snails *Segmentina nitida* and *Anisus vorticulus*. *English Nature Research Reports* no. 229.

Kuiper, J. G. J., 1949. *Pisidium pseudosphaerium* Favre in England. *J. Conchol., Lond.* 23: 27–32.

Leersnyder, M. de & Hoestlandt, H., 1958. Extension du gastropode méditerranéen *Cochlicella acuta* (Müller) dans le sud-est de l'Angleterre. *J. Conchol., Lond.* 24: 253–264.

Light, J., 1986. *Paludinella littorina* living along the Fleet, Dorset. *J. Conchol., Lond.* 32: 260.

———, 1998. *Paludinella littorina* (delle Chiaje 1828) at Brixham as a member of the upper shore crevice fauna. *Conchologists' Newsletter* no. 146: 62–64.

Lister, M., 1678. *Historiae animalium Angliae tres tractatus*. London.

———, 1694. *Exercitatio anatomica. In qua de cochleis, maxime terrestribus & limacibus, agitur.* London.

Lloyd-Evans, L., 1975. The biogeography of snails in Yorkshire. *Naturalist* 100: 1–12.

Long, D. C., 1968. *Planorbis vorticulus* Troschel in Suffolk East. *Conchologists' Newsletter* no. 27: 71.

———, 1970. *Abida secale* (Draparnaud) in the north Cotswolds. *J. Conchol., Lond.* 27: 117–120.

———, 1974. A *Vertigo lilljeborgi* site in Wigtownshire. *Conchologists' Newsletter* no. 49: 375–376.

———, 1986. A second site for *Lauria sempronii* living in Britain. *J. Conchol., Lond.* 32: 201–202.

Lousley, J. E. (ed.), 1951. *The study of the distribution of British plants*. Botanical Society of the British Isles: Arbroath.

Lucey, J., 1993. The distribution of *Margaritifera margaritifera* (L.) in southern Irish rivers and streams. *J. Conchol., Lond.* 34: 300–310.

———, McGarrigle, M. L. & Clabby, K. J., 1992. The distribution of *Theodoxus fluviatilis* (L.) in Ireland. *J. Conchol., Lond.* 34: 91–101.

Macan, T. T., 1949. *A key to the British fresh- and brackish-water gastropods*. Freshwater Biological Association: Ambleside.

McCarthy, T. & Fitzgerald, J. 1997. The occurrence of the zebra mussel *Dreissena polymorpha* (Pallas, 1771), an introduced befouling freshwater bivalve in Ireland. *Ir. nat. J.* 25: 413–416.

McMillan, N. F., 1955. The range of *Planorbarius corneus* (L.) in the British Isles. *J. Conchol., Lond.* 24: 63–65.

────── & Stelfox, A. W., 1962. *Viviparus viviparus* (L.) in Ireland. *J. Conchol., Lond.* 25: 117–121.

Makings, P., 1959. *Agiolimax caruanae* Pollonera new to Ireland. *J. Conchol., Lond.* 24: 354–356.

Marriott, D. K. & Marriott, R. W., 1982a. *Vertigo alpestris* Alder in Perthshire, Scotland. *J. Conchol., Lond.* 31: 135.

────── & ──────, 1982b. The occurrence of *Vertigo angustior* in North Lancashire. *J. Conchol., Lond.* 31: 72.

────── & ──────, 1984. *Vertigo alpestris* Alder in Scotland. *J. Conchol., Lond.* 31: 388–389.

Marriott, R. W. & Marriott, D. K., 1988. *Vertigo modesta*, a snail new to the British Isles. *J. Conchol., Lond.* 33: 51–52.

Meiklejohn, R. G., 1972. *Anodonta anatina* (Linné) in Caithness (v.c.109). *Conchologists' Newsletter* no. 41: 262.

Meredith, H., 1994. *Theba pisana* (Müller) now widespread along part of the north Cornish coast. *J. Conchol., Lond.* 34: 394–395.

Merrett, C., 1666. *Pinax rerum naturalium Britannicarum*. London. (2nd ed., 1667).

Montagu, G., 1803. *Testacea Britannica*. 2 vols. Romsey & London.

──────, 1808. *Supplement to Testacea Britannica*. Exeter & London.

Morton, J., 1712. *The natural history of Northamptonshire*. London.

Muus, B. J., 1963. Some Danish Hydrobiidae with the description of a new species, *Hydrobia neglecta*. *Proc. malac. Soc. Lond.* 35: 131–138.

Naggs, F., 1983a. *Perforatella*: the helicid snail newly recorded in Britain and other genera commonly confused with *Trichia*. *J. Conchol., Lond.* 31: 201–206.

──────, 1983b. *Perforatella*, a genus of Palaearctic snails newly recognised in Britain, is found in the London area. *London Nat.* no. 62: 59.

──────, 1985. Some preliminary results of a morphometric multivariate analysis of the *Trichia* (Pulmonata: Helicidae) species groups in Britain. *J. nat. Hist.* 19: 1217–1230.

Norris, A., 1976. *Truncatellina cylindrica* (Férussac) in Yorkshire. *Naturalist* 101: 25–27.

──────, 1982. Notes on Yorkshire Mollusca – 5. *Ferrissia wautieri* (Mirolli 1960) a freshwater limpet, new to Yorkshire. *Naturalist* 107: 59–60.

────── & Colville, B., 1974. Notes on the occurrence of *Vertigo angustior* Jeffreys in Great Britain. *J. Conchol., Lond.* 28: 141–154.

────── & Pickrell, D. G., 1972. Notes on the occurrence of *Vertigo geyeri* Lindholm in Ireland. *J. Conchol., Lond.* 27: 411–417.

Økland, J., 1990. *Lakes and snails*. Oegstgeest.

Økland, K. A., 1979. Sphaeriidae of Norway: a project for studying ecological requirements and geographical distribution. *Malacologia* 18: 223–226.

Oldham, C., 1909. *Limax tenellus* in Buckinghamshire and Hertfordshire. *J. Conchol., Lond.* 12: 283.

──────, 1922a. *Paludestrina confusa* (Frauenfeld) in the Waveney valley. *J. Conchol., Lond.* 16: 324–325.

──────, 1922b. *Limax tenellus* in Gloucester West, Hereford and Montgomery. *J. Conchol., Lond.* 16: 276.

──────, 1929a. *Agriolimax laevis* (Müll.) a woodland slug. *J. Conchol., Lond.* 18: 318–319.

──────, 1929b. *Cepaea hortensis* (Mueller) and *Arianta arbustorum* (L.) on blown sand. *Proc. malac. Soc. Lond.* 18: 144–146.

──────, 1933. Notes on *Hygromia revelata* (Mich.). *J. Conchol., Lond.* 19: 359–360.

──────, 1938. *Pisidium conventus* in Westmorland. *J. Conchol., Lond.* 21: 51.

Paul, C. R. C., 1974. *Azeca* in Britain. *J. Conchol., Lond.* 28: 155–172.

──────, 1975. *Columella* in the British Isles. *J. Conchol., Lond.* 28: 371–383.

Perring, F. H. & Walters, S. M. (eds), 1962. *Atlas of the British Flora*. London.

Phillips, R. A., 1908. *Vertigo moulinsiana*, Dupuy. An addition to the Irish fauna. *Ir. Nat.* 17: 89–93.

──────, 1914. *Helicigona lapicida* in Ireland. *Ir. Nat.* 23: 37–38.

──────, 1916. On two species of *Pisidium* (fossil) new to Ireland. *Ir. Nat.* 25: 101–105.

──────, 1935. *Vertigo genesii* Gredler in central Ireland. *J. Conchol., Lond.* 20: 142–145.

────── & Watson, H., 1930. *Milax gracilis* (Leydig) in the British Isles. *J. Conchol., Lond.* 19: 65–93.

Philp, E. G., 1984. *Perforatella rubiginosa* (Schmidt) and *Clausilia dubia* Drap. in Kent. *Conchologists' Newsletter* no. 88: 153.

──────, 1987. *Tandonia rustica* (Millet), a slug new to the British Isles. *J. Conchol., Lond.* 32: 302.

Platts, E., 1977. The land winkle *Pomatias elegans* (Müller) confirmed as an Irish species. *Ir. nat. J.* 19: 10–12.

────── & Speight, M. C. D., 1988. The taxonomy and distribution of the Kerry slug *Geomalacus maculosus* Allman, 1843 (Mollusca: Arionidae) with a discussion of its status as a threatened species. *Ir. nat. J.* 22: 417–430.

Pollard, E., 1974. Distribution maps of *Helix pomatia* L. *J. Conchol., Lond.* 28: 239–242.

Ponder, W. F., 1988. *Potamopyrgus antipodarum*–a molluscan coloniser of Europe and Australia. *J. moll. Stud.* 54: 271–285.

Praeger, R. L., 1896. On the botanical subdivision of Ireland. *Ir. Nat.* 5: 29–38.

Preece, R. C., 1977. Fossil *Helicopsis striata* (Müller) and *Trochoidea geyeri* (Soós) from the Isle of Wight. *Proc. Isle of Wight nat. hist. arch. Soc.* 6: 608–609.

──────, 1980. *An atlas of the non-marine Mollusca of the Isle of Wight*. County Museum service: Sandown.

Preece, R.C. 1992. *Cochlicopa nitens* (Gallenstein) in the British Late-glacial and Holocene. *J. Conchol., Lond.* **34**: 215–224.

——— & Holyoak, D. T. (in preparation). *The Quaternary history of non-marine Mollusca.* London.

——— & Robinson, J. E., 1984. Late Devensian and Flandrian environmental history of the Ancholme valley, Lincolnshire: molluscan and ostracod evidence. *J. Biogeogr.* **11**: 319–352.

——— & Willing, M. J., 1984. *Vertigo angustior* living near its type locality in south Wales. *J. Conchol., Lond.* **31**: 340.

——— & Wilmot, R. D., 1979. *Marstoniopsis scholtzi* (A. Schmidt) and *Ferrissia wautieri* (Mirolli) from Hilgay, Norfolk. *J. Conchol., Lond.* **30**: 135–139.

———, Burleigh, R., Kerney, M. P. & Jarzembowski, E. A., 1983. Radiocarbon age determinations of fossil *Margaritifera auricularia* (Spengler) from the River Thames in west London. *J. arch. Sci.* **10**: 249–257.

Quick, H. E., 1933. The anatomy of British Succineae. *Proc. malac. Soc. Lond.* **20**: 295–318.

———, 1947. *Arion ater* (L.) and *A. rufus* (L.) in Britain and their specific differences. *J. Conchol., Lond.* **22**: 249–261.

———, 1952. Rediscovery of *Arion lusitanicus* Mabille in Britain. *Proc. malac. Soc. Lond.* **29**: 93–101.

———, 1954. *Cochlicopa* in the British Isles. *Proc. malac. Soc. Lond.* **30**: 204–213.

———, 1957. *Vitrina major* (Férussac) in Surrey, with some remarks on the species. *J. Conchol., Lond.* **24**: 235–237.

———, 1960. British slugs (Pulmonata: Testacellidae, Arionidae, Limacidae). *Bull. Brit. Mus. (Nat. Hist.), Zool.* **6**: no. 3.

Redshaw, E. J. & Norris, A., 1974. *Sphaerium solidum* (Normand) in the British Isles. *J. Conchol., Lond.* **28**: 209–212.

Reeve, L. A., 1863. *The land and freshwater mollusks indigenous to, or naturalized in, the British Isles.* London.

Rimmer, R., 1880. *The land and freshwater shells of the British Isles.* London.

Roebuck, W. D., 1881. Proposed system of conchological locality-records. *J. Conchol., Lond.* **3**: 138–140.

———, 1921. Census of the distribution of British land and freshwater Mollusca. *J. Conchol., Lond.* **16**: 165–212. (Text by A. E. Boycott).

Scharff, R. F., 1892. The Irish land and freshwater Mollusca. *Ir. Nat.* **1**: 45–47; 65–67; 87–90; 105–109; 135–138; 149–153; 177–181.

Sparks, B. W., 1953. The former occurrence of both *Helicella striata* (Müller) and *H. geyeri* (Soós) in England. *J. Conchol., Lond.* **23**: 372–378.

Stelfox, A. W., 1911. A list of the land and freshwater mollusks of Ireland. *Proc. R. Ir. Acad.* B **29**: 65–164.

———, 1918. The *Pisidium* fauna of the Grand Junction Canal in Herts. and Bucks. *J. Conchol., Lond.* **15**: 289–304.

———, 1958. A short history of the known occurrences of *Helicella gigaxi* (L. Pfeiffer) in Ireland. *J. Conchol., Lond.* **24**: 283–284.

——— & Phillips, R. A., 1925. *Vertigo genesii* Gredler in Ireland. *J. Conchol., Lond.* **17**: 236–240.

———, Kuiper, J. G. J., McMillan, N. F. & Mitchell, G. F., 1972. The Late-glacial and Post-glacial Mollusca of the White Bog, Co. Down. *Proc. R. Ir. Acad.* B **72**: 185–207.

Step, E., 1901. *Shell life.* London.

Stratton, L. W., 1954. A new locality for *Hygromia limbata* (Draparnaud). *J. Conchol., Lond.* **24**: 20.

Stubbs, A. G., 1907. *Illustrated index of British freshwater shells.* Leeds.

Swanton, E. W., 1906. *A pocket guide to the British non-marine Mollusca.* Lockwood.

Tate, R., 1866. *A plain and easy account of the land and fresh-water mollusks of Great Britain.* London.

Taylor, J. W., 1894–1921. *Monograph of the land & freshwater Mollusca of the British Isles.* 3 vols + 3 pts (unfinished). Leeds.

——— & Roebuck, J. W., 1885. Census of the authenticated distribution of British land and freshwater Mollusca. *J. Conchol., Lond.* **4**: 319–336.

Thompson, W., 1840. Catalogue of the land and freshwater Mollusca of Ireland. *Ann. Mag. nat. Hist.* **6**: 16–34; 109–126; 194–206.

Turner, A. H., 1950. *Laciniaria biplicata* (Montagu) in Somerset. *J. Conchol., Lond.* **23**: 116.

Turton, W., 1831. *A manual of the land and fresh-water shells of the British Islands.* London.

Van der Velde, G. & Roelofs, J. G. M., 1977. *Ferrissia wautieri* (Gastropoda, Basommatophora) nieuw voor Nederland. *Basteria* **41**: 73–80.

Verdcourt, B., 1982. The occurrence of *Perforatella (Monachoides) rubiginosa* (Schmidt) in the British Isles. *Conchologists' Newsletter* no. 83: 46–48.

Walker, D. *et al.*, 1991. Two new British records for *Myxas glutinosa* (Müller). *J. Conchol., Lond.* **34**: 39.

Walker, G., 1784. *Testacea minuta rariora.* London.

Walters, S. M., 1954. The distribution maps scheme. *Proc. bot. Soc. Br. Isl.* **1**: 121–130.

Waterston, R., 1934. Occurrence of *Amnicola taylori* (E. A. Smith) and *Bithynia leachii* (Sheppard) in Scotland. *J. Conchol., Lond.* **20**: 55–56.

Watson, H. & Verdcourt, B., 1953. The two British species of *Carychium*. *J. Conchol., Lond.* **23**: 306–324.

Welch, J. M. & Pollard, E., 1975. The exploitation of *Helix pomatia*. *Biological Conservation* **8**: 155–160.

Wiktor, A. & Norris, A., 1982. The synonymy of *Limax maculatus* (Kaleniczenko 1851) with notes on its European distribution. *J. Conchol., Lond.* **31**: 75–77.

Williams, J. W., 1888. *The shell-collector's handbook for the field.* London.

Willing, M. J. & Killeen, I. J., 1998. The freshwater snail *Anisus vorticulus* in ditches in Suffolk, Norfolk and West Sussex. *English Nature Research Reports* no. 287.

Winterbourn, M. J., 1972. Morphological variation of *Potamopyrgus jenkinsi* (Smith) from England and a comparison with the New Zealand species, *Potamopyrgus antipodarum* (Gray). *Proc. malac. Soc. Lond.* **40**: 133–145.

Woodward, B. B., 1913. *Catalogue of the British species of Pisidium (recent & fossil)*. British Museum (Natural History): London.

Young, M. & Williams, J., 1983. The status and conservation of the freshwater pearl mussel *Margaritifera margaritifera* (Linn.) in Great Britain. *Biological Conservation* **25**: 35–52.

Zoer, J. A. & Visser, G., 1972. Description of some finding-spots of *Geomalacus maculosus* in S.W. Ireland. *Conchologists' Newsletter* no. 42: 267–268.

# INDEX

Synonyms are shown in *italics*

Abida secale 102
Acanthinula aculeata 110
*Acanthinula lamellata* 111
Acicula fusca 43
acicula, Cecilioides 168
*Acme lineata* 43
*Acmea subcylindrica* 38
Acroloxus lacustris 74
acronicus, Gyraulus 65
aculeata, Acanthinula 110
acuta, Cochlicella 186
acuta, Physa 50
Aegopinella nitens 142
　nitidula 142
　pura 141
aginnica, Cernuella 181
agreste, Deroceras 162
*Agriolimax agrestis* 162, 163
　*caruanae* 164
　*laevis* 161
　*reticulatus* 163
albus, Gyraulus 66
alderi, Euconulus 167
alliarius, Oxychilus 145
alpestris, Vertigo 100
*Amnicola confusa* 35
　*taylori* 37
amnicum, Pisidium 219
anatina, Anodonta 212
*anatina, Hydrobia* 35
*Ancylastrum fluviatile* 72
Ancylus fluviatilis 72
*Ancylus lacustris* 74
anglica, Leiostyla 104
angustior, Vertigo 101
Anisus leucostoma 60
　vortex 61
　vorticulus 62
Anodonta anatina 212
　cygnea 211
*Anodonta elongata* 213
　*minima* 213
　*piscinalis* 212
antipodarum, Potamopyrgus 36
antivertigo, Vertigo 92
Aplexa hypnorum 48
*arborum, Limax* 159
arbustorum, Arianta 200
*arctica, Vertigo* 96
arenaria, Catinella 75

Arianta arbustorum 200
Arion ater 120
　circumscriptus *agg.* 124
　circumscriptus *seg.* 125
　distinctus 130
　fasciatus 127
　flagellus 122
　hortensis *agg.* 128
　hortensis *seg.* 129
　intermedius 132
　lusitanicus 121
　owenii 131
　rufus 120
　silvaticus 126
　subfuscus 123
*Arion fasciatus* 124
　*lusitanicus* 122
　*minimus* 132
*Armiger crista* 67
Ashfordia granulata 190
aspera, Columella 87
aspersa, Helix 205
Assiminea grayana 41
*Assiminea littorina* 42
ater, Arion 120
auricularia, Lymnaea 55
auricularia, Margaritifera 208
*Auriculinella bidentata* 47
Azeca goodalli 80
*Azeca menkeana* 80
　*tridens* 80

Balea biplicata 173
　perversa 174
barbara, Cochlicella 187
*barbara, Cochlicella* 186
Bathyomphalus contortus 63
bidentata, Clausilia 171
bidentata, Leucophytia 47
biplicata, Balea 173
Bithynia leachii 40
　tentaculata 39
Boettgerilla pallens 153
*Boettgerilla vermiformis* 153
Bradybaena fruticum 178
*britannica, Truncatellina* 90
budapestensis, Tandonia 152
*burnetti, Lymnaea* 56
*Bythinella scholtzi* 37
　*steinii* 37

callicratis, Truncatellina 90

Candidula gigaxii 180
　intersecta 179
cantiana, Monacha 189
*caperata, Helicella* 179
caputspinulae, Paralaoma 115
carinatus, Planorbis 59
cartusiana, Monacha 188
*caruanae, Deroceras* 164
Carychium minimum 44
　tridentatum 45
casertanum, Pisidium 220
Catinella arenaria 75
Cecilioides acicula 168
cellarius, Oxychilus 144
Cepaea hortensis 204
　nemoralis 203
Cernuella aginnica 181
　neglecta 181
　virgata 181
cinctella, Hygromia 193
cinereoniger, Limax 155
*cinereum, Pisidium* 220
circumscriptus *agg.*, Arion 124
circumscriptus *seg.*, Arion 125
Clausilia bidentata 171
　dubia 172
*Clausilia biplicata* 173
　*cravenensis* 172
　*laminata* 169
　*rolphii* 170
　*rugosa* 171
*clessiniana, Ferrissia* 73
Cochlicella acuta 186
　barbara 187
*Cochlicella barbara* 186
　*ventricosa* 187
Cochlicopa lubrica 81
　lubricella 82
　nitens 83
　repentina 81
*Cochlicopa minima* 82
Cochlodina laminata 169
Columella aspera 87
　columella 88
　edentula *agg.* 85
　edentula *seg.* 86
columella, Columella 88
complanata, Pseudanodonta 213
complanatus, Hippeutis 68
*complanatus, Planorbis* 58, 68
confusa, Mercuria 35
contectus, Viviparus 26

258

# INDEX

contortus, Bathyomphalus 63
contracta, Vitrea 138
conventus, Pisidium 221
corneum, Sphaerium 214
corneus, Planorbarius 70
costata, Vallonia 107
*cravenensis, Clausilia* 172
crista, Gyraulus 67
cristata, Valvata 27
*Cryptomphalus aspersus* 205
crystallina, Vitrea 137
*Cyclostoma elegans* 30
cylindracea, Lauria 105
cylindrica, Truncatellina 89
cygnea, Anodonta 211

*denticulata, Ovatella* 46
Deroceras agreste 162
   laeve 161
   panormitanum 164
   reticulatum 163
*Deroceras caruanae* 164
*diaphana, Vitrea* 136
dilatatus, Menetus 71
Discus ruderatus 117
   rotundatus 118
distinctus, Arion 130
draparnaudi, Oxychilus 143
Dreissena polymorpha 236
dubia, Clausilia 172
*durrovensis, Margaritifera* 207

edentula *agg.*, Columella 85
edentula *seg.*, Columella 86
elegans, Pomatias 30
elegans, Trochoidea 183
*elegans, Oxyloma* 78
*elegans, Succinea* 79
*elongata, Anodonta* 213
*ericetorum, Helix* 182
Euconulus alderi 167
   fulvus *agg.* 165
   fulvus *seg.* 166
Ena montana 112
   obscura 113
*Euparypha pisana* 202
excavatus, Zonitoides 147
excentrica, Vallonia 109

fasciatus, Arion 127
*fasciatus, Arion* 124
*fasciatus, Viviparus* 26
Ferrissia wautieri 73
*Ferrissia clessiniana* 73
flagellus, Arion 122
flavus, Limax 156
fluviatilis, Ancylus 72
fluviatilis, Theodoxus 24
*fontanus, Planorbis* 68
fontinalis, Physa 49
*Fruticicola fruticum* 178
fruticum, Bradybaena 178
fulvus *agg.*, Euconulus 165
fulvus *seg.*, Euconulus 166
fusca, Acicula 43
*fusca, Hygromia* 191

*Galba truncatula* 51
gagates, Milax 149

genesii, Vertigo 98
*genesii, Vertigo* 99
Geomalacus maculosus 119
geyeri, Trochoidea 184
geyeri, Vertigo 99
gigaxii, Candidula 180
*glaber, Planorbis* 64
glabra, Lymnaea 52
glutinosa, Myxas 57
*Goniodiscus rotundatus* 118
goodalli, Azeca 80
*gracilis, Milax* 152
granulata, Ashfordia 190
grayana, Assiminea 41
*grossui, Limax* 157
Gyraulus acronicus 65
   albus 66
   crista 67
   laevis 64
gyrina, Physa 50

haliotidea, Testacella 176
hammonis, Nesovitrea 139
Heleobia stagnorum 34
Helicella itala 182
*Helicella caperata* 179
   *elegans* 183
   *geyeri* 184
   *gigaxii* 180
   *heripensis* 180
   *striata* 185
   *virgata* 181
Helicigona lapicida 201
Helicodiscus singleyanus 116
Helicodonta obvoluta 199
Helicopsis striata 185
Helix aspersa 205
   pomatia 206
*Helix aculeata* 110
   *arbustorum* 200
   *cantiana* 189
   *caperata* 179
   *cartusiana* 188
   *ericetorum* 182
   *fruticum* 178
   *granulata* 190
   *hispida* 197
   *hortensis* 204
   *itala* 182
   *lamellata* 111
   *lapicida* 201
   *limbata* 194
   *nemoralis* 203
   *obvoluta* 199
   *pisana* 202
   *revelata* 198
   *rotundata* 118
   *rufescens* 195
   *rupestris* 84
   *striolata* 195
   *terrestris* 183
   *umbilicata* 84
   *virgata* 181
helveticus, Oxychilus 146
henslowanum, Pisidium 228
*heripensis, Helicella* 180
heterostropha, Physa 50
*hibernica, Vitrina* 134
hibernicum, Pisidium 230

Hippeutis complanatus 68
hispida, Trichia 197
hortensis *agg.*, Arion 128
hortensis *seg.*, Arion 129
hortensis, Cepaea 204
*Hyalinia alliaria* 145
   *cellaria* 144
   *crystallina* 137
   *fulva* 165
   *helvetica* 146
   *lucida* 143
   *nitidula* 142
   *pura* 141
   *radiatula* 139
   *rogersi* 146
Hydrobia neglecta 32
   ulvae 33
   ventrosa 31
*Hydrobia anatina* 35
   *jenkinsi* 36
   *similis* 35
   *stagnalis* 33
   *stagnorum* 34
Hygromia cinctella 193
   limbata 194
*Hygromia fusca* 191
   *hispida* 197
   *liberta* 196
   *revelata* 198
   *rufescens* 195
   *striolata* 195
   *subrufescens* 191
   *subvirescens* 198
hypnorum, Aplexa 48

*insularis, Milax* 149
intermedius, Arion 132
intersecta, Candidula 179
*involuta, Lymnaea* 56
itala, Helicella 182

*jenkinsi, Potamopyrgus* 36

*Laciniaria biplicata* 173
lacustre, Musculium 218
lacustris, Acroloxus 74
laeve, Deroceras 161
laevis, Gyraulus 64
lamellata, Spermodea 111
laminata, Cochlodina 169
lapicida, Helicigona 201
Lauria cylindracea 105
   sempronii 106
*Lauria anglica* 104
leachii, Bithynia 40
Lehmannia marginata 159
   valentiana 160
Leiostyla anglica 104
*Leuconia bidentata* 47
Leucophytia bidentata 47
leucostoma, Anisus 60
*liberta, Hygromia* 196
lilljeborgi, Vertigo 97
lilljeborgii, Pisidium 229
*Limacus flavus* 156
   *maculatus* 157
Limax cinereoniger 155
   flavus 156
   maculatus 157

INDEX

Limax maximus 154
Limax arborum 159
   *grossui* 157
   *marginatus* 159
   *poirieri* 160
   *pseudoflavus* 157
   *tenellus* 158
   *valentianus* 160
limbata, Hygromia 194
*lineata, Acme* 43
*lineata, Segmentina* 69
*littorea, Paludinella* 42
littorina, Paludinella 42
lubrica, Cochlicopa 81
lubricella, Cochlicopa 82
*lucidus, Oxychilus* 143
lusitanicus, Arion 121
*lusitanicus, Arion* 122
Lymnaea auricularia 55
   glabra 52
   palustris 53
   peregra 56
   stagnalis 54
   truncatula 51
*Lymnaea burnetti* 56
   *glutinosa* 57
   *involuta* 56
   *ovata* 56

Macrogastra rolphii 170
macrostoma, Valvata 28
maculatus, Limax 157
maculosus, Geomalacus 119
major, Phenacolimax 135
Malacolimax tenellus 158
*Margaritana margaritifera* 207
Margaritifera auricularia 208
   margaritifera 207
*Margaritifera durrovensis* 207
margaritifera, Margaritifera 207
marginata, Lehmannia 159
*marginata, Pupa* 103
marginatus, Milax 151
marginatus, Planorbis 58
*Marpessa laminata* 169
Marstoniopsis scholtzi 37
maugei, Testacella 175
maximus, Limax 154
Menetus dilatatus 71
menkeana, *Azeca* 80
Mercuria confusa 35
*Merdigera obscura* 113
*micropleurum, Pleuropunctum* 115
Milax gagates 149
   nigricans 149
*Milax budapestensis* 152
   *gracilis* 152
   *insularis* 149
   *marginatus* 151
   *rusticus* 151
   *sowerbyi* 150
milium, Pisidium 224
*minima, Anodonta* 213
*minima, Cochlicopa* 82
minimum, Carychium 44
*minimus, Arion* 132
modesta, Vertigo 96
moitessierianum, Pisidium 233
Monacha cantiana 189
   cartusiana 188

*Monacha granulata* 190
montana, Ena 112
moulinsiana, Vertigo 95
muscorum, Pupilla 103
Musculium lacustre 218
   transversum 217
myosotis, Ovatella 46
Myxas glutinosa 57

*nautileus, Planorbis* 67
neglecta, Cernuella 181
neglecta, Hydrobia 32
nemoralis, Cepaea 203
*Neritina fluviatilis* 24
Nesovitrea hammonis 139
   petronella 140
nigricans, Milax 149
nitens, Aegopinella 142
nitens, Cochlicopa 83
nitida, Segmentina 69
nitidula, Aegopinella 142
nitidum, Pisidium 231
nitidus, Zonitoides 148

oblonga, Succinea 76
obscura, Ena 113
obtusale, Pisidium 223
obvoluta, Helicodonta 199
*Omphiscola glabra* 52
*ovale, Sphaerium* 217
*ovata, Lymnaea* 56
Ovatella myosotis 46
*Ovatella denticulata* 46
owenii, Arion 131
Oxychilus alliarius 145
   cellarius 144
   draparnaudi 143
   helveticus 146
*Oxychilus lucidus* 143
Oxyloma pfeifferi 78
   sarsi 79
*Oxyloma elegans* 78

pallens, Boettgerilla 153
*pallidum, Sphaerium* 217
*Paludestrina confusa* 35
   *stagnalis* 33
   *taylori* 37
   *ventrosa* 31
Paludina contecta 26
   *vivipara* 25
Paludinella littorina 42
*Paludinella littorea* 42
palustris, Lymnaea 53
panormitanum, Deroceras 164
Paralaoma caputspinulae 115
*parvulum, Pisidium* 233
pellucida, Vitrina 133
peregra, Lymnaea 56
Perforatella rubiginosa 192
   subrufescens 191
*Peringia ulvae* 33
personatum, Pisidium 222
perversa, Balea 174
petronella, Nesovitrea 140
pfeifferi, Oxyloma 78
Phenacolimax major 135
Physa acuta 50
   fontinalis 49
   gyrina 50

   heterostropha 50
*Physa hypnorum* 48
*Phytia myosotis* 46
pictorum, Unio 209
pisana, Theba 202
piscinalis, Valvata 29
*piscinalis, Anodonta* 212
Pisidium amnicum 219
   casertanum 220
   conventus 221
   henslowanum 228
   hibernicum 230
   lilljeborgii 229
   milium 224
   moitessierianum 233
   nitidum 231
   obtusale 223
   personatum 222
   pseudosphaerium 225
   pulchellum 232
   subtruncatum 226
   supinum 227
   tenuilineatum 234
   vincentianum 235
*Pisidium cinereum* 220
   *parvulum* 233
   *stewarti* 235
   *torquatum* 233
Planorbarius corneus 70
Planorbis carinatus 59
   planorbis 58
*Planorbis acronicus* 65
   *albus* 66
   *complanatus* 58, 68
   *contortus* 63
   *corneus* 70
   *crista* 67
   *dilatatus* 71
   *fontanus* 68
   *glaber* 64
   *laevis* 64
   *leucostoma* 60
   *marginatus* 58
   *nautileus* 67
   *spirorbis* 60
   *umbilicatus* 58
   *vortex* 61
   *vorticulus* 62
planorbis, Planorbis 58
plebeia, Trichia 196
*Pleuropunctum micropleurum* 115
*poirieri, Limax* 160
polymorpha, Dreissena 236
pomatia, Helix 206
Pomatias elegans 30
Ponentina subvirescens 198
Potamopyrgus antipodarum 36
*Potamopyrgus jenkinsi* 36
*Pseudamnicola confusa* 35
Pseudanodonta complanata 213
*Pseudanodonta rothomagensis* 213
*pseudoflavus, Limax* 157
pseudosphaerium, Pisidium 225
*Pseudotrichia rubiginosa* 192
*Pseudunio auricularia* 208
pulchella, Vallonia 108
*pulchella, Valvata* 28
pulchellum, Pisidium 232
Punctum pygmaeum 114
*Punctum pusillum* 115

# INDEX

*Pupa anglica* 104
  *marginata* 103
  *secale* 102
  *umbilicata* 105
Pupilla muscorum 103
pura, Aegopinella 141
pusilla, Vertigo 91
*pusilla, Toltecia* 115
putris, Succinea 77
pygmaea, Vertigo 94
pygmaeum, Punctum 114
Pyramidula rupestris 84
*Pyramidula rotundata* 118
pyrenaicus, Semilimax 134

*radiatula, Retinella* 139
*Radix auricularia* 55
  *peregra* 56
repentina, Cochlicopa 81
reticulatum, Deroceras 163
*Retinella nitidula* 142
  *pura* 141
  *radiatula* 139
*revelata, Hygromia* 198
rivicola, Sphaerium 215
*rogersi, Hyalinia* 146
rolphii, Macrogastra 170
*rothomagensis, Pseudanodonta* 213
rotundatus, Discus 118
rubiginosa, Perforatella 192
ruderatus, Discus 117
*rufescens, Hygromia* 195
rufus, Arion 120
rugosa, Clausilia 171
rupestris, Pyramidula 84
rustica, Tandonia 151

*Sabanaea ulvae* 33
sarsi, Oxyloma 79
scholtzi, Marstoniopsis 37
scutulum, Testacella 177
secale, Abida 102
Segmentina nitida 69
*Segmentina complanata* 68
  *lineata* 69
Semilimax pyrenaicus 134
*Semisalsa stagnorum* 34
sempronii, Lauria 106
silvaticus, Arion 126
*similis, Hydrobia* 35
singleyanus, Helicodiscus 116
*sinuatus, Unio* 208
solidum, Sphaerium 216
sowerbyi, Tandonia 150
Spermodea lamellata 111
Sphaerium corneum 214
  rivicola 215
  solidum 216
*Sphaerium lacustre* 218
  *ovale* 217

  *pallidum* 217
  *transversum* 217
*spirorbis, Planorbis* 60
stagnalis, Lymnaea 54
*stagnalis, Hydrobia* 33
*Stagnicola glabra* 52
  *palustris* 53
stagnorum, Heleobia 34
*steinii, Bythinella* 37
*stewarti, Pisidium* 235
striata, Helicopsis 185
striolata, Trichia 195
subcylindrica, Truncatella 38
subfuscus, Arion 123
subrimata, Vitrea 136
subrufescens, Perforatella 191
substriata, Vertigo 93
subtruncatum, Pisidium 226
subvirescens, Ponentina 198
Succinea oblonga 76
  putris 77
*Succinea arenaria* 75
  *elegans* 79
  *pfeifferi* 78
supinum, Pisidium 227

Tandonia budapestensis 152
  rustica 151
  sowerbyi 150
*taylori, Amnicola* 37
tenellus, Malacolimax 158
tentaculata, Bithynia 39
tenuilineatum, Pisidium 234
*terrestris, Helix* 183
Testacella haliotidea 176
  maugei 175
  scutulum 177
Theba pisana 202
*Theba cantiana* 189
  *cartusiana* 188
Theodoxus fluviatilis 24
*Toltecia pusilla* 115
*torquatum, Pisidium* 233
transversum, Musculium 217
Trichia hispida 197
  plebeia 196
  striolata 195
*Trichia subvirescens* 198
*tridens, Azeca* 80
tridentatum, Carychium 45
Trochoidea elegans 183
  geyeri 184
Truncatella subcylindrica 38
*Truncatella truncatula* 38
Truncatella callicratis 90
  cylindrica 89
*Truncatellina britannica* 90
truncatula, Lymnaea 51
*truncatula, Truncatella* 38
tumidus, Unio 210

ulvae, Hydrobia 33
*umbilicata, Helix* 84
*umbilicata, Pupa* 105
*umbilicatus, Planorbis* 58
Unio pictorum 209
  tumidus 210
*Unio margaritifer* 207
  *sinuatus* 208

valentiana, Lehmannia 160
Vallonia costata 107
  excentrica 109
  pulchella 108
Valvata cristata 27
  macrostoma 28
  piscinalis 29
*Valvata pulchella* 28
*ventricosa, Cochlicella* 187
ventrosa, Hydrobia 31
*Ventrosia ventrosa* 31
*vermiformis, Boettgerilla* 153
Vertigo alpestris 100
  angustior 101
  antivertigo 92
  genesii 98
  geyeri 99
  lilljeborgi 97
  modesta 96
  moulinsiana 95
  pusilla 91
  pygmaea 94
  substriata 93
*Vertigo arctica* 96
  *britannica* 90
  *cylindrica* 89
  *edentula* 85
  *genesii* 99
vincentianum, Pisidium 235
virgata, Cernuella 181
Vitrea contracta 138
  crystallina 137
  subrimata 136
*Vitrea diaphana* 136
Vitrina pellucida 133
*Vitrina hibernica* 134
  *major* 135
  *pyrenaica* 134
Viviparus contectus 26
  viviparus 25
*Viviparus fasciatus* 26
viviparus, Viviparus 25
vortex, Anisus 61
vorticulus, Anisus 62

wautieri, Ferrissia 73

*Zenobiella subrufescens* 191
Zonitoides excavatus 147
  nitidus 148

# NOTES AND ADDENDA

NOTES AND ADDENDA

# NOTES AND ADDENDA